Trophic Relationships in Inland Waters

Developments in Hydrobiology 53

Series editor
H. J. Dumont

Trophic Relationships in Inland Waters

Proceedings of an International Symposium held in Tihany (Hungary),
1–4 September 1987

Edited by
P. Biró and J. F. Talling

Reprinted from Hydrobiologia, vol. 191 (1990)

Kluwer Academic Publishers

Dordrecht/Boston/London

Library of Congress Cataloging-in-Publication Data

Trophic relationships in inland waters : proceedings of an
  international symposium, held at Tihany (Hungary), September 1-4,
  1987 / edited by P. Biró and J.F. Talling.
        p.   cm. -- (Developments in hydrobiology ; 53)
    Proceedings of the International Symposium on Trophic
  Relationships in Inland Waters, held Sept. 1-4, 1987 at the Balaton
  Limnological Research Institute, and sponsored by the Hungarian
  National Water Authority and the Warmwater Fish Hatchery.
    ISBN-13:978-94-010-6695-2   e-ISBN-13:978-94-009-0467-5
    DOI: 10.1007/978-94-009-0467-5

    1. Freshwater ecology--Europe--Congresses. 2. Lake ecology-
  -Europe--Congresses. 3. Food chains (Ecology)--Europe--Congresses.
  I. Biró, P. (Pál) II. Talling, J. F. (John Francis)
  III. International Symposium on Trophic Relationships in Inland
  Waters (1987 : Balaton Limnological Research Institute)
  IV. Hungary. Országos Vízügyi Hivatal. V. Warmwater Fish Hatchery
  (Százhalombatta, Hungary) VI. Series.
  QH541.5.F7T76   1989
  574.5'2632--dc20   89-15613

ISBN-13:978-94-010-6695-2

Printed on acid-free paper

Kluwer Academic Publishers incorporates the publishing programmes of Dr W. Junk Publishers, MTP Press, Martinus Nijhoff Publishers, and D. Reidel Publishing Company.

Distributors

for the United States and Canada: Kluwer Academic Publishers, 101 Philip Drive, Norwell, MA 02061, U.S.A.
for all other countries: Kluwer Academic Publishers Group, P.O. Box 322, 3200 AH Dordrecht, The Netherlands

Copyright

V

Contents

Foreword

The International Symposium on Trophic Relationships in Inland Waters, held from 1st–4th September 1987, at the Balaton Limnological Research Institute of the Hungarian Academy of Sciences, Tihany (Hungary), was intended to give an insight into current research on limnology of inland waters. The meeting was organized on the occasion of the 60th anniversary of the Institute in order to promote the exchange of ideas and discussion of new results.

Papers presented during the Symposium dealt with four main topics: (1) Interactions of inorganic nutrients, primary producers and bacteria, (2) Interactions between primary and secondary producers, (3) Trophic relationships between plankton and fish, (4) Studies on complex trophic systems. Participants from 18 countries presented 40 oral lectures and 15 posters, that reviewed the structure and functioning of inland water ecosystems from different aspects. Since in such functioning nutrients are main forcing factors, the pathways of nutrients, as well as trophic connections, are widely studied nowadays. The papers of these proceedings thus present a series of different approaches to the main results of current limnological research in this very important field.

The structure of these proceedings was somewhat altered when papers were ranked into three main groups: (1) Long-term changes, (2) Annual and seasonal cycles and (3) Short-term changes and pilot scale operations. The sequence of papers within these groups follows the four main subjects discussed during the Symposium.

During the meeting, a special workshop was held on biomanipulation, in which participants discussed the possibility of controlled physico-chemical cycles and biological processes from the viewpoint of management practice. It is hoped by the editors that the proceedings presented here will stimulate further research and a greater exchange of information in this context, because the content of this volume is more selective than comprehensive.

Our special gratitude is due to the Hungarian Academy of Sciences for the financial support of the Symposium, as well as to co-sponsors, the Hungarian National Water Authority (Budapest) and the Warmwater Fish Hatchery (Százhalombatta). We wish to extend our sincere thanks to all participants and contributors, and to all our colleagues for their essential help in many ways. Dr H. L. Golterman and Prof. M. Gophen deserve special mention for their assistance. Moreover we wish to thank the Director of the Institute, Prof. J. Salánki, and all the colleagues who were so helpful during the organization. Mr T. Forró, Dr S. Herodek, Dr J. E. Ponyi, Dr N. P.-Zánkai, Dr I. Tátrai, Dr L. Vörös and all the staff members of the Department of Hydrobiology are kindly acknowledged for their assistance.

We are extremely grateful to the referees who worked diligently in improving the scientific quality of the manuscripts. We also thank Dr G. Paulovits for his cooperation in the compilation of the General Index.

THE EDITORS

List of participants

ABAFFY-BOTHÁR Anna, Hungarian Danube Research Station of the HAS, H-2131 Göd, Hungary

BARICA Jan M., National Water Research Institute, P.O. Box 5050, Burlington, Ontario L7R 4A6, Canada

BERCZIK Árpád, Institute of Ecology and Botany of the HAS, H-2163 Vácrátót, Alkotmány u. 2-6, Hungary

BERGSTRAND Eva, Institute of Freshwater Research, S-170 11 Drottningholm, Sweden

BIRÓ Péter, Balaton Limnological Research Institute of the HAS, H-8237 Tihany, Hungary

B.-MUSKÓ Ilona, Balaton Limnological Research Institute of the HAS, H-8237 Tihany, Hungary

DAWIDOWICZ Piotr, Department of Hydrobiology, Institute of Zoology, University of Warsaw, PL-00-046 Warsaw, Nowy Swiat 67, Poland

DE BERNARDI Riccardo, CNR Istituto Italiano di Idrobiologia, Largo Vittorio Tonolli 50/52, I-28048 Pallanza, Italy

Mrs DE BERNARDI Alessandra, CNR Istituto Italiano di Idrobiologia, Largo Vittorio Tonolli 50/52, I-28048 Pallanza, Italy

DÉVAI György, Department of Ecology, L. Kossuth University, H-4010 Debrecen, Hungary

DJUKIC Nada, Institute of Biology, Faculty of Sciences, Ilije Djuricica 6, 21 000 Novi Sad, Yugoslavia

DOKULIL Martin, Institut für Limnologie, Gaisberg 116, A-5310 Mondsee, Austria

EJSMONT-KARABIN Jolanta, Institute of Ecology PAS, Hydrobiological Station, ul. Lesna 13, 11-730 Mikołajki, Poland

ENTZ Béla, H-8237 Tihany, Váralja út, Hungary

FEKETE Gábor, Institute of Ecology and Botany of the HAS, H-2163 Vácrátót, Alkotmány u. 2-6, Hungary

FORRÓ Tibor, Balaton Limnological Research Institute of the HAS, H-8237 Tihany, Hungary

GONS Herman J., Limnological Institute, Rijksstraatweg 6, 3631 AC Nieuwersluis, The Netherlands

GOLTERMAN Han L., Station Biologique de la Tour du Valat, Le Sambuc, F-13200 Arles, France

GOPHEN Moshe, Kinneret Limnological Laboratory, P.O. Box 345, Tiberias, Israel

G.-TÓTH László, Balaton Limnological Research Institute of the HAS, H-8237 Tihany, Hungary

GULATI Ramesh D., Limnological Institute, Rijksstraatweg 6, 3631 AC Nieuwersluis, The Netherlands

GYORFFY Béla, 23 000 Zrenjanin, B. Buhe 2/5, Yugoslavia

HANSEN Anne-Mette, Institute of Biology, Odense University, Campusvej 55, 5230 Odense M, Denmark

HEBAUER Franz, Wagnerstrasse 4, D-8360 Deggendorf, FRG

HERODEK Sándor, Balaton Limnological Research Institute of the HAS, H-8237 Tihany, Hungary

HERZIG Alois, Institut für Limnologie, Gaisberg 116, A-5310 Mondsee, Austria

HERZIG Barbara, Naturhistorisches Museum, Fischsammlung, A-1014 Wien, Postfach 417, Austria

ILMAVIRTA Veijo, Maj and Tor Nessling Foundation, P. Hesperiankatu 3 A 4, SF-00260 Helsinki 26, Finland

ISTVÁNOVICS Vera, Balaton Limnological Research Institute of the HAS, H-8237 Tihany, Hungary

JENSEN Henning Skovgaard, Institute of Biology, Odense University, Campusvej 55, 5230 Odense M, Denmark

JEPPESEN Erik, National Agency of Environmental Protection, The Freshwater Laboratory, Lysbrogade 52, 8600 Silkeborg, Denmark

KEREKES Joseph, Canadian Wildlife Service, c/o Biology Department, Dalhousie University, Halifax, Nova Scotia B3H 4J1, Canada

KISS Keve Tihamér, Hungarian Danube Research Station of the HAS, H-2131 Göd, Hungary

LAMMENS Eddy H. R. R., Limnological Research Institute, Tjeukemeer Laboratory, De Akkers 47, 8536 VD Oosterzee, The Netherlands

LEVENTER Haim, Mekoroth Water Co., Mekoroth, Jordan District, Nazareth Ilit 17105, Israel

MALETIN Stevan, Institute of Biology, Faculty of Sciences, Ilije Djuricica 6, 21 000 Novi Sad, Yugoslavia

MALTHUS Tim J., Limnological Institute, Rijksstraatweg 6, 3631 AC Nieuwersluis, The Netherlands

MEIJER Marie-Louise, Institute for Inland Water Management and Waste Water Treatment, Postbox 17, 8200 AA Lelystad, The Netherlands

MÜLLER German, Institut für Sedimentforschung der Universität Heidelberg, P.O. Box 103020, D-6900 Heidelberg, FRG

Mrs MÜLLER Renate, Institut für Sedimentforschung der Universität Heidelberg, P.O. Box 103020, D-6900 Heidelberg, FRG

NAKAMOTO Nobutada, Shinshu University, Department of Applied Biology, Ueda, Nagano 386, Japan

NILSSON Nils Arvid, Institute of Freshwater Research S-170 11 Drottningholm, Sweden

OZIMEK Teresa, Department of Hydrobiology, Institute of Zoology, University of Warsaw, PL-00-046 Warsaw, Nowy Swiat 67, Poland

PADISÁK Judit, Botanical Department of the Hungarian Natural History Museum, H-1475 Budapest, P.O. Box 222, Hungary

PARPAROV A. S., Sevan Hydrobiological Station, Kirov 186, Sevan, Armenian SSR, USSR 378610

PAULOVITS Gábor, Balaton Limnological Research Institute of the HAS, H-8237 Tihany, Hungary

PERROW Martin R., School of Biological Sciences, University of East Anglia, NR4 7TJ Norwich, England

PETERSEN Annette, National Agency of Environmental Protection, The Freshwater Laboratory, Lysbrogade 52, 8600 Silkeborg, Denmark

PETTERSSON Kurt, Erken Laboratory, Institute of Limnology, University of Uppsala, PL 4200, S-76100 Norrtalje, Sweden

PONYI Jeno E., Balaton Limnological Research Institute of the HAS, H-8237 Tihany, Hungary

PUJIN Vlasta, Institute of Biology, Faculty of Sciences, Ilije Djuricica 6, 21 000 Novi Sad, Yugoslavia

P.-ZÁNKAI Nóra, Balaton Limnological Research Institute of the HAS, H-8237 Tihany, Hungary

RAAT Alexander, Organisatie ter Verbetering van de Binnenvisserij, P.B. 433, 3430 AK Nieuwegein, The Netherlands

RIJKEBOER Machteld, Limnological Institute, Rijksstraatweg 6, 3631 AC Nieuwersluis, The Netherlands

RUŽICKA Libor, Fisheries Research Institute, 252 66 Libcice nad Vltavou, Czechoslovakia

RUŽICKOVÁ Jana, Fisheries Research Institute, 252 66 Libcice nad Vltavou, Czechoslovakia

SALÁNKI János, Balaton Limnological Research Institute of the HAS, H-8237 Tihany, Hungary

SØNDERGAARD Martin, Botanical Institute, University of Aarhus, National Agency of Environmental Protection, Lysbrogade 52, 8600 Silkeborg, Denmark

SORTKJAER Ole, National Agency of Environmental Protection, The Freshwater Laboratory, Lysbrogade 52, 8600 Silkeborg, Denmark

STANCZYKOWSKA-PIOTROWSKA Anna, Agricultural-Pedagogical University, 08-110 Siedlce, ul. Prusa 12, Poland

SZÁSZ Erzsébet, Balaton Limnological Research Institute of the HAS, H-8237 Tihany, Hungary

SZILÁGYI Ferenc, Research Center for Water Resources Development (VITUKI), H-1453 Budapest, Pf. 27, Hungary

SZÖLLÖSI Gyula, Institute of Civil Engineering of Vojvodina Region, Palicki put., YU-24 000 Subotica, Yugoslavia

TALLING Jack F., Freshwater Biological Association, Ambleside, Cumbria LA22 OLP, England

Mrs TALLING Ida, Freshwater Biological Association, Ambleside, Cumbria LA22 OLP, England

TÁTRAI István, Balaton Limnological Research Institute of the HAS, H-8237 Tihany, Hungary

TELTSCH Benjamin, Mekoroth Water Co., Mekoroth, Jordan District, Nazareth Ilit 17105, Israel

VAN DONK Ellen, Provinciale Waterstaat, Galileilaan 15, P.B. 80300, 3508 TH Utrecht, The Netherlands

VAN HUET Harry, Limnological Research Institute, Tjeukemeer Laboratory, De Akkers 47, 8536 VD Oosterzee, The Netherlands

VAN LIERE Louis, Limnological Institute, Rijksstraatweg 6, 3631 AC Nieuwersluis, The Netherlands

WEISSE Thomas, Limnological Institute, University of Constance, Postfach 5560, D-7750 Konstanz, FRG

WICKSTRØM Håkan, Institute of Freshwater Research, S-170 11 Drottningholm, Sweden

WINFIELD J. Ian, Freshwater Laboratory, University of Ulster, Traad Point, Drumenagh, Magherafelt, Co. Londonderry BT45 6LR, Northern Ireland

WOYNÁROVICH Elek, H-1011 Budapest, Attila u. 121, Hungary

ZLINSZKY János, Balaton Limnological Research Institute of the HAS, H-8237 Tihany, Hungary

Participants in the International Symposium on Trophic Relationships in Inland Waters, Tihany, Hungary

1. E. Jeppesen
2. M. Søndergaard
3. V. Istvánovics
4. L. G.-Tóth
5. T. Malthus
6. L. Vörös
7. N. A. Parparov
8. E. Lammens
9. A. Raat
10. P. Dawidowicz
11. M. Perényi
12. J. F. Talling
13. M. Dokulil
14. G. Paulovits
15. B. A. G. Entz
16. H. L. Golterman
17. J. Zlinszky
18. A. Abaffy-Bothár

19. F. Hebauer
20. A. Herzig
21. B. Herzig
22. R. D. Gulati
23. S. Maletin
24. V. Ilmavirta
25. K. T. Kiss
26. B. Teltsch
27. N. P.-Zánkai
28. E. Woynárovich
29. I. B.-Muskó
30. T. Ozimek
31. O. Sortkjaer
32. A. Stanczykowska-Piotrowska
33. H. Wickstrøm
34. Hesham M. Shafik
35. H. Van Huet

36. R. De Bernardi
37. I. Tátrai
38. E. Bergstrand
39. K. Pettersson
40. J. Padisák
41. J. Ejsmont-Karabin
42. N. Djukic
43. Gy. Szöllösi
44. M. L. Mejer
45. Th. Weisse
46. A. M. Hansen
47. H. S. Jensen
48. A. Petersen
49. M. Gophen
50. H. J. Gons
51. H. Leventer
52. G. Müller
53. J. Ružicková

54. Gy. Dévai
55. L. Ružicka
56. J. Kerekes
57. Á. Berczik
58. N. Nakamoto
59. G. Fekete
60. J. Salánki
61. V. Pujin
62. P. Biró
63. E. Van Donk
64. B. Györffy
65. M. R. Perrow
66. M. Rijkeboer
67. L. Van Liere
68. I. Winfield

Opening address

János Salánki
Director of the Balaton Limnological Research Institute of the Hungarian Academy of Sciences, Tihany

Ladies and Gentlemen, Dear Colleagues,

It is a great pleasure and honour to greet all of you arriving from 18 different countries at the Symposium on Trophic Relationships in Inland Waters, here, at the Balaton Limnological Research Institute of the Hungarian Academy of Sciences. One of the main interests both of classical and modern hydrobiology concerns the interrelationship between aquatic organisms and the physical and chemical environment, and also of different organisms and species at the community level. Bacteria, plants and animals – which are both consumers and prey in the water ecosystem – are interconnected and interdependent in many and various ways. Trophic relations are factors determining not only the life, abundance or rarity of individuals, but also the dynamics and diversity of species. Although their basic characteristics are rather general, they show a lot of distinct specificities in each environment and undergo variations when the environment changes.

Due to human impacts in the catchment area, significant alterations have occurred in the life of Lake Balaton during the past 20 years. The nutrient runoff from agriculture and effluent from municipal sewage plants have increased, causing an acceleration of eutrophication; the trophic relationships in the lake have also changed. The research of the Hydrobiological Department of the Institute deals with these questions. Since similar problems are in the forefront of limnological research at many places all around the world, the main reason to organize this Symposium was to promote exchange of information and discussion in this specific topic.

As you certainly know, our lake with its 77 km length, roughly 600 km^2 surface area and mean depth of 3 meters is the largest shallow lake in Europe. Limnological research of Lake Balaton looks back over about a century. Following sporadic investigations conducted in the early decades of this period, more concentrated scientific research started with the creation of a small biological station at Révfülöp in 1925. After the opening of this Institute Balaton research was conducted here continuously and in the last decade it became the main task of the Institute. Since the existence of the special research place several hundred research papers were published about Balaton-biology, until 1977 in a separate year-book, the Annales Instituti Biologici Tihany, and afterwards in various international and national journals, which makes the results more easily available for the scientific community of the world.

Due to the activity of the hydrobiologists of the Institute, of whom Béla Hankó, Géza Entz, Olga Sebestyén, Elek Woynárovich, Béla Entz and Lajos Felföldy should be mentioned as former leading scientists, the Lake became one of the best studied inland waters not only in Europe but in the world. More than 1200 algal species, around 750 other benthic and planktonic organisms and 44 fish species have been described from Lake Balaton, and a number of varied specific problems have also been studied. Recently the main research topics of the Institute have concerned eutrophication, feeding, population dynamics and functioning of aquatic animals and fish populations, as well as the accumulation and effect of environmental pollutants in Lake Balaton. The newest results of the relevant topics will be presented at this Symposium. Dear Colleagues, our Institute – called Biological Research Institute before 1982 – was opened 60 years ago, in 1927, in the presence of the participants of the 10th International Zoological

Congress. Since then several international meetings have been held here, concerning both different topics of lake research and those of invertebrate neurobiology which is the research field of our second department, the Department of Experimental Zoology. The present meeting is one of the events commemorating the 60th anniversary of the Institute and you are particularly warmly welcomed here on this occasion.

I hope very much that you will find this meeting profitable, and will enjoy all the advantages of gatherings of such size. I wish you all interesting lectures, posters and useful discussions for the next four days. I also hope that you will have the opportunity to get an insight not only into the life of Lake Balaton, but also into the life of the people living around here, and enjoy its beauties. As the Director of the Institute, I welcome you once again and declare the Symposium opened.

Welcome address

Gábor Fekete
Corresponding member of the Hungarian Academy of Sciences

Ladies and Gentlemen,

On behalf of the Department of Biological Sciences of the Hungarian Academy of Sciences I would like to welcome all the participants to the International Symposium on 'Trophic Relationships in Inland Waters'. We are glad that this conference is being held in Hungary. There is no doubt that the topics to be discussed in the forthcoming days are of great importance.

The description, detection and modelling of trophic relationships have always represented a central issue of hydrobiology. The methodology developed around the concept of trophic relations is an integral part of this branch of science. As a specialist of terrestrial ecology, I am always deeply impressed that problems encountered in hydrobiology allow for an easier investigation and interpretation of trophic relationships than in terrestrial ecology. I think that this difference is only partly explained by differences in organisms; the freshwater medium is a much more important aspect to consider.

Under the pressure of worldwide nutritional problems, in the sixties the International Biological Programme was launched. It is well-known that in this programme main interest was focused on biological production, productivity, and the underlying ecological processes. The relatively new views of production biology emphasized the importance of nutritional relationships and of trophic structures. At that time, terrestrial ecologists made an attempt to follow closely the lines developed by hydrobiologists. Recently, the trophic aspect of ecosystems is still of primary concern. However, the formulation of problems may be a bit changed. There are research projects of increasing number in which the undesirable change or transformation of communities and ecosystems requires the detailed study of trophic relationships. Whereas some years ago the term 'eutrophication' was restricted to hydrobiology, it is now adopted in terrestrial ecology as well. A consequence of intensive agriculture near natural stands of communities is – as Professor Salánki has already mentioned – that fertilizers are washed into the environment. As a result, the original species composition is changed: sensitive and rare species disappear owing to the competitive effect of species that utilize the increased phosphorus and nitrogen content more effectively. Some nitrophilous species, not present otherwise, also become established due to eutrophication. These changes are responsible for the decrease of diversity. There is a new ecological programme in Hungary to measure the resistance of terrestrial communities against weeds. Also, a current task is to characterize the species according to their nutrient requirements. In plant communities, within the same trophic level, questions concerning the niche of populations, the niche space of communities and its partitioning become in our days really significant. Nutrients have a primary role in specifying the niche space and competition is often the manifestation of trophic relationships.

Terrestrial ecology and hydrobiology have several fields of interest in common. The phytosociology, distribution, and moreover the productivity of aquatic macrophyte communities are analyzed with standard methods of vegetation science. There is very little information on the events along the boundary of terrestrial and aquatic ecosystems – for instance, on the main features of nutrient dynamics. It is the theoretical basis that forms a powerful link between hydrobiology and terrestrial ecology. Hydrobiologists contributed much to the development of theoretical ecology. The best examples are the system models describing nutrient cycling in ecosystems. I am happy to emphasize this example here in Tihany, because the well-known Balaton Eutrophication Model was developed and verified by the scientists of this Institute in cooperation with other institutions.

My personal experience is that hydrobiologists look for and find the balance between theory and practice. Looking at the programme of this symposium, I anticipate that the results to be presented here will provide a scientific basis for further practical measures. I am convinced that the speakers, and their papers, guarantee that this conference will be a memorable contribution on the occasion of the sixtieth anniversary of the Balaton Limnological Research Institute.

I wish you a successful meeting and hope that all participants will benefit from the papers and personal contacts.

Hydrobiologia **191**: 1–8, 1990.
P. Biró and J. F. Talling (eds), Trophic Relationships in Inland Waters.
© 1990 *Kluwer Academic Publishers.*

Trophic status and the pelagic system in Lago Maggiore

R. de Bernardi, G. Giussani, M. Manca & D. Ruggiu
C.N.R. Istituto Italiano di Idrobiologia, 28048 Pallanza, Italy

Key words: Lago Maggiore, trophic evolution, plankton dynamics, eutrophication, lake recovery

Abstract

Long-term data on the main limnological parameters enabled a description of recent trophic evolution of Lago Maggiore (N. Italy). The results allow a better definition of the conditions of lake eutrophication and recovery. Moreover, they identify some important phenomena that show resilience in the lake's biota. The results have important implications for lake management and recovery.

Introduction

Lago Maggiore is one of the largest Italian lakes. Located in Northern Italy, it represents an important water resource for the densely inhabited region close to Milan. The main morphometric and hydrographic characteristics of this lake are reported in Table 1. Some figures from this table demonstrate the lake's importance: its volume represents about one-fifth of the total amount of lake water in Italy, and it contributes, via the Ticino River, a fifth of the discharge of Italy's largest river, the Po. The drainage area of Lago Maggiore (6599 km²) is very large relative to the lake's surface area (ratio 31 : 1) and, although most of the basin lies above 1000 m a.s.l., about 660 000 people inhabits this area (100 ind km^{-2}). With a maximum depth of 370 m and an average depth of 177.4 m, the lake is holo-oligomictic. As demonstrated by Piontelli & Tonolli (1964), the water renewal time is 14.5 years and so the lake tends to accumulate solutes. As a consequence, in recent decades, the lake has undergone some important changes, the most important being variations in its trophic status.

Because of its importance as a water resource and as a consequence of the deterioration in its quality, the lake has been the object of a series of limnological studies since the early 50's. These

Table 1. Morphometric characteristics of Lago Maggiore.

Catchment area (km²)	6599
Maximum altitude of catchment (m.a.s.l.)	4633
Mean altitude of catchment (m.a.s.l.)	1283
Ice cover in catchment area (km²)	7.3
Lake area (km²)	212.5
Catchment area/lake area	31.1
Lake length (km)	54
Length along the thalweg (km)	66
Maximum breadth (km)	10
Mean breadth (km)	3.94
Shoreline length (km)	170
Development of shoreline (DL)*	3.07
Volume (10⁶ m³)	37.502
Maximum depth (m)	370
Mean depth (m)	177.4
Depth of cryptodepression (m)	177
Mean depth/maximum depth	0.48
Development of volume	1.44

* Ratio of the length of shoreline to the length of the circumference of the circle of area equal to that of the lake.

have progressively intensified in the last ten years and the most relevant parameters for the assessment of water quality evolution have been analyzed. In this paper we will consider the long-term evolution of those parameters that better illustrate the recent changes in the lake's conditions.

Phosphorus loading

Starting from 1978, the phosphorus load to the lake has been measured by regularly sampling the most important tributaries (Mosello & Ruggiu, 1983). The results (Table 2) show that the total amount of phosphorus carried to the lake increas-

Table 2. Total phosphorus load from 1978 to 1985.

Year	P (t yr^{-1})
1978	370
1979	450
1980	380
1981	340
1982	280
1983	250
1984	260
1985	260

ed progressively until 1979 (when it reached a maximum of about 450 t yr^{-1}) and then started to decline, reaching the present level of about 250 t yr^{-1}.

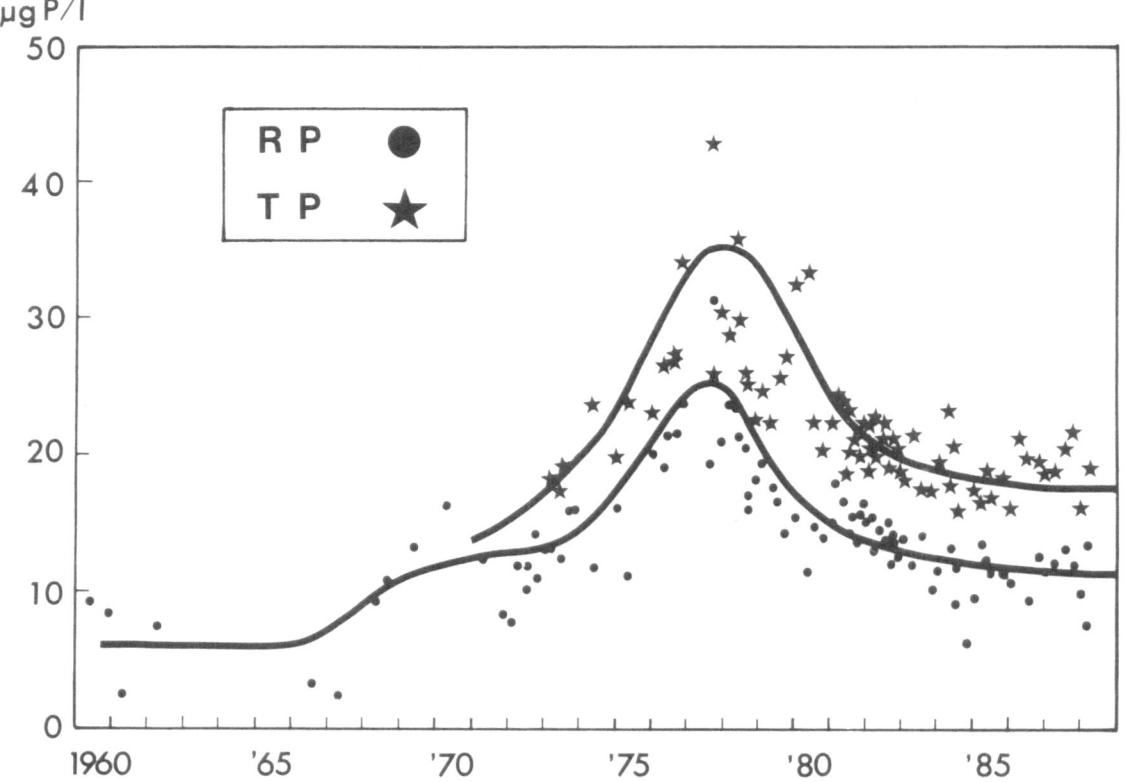

Fig. 1. Recent trends of phosphorus concentrations in the open waters of Lago Maggiore. Average values on the water column at the winter overturn.

This reduction largely reflects three changes in human activity and anthropogenic phosphorus load:

a) the implementation of sewage treatment plants that presently treat about 30% of the whole population in the near-shore area;

b) a gradual reduction down to 2.5% phosphorus in detergents, as required by recent national laws;

c) negative demographic and socio-economic trends in the drainage area of the lake in the last years.

Phosphorus and nitrogen concentration

Although total phosphorus concentrations are not consistently available before 1975, estimates of reactive phosphorus concentration in the water column at winter overturn provide a valid index of trophic status.

Starting from the early 60's, reactive phosphorus concentration (Fig. 1) progressively increased from values below $10 \mu g l^{-1}$ up to $30 \mu g l^{-1}$ in 1977, then gradually fell to the present value of $17 \mu g l^{-1}$.

Total phosphorus shows a similar trend with a maximum of about $37 \mu g l^{-1}$ in 1977 and a present value of about $20 \mu g l^{-1}$

Over the same period, nitrate-nitrogen (Fig. 2) presents a different dynamics. It progressively increased to $800 \mu g l^{-1}$ in 1977 and then maintained a steady-state (Fig. 2). These data suggest that phosphorus, rather than nitrogen, is more likely to limit productivity in Lago Maggiore.

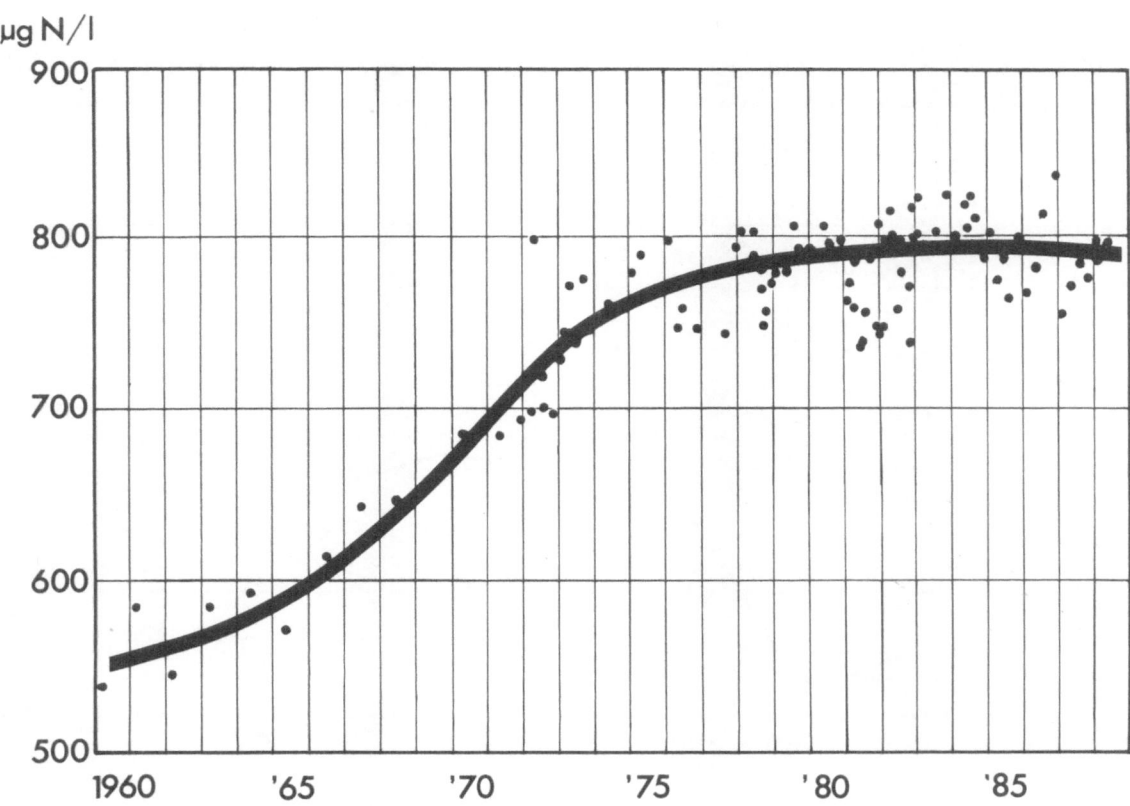

Fig. 2. Recent trends of $NO_3 - N$ concentrations in the open waters of Lago Maggiore. Average values in the water column at winter overturn.

4

Phytoplankton community

Primary productivity increased three-fold in the 60's, presenting a long term pattern similar to that described for phosphorus (Fig. 3). However, the same figure also shows that neither primary productivity nor chlorophyll *a* concentration declined in the last ten years, despite a reduction in phosphorus availability. Figure 4 shows the trends of total phosphorus and chlorophyll *a* concentrations in the upper 20 m from 1979 to 1986. Despite wide seasonal fluctuations, a decreasing trend in P concentration is clearly evident. This is not paralleled by a similar trend in chlorophyll *a* concentration. In fact, statistical analysis shows that, in the most recent years, algal biomass, as expressed by chlorophyll *a*, is not correlated with either P load or P concentration (Sas & Vermij, 1987).

Data from the 50's, prior to the phase of intense eutrophication, show that the community structure was characterized by low biomass, high diversity, and by species indicative of oligo-meso-trophy (de Bernardi *et al.*, 1987). This changed during the 60's and, since then, the community has been characterized by higher biomass, lower diversity, and a much greater importance of the blue-greens (especially *Oscillatoria rubescens*) as well as diatoms. Figure 5 shows the trends in the biomass of phytoplankton and blue-greens in recent years.

Zooplankton community

Data on zooplankton populations are available from the beginning of this century, and are useful at least for comparing species composition and community structure. The analysis of available data reveals progressive and marked changes in

Fig. 3. Recent trends of primary production (——) and annual average of chlorophyll *a* concentrations (······) in the euphotic zone of Lago Maggiore.

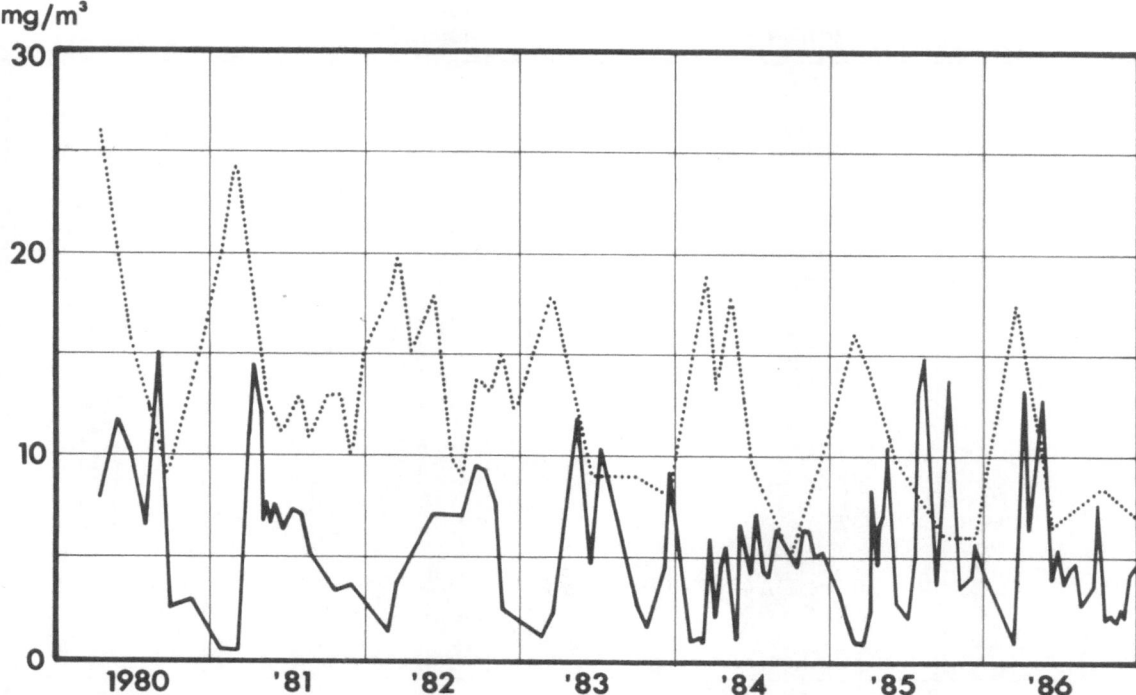

Fig. 4. Recent trends of total phosphorus (·······) and chlorophyll *a* (——) in the upper 20 m of Lago Maggiore.

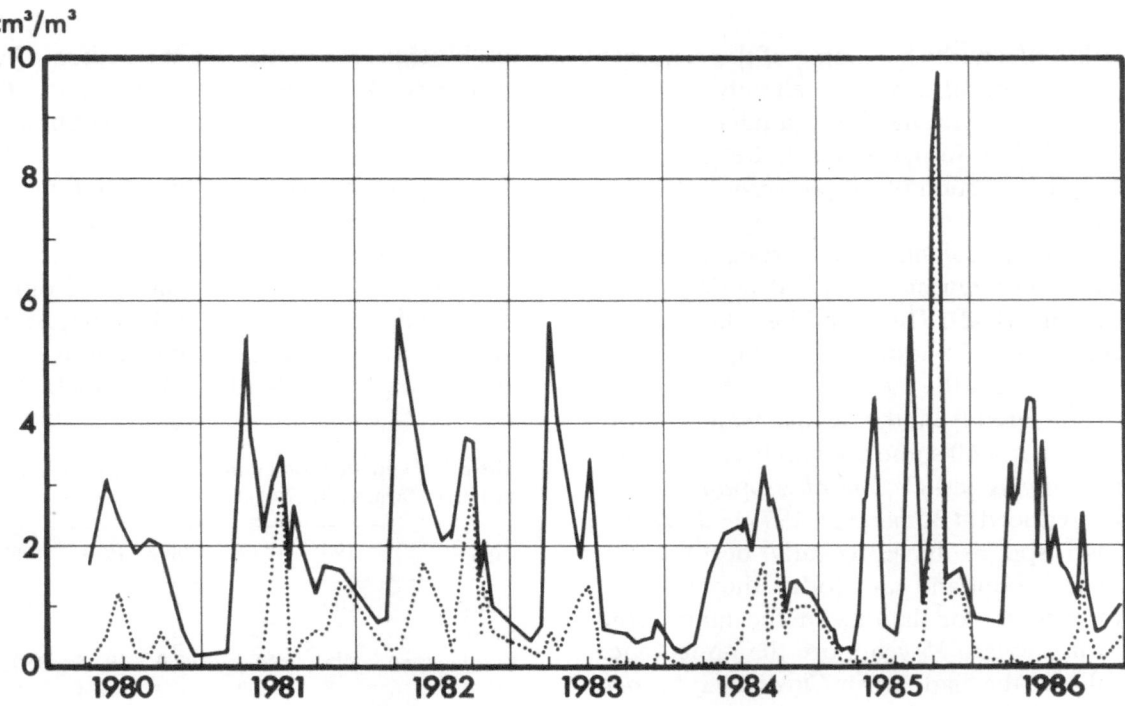

Fig. 5. Recent trends of biomass (as cell volume) of total phytoplankton (——) and blue-green algae (·······) in the upper 20 m of Lago Maggiore.

6

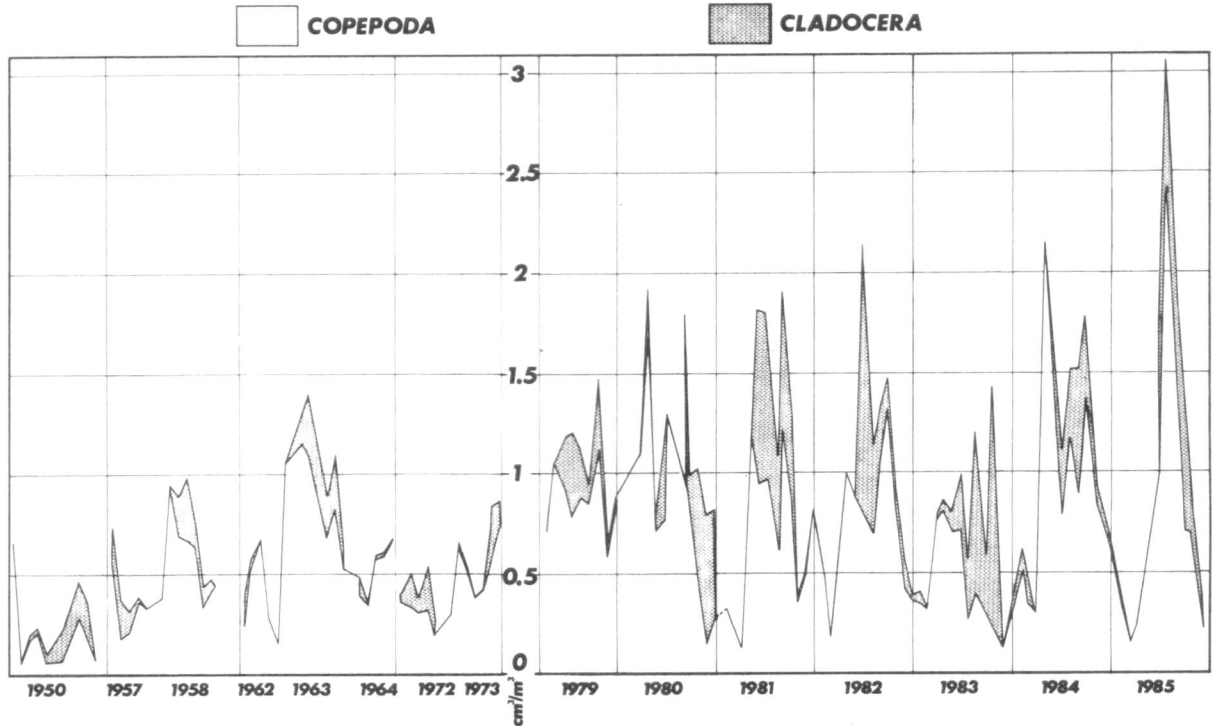

COPEPODA   CLADOCERA

Fig. 6. Recent trend of biomass (as volume) of herbivorous crustacea in the upper 50 m of Lago Maggiore.

zooplankton from the beginning of the century to 1985. This confirms a pattern already shown by Tonolli (1962), de Bernardi & Canali (1975), and Bonacina (1977). In particular, there was a relatively heavy reduction of *Diaphanosoma brachyurum* and a smaller reduction of *Mesocyclops leuckarti*, counterbalanced by a strong increase of *Daphnia hyalina* during a 'critical period' in the 50's (Tonolli, 1962). This trend has subsequently been maintained. In addition, the copepod *Heterocope saliens* and the cladoceran *Sida cristallina* have disappeared from the pelagic waters. Moreover, in the early 60's, the copepods *Eudiaptomus padanus padanus* and *Cyclops abyssorum* were noticeably reduced (Bonacina, 1977). However, these two species have recently become very much more abundant, and today they form the most important populations in the lake. Other changes in the last 35 years are the appearance, in the 60's, of the cladoceran *Chydorus sphaericus* in the open waters, with a progressive increase in density over the last 20 years, and the occurrence

of 'blooms' of species like the rotifers *Conochilus unicornis*, *Kellicottia longispina* and *Keratella cochlearis* starting in the 70's. These changes in the zooplankton community cannot always be satisfactorily or consistently explained in terms of modifications of lake trophy alone.

Focusing on herbivores (Fig. 6), two distinct periods can be identified: the first from 1950 to the middle of 70's, in which the biomass of herbivorous crustaceans remained almost constant; the second (1979–1985) in which herbivore

Table 3. Estimated total annual yield of pelagic fish species in Lago Maggiore (t yr^{-1}).

Species	1965	1979	1980	1981	1982	1983
Bleak	100	58.3	35	75	111	99
Shad	37	–	–	–	–	–
Coregonus sp.	75	51	97.4	193.5	370.5	384.3
Trout	5	2.2	4.2	8.4	6.3	6.2
Total	217	111.5	136.6	276.9	487.8	489.5

biomass reached higher values. The maximum, of $3 \text{ cm}^3 \text{ m}^{-3}$, was observed in 1985. It must be noticed that the higher herbivore biomass attained in the last decade is supported by a phytoplankton community similar to that present in the 70's.

Fish community

The evaluation of changes that have been occurred in the fish population of Lago Maggiore in recent decades is very difficult. The introduction of exotic species and selective exploitation of particular species have drastically affected the original structure of the fish community. In particular, the very successful introduction of *Coregonus* 'bondella' in the early 50's strongly influenced the dynamics of the entire native fish community. The landlocked shad (*Alosa fallax lacustris*) and the coregonine *Coregonus* 'lavarello' almost completely disappeared in the years immediately following this introduction. However, it is possible to observe an overall increase in the total catch of pelagic species (Table 3), mainly due to the newly introduced 'bondella'. After a decline at the end of the 70's, there has been a progressive increase in fish production in the last few years.

Discussion

These data provide a good picture of recent trends in the trophic status of Lago Maggiore. Two distinct periods emerge: the first one of rapid eutrophication (early 60's to late 70's), and the second of moderate oligotrophication (early 80's to the present). However, the biota has not responded similarly to similar abiotic water characteristics in these periods. During the eutrophication phase, the biotic communities at all trophic levels increased rapidly in standing stock and productivity and the eutrophication of Lago Maggiore was in a very good agreement with general models of eutrophication. Indeed, data from the end of the 70's fits the Vollenweider-OECD model perfectly (OECD, 1982; Mosello & Ruggiu, 1985).

On this basis, predictions were made in the early 80's about the possible recovery of the lake, given different levels of reduction of phosphorus load. According to these predictions, a 50% reduction should have provoked a decline in the trophic status of the lake from meso-eutrophy to the upper limit of oligotrophy.

However, data collected in the last eight years have not confirmed this prediction. During the phase of moderate oligotrophication, the total phosphorus load to the lake and, consequently, the phosphorus concentration in the lake, have been halved. Nevertheless, no corresponding declines in the standing stocks of the biota were observed.

It is thus evident that the biota of the lake are somehow resilient to reduction in their nutrient supply. No data are at present available on the dynamic activities of the different biological communities. However, it is clear that they are more efficiently utilizing and recycling their nutrients, because a smaller resource base is supporting undiminished (and possibly even larger) plant and animal communities.

In this situation in which the phosphorus concentration is considerably reduced, the biotic communities seem to be particularly sensitive to biotic interactions (predation, competition) and meteorological features. Strong evidence for this phenomenon has been collected in the last few years (Ruggiu & Panzani, 1985, de Bernardi *et al.*, 1987). Such resilience seems to be common for lakes of different typology, subjected to restoration through nutrient reduction (Sas & Vermij, 1987). However, the basic mechanisms determining this process are largely unknown. This gap in our knowledge of the effects of a fundamental tool for lake management should hold considerable interest for both pure and applied research.

Acknowledgement

We express our grateful thanks to Prof. Robert H. Peters for his suggestions and English revision.

8

References

Bonacina, C., 1977. Lo zooplancton del Lago Maggiore: situazione presente e modificazioni a lungo termine della struttura comunitaria. Mem. Ist. Ital. Idrobiol. 34: 79–120.

de Bernardi, R. & S. Canali, 1975. Population dynamics of pelagic cladocerans in Lago Maggiore. Mem. Ist. Ital. Idrobiol. 32: 365–392.

de Bernardi, R., G. Giussani, M. Manca & D. Ruggiu, 1987. Long-term dynamics of plankton communities in Lago Maggiore (N. Italy). Verh. int. Ver. Limnol. 23: 729–733.

Mosello, R. & D. Ruggiu, 1983. Apporti e concentrazioni di nutrienti, fitoplancton e livello trofico del Lago Maggiore. In: Commissione Internazionale per la Protezione delle Acque Italo-Svizzere, Rapporto Quinquennale 1978–1982. Relazione Conclusiva sulle Indagini Limnologiche Relative al Lago Maggiore: 39–64.

Mosello, R. & D. Ruggiu, 1985. Nutrient load, trophic conditions and restoration prospects of Lake Maggiore. Int. Revue ges. Hydrobiol. 70: 63–75.

OECD, 1982. Eutrophication of waters – Monitoring, assessment and control., Paris, 1982: 154 pp.

Piontelli, R. & V. Tonolli, 1964. Il tempo di residenza delle acque lacustri in relazione ai fenomeni di arricchimento in sostanze immesse, con particolare riferimento al Lago Maggiore. Mem. Ist. Ital. Idrobiol. 17: 247–266.

Ruggiu, D. & P. Guida, 1987. Indagini sul fitoplancton. In: Commissione Internazionale per la Protezione delle Acque Italo-Svizzere, Ricerche sull'Evoluzione del Lago Maggiore. Aspetti di Limnologia., Campagna 1986: 64–70.

Sas, H. & S. Vermij, 1987. Eutrophication management in international perspective. Third Interim Report IMSA: 312 pp.

Tonolli, V., 1962. L'attuale situazione del popolamento planctonico del Lago Maggiore. Mem. Ist. Ital. Idrobiol. 15: 81–133.

Hydrobiologia **191**: 9–14, 1990.
P. Biró and J. F. Talling (eds), Trophic Relationships in Inland Waters.
© 1990 *Kluwer Academic Publishers.*

The spring development of phytoplankton in Lake Erken: species composition, biomass, primary production and nutrient conditions – a review

Kurt Pettersson
Uppsala University, Institute of Limnology, Erken Laboratory, Norr Malma 4200, S-761 73 Norrtälje, Sweden

Key words: phytoplankton, primary production, phosphorus, nitrogen

Abstract

In Lake Erken climatic factors such as duration of ice cover, snow-depth and insolation govern the phytoplankton development and the species composition during the spring, with significant variations from year to year.

Generally the small diatom, *Stephanodiscus hantzschii var. pusillus* creates a conspicuous peak at ice-break. In some years motile dinoflagellates start to develop under the ice already in early March, which results in a much longer spring bloom.

The highest biomasses were recorded in 1954–1955 with values up to 11 mg l^{-1} of fresh weight. The chlorophyll *a* concentrations have at most reached an epilimnetic average of 30 μg l^{-1}.

The primary production reached a maximum value of 2200 mg C m^{-2} d^{-1} in 1955 and the average production for two months during the spring varied from 30 to 64 mg C m^{-3} d^{-1}.

Concerning nutrients, phosphorus was shown to be the limiting nutrient at the end of the spring bloom. This fact was confirmed by orthophosphate concentrations, algal surplus phosphorus content and alkaline phosphatase activity, as well as estimations of inorganic N : P and C : P ratios and nutrient enrichment experiments.

Introduction

From the early fifties and onwards the spring development of phytoplankton and its conditions have been a central theme for limnological research at the Erken laboratory. Already in 1954 and 1955 the biomass development and the primary production were measured intensively (Rodhe *et al.*, 1958). Nauwerck (1963) continued with a year-round study in 1957 of phytoplankton and zooplankton biomass, species composition

and production. Pechlaner visited the laboratory in 1960 in order to follow the outburst of phytoplankton during the spring in relation to nutrient conditions and climatic factors (Pechlaner, 1970). During the 1970's considerable effort was made to estimate the phosphorus situation during the spring, and this work was continued during the early 1980's (Boström & Pettersson, 1977; Pettersson, 1979a, b; 1980, 1985; Eriksson & Pettersson, 1984). The nitrogen cycle was also followed (Boström, 1981; Tirén & Boström, 1983).

In 1983 Bell and Kuparinen assessed phytoplankton and bacterioplankton production during early spring (Bell & Kuparinen, 1984). The information given in these papers is here summarized and some new results from more recent years are included.

Lake Erken

Lake Erken (59° 51' N, 18° 35' E) is situated 11.1 m above sea level and has an area of 24 km². The mean depth is 9 m and the water residence time about 7 years (Widell, 1970).

The lake is regularly covered with ice for 16 to 18 weeks from late December/early January to late April/early May. After the spring overturn the summer stratification starts to develop in late May.

The conductivity (at 20 C) in the epilimnion is about 27 mS m^{-1} and the pH within 7.9–8.7. The alkalinity of the lake has according to yearly averages decreased from 1.8 mequ l^{-1} in the 1960's to 1.4 mequ l^{-1} twenty years later as shown in Fig. 1 (Rodhe, 1981).

Nitrate is accumulated in the whole water column during autumn and winter and reaches a level of around 300 μg N l^{-1}. Ammonia, however, is not accumulated and the concentration is less than 10 μg N l^{-1}, except at certain deep sites. Phosphate has a concentration of around 10 μg P l^{-1} during winter, but is rapidly taken up by the phytoplankton in spring.

About a month before ice-break the transparency is about 10 m and then it decreases to 3 m as a minimum during the spring peak of phytoplankton.

A common sampling point situated above the deepest point has been used by all investigators of the phytoplankton spring bloom.

In some cases comparisons cannot be made due to differences in technique and approach. For

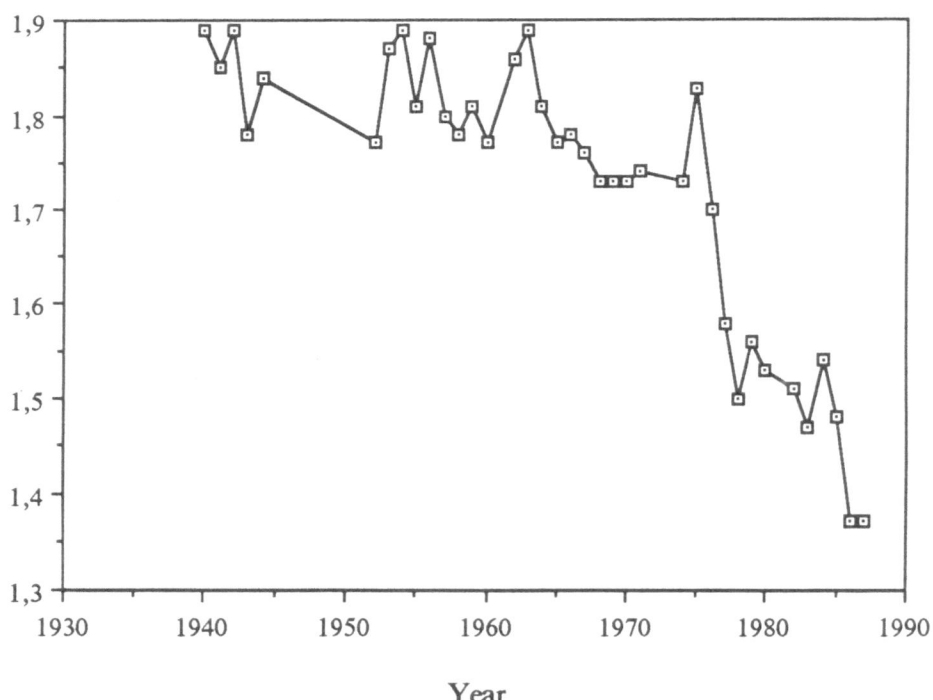

Fig. 1. The alkalinity (yearly means) in Lake Erken during the last forty years.

11

the determination of primary production several incubation times have been used as well as 'reduced series' with measurements at one or two depths. The earliest chlorophyll measurements were made with a fluorometric technique, which gave too high values. The methodological problems in chemical analyses used at the Erken laboratory are discussed by Boström & Pettersson (1973).

Results and discussion
Algal abundance

Stephanodiscus hantzschii var. *pusillus* dominated the spring outburst of phytoplankton during most of the investigation period (Rodhe *et al.*, 1958; Nauwerck, 1963; Pechlaner, 1970; Boström, 1981; Bell & Kuparinen, 1984). Regular winters with a snow- and ice-cover, which efficiently extinguishes the light, create conditions favourable for this small-size (150 μm^3 – Nauwerck, 1963; 85-235 μm^3 – Blomqvist, pers. comm.) diatom to develop exponentially around ice-break, when the light conditions improve.

Atypical climatic conditions, however, affect the species composition of phytoplankton during the spring severely. In the winter of 1973, with only 10 weeks of ice-cover and almost no snow, autumnal diatoms could persist and the large species *Stephanodiscus rotula* (Ehr.) Grun. (800-4000 μm^3 – Blomqvist, pers. comm.) and *Asterionella formosa* (280-1000 μm^3 – Blomqvist, pers. comm.) were dominant already in early March; the biomass maximum was reached on 1 April (Boström & Pettersson, 1977).

In 1979 the ice-cover lasted for 21 weeks, but the light conditions improved somewhat in early March due to a snow-melting period. This year motile dinoflagellates developed just under the ice and intensive vertical migration was shown (Pettersson, 1985). Similar migrations were well followed by Nauwerck (1963) in 1957. The dinoflagellates *Woloszynskia ordinata* and *Peridinium aciculiferum* were dominant during the entire spring development in 1979, a phenomenon not recorded before. In 1984 *Melosira islandica* ssp.

helvetica was more abundant than in earlier years (Eriksson, pers. comm.). A similar increase in number for this species was reported by Pechlaner (1970) for the spring of 1960, although the biomass value of 1984 (1.1 mg l^{-1}) was not reached.

The species composition of phytoplankton affects the form of the spring peak as shown in Fig. 2. It is a well-known fact that the sedimentation of diatoms is very rapid; at the traditional outburst of diatoms during the spring in Lake Erken, they have decreased significantly a few days after the peak.

The average chlorophyll *a* concentrations in the epilimnion have never exceeded 30 $\mu g l^{-1}$ and the

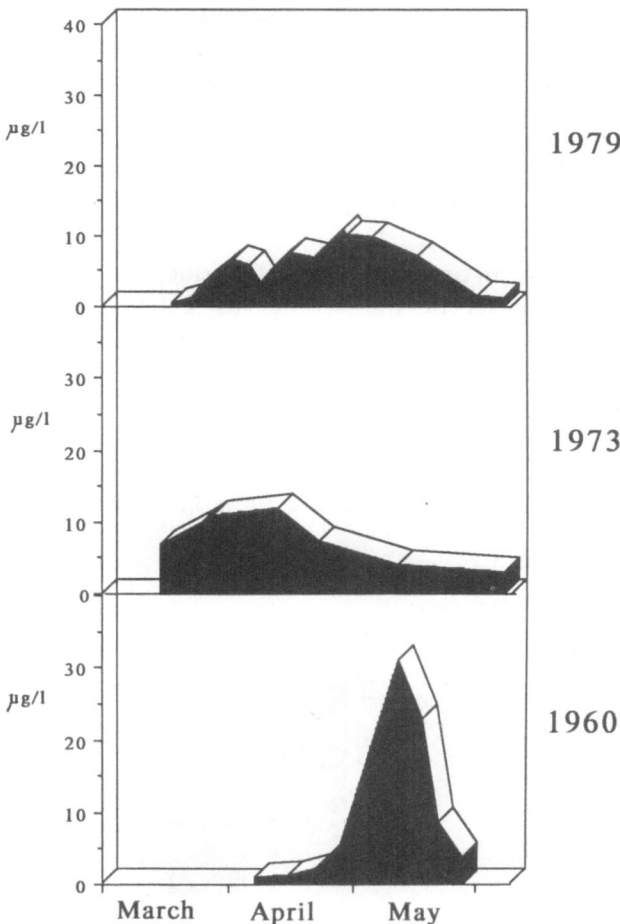

Fig. 2. Temporal change in the average (0–10 m) phytoplankton biomass (expressed as μg chl *a* l^{-1}) in Lake Erken during springs with different species composition.

range for recorded maxima is 11 to 30 μg l^{-1}. The reported fresh weight maxima vary from 4 to 11 mg l^{-1}. However, in thin layers beneath the ice up to 75 μg chl a l^{-1} was measured in 1980 (unpubl. data). The fresh weights recorded by Rodhe *et al.* (1958) for the springs of 1954 and 1955 are still the highest reported for Lake Erken.

Primary production

The primary production has been measured during several springs from 1954 onwards with the ^{14}C-method (Rodhe *et al.*, 1958; Nauwerck, 1963; Pechlaner, 1970; Boström & Pettersson, 1977; Eriksson & Pettersson, 1984; Bell & Kuparinen, 1984). The maximum production rates of the different springs are presented in Fig. 3. The highest primary production rate yet measured was recorded in May 1955, as 2200 mg C m^{-2} d^{-1}. However, a value of 2700 mg C m^{-2} d^{-1} was calculated by Boström & Pettersson (1977) as a maximum for the spring of 1960.

The mean probable primary productions during two months of the spring, as calculated by Boström & Pettersson (1977), are shown in Fig. 4. There was a twofold variation in average production.

Bell & Kuparinen (1984) described Lake Erken as a batch culture during the spring outburst of phytoplankton – a good comparison since the

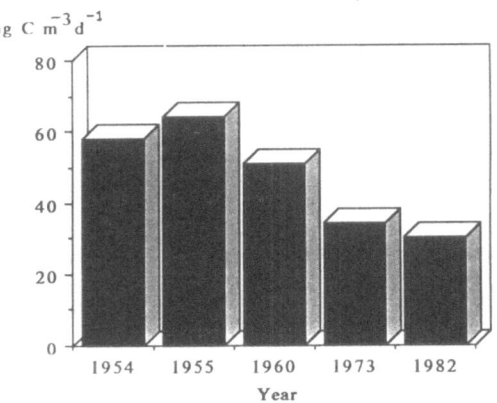

Fig. 4. Comparison of the mean probable primary production (see Boström & Pettersson, 1977) in Lake Erken during two months in the springs of five years.

grazing from zooplankton is negligible (Nauwerck, 1963; Pechlaner, 1970). They also estimated the bacterioplankton production to 1.2–1.7 mg C m^{-3} d^{-1} during the spring of 1983 from ^{3}H-thymidine incorporation. Under the assumption of similar bacterioplankton production throughout the water column this corresponds to 11–15 mg C m^{-2} d^{-1}. The gross bacterioplankton production (production plus respiration) was 20% of gross primary production, when calculated per square meter of surface area.

Nutrients

The nutrient supply and availability during the spring development of phytoplankton were followed intensively from 1975 to 1984.

Nitrogen cycling was studied by determination of nitrate, nitrite and ammonia concentrations regularly and by laboratory experiments on nitrification and denitrification (Boström, 1981; Tirén & Boström, 1983). Nitrate has a markedly regular seasonal variation in Lake Erken. The formation of nitrate by nitrification occurs mainly in the surface sediment and nitrate is accumulated during the turnover period in autumn with low primary production. Under these conditions the denitrification process is inhibited. The concentration of nitrite is negligible during the whole year. Ammonia is never accumulated in high con-

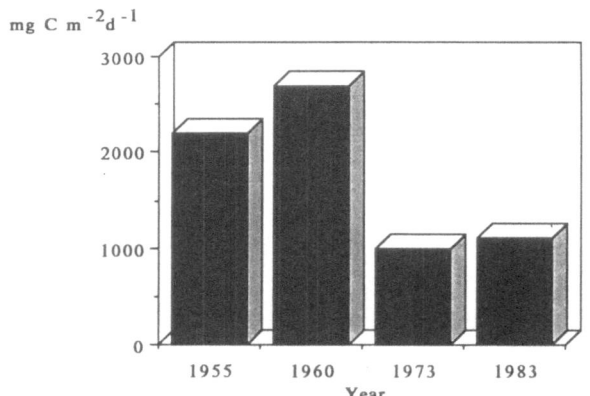

Fig. 3. The maximum primary production (recorded or calculated) for one day in Lake Erken during four springs.

centrations in the whole water column. Inflowing water occasionally causes high concentrations of $100–200 \, \mu g \, NH_4–N \, l^{-1}$ (Boström, 1981).

In order to investigate the phosphorus conditions, measurements of total phosphorus, particulate phosphorus and enzymatically determined orthophosphate (Pettersson, 1979a) were combined with determinations of biomass-related physiological parameters such as surplus phosphorus content and specific alkaline phosphatase activity (for methods and results see Pettersson, 1979b, 1980, 1985). Laboratory experiments on nutrient enrichment were performed during 1982–1984 with a completely randomized factorial design (Eriksson & Pettersson, 1984; Eriksson, in prep.).

The results presented in the above mentioned papers can be summarized as follows:

The orthophosphate concentration decreased from about $10 \, \mu g \, P \, l^{-1}$ to below the detectable level ($0.1 \, \mu g \, P \, l^{-1}$) by late April, after a rapid uptake by phytoplankton starting in late March/early April.

A luxury uptake of phosphorus was shown by high surplus phosphorus content in the particulate matter. The maximum content registered, $10 \, \mu g \, P$ per mg C, appeared in March 1979, when the dinoflagellates dominated the biomass beneath the ice. This figure was twice as high as any previously measured value.

Surplus phosphorus can also be expressed as concentration in the water, to be compared with orthophosphate, particulate phosphorus and total phosphorus. In 1979 and 1980, $12 \, \mu g \, P \, l^{-1}$ was measured as surplus phosphorus beneath the ice in late March due to a dense dinoflagellate bloom. In 1983 and 1984 more moderate concentrations, around $3.5 \, \mu g \, P \, l^{-1}$, were registered in the middle of April, before the phytoplankton outburst. This was in agreement with the results obtained during the years 1975 to 1978. However, surplus phosphorus determination turned out to be useful in showing luxury phosphorus storage by phytoplankton before the logarithmic growth began around ice-break. When related to particulate phosphorus, surplus phosphorus decreased from more than 30% to 15% in 1979. The minimum

values for the specific surplus phosphorus content, reached in May, were about $0.4 \, \mu g \, P$ per mg C, i.e. one order of magnitude less than in March/April.

As recommended by several authors (see Pettersson, 1985), alkaline phosphatase activity was used as an indicator of phosphorus deficiency in phytoplankton. In Lake Erken, measurements of the alkaline phosphatase activity were performed during the spring of the years 1975 to 1984. The specific activity increased in 1979 from 0.4 nmol $(mg \, C)^{-1} \, min^{-1}$ in March and April to 7 nmol $(mg \, C)^{-1} \, min^{-1}$ in late May. This picture was also valid for the years before as well as after. Also, as shown in Eriksson & Pettersson (1984), the actual activity (nmol $l^{-1} \, min^{-1}$) increased drastically after the spring peak of phytoplankton in 1982 (see also Pettersson, 1980). In the latter paper it was also noticed that the proportion of free enzymes was high, 40% to 70% of the total activity, when the phosphorus deficiency was severe in late May.

In 1976 it was possible to calculate the Michaelis constant of the alkaline phosphatases, which decreased to a minimum of $0.3 \, \mu mol \, l^{-1}$ in May, when the algae were phosphorus-limited. It was concluded that the phytoplankton compensated the phosphorus deficiency with an increase in enzyme production, liberation of free exoenzymes and with an improved ability to use low substrate concentrations (Pettersson, 1980).

Additional information from N:P and C:P ratios as well as nutrient enrichment experiments confirmed the statement that phosphorus limitation ended the spring outburst of phytoplankton in Lake Erken (Pettersson, 1980, 1985; Eriksson & Pettersson, 1984). In some years also nitrate was depleted, but generally phosphorus is the controlling factor. Pechlaner (1970) stated that depletion of inorganic nutrients was the obvious cause for the end of the spring outburst. He also concluded that silicon did not decrease enough to limit the growth of diatoms. This was verified by all later investigations. The concentration of molybdate reactive silicon very rarely decreased below $0.5 \, mg \, Si \, l^{-1}$. The concentrations of micronutrients have not been followed in Lake Erken.

14

However, selenium has been studied and shown to have significant effect on algal growth in laboratory enrichment experiments with natural algal assemblages (Eriksson & Pettersson, 1984) as well as algal cultures (Eriksson, 1982).

Conclusion

The spring development of phytoplankton is an event which has been followed more or less intensively on a number of occasions in Lake Erken during the last thirty years.

The form of the spring bloom varies, mainly due to climatic factors such as duration of ice- and snow-cover and thus light conditions at the end of the winter stratification affecting the species composition. A sharp phytoplankton peak with the small diatom *Stephanodiscus hantzschii var. pusillus* as the dominant species is most often occuring, but exceptions are not uncommon.

Studies of the nutrient conditions, including usage of physiological indicators of algal phosphorus status, have shown phosphorus to be the factor ending the spring peak. The studies of the pelagial community in Lake Erken are to be continued; emphasis will be put on improvements in primary production estimates, automatic continuous measurement of *in vivo* fluorescence and determination of phosphorus turnover.

References

Bell, R. J. & J. Kuparinen, 1984. Assessing phytoplankton and bacterioplankton production during early spring in Lake Erken, Sweden. Appl. Envir. Microbiol. 48: 1221–1230.

Boström, B., 1981. Factors controlling the seasonal variation of nitrate in Lake Erken. Int. Revue ges. Hydrobiol. 66: 821–836.

Boström, B. & K. Pettersson, 1973. Fysikaliska och kemiska data från Erken sedan 1930-talet. Scripta Limnologica Upsaliensia 344, 26 pp. (in Swedish).

Boström, B. & K. Pettersson, 1977. The spring development of phytoplankton in Lake Erken. Freshwat. Biol. 7: 327–335.

Eriksson, C., 1982. Effects of selenium and phosphorus on growth of the diatom *Stephanodiscus Hantzschii v pusillus*. In H. Håkansson (ed.) Rapport från diatomésymposium. CODEN. LUNBDS/(NBGK-7022)/1-120(1982) pp. 83–102.

Eriksson, C. & K. Pettersson, 1984. The physiological status of spring phytoplankton in Lake Erken – field measurements and nutrient enrichment experiments. Verh. int. Ver. Limnol. 22: 743–749.

Nauwerck, A., 1963. Die Beziehungen zwischen Zooplankton und Phytoplankton im See Erken. Symb. Bot. Ups. 17: 5, 163 pp.

Pechlaner, R., 1970. The phytoplankton spring outburst and its conditions in Lake Erken (Sweden). Limnol. Oceanogr. 15: 113–130.

Pettersson, K., 1979a. Enzymatic determination of orthophosphate in natural waters. Int. Revue ges. Hydrobiol. 64: 585–607.

Pettersson, K., 1979b. Orthophosphate, alkaline phosphatase activity and algal surplus phosphorus in Lake Erken. Acta Upsaliensia 505, 15 pp.

Pettersson, K., 1980. Alkaline phosphatase activity and algal surplus phosphorus as phosphorus-deficiency indicators in Lake Erken. Arch. Hydrobiol. 89: 54–87.

Pettersson, K., 1985. The availability of phosphorus and the species composition of the spring phytoplankton in Lake Erken. Int. Revue ges. Hydrobiol. 70: 527–546.

Rodhe, W., 1981. Tänkbara åtgärder i recipienter, sjöar – biologiska metoder. In Kontakkonferens om försurning, Elmia Jönköping September 1981, National Swedish Environmental Protection Board, 4 pp. (in Swedish).

Rodhe, W., R. A. Vollenweider & A. Nauwerck, 1958. The primary production and standing crop of phytoplankton. In A. A. Buzzati-Traverso (ed.), Perspectives in Marine Biology. Univ. Calif. Press, Berkeley: 299–322.

Tirén, T. & B. Boström, 1983. Simultaneous nitrification-denitrification in Lake Erken sediments. In T. Tirén, On denitrification in lakes (thesis). Acta universitatis Upsaliensis 673, 15 pp.

Widell, A., 1970. Närsalterna i sjön Erken sommaren 1965. Scripta Limnologica Upsaliensia 250, 24 pp. (in Swedish).

Hydrobiologia **191**: 15–21, 1990.
P. Biró and J. F. Talling (eds), Trophic Relationships in Inland Waters.
© 1990 *Kluwer Academic Publishers.*

Some characteristics of the community of autotrophs of Lake Sevan in connection with its eutrophication

A.S. Parparov
Sevan Hydrobiological Station of the Academy of Sciences of Armenian SSR, Kirov 186, Arm. SSR, USSR 378610

Key words: Lake Sevan, water level lowering, eutrophication, phytoplankton, primary productivity

Abstract

Eutrophication of Lake Sevan caused by the artificial lowering of water level was accompanied by changes in the structure and dynamics of the planktonic communities. A dominance of diatoms up to 1983 was changed to that of green algae in the last years. Primary production of plankton rose and then decreased in the process of eutrophication. The annual average primary production in 1982–1986 – $250 \text{ g C m}^{-2} \text{ yr}^{-1}$ – is evidently close to the steady state production under the present morphometry of the lake. The activity coefficient of phytoplanktonic photosynthesis changed within relatively narrow limits, in spite of significant changes in the concentrations of major nutrients and in the structure and productivity of the phytoplankton.

Introduction

One of the principal concepts of modern limnology concerns the determining influence of the community of primary producers on the functioning of following links in a trophic chain and on the trophic status of a whole water-body (Winberg, 1960; Rossolimo, 1977). Studies on Lake Sevan enable us to trace structural and functional changes of autotrophic communities in one of the largest mountain lakes that has undergone eutrophication after the artificial changing of lake morphometry (Oganesian & Parparov, 1983).

Published data about the composition and biomass of algae (Stroykina, 1953; Meshkova, 1975; Nikulina & Mnatsakanyan, 1984), and results of the authors' studies on primary production and chlorophyll *a* concentration of the plankton, are analysed below.

The studied area

Lake Sevan – the largest waterbody of the Caucasus mountain area – is situated in north-west Armenia at the altitude 1897.5 m a.s.l. (Fig. 1). The lake consists of two morphometrically different parts – the deeper Minor and the comparatively shallow Major Sevan. Some morphometric and limnological characteristics of the lake are given in Table 1. Before an artificial increase of outflow the level of the lake was at the altitude 1916 m a.s.l. Signs of eutrophication – water-blooms of the blue-green *Anabaena* – were first observed in 1964. Eutrophication of Lake

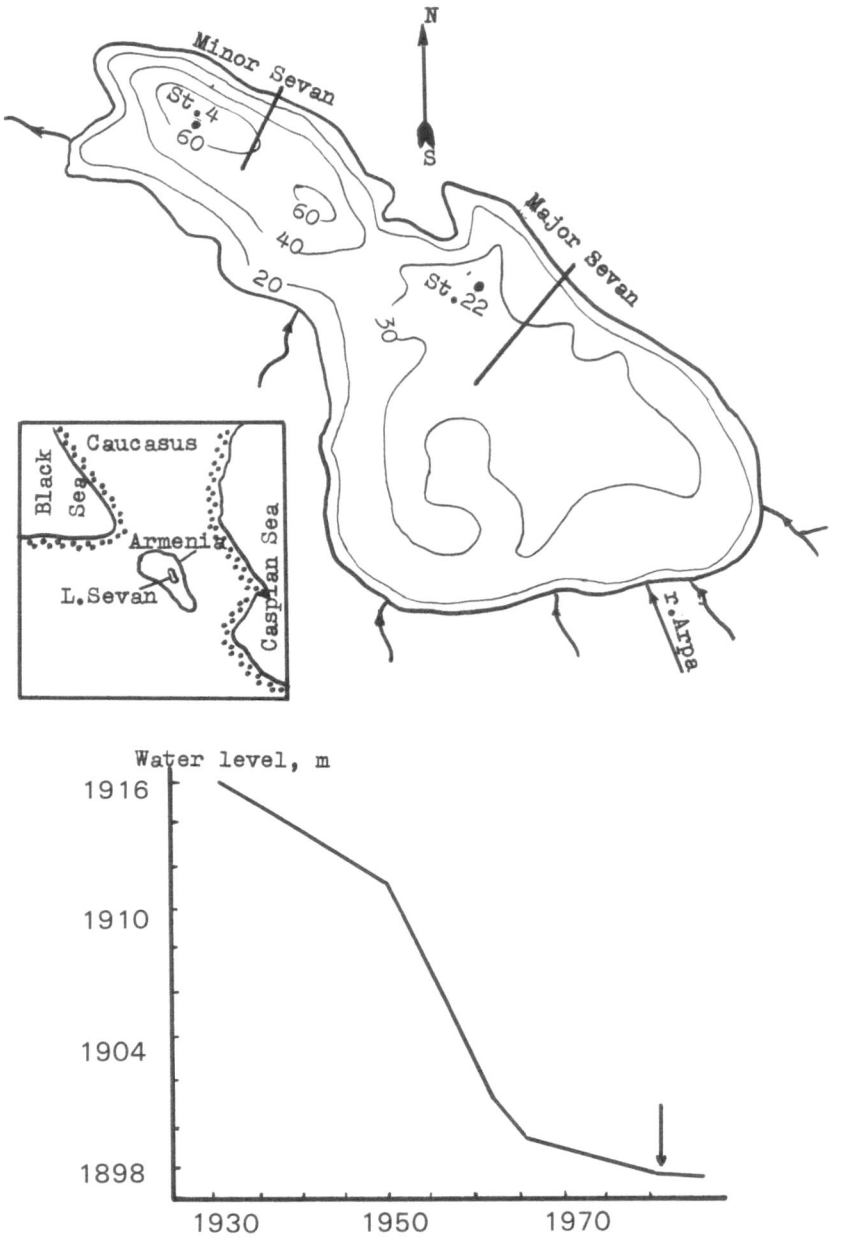

Fig. 1. (Above) Lake Sevan, its location in Armenia, contour lines of depth and sampling locations; (below) dynamics of the lowering of water level, with arrow showing incoming of the River Arpa.

Sevan was accompanied by changes in its original hydrochemical regime which evidently increased the concentration of soluble inorganic nitrogen and decreased the concentration of phosphorus (Parparova, 1985).

It was shown that the artificial raising of water level would be a fundamental way of retarding the eutrophication of Lake Sevan (Oganesian & Parparov, 1983). To raise the water level additional water from the River Arpa was channeled into the lake along a 48-km long tunnel since early 1981; it enabled the water level to be stabilized.

Table 1. Some morphometric and limnological indices of Lake Sevan before water level lowering (1) and at present (2). Morphometry after Gyozalyan (1984), hydrochemistry after Parparova (1985).

Indices	Minor Sevan		Major Sevan	
	1	2	1	2
Surface area, km^2	384	328	1032	916
Volume, km^3	19.5	12.8	39.0	20.9
Average depth, m	50.9	39.2	37.7	23.6
Transparency (Secchi depth), m	14.3	3.5*	14.3	2.8*
pH	9.2	8.6–8.8	9.2	8.6–8.8
Oxygen near bottom at the end of stratification, mg l^{-1}	5.0	0.0	4.6	0.0
Inorganic nitrogen, mg l^{-1}	0.00	0.10*	0.00	0.10*
Inorganic phosphorus, mg l^{-1}	0.320	0.025*	0.320	0.025

* Average of 1982–1983.

Material and methods

Primary production of plankton was measured in 1976–1986 with the oxygen light-and-dark bottle method (Winberg, 1960) on two typical pelagic stations of Minor (depth 60 m) and Major (depth 28 m) Sevan. Concentrations of chlorophyll *a* (Chl) was determined spectrophotometrically (SCOR-UNESCO, 1966) (in 1982–1983 Chl was determined with the same method by L.O. Glooshchenko). Studies in 1976–1977 (Parparov, 1983) and 1986 showed that those stations are representative for the pelagial of both parts of the lake.

Phytoplankton data of Stroykina (1953), Meshkova (1975), Kazaryan (1979), and Nikulina & Mnatsakanyan (1984) were also used in regression calculations. A conversion factor ratio of 1 : 500 between chlorophyll *a* and total biomass (fresh weight) of algae was employed (Boulyon, 1983).

Results and discussion

Submerged macrophytes and phytoplankton were the main autotrophs in Lake Sevan before lowering of the water level, when the annual biomass of macrophytes (wet weight) was about $6 \cdot 10^5$ t (Gambaryan, 1984). Assuming that the P/B coefficient of macrophytes = 1, the estimated production of macrophytes was 40 g C m^{-2} yr^{-1}. Since then the biomass of macrophytes decreased twentyfold (Gambaryan, 1984) and the main producer in Lake Sevan became phytoplankton.

The first systematic studies of primary production by the plankton of Lake Sevan were made in 1959 (Gambaryan, 1968). Our calculations on the data of Gambaryan (oxygen method) give a production estimate of 90 g C m^{-2} yr^{-1}. This value is considered characteristic for the period before the lowering of water level.

Up to 1961–1966 the annual average biomass of phytoplankton (Figs. 2a, b) decreased only slightly in spite of summer water-blooms of blue-green algae after 1964 (Legovich, 1968). From 1976 the annual average biomass of algae exceeded earlier values by one order of magnitude.

Diatoms were the most abundant group of algae in the period from 1947 to 1982, and a close relation then existed between total biomass of phytoplankton (B) and biomass of diatoms ($r = +0.97$; $n = 108$).

From 1983 green algae became the most abundant group of algae in all depths studied (Fig. 3), and there existed an inverse relation between biomass of diatoms and green algae: $r = -0.72$ ($n = 36$) and $r = -0.79$ ($n = 38$) for Minor and Major Sevan, respectively.

18

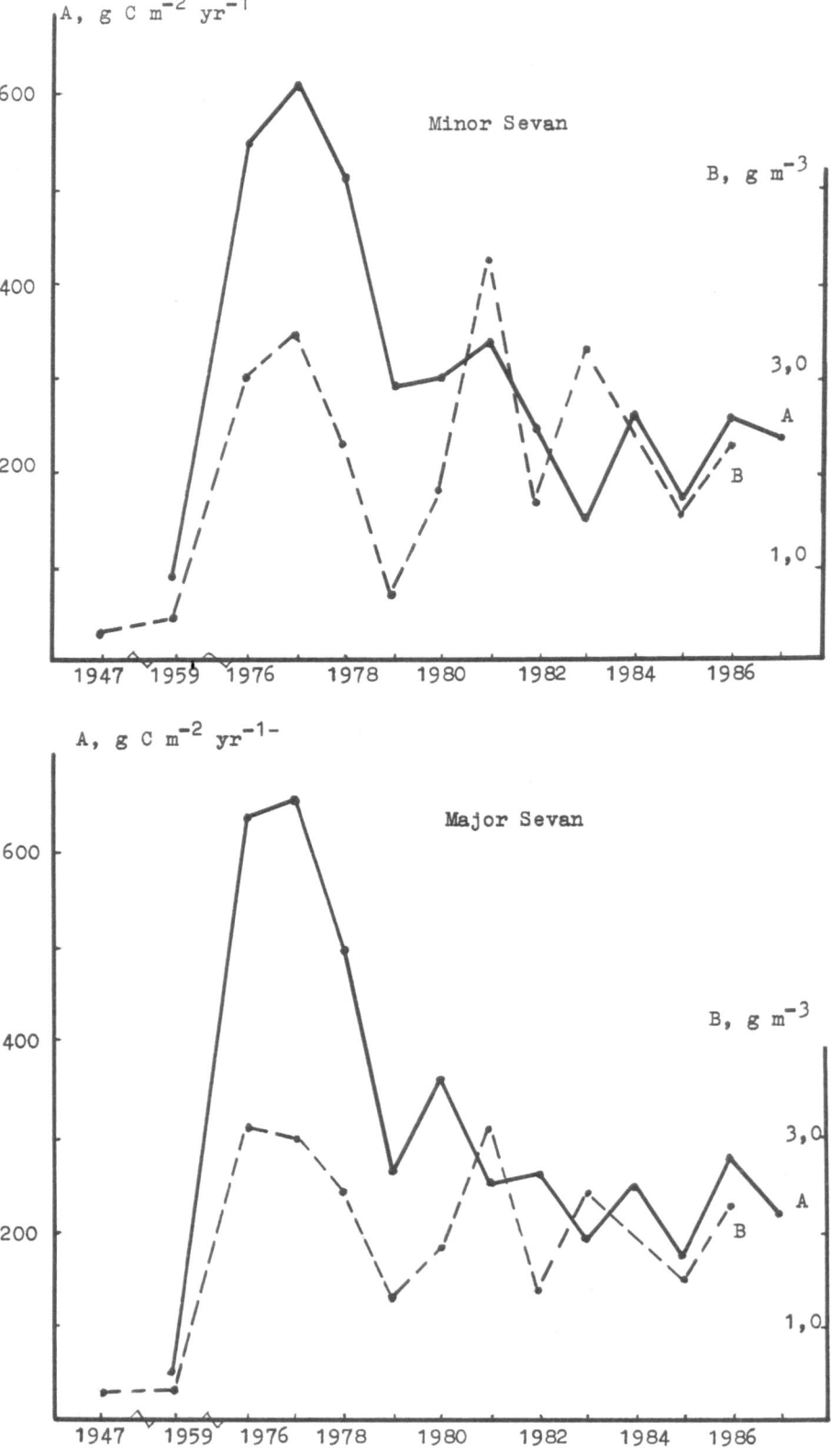

Fig. 2. Dynamics of primary production (A) and biomass (B) of phytoplankton in (a) Minor and (b) Major Sevan.

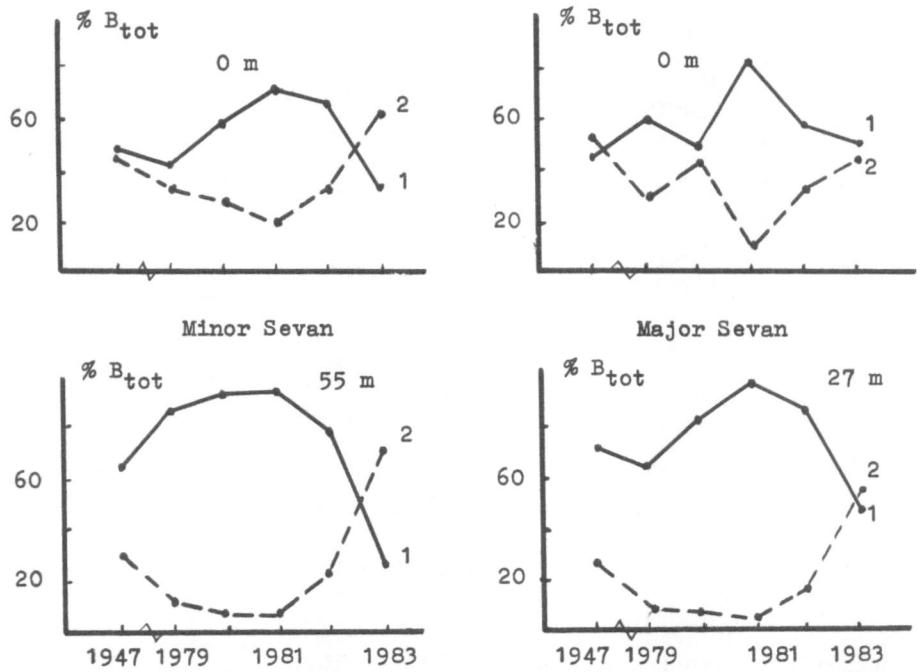

Fig. 3. Dynamics of biomass of diatoms (1) and green (2) algae in% of total biomass (B_{tot}) of phytoplankton in surface and deep water of Minor and Major Sevan.

Blue-green algae – *Anabaena* and *Aphanizomenon*, which caused intense water blooms in 1974–1978, decreased sharply. The maximum biomass of algae for the summer period in 1976–1977 was 18.0 g m^{-3}, but dropped to 0.5–1.0 g m^{-1} in 1982–1986.

Considering the dynamics of total primary production in 1976–1986, one can distinguish a highly productive period 1976–1978 and a fairly low productive period 1982–1986 (Figs. 2a, b).

Primary production values of the highly productive period are typical for hyper-eutrophic lakes of the USSR (Winberg, 1960; Boulyon, 1983), but the period 1982–1986 can be considered mesotrophic (taking into account the all-year vegetation of algae in Lake Sevan and the fairly large depth of the euphotic zone – about 10 m). Productivity during different seasons decreased approximately equally: in spring from 4.5 g O$_2$ m^{-2} d^{-1} in 1976–1977 to 2.3 g O$_2$ m^{-2} d^{-1} in 1983–1984, and in summer from 7.4 to 2.9 g O$_2$ m^{-2} d^{-1}.

The decrease of productivity was not completely inverse but it was accompanied by some hysteresis. In spite of the decrease of productivity there was no increase in transparency or in the concentration of oxygen near the bottom. The period of rising and then diminishing productivity (1975–1980) was characterised by frequent and unpredictable changes of dominant species of algae in the spring phytoplankton, which means an unstable state of the community of autotrophs. The character of present changes of productivity oscillating about 250 g C m^{-2} yr^{-1}, and the relative stability of community structure of phytoplankton, enables us to consider the present state as close to stable.

It is interesting to consider changes of the activity coefficient (AC, mg C mg Chl^{-1} d^{-1}) under pronounced changes of productivity (Fig. 4). Comparison of data in Figs. 4 and 2 shows that there is a tendency for simultaneous decrease of annual average productivity and activity coefficient ($r = 0.55$, $P = 0.05$). It is clear that, in both parts of the lake, changes of annual average activity coefficient were small in spite of

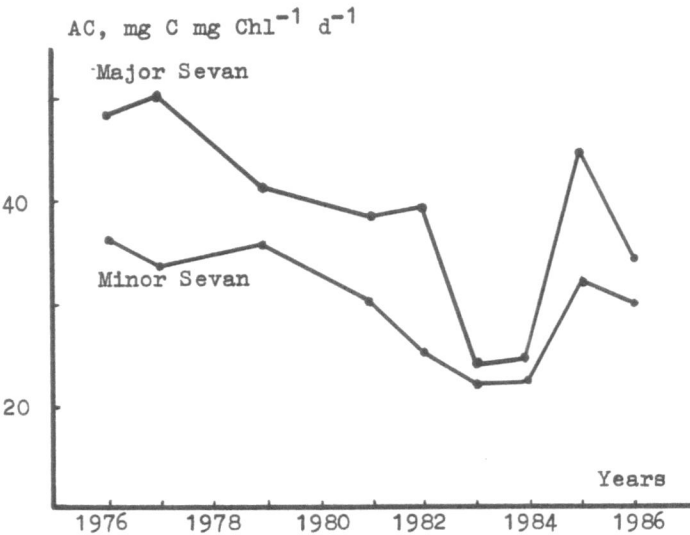

Fig. 4. Dynamics of the annual average activity coefficient (AC) in Lake Sevan.

considerable changes of productivity and regime of major nutrients. For the whole period of studies the mean annual activity coefficient was 29.9 ± 4.1 in Minor and 38.3 ± 7.2 mg C mg Chl^{-1} d^{-1} in Major Sevan. These values are close to those commonly recorded in the literature (Boulyon, 1983).

Positive non-linear correlation was revealed between the activity coefficient and water temperature (T, °C):

$$\ln AC = (2.644 \pm 0.252) + (0.086 \pm 0.19)T$$

$$r = 0.65, \ n = 105 \tag{1}$$

This equation may be rewritten

$$AC = (14.1 \pm 1.5)e^{(0.086 \pm 0.019)T} \tag{2}$$

where the coefficient (14.1 ± 1.5) estimates the activity coefficient under lower water temperatures. The temperature coefficient of Vant-Hoff (Q_{10}) is 2.36 ± 0.45. Higher values of AC in Major Sevan may be explained by higher summer temperature of surficial water layers in the shallower part of the lake.

Conclusion

Primary production of plankton first rose and then decreased in the process of eutrophication of Lake Sevan. At the present time (1987) the level of productivity is about twice that before the lowering of lake water level. The increase and decrease of primary production characterise the transition processes of ecosystem response to changes in morphometry. The average value (1982–1986) of primary production of plankton – 250 g C m^{-2} yr^{-1} – is evidently close to stationary (at the present morphometry of the lake).

The activity coefficient of photosynthesis by plankton (annual average) of Lake Sevan changed within comparatively narrow limits in spite of significant changes in the regime of major nutrients and in the structure and productivity of phytoplankton. This enables the use of an average AC value to predict and express estimations of primary productivity in Lake Sevan.

References

Boulyon, V. V., 1983. Primary production of plankton of internal waterbodies. 'Nauka', Leningrad, 149 pp. (in Russian).

Gambaryan, M. Ye., 1968. Microbiological studies on Lake Sevan. Yerevan, 165 pp. (in Russian).

Gambaryan, P. P., 1984. Macrophytes of Lake Sevan. In: Limnology of alpine lakes. Yerevan, 55–56 (in Russian).

Gyozalyan, M. G., 1984. Some modern morphometric and hydrologic parameters of Minor and Major Sevan. In: Limnology of alpine lakes. Yerevan, 57–58 (in Russian).

Kazaryan, H. G., 1979. Data for studying the phytoplankton of Lake Sevan. In: Ecology of hydrobionts of Lake Sevan. Yerevan, 75–87 (in Russian).

Legovich, N. A., 1968. Changes of structure of the phytoplankton of Lake Sevan affected by lowering of its level. Biol. J. Armenia 21 (12): 31–41 (in Russian).

Meshkova, T. M., 1975. Objectives of development of zooplankton in Lake Sevan. Yerevan, 276 pp. (in Russian).

Nikulina, V. N. & A. T. Mnatsakanyan, 1984. Phytoplankton of Lake Sevan in 1979–1983. In: Experimental and field studies of hydrobionts of Lake Sevan. Yerevan, 18–43 (in Russian).

Oganesian, R. O., 1978. Present state of Lake Sevan (Armenia). Verh. int. Ver. Limnol., 20, 1103–1104.

Oganesian, R. O. & A. S. Parparov, 1983. Ecological aspects of the problem of Lake Sevan. In: Production processes in the ecosystem of Lake Sevan. Yerevan, 5–13 (in Russian).

Parparov, A. S., 1983. Some features of the study of primary productivity of Lake Sevan. In: Production processes in the ecosystem of Lake Sevan. Yerevan, 14–50 (in Russian).

Parparova, R. M., 1985. Peculiarities of the cycle of phosphorus in Lake Sevan on a background of the changes of its hydrochemical regime in relation with anthropogenic influence. Avtoref. diss. Rostov-on-Don, 24 pp. (in Russian).

SCOR-UNESCO, 1966. Rept. Work. Group 17: 9–18.

Rossolimo, L. L., 1977. Changing of limnetic ecosystems under the influence of anthropogenic factors. 'Nauka', Moscow, 245 pp. (in Russian).

Stroykina, V. G., 1953. Phytoplankton of the pelagial of Lake Sevan. Proc. Sevan Hydrobiol. Sta. Yerevan 13: 171–212 (in Russian).

Winberg, G. G., 1960. Primary production of waterbodies. Minsk, 329 pp. (in Russian).

Hydrobiologia **191**: 23–27, 1990.
P. Biró and J. F. Talling (eds), Trophic Relationships in Inland Waters.
© 1990 *Kluwer Academic Publishers.*

Aspects of the ecology of a filamentous alga in a eutrophicated lake

Teresa Ozimek
Department of Hydrobiology, Institute of Zoology, University of Warsaw, Nowy Swiat 67, 00-046
Warsaw, Poland

Key words: lake littoral, submerged macrophytes, filamentous green algae, utilization of phosphorus and nitrogen

Abstract

In the eutrophic Lake Mikołajskie macrophytes disappearing from the deeper parts of the littoral are replaced by *Vaucheria dichotoma* Ag. This forms a belt at 3.0–4.5 m to which depth only 1% of surface light penetrates. The zone of *V. dichotoma* has a layered structure. Some filaments are covered in mud and receive no light, but are alive and photosynthesize when transferred to light. *V. dichotoma* prefers fertile environments with a high content of phosphate-phosphorus and ammonium-nitrogen. It is evergreen and its biomass changes relatively little during a year.

Introduction

Many years of observations on Polish eutrophic lakes in the Masurian Lakeland (North Poland) show that as submerged macrophytes disappear from deeper parts of the littoral, they are replaced by *Vaucheria dichotoma* Ag. (Ozimek, unpub. data). For example such situations have taken place in lakes Mikołajskie, Juno, Wobel and Niegocin. These lakes are highly eutrophic (Kajak & Zdanowski, 1983). According to Lakatos (1976, 1978) *V. dichotoma* can be the dominant species in the littoral flora of eutrophic lakes.

The aim of this paper is to determine the role of *V. dichotoma* in the functioning of the lake's littoral. This has been done by analysing the distribution of *V. dichotoma*, its biomass, photosynthetic activity, content of chlorophyll *a*, ability to utilize phosphorus and nitrogen, and the influence of light and different levels of nutrients on its growth.

Material and methods

Studies were made in 1974–1985 in Lake Mikołajskie (North Poland). The lake is eutrophic, holomictic and has a surface area of 460 ha, a mean depth of 11 m and maximum of 27.8 m. The area occupied by submerged macrophytes was 44 ha in 1971 and only 30 ha in 1980 (Ozimek & Kowalczewski, 1984).

V. dichotoma, growing beyond the zone of submerged macrophytes, was the object of the investigations. The Bernatowicz type of sampler (0.16 m² sampling area) (Bernatowicz, 1960) was used to obtain data on *V. dichotoma* distribution and biomass. In 1975 and 1980 some 150 samples were collected from 50 sites in the littoral of Lake Mikołajskie.

Undisturbed sediment cores, together with a near-bottom water layer, were sampled from 5 sites of littoral sediment (at a depth of 3–4 m) into plexiglass cylinders attached to a tubular bottom sampler of the Kajak type (Kajak *et al.*, 1965). Each cylinder contained a sediment core about 15 cm in height and approximately 0.5 dm³ of near-bottom water. Interstitial water was obtained by centrifugation of sediment at 5000 r.p.m. The organic matter content in sediments, and phosphate-phosphorus, ammonium-nitrogen and nitrate-nitrogen in the interstitial, near-bottom water and surface water and in aquarium cultures, were analysed according to Hermanowicz *et al.* (1976). Each type of water was filtered through glass-fibre filter Whatman GF/C before determination.

The light measurements were made using photometer LI 185A.

Photosynthetic activity was measured by the oxygen (Winkler method) light and dark bottle technique (Vollenweider, 1969) under laboratory conditions (temperature, 18 °C; time of exposure, 4 h). The photosynthetic activity measurements were made using filaments of *V. dichotoma* from the mud surface, from mud 1–5 cm depth, and filaments kept 2 weeks in an aquarium without light. For each treatment 5 light and 5 dark bottles were used.

Laboratory experiments were carried out on the

Table 1. Content of phosphate-phosphorus, ammonium-nitrogen and nitrate-nitrogen in water of Lake Mikołajskie (means ± SD, July 1985). Means and SD are based on samples from five sites in the littoral.

Water-type	PO_4-P mg l^{-1}	NH_4-N mg l^{-1}	NO_3-N mg l^{-1}
Interstitial	0.674 ± 0.10	10.5 ± 0.05	0.08 ± 0.01
Near-bottom	0.222 ± 0.05	3.3 ± 0.02	0.05 ± 0.01
Surface	0.018 ± 0.00	0.1 ± 0.00	0.02 ± 0.00

growth rate of *V. dichotoma* in different levels of nutrient content. *V. dichotoma* (5 g fresh weight) was cultivated in 1 l of filtered surface water, in 0.5 l of filtered surface water mixed with 0.5 l of interstitial water, and 1 l filtered surface water and 250 ml of mud. Five replicates were set up of each treatment.

Laboratory experiments on the utilization of phosphate-phosphorus, ammonium-nitrogen and nitrate-nitrogen were carried out. *V. dichotoma* (20 g fresh weight) was cultivated in 1 l of surface water mixed with interstitial water. In initial water and after 24 h in control water and in water with *V. dichotoma*, phosphate-phosphorus and am-

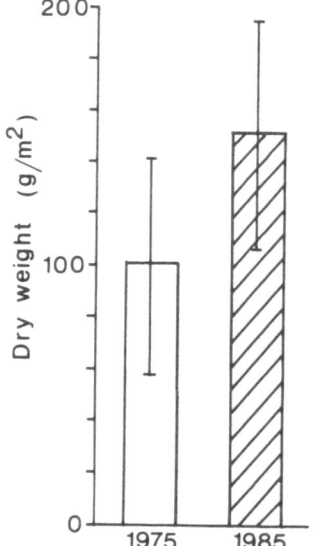
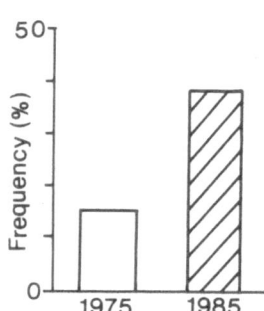

Fig. 1. Biomass (mean ± SD) and frequency of *V. dichotoma* in Lake Mikołajskie. (Frequency = % of occurrence in the total number of sites).

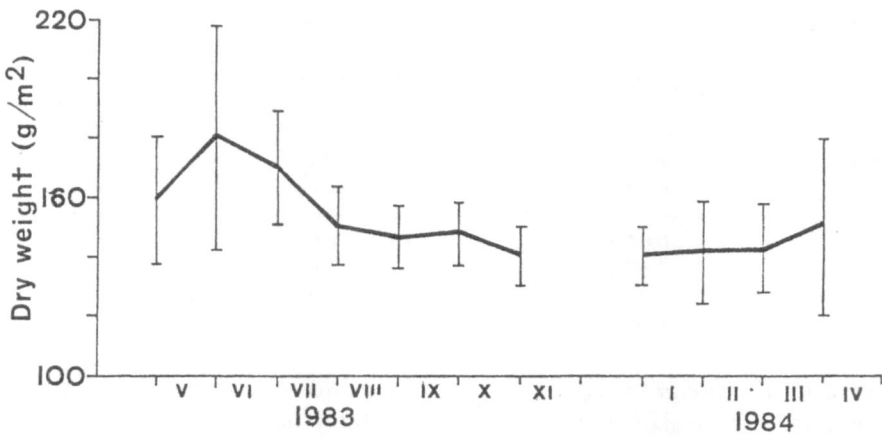

Fig. 2. Changes of biomass of *V. dichotoma* (means ± SD) during the year in Lake Mikołajskie (typical data, from one site).

monium-nitrogen were analysed. The experiments were in 5 replications.

Results

V. dichotoma was noted in deeper parts of the littoral of Lake Mikołajskie for the first time in 1974, as isolated patches. In 1985 it was very common in the deeper littoral. *V. dichotoma* occurs at a depth of 3.0–4.5 m on a muddy substratum with a high organic matter content of about 20% of dry weight. The interstitial and near-bottom waters have concentrations of phosphate-phosphorus, ammonium-nitrogen and nitrate-nitrogen many times higher than in the surface water (Table 1).

In shallower parts of the littoral (0.1–2.5 m) single filaments of *V. dichotoma* occur only sporadically. The spread of *V. dichotoma* at the depth of 3.0–4.5 m has changed over 10 years.

Frequency of occurrence has increased almost 3 times, but mean dry weight · m $^{-2}$ has not changed significantly (Fig. 1). *V. dichotoma* is evergreen and its biomass does not vary much over the year (Fig. 2).

The *V. dichotoma* zone has layers: some filaments are deep in the mud and receive no light, some occur on the mud surface. In summer only 1% of surface light penetrates to the filaments growing on the mud surface.

Filaments isolated from mud (at 1–5 cm depth) and those occurring on the mud surface vary in chlorophyll *a* content and photosynthetic activity (Table 2). The differences are statistically significant at the level of significance $P = 0.01$.

Filaments in mud are alive and photosynthesize when transferred to light, but their photosynthetic activity is lower than that of filaments on the mud

Table 2. Content of chlorophyll *a* and photosynthetic activity (mean ± SD) of *V. dichotoma* growing under different light regimes. Means and SD are based on five replicates.

Filaments origin	Chlorophyll *a* $\mu g\ g^{-1}$ d.w.	Photosynthetic activity mg O_2 mg^{-1} chl. *a*
From surface of mud	10.6 ± 2.0	0.40 ± 0.05
From mud 1–5 cm depth	4.2 ± 0.5	0.25 ± 0.05
Kept 2 weeks in aquarium	5.6 ± 1.0	0.26 ± 0.06

Table 3. Increments of *V. dichotoma* dry weight (means ± SD) under different nutritional conditions (July 1985, laboratory experiment). Means and SD are based on five replicates.

Aquarium of 1 l volume	Initial d.w. (mg)	D.w. after 2 weeks (mg)	Increments in % of initial d.w.
Surface water	0.25	0.29 ± 0.05	16
Surface water + interstitial	0.25	0.34 ± 0.10	36
Surface water + mud	0.25	0.51 ± 0.10	104

Table 4. Removal from water of phosphate-phosphorus, ammonium-nitrogen and nitrate-nitrogen by *V. dichotoma*. Means and SD are based on five replicates.

Content in water in mg l^{-1}				Reduction by 1 g d.w. fo *V. dichotoma* in %
	Initial water	After 24 h, control water	After 24 h, water with *V. dichotoma*	
PO$_4$-P	0.05	0.04 ± 0.01	0.020 ± 0.005	50
NH$_4$-N	3.25	3.13 ± 0.50	0.180 ± 0.039	94
NO$_3$-N	0.05	0.05 ± 0.00	0.050 ± 0.000	0

surface. Filaments of *V. dichotoma* kept for a fortnight without light photosynthesize when transferred to light, and their photosynthetic activity is similar to that of filaments from the mud (Table 2). The laboratory experiment showed that *V. dichotoma* had a 6-times greater increase in biomass in water with addition of mud than in surface water with lower nutrient content (Table 3).

. *V. dichotoma* plays a significant part in nitrogen and phosphorus cycling (Table 4). After 24 h *V. dichotoma* (20 g fresh weight, i.e. 1 g dry weight) removed 0.02 mg of phosphate-phosphorus (50% of values in control water) and 2.95 mg of ammonium-nitrogen (94% of content in control water). *V. dichotoma* uses mainly ammonium as a source of nitrogen.

Discussion

Over the last 20 years Lake Mikołajskie has became a highly eutrophic lake (Kajak, 1978; Gliwicz *et al.*, 1980). The depth range of submerged macrophytes decreased by 2 m in the years 1971–1980 (Ozimek & Kowalczewski, 1984). This has been connected primarily with deteriorating light conditions in deeper parts of the littoral. But it may also be due to the enrichment of littoral sediments in organic matter, which, according to Barko & Smart (1983), inhibits the growth of some species of submerged macrophytes. At present they have been replaced by *V. dichotoma*, the frequency of which increased along with the disappearance of submerged macrophytes in Lake Mikołajskie. It is interesting to find out whether *V. dichotoma* occupies a habitat no longer occupied by macrophytes, or whether it has displaced the macrophytes from deeper parts of the littoral. The results available suggest that *V. dichotoma* accelerates the disappearance of macrophytes, but in order to test this hypothesis a series of field and laboratory experiments must be conducted. Macrophytes have also been replaced by *V. dichotoma* in other eutrophic and hypertrophic lakes of the Mazurian Lakeland (Ozimek, unpubl. data). According to the literature (Lewin, 1962; Lakatos, 1975) and the present results, *V. dichotoma* may grow under conditions of low light intensity. The alga may also live without access to light (as confirmed by laboratory experiments), possibly heterotrophically using dissolved organic matter.

V. dichotoma prefers fertile environments, attaining there a biomass density several times higher than macrophytes previously colonizing these parts of the littoral (Ozimek & Kowalczewski 1984).

V. dichotoma uses considerable amounts of nutrients from interstitial and near-bottom waters, thus limiting their transfer to higher water layers, where they could be used by phytoplankton. It forms a compact turf, fixing the sediments and protecting them from being resuspended during the spring and autumn circulation.

Acknowledgements

I gratefully acknowledge Miss. A. Hankiewicz who helped me in chemical analyses.

References

Barko, J. W. & R. M. Smart, 1983. Effects of organic matter additions to sediment on the growth of aquatic plants. J. Ecol. 71: 161–175.

Bernatowicz, S. 1960. Metody badania roslinnosci naczyniowej w jeziorach. Roczn. Nauk roln. B. 5: 61–78.

Gliwicz, Z. M., A. Kowalczewski, T. Ozimek, E. Pieczynska, A. Prejs, K. Prejs & J. I. Rybak, 1980. An evaluation of the degree of eutrophication of Great Mazurian Lakes. Wydawnictwa Akcydensowe, Warszawa, 100–103.

Golterman, H. L. & R. S. Clymo, 1969. Methods for chemical analysis of freshwaters, IBP Handbook 8, Oxford, 172 pp.

Hermanowicz, W., W. Dozanska, J. Dojlido & B. Koziorowski, 1976. Fizycznochemiczne badania wody i sciekow. Arkady, Warszawa, 847 pp.

Kajak, Z. 1978. The characteristics of a temperate eutrophic, dimictic lake (Lake Mikołajskie, Northern Poland). Int. Revue ges. Hydrobiol. 63: 451–480.

Kajak, Z., K. Kacprzak & R. Polkowski, 1965. Chwytacz rurowy do pobierania prob dna. Ekol. pol. B. 11: 159–165.

Kajak, Z. & B. Zdanowski, 1983. Ecological characteristics of lakes northeastern Poland versus their trophic gradient. Ekol. pol. 31: 239–256.

Lakatos, G. 1976. On the photosynthesis and chlorophyll efficiency by *Vaucheria dichotoma* Agh. in Lake Velence, Hungary. Acta Bot. Acad. Scient. Hungar. 22: 381–391.

Lakatos, G., 1978. The phenomenon and significance of benthonic eutrophication in Lake Velence, Hungary. Acta Biol. Debrecina 15: 147–168.

Lewin, R. A., 1962. Physiology and biochemistry of algae. Academic Press, N.Y., Lond. 703 pp.

Ozimek, T. & A. Kowalczewski, 1984. Long-term changes of the submerged macrophytes in eutrophic Lake Mikołajskie (North Poland). Aquat. Bot. 19: 1–11.

Vollenweider, R. A., 1969. A manual on methods for measuring primary production in aquatic environments, IBP Handbook 12, Oxford & Edinburgh, 213 pp.

Hydrobiologia **191**: 29–37, 1990.
P. Biró and J. F. Talling (eds), Trophic Relationships in Inland Waters.
© 1990 *Kluwer Academic Publishers.*

The relation of biotic and abiotic interactions to eutrophication in Tjeukemeer, The Netherlands

Eddy H.R.R. Lammens
Limnological Institute, De Akkers 47, 8536 VD Oosterzee, The Netherlands

Key words: total-P, chlorophyll *a*, *Daphnia hyalina*, temperature, predation, smelt, bream, hydrology

Abstract

During the summer months of 1974–1985 chlorophyll-*a* and total P concentration, biomass of *Daphnia hyalina*, smelt *Osmerus eperlanus*, bream *Abramis brama* and pikeperch *Stizostedion lucioperca*, water temperature and water intake from lake IJsselmeer were monitored in Tjeukemeer. During this period there were manipulations with the bream and pikeperch stocks as a consequence of the termination of a gill-net fishery in 1977, and larval smelt immigrated each year from the large lake IJsselmeer and contributed largely to the yearly smelt recruitment.

The correlation matrix of the nine variables mentioned above showed a positive correlation between bream and chlorophyll-*a*, but surprisingly a negative one between smelt and chlorophyll-*a*. The latter can only be explained when smelt is the dependent variable. In a multi-linear regression there was a negative effect of temperature, chlorophyll *a* and pikeperch on smelt and a positive effect of water intake. *Daphnia hyalina* was negatively influenced by the biomass of smelt and the water intake of lake IJsselmeer. The positive relation of *Daphnia hyalina* and chlorophyll-*a* was probably related to better survival chances of *D. hyalina* in an *Oscillatoria*-rich environment when smelt is the most important predator. An increasing biomass of bream coincided with higher total-P levels and probably contributed to higher chlorophyll-*a* levels.

Introduction

Interaction between trophic levels has drawn much attention since eutrophication has become a world-wide problem. An overload of nutrients causes severe algal blooms, transparency decreases and macrophytes disappear. Enclosure experiments (Anderson *et al.*, 1978; Tatrai & Istvanovics, 1986) and whole lake experiments (Stenson *et al.*, 1978; Benndorf *et al.*, 1984; Shapiro & Wright, 1984; Bernardi, 1986) point to the importance of the higher trophic levels (zooplankton and fish) in the process of

eutrophication (for a review see McQueen & Post, 1986). Lake Tjeukemeer in the northern part of the Netherlands has experienced this eutrophication process too and has now turned into a system dominated by blue-greens (Moed & Hoogveld, 1982) and bream and pikeperch (Vijverberg & Van Densen, 1984; Lammens *et al.*, 1985; Lammens, 1986).

Tjeukemeer is part of the Frisian lake district and is interconnected with lake IJsselmeer in the central part of the Netherlands (Fig. 1). This connection provides a permanent experimental situation in which each year large amounts of

30

Fig. 1. The location of Tjeukemeer in the Frisian lake district and The Netherlands. Lake IJsselmeer is also shown.

larval smelt are passively transported from IJsselmeer to the Frisian lakes, where they serve as the most important food for pikeperch (Vijverberg & Van Densen, 1984). The development of this smelt population varies strongly from

year to year and gives the opportunity to study its effect on eutrophication.

Apart from this yearly injection of smelt larvae, the fishery management changed during this twelve-year study (1974–1985) by the ending of the gill-net fishery for pikeperch in 1977. This fishery removed yearly about 70% of the pikeperch > 42 cm (Goldspink & Banks, 1975) and a large part of the bream population > 28 cm. Since 1977 the pikeperch and bream populations have changed significantly with respect to the larger length-classes (Lammens, 1986). This change in bream and pikeperch biomass is another variable influencing the eutrophication process. Apart from these effects, temperature and water intake were important abiotic variables as the IJsselmeer water contains different nutrient concentrations compared to the drainage water from the surrounding agricultural areas (Fig. 2).

In order to explain the development of the phytoplankton (as a function of chlorophyll *a*) and the zooplankton (*D. hyalina*) during this period, multivariate analyses were performed with the summer values of chlorophyll *a* and biomass of *Daphnia hyalina* as dependent variables. As independent variables the biomass of the smelt, bream and pikeperch population, temperature, yearly water intake from IJsselmeer and total P concentration were chosen.

Study Area

Tjeukemeer is a shallow (depth 1.5–2 m), hypertrophic and turbid (Secchi disc 25–35 cm) freshwater lake with a surface area of 21.5 km². The total P concentration amounts to 0.25 mg l^{-1} in summer and to 0.3 mg l^{-1} in winter, whereas total

N amounts to 3 and 6 mg l^{-1} respectively. The inshore zone is poorly developed, the index of shore-line development being 1.25 (Hutchinson, 1957). The only vegetation is some reed. Most of the primary production is accounted for by the blue-green *Oscillatoria agardhii* (Moed & Hoogveld, 1982) and most of the secondary production by cladocerans and copepods, of which *Daphnia hyalina* is most productive (Vijverberg & Richter, 1982a, b). The biomass and production of chironomids is very small (Beattie, 1982). The fish community is dominated by bream (*Abramis brama*) and pikeperch (*Stizostedion lucioperca*), whereas smelt (*Osmerus eperlanus*) is present in varying amounts (Lammens, 1986). As the lake is shallow and exposed to wind, summer stratification does not occur and in consequence zooplankton is almost homogeneously distributed (Nie *et al.*, 1978; Nie & Vijverberg, 1985). The lake is part of an interconnected system of lakes, that receives water from the nearby lake IJsselmeer during the summer and from the surrounding polders in winter, when precipitation exceeds evaporation (Fig. 1). The management of the hydrological regime is designed to control the water level which is required for agricultural purposes. Chemistry and hydrology have been described by DeHaan (1982), DeHaan & Moed (1984) and Leenen (1982).

Materials and methods

Fish and zooplankton were collected during a period of 12 years (1974–1985). The fish were caught once a month in summer with a small mesh trawl (5.5 mm cod end) at five stations. Standard hauls were made with a speed of 1 m s^{-1} during 10 min. Length-frequency distribution and total number of each species were determined. Yearly catches of bream are given as an average for the whole summer, that is 4 times 5 separate catches. However, because smelt (*Osmerus eperlanus*) is not caught efficiently with this trawl until the second half of the summer, only the September catches of this species were used.

Zooplankton was sampled at the same time

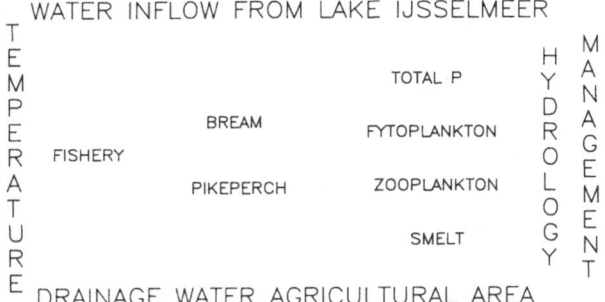

Fig. 2. A schematic representation of variables influencing eutrophication in Tjeukemeer. In the central part the most important biotic variables are shown, in the periphery the abiotic variables.

32

and place as the trawling. Using a 5-l Friedinger sampler with a cylinder of 60 cm height, two samples were taken at each of the five stations, one just below the water surface and the other just above the bottom. The samples were concentrated by filtering through a plankton gauze of 120 μm mesh and preserved in 4% formalin. The combined sample was sub-sampled with a whirling vessel after Kott (1953) and organisms were identified, counted and their lengths measured. The length of the cladocerans was measured from the top of the head or, if a helmet was present, from the base of the helmet to the base of the tail spine.

Chlorophyll a and total P concentrations were determined fortnightly in a mixed sample from ten stations. Chlorophyll a was determined according to Moed (1973) and the concentration was calculated after Nusch & Palme (1975). Total P was analysed after hydrolysis to ortho-phosphate for

2.5 h at 110 °C in acidic (3N H_2SO_4) persulphate (5%). Ortho-phosphate was analysed colorimetrically (Murphy & Riley, 1962) either manually or using an auto-analyser (Technicon AA II). Water temperature was measured weekly from three whole-column samples. The inflow of IJsselmeer water was recorded by the Department of Public Works in the province of Friesland.

Results and discussion

During the period 1974–1985 significant changes occurred in the trophic levels of the Tjeukemeer ecosystem. Chlorophyll a and biomass of *Daphnia hyalina* and bream increased significantly ($p < 0.05$), whereas there was a tendency of smelt *Osmerus eperlanus* to decrease and total P levels to increase ($p < 0.2$) (Fig. 3, Table 1). In the same period there were two major management

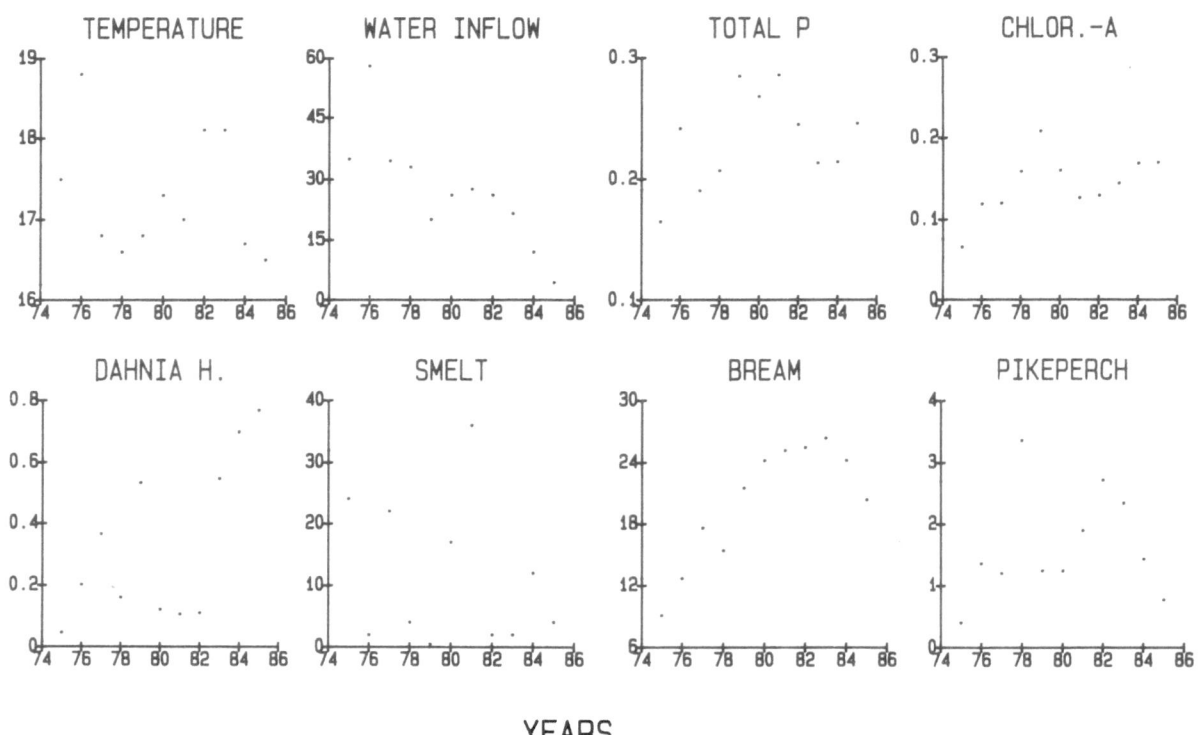

YEARS

Fig. 3. Temperature (°C), water intake from lake IJsselmeer (10^7 m^3 · year^{-1}), total-P (mg · l^{-1}), chlorophyll a (mg · l^{-1}), biomass of *Daphnia hyalina* (mg dry wt · l^{-1}), and minimum biomass of smelt, bream and pikeperch (g · m^{-2}) are plotted to the corresponding years.

Table 1. Simple correlations of the most important variables.

	Years	Temperature	Water-intake	Total-P	Chlor.-*a*	*D. hyalina*	Smelt	Bream
Years								
Temperature	− 0.0097							
Water-intake	− 0.7465**	0.5554						
Total-P	0.4473	0.0902	− 0.2110					
Chlorophyll-*a*	0.6379*	− 0.1716	− 0.5156	0.6405*				
D. hyalina	0.6973*	− 0.1919	− 0.6509*	0.1597	0.6710*			
Smelt	− 0.3988	− 0.3978	0.0979	− 0.2054	− 0.6133*	− 0.4983		
Bream	0.8237***	0.0637	− 0.5730*	0.6116*	0.6040*	0.4045	− 0.2142	
Pikeperch	− 0.0296	− 0.0456	0.0894	− 0.0490	− 0.0186	− 0.3264	− 0.0972	0.1646

* $p < 0.05$
** $p < 0.01$
*** $p < 0.005$

changes regarding the fishery of bream and pikeperch on the one hand and the yearly water import from lake IJsselmeer on the other. Both changes affected directly and indirectly the intermediate trophic levels and make them difficult to interpret (Fig. 2). The high correlations between the different trophic levels (Table 1) are markedly changed when they are analysed in a multilinear regression. The variables chlorophyll *a* and *Daphnia hyalina* are treated as dependent variables in relation to temperature, water input, total-P, and biomass of smelt, bream and pikeperch as independent variables.

Chlorophyll *a* as dependent variable

In the correlation matrix chlorophyll *a* shows significantly negative correlations to water input from lake IJsselmeer and to smelt biomass and significantly positive correlations to total-P, *Daphnia hyalina* and bream biomass (see also Fig. 4). These five variables, however, also show significant correlations to each other, which causes difficulties in interpretation. A possible way to correct for so called auto-correlations may be a multilinear regression; however, one should be careful in the choice of the variables. If, for instance, temperature, water-input, smelt, bream and pikeperch are used as independent variables and chlorophyll *a* as a dependent variable, then smelt, pikeperch and temperature contribute

significantly and negatively to the chlorophyll *a* level, whereas water input and bream contribute positively (Table 2). This combination of variables gives the highest explanation of the chlorophyll *a* levels, but it is probably not a realistic explanation. If, instead, smelt is used as dependent variable and chlorophyll *a*, temperature, water intake, bream and pikeperch as independent variables, then the same high explanation of variation (89%) is found, which seems much more realistic, since smelt is a visual feeder (chlorophyll *a*), avoids high temperatures (Densen & Hadderingh, 1982), is largely imported together with the water intake, is preyed upon by pikeperch (van Densen, 1985) and competes with bream for food (Lammens *et al.*, 1985). Therefore the choice of the independent variables must be made carefully.

Other independent variables are total-P concentration and the density of *Daphnia hyalina*. The combination of both variables explains almost 70% of the variation in chlorophyll *a* concentration. However, the relation to *Daphnia hyalina* is positive and therefore it seems more likely that the latter is a dependent variable.

Daphnia hyalina as dependent variable

The biomass of *Daphnia hyalina* was significantly positively correlated to the chlorophyll *a* concentration and significantly negatively to the water

34

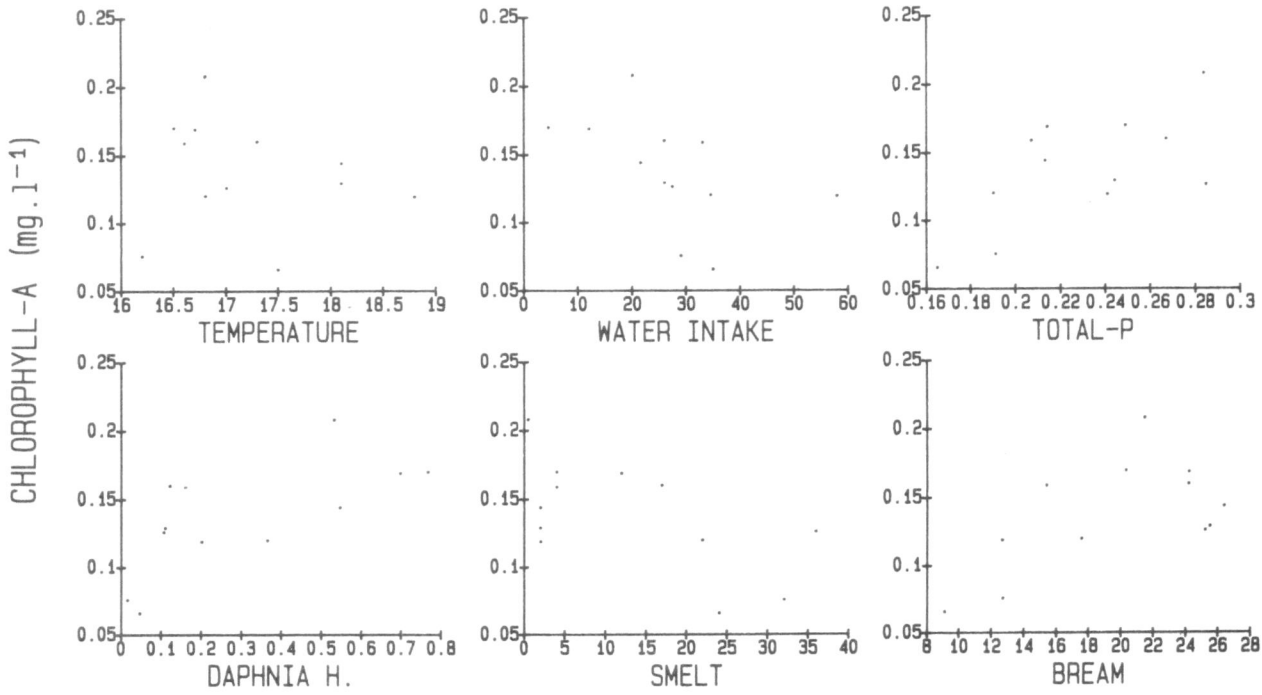

Fig. 4. Chlorophyll *a* concentration (mg · l⁻¹) plotted to temperature, water intake (10^7 m³ · year⁻¹), total-P (mg · l⁻¹), biomass of *Daphnia hyalina* (mg dry wt · l⁻¹) and minimum biomass of smelt and bream (g wet wt · m⁻²).

input from lake IJsselmeer, when simple correlations are calculated (see also Fig. 5). The correlation to smelt was negative for $p < 0.11$ and to bream positive for $p < 0.2$. When smelt, pikeperch and water intake from lake IJsselmeer are combined, then the contributions of both smelt and water intake are negatively significant (Table 3). About 60% of the total variation in the biomass of *Daphnia hyalina* can be explained by this combination. The contribution of chlorophyll *a* to the variation of *Daphnia hyalina* is in most combinations not significant, and each of these combinations explains a much lower per-

centage of the total variation. The positive correlation of *Daphnia hyalina* and chlorophyll *a* in the simple correlation matrix must be coincidence because chlorophyll *a* is negatively correlated to smelt and water-import. So the biomass of *Daphnia hyalina* is relatively less affected by smelt predation when the chlorophyll *a* concentration increases, and this occurs in periods when only small amounts of IJsselmeer water are taken in. This explanation seems much more likely than an indirect stimulation of blue-greens by selective feeding of *D. hyalina* (Lammens, 1988).

Table 2. Multiple regression analysis with chlorophyll *a* as dependent variable. Three combinations are shown.

Temperature	Water-intake	Total-P	*D. hyalina*	Smelt	Bream	Pikeperch	P-value	Adjusted R^2
0.0018(−)	0.0180(+)			0.0003(−)	0.0013(+)	0.0176(−)	0.0011	0.8942
		0.0110(+)	0.0078(+)				0.0013	0.6848
0.6887(−)	0.9198(−)				0.2311(+)	0.7121(−)	0.3600	0.1003

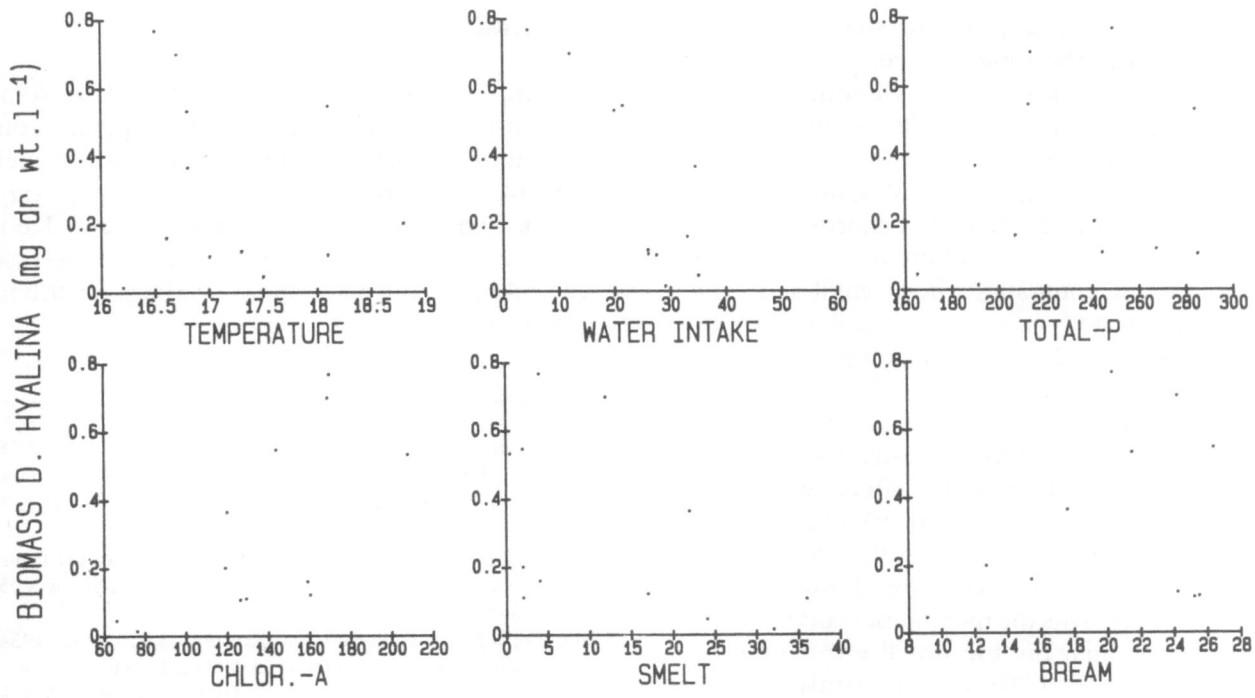

Fig. 5. Biomass of *Daphnia hyalina* (mg dry wt·l⁻¹) plotted to temperature (°C), water intake (10⁷ m³·year⁻¹), total-P (mg·l⁻¹), chlorophyll *a* concentration (mg·l⁻¹), and minimum biomass of smelt and bream (g wet wt·m⁻²).

The most probable explanation

During the years 1974–1985 the bream biomass has markedly increased, most probably as a result of the termination of the gill-net fishery in 1977, since only the specimens larger than 30 cm contributed to the biomass increase (Lammens, 1986). This increase could be responsible for the change in total-P concentration, because of the partly benthivorous feeding behaviour of this fish. Phytoplankton is consequently influenced, but consists predominantly of large inedible threads. This creates increasingly unfavourable conditions for the feeding of smelt. The threads may clog the gill-rakers or limit the visual field. The negative relation between the biomass of smelt and chlorophyll *a* points to this. The decreasing feeding efficiency of this fish gives *Daphnia hyalina* better chances of survival. On the other hand the decreasing water intake from lake IJsselmeer during this period also contributed to the changing chlorophyll *a* concentration and biomass of *Daphnia hyalina*, as phytoplankton and *Daphnia hyalina* are more poorly represented in this lake. So both the

Table 3. Multiple regression analysis with *Daphnia hyalina* as dependent variable. Five combinations are shown.

Temperature	Water-intake	Total-P	Chlorophyll-*a*	Smelt	Bream	Pikeperch	P-value	Adjusted R²
	0.0167(−)			0.0386(−)		0.1350(−)	0.0117	0.6059
	0.0170(−)			0.0640(−)			0.0093	0.5285
	0.1357(−)		0.1041(+)				0.0147	0.4828
	0.0900(−)		0.5380(+)	0.2998(−)			0.0319	0.4956
			0.0908(+)	0.6631(−)			0.0496	0.3433

36

change in fishery and in hydrology management contributed to the total change.

This situation seems to be essentially different from experiments described in the literature (Andersson *et al.*, 1978; Benndorf *et al.*, 1984; Shapiro & Wright, 1984; McQueen & Post, 1986). The general idea is that a strong predation pressure on the zooplankton changes the zooplankton community into small individuals and species and thus stimulates eutrophication because of the relieved predation pressure on the phytoplankton. Top-down effects are often obvious up to the nutrient level. However, most of these experiments were short-term experiments. In Tjeukemeer the top-down effects were limited to fish (smelt) and large zooplankton. Here smelt probably feeds more efficiently when chlorophyll *a* levels are relatively low. However, if the smelt biomass was the most important variable in this system then the top-down effect may have been quite different. The independently increasing bream population and decreasing water intake from lake IJsselmeer (containing less nutrients than the drainage water from the surrounding polders) overcompensated the effect of planktivorous feeding by smelt. The effect of benthivorous fish was already known from earlier work (Andersson *et al.*, 1978; Brabant *et al.*, 1986; Tatrai & Istvanovics, 1986).

It is clear that both bottom-up and top-down effects are present in Tjeukemeer. However the bottom-up effects are more important than the top-down effects, due to the large impact of hydrology on this lake, the interfering bream biomass and the smelt as most important planktivore. Biotic and abiotic interactions are interrelated in a very complex way, because the hydrology connects IJsselmeer, Tjeukemeer and surrounding polders (Huet, this volume) and causes an exchange of nutrients, phytoplankton, zooplankton and fish larvae. It is impossible to deal with all these interactions, but in trying to do so I have shown the complexity of this system and the difficulties one will meet in attempting to reduce eutrophication.

Acknowledgements

I thank Henk de Haan and Jan Moed for some of their unpublished results of chlorophyll *a* concentration, total P and temperature and for their valuable comments on the manuscript; Aafje Frank-Landman, Bertus Lemsma, and Koos Swart are acknowledged for assistance in the field and Sikko Parma and Koos Vijverberg for critical comments.

References

Andersson, G., H. Berggren, G. Cronberg & C. Gelin, 1978. Effect of planktivorous and benthivorous fish on organisms and water chemistry in eutrophic lakes. Hydrobiologia 59: 9–15.

Beattie, M., 1982. Distribution and production of the larval chironomid populations in Tjeukemeer. Hydrobiologia 95: 287–306.

Benndorf, J., H. Kneschke, K. Kossatz & E. Penz, 1984. Manipulation of the pelagic food web by stocking with predacious fishes. Int. Rev. ges. Hydrobiol. 69: 407–428.

Bernardi, R. de, 1986. Aquatic food chains manipulation and its effect on water quality. EC report. Ist. Ital. Idrobiol.

Brabrand A., B. Faafeng & J. P. M. Nilssen, 1986. Juvenile fish and invertebrate predators: delaying the recovery phase of eutrophic lakes by suppression of efficient filter feeders. J. Fish Biol. 29: 99–107.

DeHaan, H., 1982. Physico-chemical environment in Tjeukemeer with special reference to speciation of algal nutrients. Hydrobiologia 95: 205–221.

DeHaan, H. & J. R. Moed, 1984. Phosphorus, nitrogen and chlorophyll-a concentrations in a typical Dutch polder lake, Tjeukemeer in relation to its water regime between 1968 and 1982. Wat. Sci. Technol. 17, 733–743.

Densen, W. L. T. van, 1985. Feeding behaviour of major 0 + fish species in a shallow, eutrophic lake (Tjeukemeer, The Netherlands). J. Appl. Ichtyol. 2: 49–70.

Densen, W. L. T. van & Hadderingh, R. H., 1982. Effects of entrapment and cooling water discharge by the Bergum Power Station on 0 + fish in the Bergumermeer. Hydrobiologia 95: 351–368.

Densen, W. L. T. van & J. Vijverberg, 1982. The relations between 0 + fish density, zooplankton size and vulnerability of pikeperch, *Stizostedion lucioperca*, to angling in the Friesian lakes. Hydrobiologia 95: 321–336.

Goldspink, C. R. & J. W. Banks, 1975. A description of the Tjeukemeer fishery together with a note upon the yield statistics between 1964 and 1970. J. Fish Biol. 7: 687–708.

Hutchinson, G. E., 1957. A treatise on limnology II. Wiley, New York.

Huet, H. J. W. J., Phosphorus eutrophication research in the lake district of south-western Friesland, The Netherlands.

Preliminary results of abiotic studies. Hydrobiologia, this volume.

Kott, P., 1953. A modified whirling apparatus for sub-sampling of plankton. Aust. J. Mar. Freshwat. Res. 4: 387–393.

Lammens, E. H. R. R., H. W. de Nie, J. Vijverberg & W. L. T. van Densen, 1985. Resource partitioning and niche shifts of bream (*Abramis brama*) and eel (*Anguilla anguilla*) mediated by predation of smelt (*Osmerus eperlanus*) on *Daphnia hyalina*. Can. J. Fish. aquat. Sci. 42: 1342–1351.

Lammens, E. H. R. R., 1986. Interactions between fishes and the structure of fish communities in Dutch shallow eutrophic lakes. Ph.D. Thesis. 100 pp.

Lammens, E. H. R. R., 1988. Trophic interactions in the shallow eutrophic lake Tjeukemeer: bottom-up and top-down effects in relation to hydrology, bioturbation and predation during the period 1974–1985. Hydrobiologica 19: 81–87.

Leenen, J. D., 1982. Hydrology of Tjeukemeer. Hydrobiologia 95: 199–203.

Lynch, M., 1979. Predation, competition and zooplankton community structure: an experimental study. Limnol. Oceanogr. 24: 253–272.

McQueen, D., J. R. Post & E. L. Mills, 1986. Trophic relationships in pelagic ecosystems. Can. J. Fish. aquat. Sci. 43: 1571–1581.

Moed, J. R. & H. L. Hoogveld, 1982. The algal periodicity in Tjeukemeer during 1968–1978. Hydrobiologia 95: 205–223.

Moed, J. R., 1973. Effect of the combined action of light and silicon depletion on *Asterionella formosa Hass*. Verh. int. Ver. Limn. 18: 1367–1374.

Murphy, J. & J. P. Riley, 1962. A modified single-solution method for the determination of phosphate in natural waters. Anal. Chim. Acta 27: 31.

Nie, H. W. de, H. J. Bromley & J. Vijverberg, 1978. Distribution patterns of zooplankton in Tjeukemeer, The Netherlands. J. Plankton Res. 2: 317–334.

Nie, H. W. de & J. Vijverberg, 1985. The accuracy of population density estimates of copepods and cladocerans, using data from Tjeukemeer (The Netherlands) as an example. Hydrobiologia 124: 3–11.

Nusch, E. & G. Palme, 1975. Biologische Methoden für die Praxis der Gewässeruntersuchung. Gwf-Wasser/Abwasser 116: 562–565.

Shapiro, J. & D. I. Wright, 1984. Lake restoration by biomanipulation: Round Lake, Minnesota, the first two years. Freshwat. Biol. 14: 371–383.

Stenson, J. A. E., T. Bohlin, L. Henrickson, B. I. Nilsson, H. G. Nyman, H. G. Oscarson & P. Larsson, 1978. Effects of fish removal from a small lake. Verh. int. Ver. Limnol. 20: 794–801.

Tatrai, I. & V. Istvanovics, 1986. The role of fish in the regulation of nutrient cycling in Lake Balaton, Hungary. Freshwat. Biol. 16: 417–424.

Vijverberg J. & A. F. Richter, 1982a. Population dynamics and production of *Daphnia hyalina* Leydig and *Daphnia cucullata* Sars in Tjeukemeer. Hydrobiologia 95: 235–259.

Vijverberg J. & A. F. Richter, 1982a. Population dynamics and production of *Acanthocyclops robustus* Sars and *Mesocyclops leuckarti* Claus. Hydrobiologia 95: 261–274.

Vijverberg J. & W. L. T. van Densen, 1984. The role of the fish in the food web of Tjeukemeer. Verh. int. Ver. Limnol. 22: 891–896.

Hydrobiologia **191**: 39–46, 1990.
P. Biró and J. F. Talling (eds), Trophic Relationships in Inland Waters.
© 1990 *Kluwer Academic Publishers.*

39

Zooplankton community changes in Lake Kinneret (Israel) during 1969–1985

M. Gophen, S. Serruya & P. Spataru
The Yigal Allon Kinneret Limnological Laboratory, (Israel Oceanographic & Limnological Research) P.O. Box. 345, Tiberias, Israel 14102

Key words: zooplankton changes, Lake Kinneret, biomass reduction, fish-zooplankton predation

Abstract

Long-term records (1969–1985) of zooplankton density in Lake Kinneret indicated significant reduction of biomass (Copepoda, Cladocera, Rotifera) and production (Copepoda, Cladocera). Nauplius and adult copepod densities decreased but those of copepodites did not change. *Mesocyclops* was suppressed more than the smaller *Thermocyclops* and males of both genera became more abundant relative to the larger females. Ratios of 'small/large' Cladocera densities became higher. Numbers of total cladocerans were stable, and therefore reduction of Cladocera grazing capacity is assumed. The abundant *Keratella* spp. were reduced. It is likely that intensification of fish visual-attack-predation pressure shifted the size-class structure towards smaller adult copepods and cladocerans. Reduction of *Keratella* spp. and copepod nauplii was probably affected by increasing pressure of fish filter-feeders. Data on fish food consumption, feeding behaviour and fisheries management suggested their direct impact on long-term changes of zooplankton in Lake Kinneret.

Introduction

Seasonal distribution and dynamics of zooplankton populations for a relatively short-term record from Lake Kinneret were presented by Gophen (1972, 1973, 1976a, 1978, 1979, 1980b, 1981b, 1984, 1986a, 1987) since 1969. Interactions between Kinneret fish and plankton were experimentally investigated by Gophen (1980a), Drenner *et al.* (1982, 1987a, b), Gophen *et al.* (1983) and Vinyard *et al.* (1988). Food sources of Kinneret fish and zooplankton and their potential impacts on the food-web structure were also described (Serruya *et al.*, 1980; Gophen & Landau, 1977; Gophen & Scharf, 1981; Gophen, 1976a, b, 1977, 1980b, 1981a, b, 1985a, b, 1986b,

1987; Spataru & Gophen, 1985a, b, c, 1986a, b, 1987a, b; Gophen & Pollingher, 1985; Gophen *et al.*, 1988; Spataru *et al.*, 1987). These data, and information on fisheries and fish stocking operations, were utilised to analyse the long-term (1969–1986) changes of the Kinneret zooplankton communities presented in this paper.

Background: zooplankton

Species composition

The planktonic copepods of Lake Kinneret are dominated by two species, *Mesocyclops ogunnus* (Van-de-Velde) and *Thermocyclops dybowskii*

(Lande); cladocerans are represented by three major genera: *Diaphanosoma brachyurum*, *Ceriodaphnia reticulata* and *C. rigaudi*, and *Bosmina longirostris*, and the Rotifera by about 20 species of which only *Keratella cochlearis* and *K. valga tropica* populations are relatively stable throughout the whole year cycle.

Body size and vulnerability

The body size of *Mesocyclops ogunnus* is larger than that of *Thermocyclops dybowskii* and females are longer than males (Gophen, 1973). Nauplii are small and probably invisible to visual particulate feeders but heavily preyed by filtrator fish (Drenner *et al.*, 1982) and by adult Cyclopidae. Body size of cladocerans follows the sequence *Diaphanosoma*, *Ceriodaphnia*, *Bosmina* (Gophen, 1973). Cladocerans are preyed by fish visually and also by filter feeders (Drenner *et al.*, 1982). Most of the rotifers are small (80–150 μm) (*Keratella* spp., *Synchaeta oblonga*, *Brachionus angularis*, *Pedalia fennica* etc.) and therefore vulnerable mostly to filter feeding adult fishes and particulate feeding young larvae (Landau *et al.*, 1988). Large rotifers (*Asplanchna* spp. and *Synchaeta pectinata*: Gophen, 1973) can probably be preyed visually (Gophen *et al.*, 1987). Vulnerability of zooplankters is also affected by their swimming behaviour. Copepods (copepodites and adults) and *Diaphanosoma* are better escapers than *Bosmina* and *Ceriodaphnia* (Drenner *et al.*, 1982).

Background: fish

Feeding habits (reviewed in Gophen 1986b, 1987)

There are 19 native and 5 exotic fish species in the lake. The food composition and feeding habits of the dominant species were studied. In spite of differences between the major food resources of fish species, most fish (adults and fingerlings) prey on zooplankton especially during summer-fall season, and ingest *Peridinium* cells during its bloom period. It was found that *S. galilaeus* digests dinoflagellates most efficiently, and the very common *Mirogrex terraesanctae* (adults, fingerlings and larvae) consume mostly zooplankton throughout the whole year cycle.

Fishery and stocking

The lake is intensively stocked by fingerlings of the exotic silver carp (*Hypophthalmichthys molitrix*), mullets (*Mugil cephalus* and *Liza ramada*) and native (*S. galilaeus* (Artedi) and *O. aureus* (Steindachner)) species. *Oreochromis* (= Tilapia) *aureus* was rare prior to its artificial stocking. Mullet fingerlings are introduced during the winter months and the others in the summer-fall season. The major food sources of all introduced and native fingerlings are zooplankton and zoobenthos.

The zooplanktivorous bleaks (*Mirogrex terraesanctae*, *Acanthobrama lissneri*) are the most abundant fish in the lake. Landau (1985) indicated that an increase in the biomass of bleaks occurred in the lake during the 1970's. Due to changes of fish populations and the stocking activity during 1960–1983, it is suggested that fish predation pressure on zooplankton in the Kinneret ecosystem has increased in recent years.

Results and discussion

Data presented in Tables 1, 2, 4 and 7 indicate significant reduction of zooplankton biomass and production during 1969–1985. Densities (organisms 1^{-1}) of *Ceriodaphnia* spp. and *Diaphanosoma* sp. were not significantly changed (Table 2). Results in Tables 3 and 7 show modification of the size-class distribution in cladoceran communities where smaller organisms became more abundant.

Structural changes of copepod populations were also observed (Tables 4, 5, 7). Adults and nauplii stages became less abundant; densities of females (*Thermocyclops* sp. and *Mesocyclops* sp.)

Table 1. Annual averages of zooplankton biomass ($g_{(w.w.)}$ m^{-2}) and total production ($g_{(w.w.)}$ m^{-2} month^{-1}) of Copepoda and Cladocera during 1969–1985.

Year	Copepoda	Cladocera	Rotifera	Total production
1969	19.3	16.5	5.9	151
1970	16.4	27.6	4.0	193
1971	14.6	24.0	1.7	179
1972	12.0	19.7	1.6	135
1973	9.6	26.2	3.0	154
1974	11.1	22.7	2.3	147
1975	9.9	17.5	2.4	115
1976	11.8	18.0	2.7	126
1977	12.5	21.1	2.1	143
1978	14.1	16.4	2.1	125
1979	14.7	16.4	2.7	124
1980	12.7	21.0	1.9	143
1981	9.2	17.3	1.6	114
1982	8.9	20.2	2.4	128
1983	7.6	12.5	1.2	91
1984	8.6	14.3	1.3	100
1985	7.0	12.2	1.3	86

Table 3. Ratio of small/large cladoceran individuals* based on annual averages of densities (animals l^{-1}) during 1975–1985.

Year	*Diaphanosoma*	*Bosmina*	*Ceriodaphnia*
1975	0.22	0.14	0.29
1976	0.25	0.47	0.61
1977	0.32	0.42	0.47
1978	0.12	0.15	0.88
1979	0.74	0.72	0.67
1980	0.93	1.15	0.57
1981	0.70	0.66	0.52
1982	0.84	1.06	0.60
1983	1.47	1.54	0.71
1984	0.75	0.92	0.78
1985	0.89	1.06	1.14

* 'small' = young: new born and 1–2 neonates; 'large' = adults: larger than 1–2 neonates.

Table 2. Annual averages of Cladocera densities (animals l^{-1}) and production ($g_{(w.w.)}$ m^{-2} month^{-1}) in Lake Kinneret during 1969–1985*.

Year	*Diaphanosoma* sp.	*Bosmina* spp.	*Ceriodaphnia* spp.	Production
1969	4.2	12.2	7.5	94
1970	10.9	19.3	10.6	147
1971	4.4	14.3	16.8	124
1972	4.4	12.1	16.6	94
1973	6.4	16.6	25.4	122
1974	7.4	14.6	20.2	108
1975	3.3	10.0	13.4	83
1976	2.0	14.3	17.1	86
1977	3.3	18.4	15.4	98
1978	2.9	9.7	13.2	81
1979	6.4	12.4	9.2	80
1980	7.7	9.9	15.1	100
1981	4.6	10.3	14.3	82
1982	4.6	10.1	21.0	98
1983	4.7	9.4	11.6	62
1984	3.5	10.2	13.2	68
1985	3.4	11.1	13.7	61

* *Moina rectirostris* (1975, 1985) and *Chydorus sphaericus* (1975, 1983–1985) were also recorded.

Table 4. Annual averages of Copepoda (adults and nauplii) densities (animals l^{-1}) and copepod production ($g_{(w.w.)}$ m^{-2} month^{-1}).

Year	Adults	Nauplii	Production
1969	16.3	125.1	57
1970	14.3	157.0	46
1971	14.1	124.9	55
1972	14.9	139.1	41
1973	12.4	146.6	32
1974	16.2	172.0	39
1975	10.7	93.3	32
1976	11.8	105.1	40
1977	11.9	114.7	45
1978	12.0	130.3	44
1979	13.3	154.5	44
1980	10.8	154.4	43
1981	8.6	92.7	32
1982	10.1	116.0	30
1983	10.0	118.9	29
1984	7.5	104.5	32
1985	10.3	66.1	25

Table 5. Annual mean density (animals l⁻¹) of adult (males and females) copepods (*Mesocyclops* sp. and *Thermocyclops* sp.) in Lake Kinneret during 1970–1985*.

Year	*Mesocyclops* sp.		*Thermocyclops* sp.	
	Males	Females	Males	Females
1970	7.4	5.5	0.1	1.3
1971	6.3	7.0	0	0.8
1972	9.1	5.2	0	0.6
1973	6.8	3.1	0.5	2.0
1974	7.5	3.1	3.0	2.6
1975	4.1	2.5	2.1	2.0
1976	4.2	6.7	0.3	0.6
1977	5.3	5.8	0.1	0.7
1978	5.2	6.5	0	0.3
1979	5.1	6.6	0.2	1.4
1980	5.5	2.8	0.5	2.0
1981	0.9	1.9	2.2	3.6
1982	3.9	2.3	1.3	2.6
1983	5.4	2.2	1.0	1.4
1984	3.0	1.0	2.1	1.4
1985	5.8	2.2	1.6	0.7

* *Arctodiaptomus gracilis* was recorded in 1985.

Table 7. Correlation coefficients (*r*) of linear regression between zooplankton population variables (annual averages) in Lake Kinneret and years (1969–1985): + = enhancement; – = reduction; significant *P* values (0.01 or 0.05) are given in parenthesis.

Parameter	*r*
Copepoda biomass	– 0.73 (0.01)
Cladocera biomass	– 0.67 (0.01)
Rotifera biomass	– 0.66 (0.01)
Total biomass	– 0.86 (0.01)
Copepoda and Cladocera production	– 0.83 (0.01)
Bosmina density	– 0.60 (0.05)
Cladocera production	– 0.74 (0.01)
Diaphanosoma: small/large	+ 0.77 (0.01)
Bosmina: small/large	+ 0.79 (0.01)
Ceriodaphnia: small/large	+ 0.66 (0.05)
Total number of large cladocerans	– 0.80 (0.01)
Total number of small cladocerans	+ 0.69 (0.05)
Nauplii density	– 0.50 (0.05)
Adult copepods (*Mesocyclops* + *Thermocyclops*) density	– 0.84 (0.01)
Copepod production	– 0.72 (0.01)
Total adult female copepods	– 0.64 (0.01)
Total adult *Mesocyclops* (♀ + ♂)	– 0.73 (0.01)
Keratella spp. density	– 0.74 (0.01)

Table 6. Annual mean density (animals l⁻¹) of *Keratella* spp., large (*Asplanchna* spp. and *Synchaeta pectinata*) and small (other species) rotifers in Lake Kinneret during 1969–1985.

Year	*Keratella* spp.	Large rotifers	Small rotifers
1969	71.7	7.2	83.5
1970	86.6	8.3	99.2
1971	23.7	3.9	40.1
1972	39.4	4.5	49.1
1973	39.1	14.9	56.3
1974	60.4	7.3	89.5
1975	61.1	9.4	89.8
1976	53.8	9.3	114.2
1977	34.8	3.6	85.5
1978	15.9	3.1	59.0
1979	38.2	5.9	103.3
1980	24.6	3.2	51.8
1981	38.9	3.3	90.6
1982	21.6	10.1	83.5
1983	15.8	4.2	26.8
1984	10.4	2.5	32.3
1985	16.6	5.2	36.4

Table 8. Annual averages of zooplankton densities (animals l⁻¹): 'good' (GE) and 'poor' (PE) escapers, small (SZ) and large (LZ) organisms. 'Zooplankton Predation Index' (ZPI = GE/PE) and SZ/LZ ratios are given (see text).

Year	PE	GE	ZPI	SZ	LZ	SZ/LZ
1969	235.5	62.5	0.3	208.6	89.4	2.3
1970	294.4	56.9	0.2	256.2	95.1	2.7
1971	200.0	63.8	0.3	165.0	98.8	1.7
1972	221.4	60.0	0.3	188.2	93.2	2.0
1973	259.8	53.5	0.2	202.9	110.4	3.9
1974	303.6	74.5	0.2	261.5	116.6	2.2
1975	224.6	45.0	0.2	191.8	77.8	2.5
1976	256.0	50.9	0.2	219.3	91.6	2.4
1977	237.6	61.9	0.3	200.2	99.1	2.0
1978	215.3	58.2	0.3	189.3	84.2	2.2
1979	285.3	66.2	0.2	257.8	91.7	2.8
1980	234.4	65.7	0.3	206.2	93.9	2.2
1981	211.2	53.7	0.3	183.3	81.6	2.2
1982	240.7	46.3	0.2	199.5	87.5	2.3
1983	170.9	49.8	0.3	145.7	75.0	1.9
1984	162.7	50.8	0.3	136.8	76.7	1.8
1985	132.5	54.1	0.4	102.5	84.1	1.2

and relative abundance of adults (males and females) of *Mesocyclops* sp. decreased.

Large bodied cladocerans and adult copepods populations, which are suitable prey for visual attack by fishes, decreased during 1969–85.

Results in Tables 6 and 7 show a density reduction of *Keratella* during 1969–1985. Other rotifers (small and large) were abundant in the lake during short periods, mostly in the winter-spring season, and their concentrations were not lowered.

Small zooplankters, *Keratella* and nauplii have become less abundant in Lake Kinneret in recent years. These small organisms are efficiently preyed on by filter feeding fish (Drenner *et al.*, 1982; Vinyard *et al.*, 1988).

Suppression of large zooplankters by fish has been documented by many authors (Hrbáček *et al.*, 1961; Brooks & Dodson, 1965; Shapiro & Wright, 1984; Carpenter *et al.*, 1985). In Lake Kinneret it was described by Drenner *et al.* (1982, 1987), Gophen (1985a) and Vinyard *et al.* (1988).

It is suggested that the observed changes in zooplankton communities were the result of an increase of predation pressure produced by particulate- and filter-feeding fish. Results in Table 8 show annual averages of the 'Zooplankton Predation Index' (ZPI) and ratios between Small (small rotifers and nauplii) (SZ) and Large (*Bosmina*, *Ceriodaphnia*, *Diaphanosoma*, copepodites and adult copepods) (LZ) zooplankters in Lake Kinneret during 1969–1985. 'ZPI' values represent the ratios between densities of 'Good Escapers' (GE) (Copepodites, adult copepods and *Diaphanosoma*) and 'Poor Escapers' (PE) (rotifers, nauplii, *Bosmina* and *Ceriodaphnia*). Annual fluctuations of visual and filter-feeding zooplanktivory can be reflected by changes of both ZPI and SZ/LZ ratios. Nevertheless, each of these two indices is not related to only one of the feeding types because of some degree of overlap. These two values partially represent particulate (ZPI) and filter (SZ/LZ) zooplanktivory. Values of ZPI and SZ/LZ show low fluctuations, 0.2–0.4 and 1.7–2.8 respectively with two exceptions of 1.2 and 3.8 in SZ/LZ values. Therefore we suggest that both types of feeding activity were intensified.

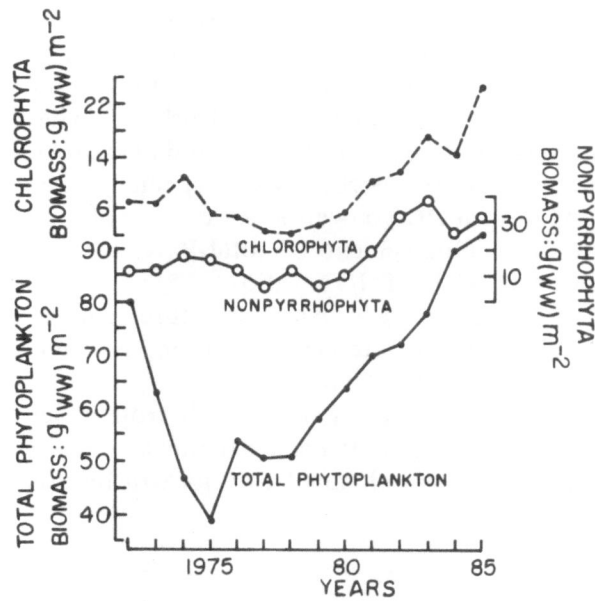

Fig. 1. Annual averages of total phytoplankton, non-Pyrrhophyta and Chlorophyta ($g_{(w.w.)} m^{-2}$) in Lake Kinneret during 1972–1985 (Pollingher, 1972–1985).

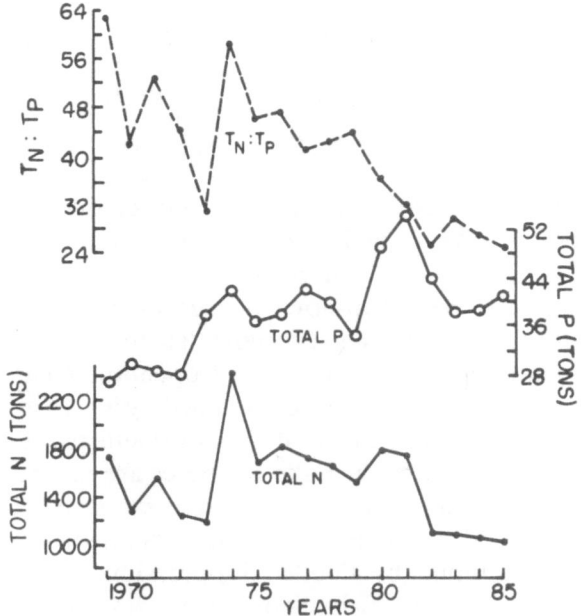

Fig. 2. Annual averages of mass contents (metric tons) of total N (TN) and total P (TP) and the TN/TP (wt/wt) ratio in the epilimnion (0–16 m) of Lake Kinneret during 1969–1985 (S. Serruya, IOLR Lake Kinneret database).

The temporal trends of phytoplankton and nutrient changes in the Kinneret epilimnion during 1969–1985 are shown in Figs. 1 and 2 respectively. Data in Fig. 1 clearly indicate an increase of total phytoplankton and of the relative biomass of non-pyrrhophytes (especially Chlorophyta) from 1975 onwards. A decrease of total-N and an enhancement of total-P resulted in a lowering of the TN/TP ratio.

Two different ecosystem structures should be considered to interpret plankton community changes in Lake Kinneret.

a) A steady-state pattern with trophic levels containing roughly the same biomass and where energy is poorly channelled to growth by organisms. This system is 'bottom-up' (or producer-controlled), set by the rate of nutrient input, and non-predatory losses are negligible (McQueen et al., 1984, 1986; Sprules, 1987).

b) 'Top-down' (or consumer-controlled) forcing where transmitted or 'cascading' trophic interactions control the productivity of lower levels and superimpose on consequences of nutrient supply (Carpenter et al., 1985). The summer-fall (May–November) conditions in the Lake Kinneret epilimnion were identified as 'steady-state'. A typical example of this pattern of carbon-balance (in mg C m^{-2} day^{-1}) for summer 1972 has been described by Serruya et al. (1980) and Gophen (1986b):

Input		Output		% of input
Phytobenthos primary productivity	35	Phytoplankton respiration	600	28
Plankton primary productivity	2036	Bacteria respiration	817	39
From drainage basin	38	Zooplankton respiration	797	38
		Fish respiration	160	8
		Outflow	34	2
		Sedimentation	85	4
Total	2109	Total	2493	119

* Excess carbon output (384 mg C m^{-2} day^{-1}) means 'organic pool' depletion.

Although densities (animals l^{-1}) of cladocerans were relatively stable during 1969–1985, reductions of zooplankton biomass and production were observed. 'Top-down' forces are likely to be relevant to the Kinneret summer 'steady-state' for which a biomass reduction of Copepoda, Cladocera and Rotifera was accompanied by increase in the relative abundance of small organisms. The winter conditions in Lake Kinneret are different because of lower temperature and consequently lower food requirements of organisms and high nutrient influx by floods. Suppression of large zooplankters by fish planktivory may increase phytoplankton biomass (Hrbáček et al., 1961; Brooks & Dodson, 1965; Shapiro & Wright, 1984). Enhancement of nanophytoplankton biomass in Lake Kinneret was recently observed (Fig. 1; Pollingher, 1972–1985, 1986). It is not excluded that increase of phosphorus content observed in lake water, accompanied by lowering of the ratio between total N and total P (wt/wt) (Fig. 2), also contributed to nanoplankton enhancement. However under such conditions, without top-down (consumer controlled) effect of fish predation, an increase of zooplankton biomass would be predicted. Because zooplankton biomass was significantly reduced, the associated cascading-effect hypothesis is likely. The enhancement of nanoplankton in Lake Kinneret (Fig. 1) might cause a deleterious effect on water quality in the major reservoir of drinking water in Israel, and therefore fishery management aimed at lowering pressure from zooplankton is recommended.

Acknowledgements

We warmly thank M. Shniper, B. Azoulay, Z. Katz and M. Hatab (skipper of R/V 'Hermona') for their technical help.

References

Brooks, J. L. & S. L. Dodson, 1965. Predation, body size and composition of plankton. Science 150: 28–35.

Carpenter, S. R., J. F. Kitchell & J. Hodgson, 1985. Cascading trophic interactions and lake productivity. Bioscience 35: 634–639.

Drenner, R. W., G. L. Vinyard, M. Gophen & S. R. McComas, 1982. Feeding behaviour of the cichlid *Sarotherodon galilaeus*: selective predation on Lake Kinneret zooplankton. Hydrobiologia 87: 17–20.

Drenner, R. W., G. L. Vinyard, K. D. Hambright & M. Gophen, 1987a. Particle ingestion by Galilee St. Peter's fish is not affected by removal of gill rakers and microbranchiospines. Trans. Am. Fish. Soc. 116: 272–276.

Drenner, R. W., K. D. Hambright, G. L. Vinyard, M. Gophen & U. Pollingher, 1987b. Experimental study of size-selective phytoplankton grazing by a filter-feeding cichlid's impact on plankton community structure. Limnol. Oceanogr. 32(5): 1138–1144.

Gophen, M., 1972. Zooplankton distribution in Lake Kinneret (Israel). Israel J. Zool. 21: 17–27.

Gophen, M., 1973. Zooplankton in Lake Kinneret. In: Lake Kinneret Data Record. (T. Berman, ed.). Israel Science Research Conservation, pp. 61–67.

Gophen, M., 1976a. Temperature dependence of food intake, ammonia excretion and respiration in *Ceriodaphnia reticulata* (Jurine) (Lake Kinneret, Israel). Freshwat. Biol. 6: 451–455.

Gophen, M., 1976b. Temperature effect on lifespan, metabolism and development time of *Mesocyclops leuckarti* (Claus). Oecologia (Berl.). 25: 271–277.

Gophen, M., 1977. Food and feeding habits of *Mesocyclops leuckarti* (Claus) in Lake Kinneret (Israel). Freshwat. Biol. 7: 513–518.

Gophen, M., 1978. Zooplankton. In: Lake Kinneret, Monographiae Biologicae, 32: 297–311. C. Serruya (ed.). Dr. W. Junk Publishers, The Hague.

Gophen, M., 1979. Sex ratio in *Mesocyclops leuckarti* (Claus) populations in Lake Kinneret (Israel). Hydrobiologia 66: 41–43.

Gophen, M., 1980a. Food resources, feeding behaviour and growth rates of *Sarotherodon galilaeum* (Linnaeus) fingerlings. Aquaculture 20: 101–115.

Gophen, M., 1980b. Artemia nauplii as a food resource for cyclopoids: extrapolation of experimental measurements to the metabolic activities of copepods in Lake Kinneret, Israel. In G. Persoone, P. Sorgeloos, O. Roels and E. Jaspers (eds.), The Brine Shrimp Artemia. Universa Press, Wetteren, Belgium. Vol. 3: 67–76.

Gophen, M., 1981a. The metabolism of adult *Mesocyclops leuckarti* populations in Lake Kinneret (Israel) during 1969–1978. Verh. int. Ver. Limnol. 21: 1568–1572.

Gophen, M., 1981b. Metabolic activity of herbivorous zooplankton in Lake Kinneret (Israel) during 1972–1977. J. Plankton Res. 3: 15–24.

Gophen, M., 1984. The impact of zooplankton status on the management of Lake Kinneret (Israel). Hydrobiologia 113: 249–258.

Gophen, M., 1985a. Effect of fish predation on size class distribution of cladocerans in Lake Kinneret. Verh. int. Ver. Limnol. 22: 3104–3108.

Gophen, M., 1985b. The management of Lake Kinneret and its drainage basin. In: Scientific Basis for Water Resources Management. Proceedings Jerusalem Symposium. September 1985. Int. Assoc. Hydrol. Soc., Publ. No. 153, pp. 127–138.

Gophen, M. 1986a. *Mesocyclops* and *Thermocyclops* populations in Lake Kinneret (Israel). Proc. 2nd Internat. Conf. on Copepoda, Ottawa, Canada, 13–17 Aug. 1984. Syllogeus No. 58: 294–300.

Gophen, M., 1986b. Fisheries management in Lake Kinneret (Israel). In: Proc. Ann. Meeting North American Lake Management Soc., Lake Geneva, Wisconsin, USA, 13–16 Nov. 1985: 327–332.

Gophen, M., 1987. Fisheries management, water quality and economic impacts: a case study of Lake Kinneret. Proc. World Conf. on Large Lakes, Mackinac Island, Michigan, USA, 18–21 May 1986 Vol. II: 5–24.

Gophen, M. & R. Landau, 1977. Trophic interactions between zooplankton and sardine (*Mirogrex terraesanctae*) populations in Lake Kinneret, Israel. Oikos 29: 166–174.

Gophen, M. & A. Scharf, 1981. Food and feeding habits of *Mirogrex* fingerlings in Lake Kinneret (Israel). Hydrobiologia 78: 3–9.

Gophen, M., R. W. Drenner & G. L. Vinyard, 1983. Cichlid stocking and the decline of the Galilee St. Peter's fish (*Sarotherodon galilaeus*) in Lake Kinneret, Israel. Can. J. Fish. aquat. Sci. 40: 983–986.

Gophen, M. & U. Pollingher, 1985. Relationships between food availability, fish predation and the abundance of the herbivorous zooplankton community in Lake Kinneret. Arch. hydrobiol. 21: 397–405.

Gophen, M., B. Azoulay & M. N. Bruton, 1988. Selective predation of Lake Kinneret zooplankton by fingerlings of *Clarias gariepinus*. Ver. int. Ver. Limnol. 23: 1763–1765.

Hrbáček, J., M. Dvornaková, V. Kořinek & L. Procházková, 1961. Demonstration of the effect of the fish stock on the species composition of zooplankton and the intensity of metabolism of the whole plankton assemb?age. Verh. int. Ver. Limnol. 14: 192–195.

Landau, R., 1985. Fish population dynamics. In: Kinneret Limnological Laboratory, Annual Report. 1984. pp. 19 (Gophen, M. ed.). (in Hebrew).

46

Landau, R., M. Gophen & P. Walline, 1988. Larval *Mirogrex terraesanctae* (Cyprinidae) of Lake Kinneret (Israel): growth rate, plankton selectivities, consumption rates and interactions with rotifers. Hydrobiologia 169: 91–106.

McQueen, D. J. & J. R. Post, 1984. Effects of planktivorous fish on zooplankton, phytoplankton and water chemistry. Lake and Reservoir Management. Proc. 4th Ann. Meeting North American Lake Management Soc., McAfee, October 1984.

McQueen, D. J., J. R. Post & E. L. Mills, 1986. Trophic relationships in freshwater pelagic ecosystems. Can. J. Fish. Aquat. Sci. 43: 1571–1581.

Pollingher, U., 1972–1985. Phytoplankton. In: Annual Reports of Kinneret Limnological Laboratory.

Pollingher, U., 1986. Phytoplankton periodicity in a subtropical lake (Lake Kinneret, Israel). Hydrobiologia 138: 127–138.

Serruya, C., M. Gophen & U. Pollingher, 1980. Lake Kinneret: carbon flow patterns and ecosystem management. Arch. Hydrobiol. 88: 265–302.

Shapiro, J. & D. I. Wright, 1984. Lake restoration by biomanipulation. Freshwat. Biol. 14: 371–383.

Spataru, P. & M. Gophen, 1985a. Feeding behaviour of silver carp *Hypophthalmichthys molitrix*. Val. and its impact on the food web in Lake Kinneret, Israel. Hydrobiologia 120: 53–61.

Spataru, P. & M. Gophen, 1985b. Food composition of the barbel *Tor canis* (Cyprinidae) and its role in the Lake Kinneret ecosystem. Envir. Biol. Fish 14: 295–301.

Spataru, P. & M. Gophen, 1985c. Food composition and feeding habits of *Astatotilapia flaviijosephi* (Lortet) in Lake Kinneret (Israel). J. Fish Biol. 26: 503–507.

Spataru, P. & M. Gophen, 1986a. Food composition of *Tristramella simonis simonis* (Gunther, 1864) in Lake Kinneret (Israel). J. Aquacult. in the Tropics 1: 111–117.

Spataru, P. & M. Gophen, 1986b. Food and feeding habits of *Capoeta damascina* (Cyprinidae) in Lake Kinneret, Israel. J. Aquacult. in the Tropics 1: 147–153.

Spataru, P. & M. Gophen, 1987a. The food and benthophagous feeding habits of *Barbus longiceps* (Cyprinidae) in Lake Kinneret (Israel). Arch. Hydrobiol. 110(3): 331–337.

Spataru, P. & M. Gophen, 1987b. Food composition of *Tristramella sacra* (Gunther, 1864) (Cichlidae) in Lake Kinneret (Israel). Israel J. Zool. 34(3/4): 183–189.

Spataru, P., Viveen, W. J. A. R. & M. Gophen, 1987. Food composition of *Clarias gariepinus* (= *C. lazera*) (Cypriniformes, Clariidae) in Lake Kinneret (Israel). Hydrobiologia 144: 77–82.

Sprules, W. G., 1987. Effect of trophic interactions on the shape of pelagic size spectra. Verh. int. Ver. Limnol. (in press).

Vinyard, G. L., R. W. Drenner, M. Gophen, U. Pollingher, D. L. Winkleman & K. D. Hambright, 1988. An experimental study of the plankton community impacts of two filter-feeding cichlids, the Galilee Saint Peter's fish (*Sarotherodon galilaeus*) and Blue *Tilapia* (*Tilapia aurea*). Can. J. Fish. aquat. Sci. 45(4): 685–690.

Hydrobiologia **191**: 47–55, 1990.
P. Biró and J. F. Talling (eds), Trophic Relationships in Inland Waters.
© 1990 *Kluwer Academic Publishers.*

The contribution of silver carp (*Hypophthalmichthys molitrix*) to the biological control of Netofa reservoirs

H. Leventer & B. Teltsch
Mekoroth Water Co., Jordan District, Central Laboratory Nazareth Ilit 17105, Israel

Key words: silver carp, biological control, algae, zooplankton, organic matter

Abstract

When silver carp were introduced into the Netofa reservoirs at an initial density of 300–4500 fish per hectare in order to control phytoplankton and zooplankton, there was a significant reduction of algae, zooplankton, and suspended organic matter; the silver carp prevents the growth of blue-green algae.

Annual yield ranged from 600 to 1500 kg per hectare. The growth of individual fish after 6 to 8 years was 6 to 15 kg per fish.

Introducing silver carp to reservoirs as a means of biological control creates a balanced ecological system in which the interspecific competition is minimal and the environmental improvements are considerable.

Silver carp and bottom-feeding fish create a positive synergism in the water-body by filtering phytoplankton and zooplankton from the water, excreting a major part of it to the bottom and enriching it with organic matter suitable for zoobenthos.

Introduction

The Israel National Water Carrier (INWC), which carries water from Lake Kinneret to the southern part of Israel, was built during the 1960s. This water system consists of open reservoirs, channels, and 250 km of pipes, supplying water for drinking and for agricultural purposes.

Lake Kinneret is rich in algae and zooplankton. The quantity of algae in the trophogenic layer ranged from 39 to 83 g m^{-2} (Berman, 1973). The dominant algae is *Peridinium cinctum*. Most of the algae, zooplankton, and detritus settles on the bottom of the Tsalmon and Netofa reservoirs.

Nuisances were caused by some species of algae (see Table 1) such as *Peridinium, Oscilla-* *toria,* and *Cyclotella,* and submerged plants, snails, clams, shrimps, insect larvae, and small fish.

When the system began to function in 1964–1968, the algae were treated with copper sulphate and chlorine. Submerged plants were treated by drying the reservoirs (see Table 1).

From 1968 to 1985, biological control was achieved by using fish (Leventer, 1981). Biological nuisances such as algae causing foul tastes and odors were removed by the fish tilapia and grey mullet. Various carp removed other nuisances: grass carp removed submerged plants, black carp removed snails and common carp removed insect larvae.

These first steps achieved successful biological

Table 1. Israel national water carrier, nuisances and control.

	NUISANCES and CONTROL / YEARS	1964	1965	1966	1967	1968	1969	1970	1971	1972	1973	1974	1975	1976	1977	1978	1979	1980	1981	1982	1983	1984
NUISANCES	PERIDINIUM	☼	☼	☼	☼	☼	☼	☼														
	OSCILLATORIA taste and odor			★	★	★																
	OTHER ALGAE	•	•	•	•	•	•	•		•	•	•		•	•	•	•	•	•	•	•	•
	SUBMERGED PLANTS		◠	◠	◠	◠	◠	◠	◠				◠				◠					
	SNAILS	▪		▪	▪	▪	▪	▪	▪	▪	▪	▪	▪	▪	▪	▪	▪					
	SHRIMPS							○	○				○	○								
	CHIRONOMIDAE																☆	☆				
	ROUGH FISH									▲	▲			▲	▲						▲	▲
CONTROL DRYING	Tsalmon		▲		▲											▲						
	Settling		▲		▲		▲				▲					▲						
	Eshkol			▲	▲	▲						▲				▲						
FISH	TILAPIA						●	●	●	●	●	●	●	●	●	●	●	●	●	●	●	●
	GREY MULLET							●	●	●	●	●	●	●	●	●	●	●	●	●	●	●
	GRASS CARP								◒	◒	◒		◒	◒		◒	◒	◒	◒	◒	◒	◒
	SILVER CARP															●	●	●	●	●	●	●
	BLACK CARP																▪	▪	▪	▪	▪	
	COMMON CARP																☆	☆	☆	☆	☆	☆
	BASS														▲	▲					▲	▲

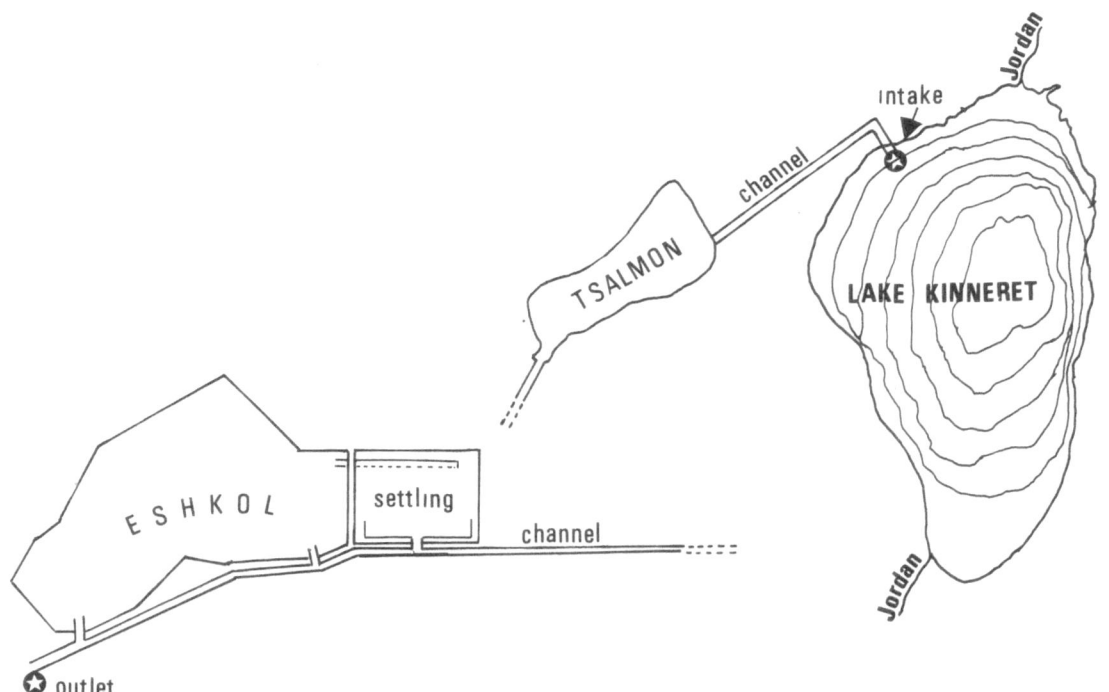

Fig. 1. The Israel National Water Carrier (INWC). The scheme includes: Channels, Tsalmon reservoir and Netofa reservoirs.

control. Then the introduction of silver and bighead carps was planned to reduce the quantity of phytoplankton and zooplankton.

There are many different opinions regarding the contribution of silver carp to biological control of algae. Opuszynski (1980) found that despite the fact that silver carp feed on algae, the quantity of algae in the water actually increases. He called this phenomenon 'Ichthyœutrophication'.

Growth of blue-green algae is a typical phenomenon in water bodies all over the world. The excessive algae growth that occurred during the 1950s in the big water reservoirs of the U.S.S.R. was controlled through the introduction of silver carp.

Summarizing the results of introducing silver carp into the reservoirs, Vovk (1974), Yefimova (1977) and Verigin (1979) noted that a carp population prevented not only algae growth, but also eutrophication of the reservoirs. Vinogradov (1979), in his investigation of the nourishment patterns of silver carp, found their food to be composed mainly of the algae *Microcystis* and *Anabaena*.

From 1979 on, the reservoirs were populated by seven species of fish, including the silver carp. This project concluded 8 years of study on the impact of silver carp on algae, zooplankton and suspended organic matter in the reservoirs.

Description of sites studied

The investigation under discussion was carried out along a section of the INWC from Lake Kinneret to the outlet of the Eshkol reservoir (Fig. 1). After the water is pumped from the lake at a depth of 12–15 m, it flows 17 km through open channels to the Tsalmon reservoir. The reservoir covers an area of 20 ha, a volume of 800 000 m³ and a maximum depth of 4.5 m. Approximately 1 million cubic meters of water flow through the reservoir daily, and the average residence time is approximately one day.

The water is pumped from the Tsalmon reservoir and flows through 16 kilometers of open channels into the Netofa reservoirs, which include a sedimentation reservoir with an area of 40 ha and a volume of 1.5×10^6 m³; and an operational reservoir (the Eshkol reservoir), with an area of 100 ha, a volume of 4.5×10^6 m³ and a maximum depth of 7 m. The residence period is between 6 and 8 days.

Both channels have concrete bases approximately 35 km long and areas of 20 ha. The flow rates range from 24 000 to 68 700 m³ h⁻¹ depending on the number of pumps in operation at Lake Kinneret. The average flow velocity is 1 m s^{-1} and the hydraulic resistance time of each channel is approximately 3.5 h.

Material and methods

The reservoirs were populated with fish from 1978–1981 by the species and quantities shown below:

Species	No. per Ha			Total no.
	Tsalmon	Settling	Eshkol	
Tilapia	1500	–	–	30 000
Mullet	–	400	700	86 000
Silver	4350	538	235	132 000
Bighead	800	125	235	44 500

Initial weights of the fish were tilapia – 10 g; mullet – 0.3 g; and silver and bighead carp – 5 g. Only tilapia fish reproduced in the Tsalmon reservoir.

Over a period of six consecutive years, weekly qualitative and quantitative analyses of algae, zooplankton and suspended matter were conducted at the intake to the INWC at Lake Kinneret and at the outlet from the Eshkol reservoir. Analyses of algae were carried out according to the 'Standard Methods For Examination of Water and Waste Water' (15th ed.; 1980). One liter of water was filtered through fine sand, which was then washed with 10 ml of water. A 1-ml glass cell Sedgewick-Rafter served to count the algae. For zooplankton analysis, 5 liters of water were filtered through an 80-micron net.

The zooplankton was concentrated in 50 ml of water and counted. Determination of dry organic suspended matter was carried out by filtering water through a GF/C filter with pore size $< 1 \mu$m followed by drying and combustion at 105 °C and 550 °C. The results were presented as $mg\,l^{-1}$ of dry matter.

Estimations of net balance involving suspended solids, channel community photosynthesis and respiration were conducted through iterative routing of two-station diel measurement of dissolved oxygen (at each channel) and through suspended matter measurement (Telsch & Ben-Zur, 1988).

Sources of algae, zooplankton and detritus

Suspended organic matter, algae, zooplankton, and detritus are found in the water from Lake Kinneret, the open channels and the reservoirs. In the large number of analyses conducted for Lake Kinneret during 1969–1984, (see Annual Reports: Kinneret Limnological Laboratory), the quantity of algae in the trophogenic layer ranged from 39 to 83 $g\,m^{-2}$. The dominant algae, *Peridinium*, decreased from 77 $g\,m^{-2}$ in 1970 to 23 $g\,m^{-2}$ in 1975 (see Table 2). Green algae constituted 6 to 16% of the algal population, except in 1974 when they constituted 27%. The blue-green algae generally appeared in low concentrations, except for 1975 when they constituted 20%. A high concentration of diatoms was found in 1982 (20%) and in 1983 (23%).

The dominant species of zooplankton present in the waters of Lake Kinneret are *Mesocyclops oqunum*, *Thermocyclops dybowskii*, and the Cladocera *Ceriodaphnia riguadi*, *Bosmina longirostris* and *Diaphanosoma brachyurum*.

Many species of algae grow on the walls and bases of the channels, especially diatoms of which *Cymbella* is the main genus. From June until November the periphytic primary production ranges from 3550 to 820 $mg\,C\,m^{-2}d^{-1}$. The channels contribute approximately 10% of the total quantity of organic suspended matter flowing through them (Teltsch & Ben-Zur 1988). Primary production in the reservoirs ranges from 60 to 300 $mg\,C\,m^{-2}d^{-1}$, and the number of crustacean zooplankters between 3 and 7 individuals l^{-1}.

Table 2. The quantity of algae in the trophogenic layers of Lake Kinneret during the years 1969–1984.

Year	Total algae $g\,m^{-2}$	*Peridinium* $g\,m^{-2}$	Other algae $g\,m^{-2}$	Dominant
1969	83	70	13	
1970	85.3	77	8.3	
1971	52.3	43	8.3	
1972	79	69	10	
1973	63	51	12	
1974	48	30	18	green algae 27
1975	39	23	16	blue-greens 20%
1976	53	42	11	
1977	50.5	45	5.5	
1978	51	39	12	
1979	58	52	6	
1980	68.3	58	10.3	
1981	69	50	19	diatoms 25%
1982	72	39	33	diatoms 23%
1983	78	39	39	
1984	90	65	25	

From: Annual Reports, Lake Kinneret Limnological Lab.

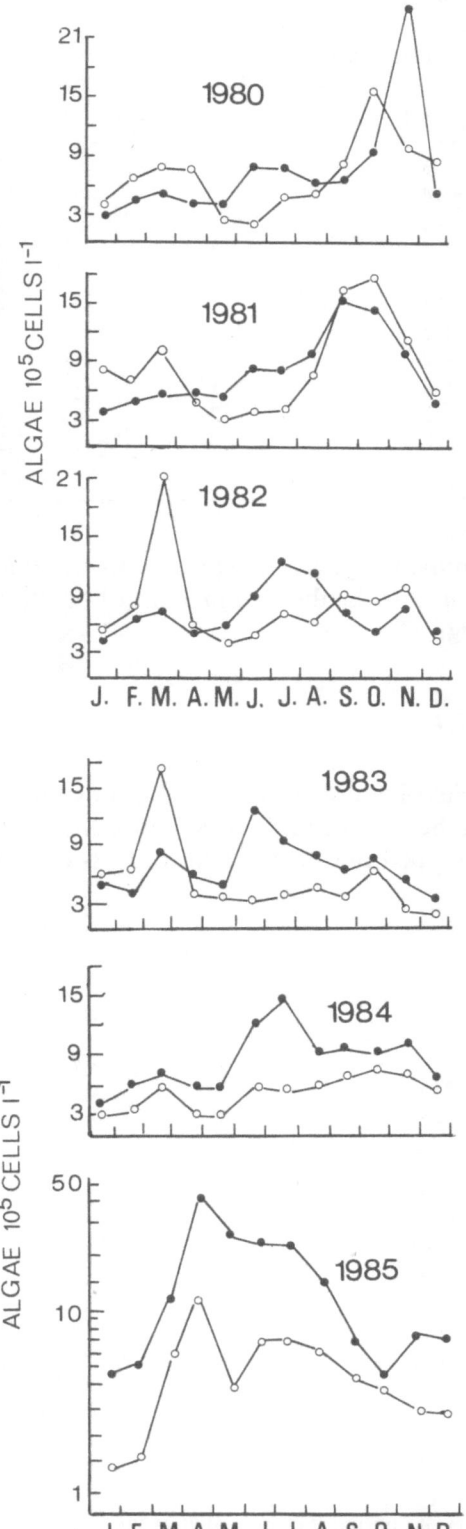

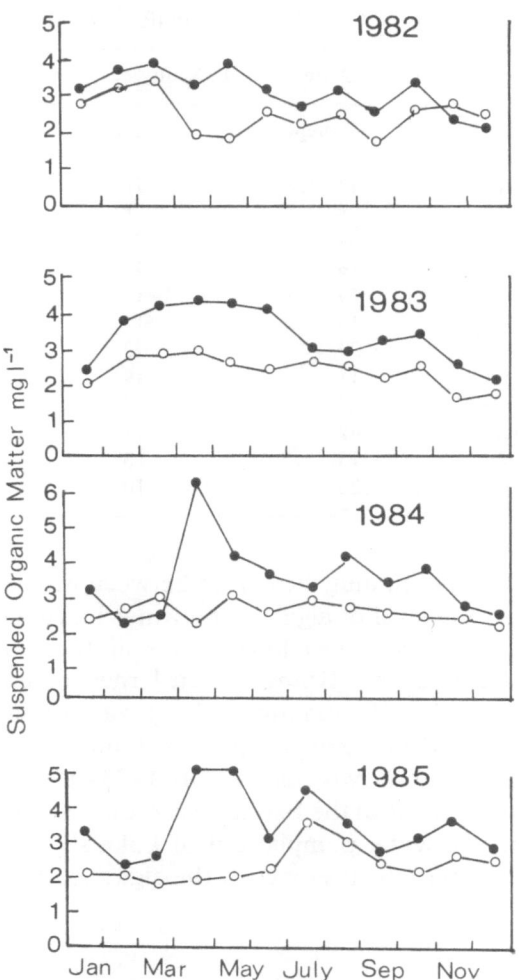

Fig. 3. The suspended organic matter, dry weight, from the intake of INWC (filled circle) and from the water of Eshkol reservoir (open circle) in the years 1982–1985.

Results

Phytoplankton

The results are presented in Fig. 2. During the years 1980–1986 the water was treated with chlorine to control *Peridinium*. The treatment, generally applied from February until June, led to a reduction in the number of algae.

Fig. 2. Number of algae in the intake of INWC (black circle) and in the outlet of Eshkol reservoir (white circle) in the years 1980–1985.

Table 3. Density of Crustacea in the outlets from Lake Kinneret and from Eshkol reservoir, 1981 (No. 1^{-1})

Month	Outlet from L. Kinneret			Outlet from Eshkol		
	Copepoda	Cladocera	Total	Copepoda	Cladocera	Total
January	15	10	25	3	1	4
February	38	28	66	4	3	7
March	34	29	63	2	2	4
April	18	16	34	3	1	4
May	18	13	31	2	1	3
June	16	11	27	3	1	4
July	37	3	40	3	1	4
August	11	19	30	2	1	3
September	20	7	37	2	1	3
October	42	11	53	2	2	4
November	15	3	18	2	0	3
December	26	10	36	1	3	4

In 1980, excluding the period between May and August, the level of algae in the water leaving the Eshkol reservoir was lower than in the water pumped from Lake Kinneret. The largest quantity of algae (24×10^5 organisms 1^{-1}) was found in October and November, and the dominant type was the diatom *Navicula*. From 1983–1985 the number of algae in the Eshkol reservoir was lower than in the water pumped out of Lake Kinneret. In addition to the presence of the alga *Peridinium*, the dominant alga in the spring was the diatom *Cyclotella*, and in the summer and fall the blue green alga *Chroococcus*.

Zooplankton

All species of zooplankton found in Lake Kinneret reach the reservoirs. In 1981, the density in the water pumped out of the lake was 18–66 indivi-

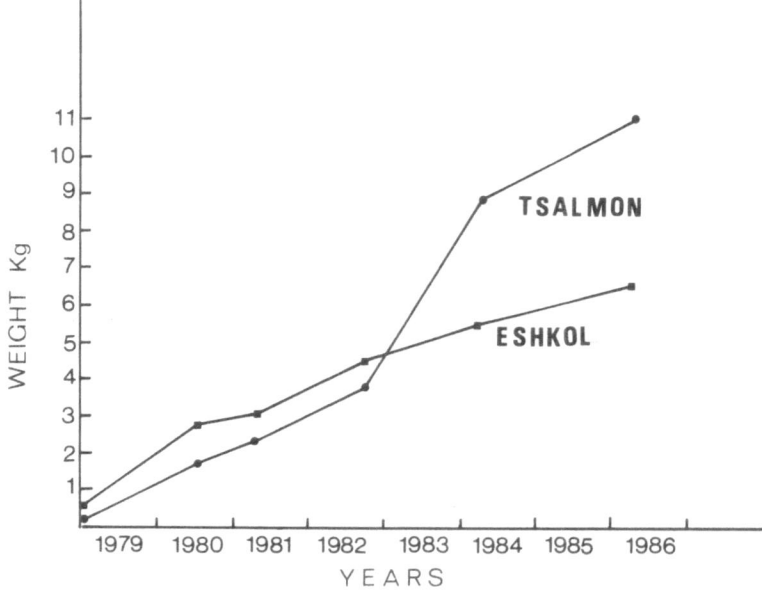

Fig. 4. The growth of silver carp in Tsalmon and Eshkol reservoirs during the years 1979–1986.

duals l^{-1}, compared to 2–7 individuals l^{-1} in the Eshkol reservoir. Similar results were obtained over the years 1982–1985 (Table 3).

Organic Suspended Matter

Most of the organic suspended matter in Lake Kinneret water consists of algae and zooplankton, whereas water leaving the Eshkol reservoir contains mainly undefined organic matter.

Figure 3 shows that from 1982 onwards the quantity of suspended matter at the outlet of the Eshkol reservoir was lower than in the water pumped from Lake Kinneret. This reduction is even higher if one takes into account the organic matter produced by and entering the water that flows through the channels.

Fish

Four groups of silver carp were introduced into the reservoirs between May 1979 and November 1981. The growth patterns of the first group in the Tsalmon and Eshkol reservoirs are compared in Fig. 4. After 7 years, the silver carp in Tsalmon reached an average weight of 11 kg compared to 6.5 kg in Eshkol.

In 1986 the reservoirs were dried, and a total of 300 000 kg of silver carp was collected. Of this amount, 240 000 kg was from the Tsalmon reservoir, and the remaining 60 000 kg from the Netofa reservoirs. After 8 years the yield from Tsalmon was 12 000 kg ha^{-1}, with 421 kg ha^{-1} from the Netofa reservoirs.

In addition to the silver carp, 400 kg ha^{-1} of tilapia was removed from the Tsalmon reservoir and 100 kg ha^{-1} of grey mullet from the Netofa reservoirs.

Discussion

This section summarizes the biological control carried out at the reservoirs of the Israel National Water Carrier over the 8-year period 1978–1986,

in order to assess the contribution of silver carp to this control.

Until 1979 the following species of fish were used for biological control: tilapia, grey mullet, common carp, grass carp and black carp. After 1979, new species of plankton-feeders were introduced – silver carp for control of phytoplankton and bighead carp for zooplankton.

Within the framework of the simultaneous biological control of the reservoirs, the reason for these new species of fish was twofold: to improve the water quality of Lake Kinneret by partially reducing the organic suspended matter; and to transfer organic suspended solids from the water to the lake bottom.

To achieve the above goals, a 'biological filter' was created, consisting of a high concentration of silver carp in the Tsalmon reservoir and a low concentration in the Netofa reservoirs. The Tsalmon reservoir can be considered a large channel with a short retention time. The reservoir was stocked with 4350 silver carp and 1500 tilapia per hectare, and over the 8-year control period reproduction of the tilapia was very high, although the fish remained small.

The yield of 12 000 kg ha^{-1} of silver carp from Tsalmon was 27 times greater than from the Netofa reservoirs, probably because the water in the Tsalmon reservoir is rich in both plankton and dissolved oxygen.

The silver carp reduced the number of algae and zooplankton at a rate that increased annually (see Fig. 2 and Table 3). The process was influenced by the biomass of the fish, which also increased.

Throughout the 8-year study period, the concentration of suspended organic matter in water flowing out of the Eshkol reservoir was smaller than in water pumped from Lake Kinneret (Fig. 3). The average annual load of dry suspended organic matter from Lake Kinneret and the channels was 1634 tons, compared to 931 tons from the Eshkol reservoir. Thus the annual reduction rate was 703 tons yr^{-1}. It can be assumed, therefore, that over the 8 years approximately 5600 tons of dry suspended organic matter had been removed by the fish and by other biological

Table 4. Quantity of suspended organic matter (tons dry weight) in the water pumped from Lake Kinneret and flowing from Eshkol reservoir, 1982–1985.

Year	Suspended organic matter in L. Kinneret water production in channels & reservoirs	Suspended organic matter in water leaving Eshkol reservoir	Reduction	
			Tons	%
1982	1570	876	694	44
1983	1606	913	693	43
1984	1643	1022	621	38
1985	1716	913	803	47
Average	1634	931	703	43

processes. During the same period, 330 tons of fish were removed from all three reservoirs, of which 300 tons were silver carp, 16 tons grey mullet, 8 tons tilapia and 6 tons other fish.

Silver carp that feed on natural food consume approximately 8 kg of dry food to produce 1 kg of weight. According to this relation, it appears that the fish removed 2500 tons of suspended organic matter over the 8-year period, although up to 80% of this food is excreted into the water and settles to the bottom. The process is a transfer of matter from the water-mass to the reservoir bottom. This enrichment of the reservoir floor by organic matter

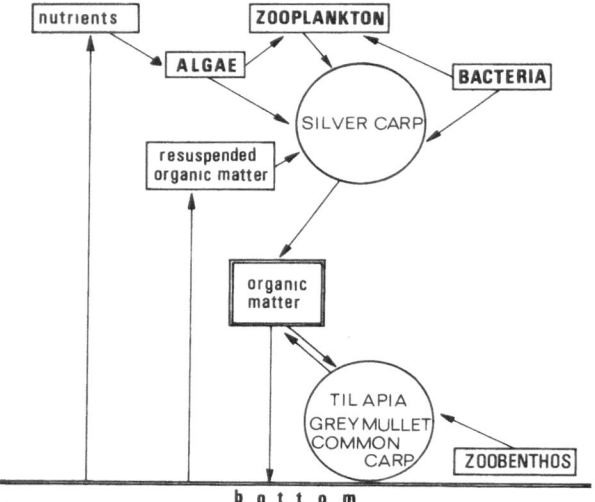

Fig. 5. Silver carp in water-bodies: the circulation of organic matter.

is a negative phenomenon, especially in drinking water reservoirs. Zoobenthos and some species of fish, that consume organic matter from the bottom of the reservoir, can reduce its concentration.

In previous research conducted in the Tsalmon reservoir from 1968–1978, it was found that over a 3-year period Tilapia aurea reduced the quantity of organic matter on the upper layer of the reservoir bottom from 19% in 1968 to 11% in 1971 (Leventer, 1972). In 1978, the concentration of organic matter in the sediment was 6%.

Yashouv (1971) indicated that the presence of silver carp in polyculture improves the growth of common carp and tilapia. He called this phenomenon 'positive synergism'. Tilapia fish in the Tsalmon reservoir grew bigger in the presence of silver carp than without them (Leventer, 1987). Hepher *et al.* (1988) found that silver carp in polyculture with bottom-feeders such as common carp, tilapia or grey mullet grow better than in monoculture.

This phenomenon can be explained by the fact that benthophage fish cause resuspension of organic matter, part of which is filtered by silver carp.

Silver carp have long been present in the Amur River region of northern China. The Chinese have bred the species in captivity, for they regard it as an efficient fish, feeding on algae. In both Eastern and Western Europe the silver carp has been adopted and raised in polyculture. Many countries have raised questions about the introduction of this fish, as whether it could live alongside the native fish, and how to prevent 'Ichthyœutrophication'.

Conventionally, silver carp are said to feed on algae. Our work and that of others shows that silver carp also feed on bacteria, zooplankton, detritus, and small particles of clay. They are able to filter from the water suspended solids up to 100 microns in size. Reich (1975) writes: 'The three species of fish, silver carp, tilapia and common carp create a balanced ecological system in which the interspecific competition is small and improvement of the environment large'.

Figure 5 shows the circulation of organic matter in reservoirs populated by silver carp, tilapia, grey mullet, and common carp. Algal populations develop on nutrients. Zooplankton feeds on small algae, while silver carp eat both small and large algae. Zooplankton and silver carp also consume bacteria. In addition to these three organic food components, the silver carp feeds on detritus, which may be its main food source.

The tilapia, grey mullet, and common carp use the excretion of silver carp both directly and indirectly. Bottom-living fish cause resuspension of detritus from sediment to water. If silver carps only are present in the reservoirs, the quantity of algae will continually increase (Milstein *et al.*, 1985). The introduction of bottom-feeders may stop this trend.

References

Annual reports. Lake Kinneret Limnological Laboratory.

Berman, T. (ed.), 1973. Lake Kinneret data record. N.C.R.D., 73 pp.

Hepher, B. & A. Milstein, H. Leventer & B. Teltsch, (in press). The effect of fish density and species combination on growth and utilization of natural foods in ponds. Aquacult. Fish. Mngmt

Leventer, H., 1973. Eutrophication control of Tsalmon Reservoirs by the *cichlid* fish *Tilapia aurea*. In S. H. Jenkins (ed.), Water Pollution Research. Pergamon: 217–229.

Leventer, H., 1981. Biological control of reservoirs by fish. Bamidgeh 33: 3–23.

Leventer, H., 1987. The contribution of silver carp *Hypophthalmichthys molitrix* to the biological control of reservoirs. Mekoroth, 106 pp.

Milstein, A., B. Hepher & B. Teltsch, 1985. Interaction between fish species and the ecological conditions in mono and polyculture ponds system. I. Phytoplankton. Aquacult. Fish. Mngmt 16: 305–317.

Moskul, A. G., 1977. Feeding of two years old silver and bighead in foraging lagoons of the Krasnodar area. J. Hydrobiol. (Trans. by A.F.S.) 13: 37–41.

Opuszynski, K., 1980. The role of fishery management in counteracting eutrophication processes. Developments in Hydrobiology (eds J. Barica & L. R. Mur) 2: 263–269.

Reich, K., 1975. Multispecies fish culture (polyculture) in Israel. Bamidgeh 27: 85–101.

APHA. Standard Methods for Examination of Water and Wastewater, 1980. New York.

Teltsch, B. & J. Ben Zur, 1988. Periphytic suspended solid production within a water distribution channel (Hebrew). Mekoroth, 105 pp.

Verigin, V. B., 1979. The role of herbivorous fishes at reconstruction of ichthyofauna under the condition of anthropogenic evolution of water bodies. Proc. Grass Carp Conference, Gainesville.

Vinogradov, V. K., 1979. Herbivorous fish breeding and rearing. EIFAC Technical Paper 35, Suppl. 1 (eds E. A. Huisman & H. Hogendoorn).

Vovk, P. S., 1974. The possibility of using the silver carp (*Hypophthalmichthys molitrix*) to increase the fish production of the Dnieper reservoirs and to decrease eutrophication. J. Ichthyol. 14: 351–358.

Yashouv, A. 1971. Interaction between the common carp (*Cyprinus carpio*) and silver carp (*Hypophthalmichthys molitrix*) in fish ponds. Bamidgeh 12: 85–92.

Yefimova, A. T. & Yu. I. Nikinovov, 1977. Prospects for the introduction of phytophagous fishes into Ivankovskoye reservoir. J. Ichthyol. 17: 634–644.

Hydrobiologia **191**: 57–66, 1990.
P. Biró and J. F. Talling (eds), Trophic Relationships in Inland Waters.
© 1990 *Kluwer Academic Publishers.*

Changes in the fish and zooplankton communities of Ringsjön, a Swedish lake undergoing man-made eutrophication

Eva Bergstrand
Institute of Freshwater Research, S-170 11 Drottningholm, Sweden

Key words: plankton feeders, grazing, eutrophication, trophic structure, zooplankton, freshwater fish

Abstract

During the 20th century Lake Ringsjön has developed from a mesotrophic to a eutrophic lake, and the phytoplankton community has changed from a rather diverse community to a monoculture of blue-green algae. The eutrophication process has accelerated during the last decade. The most important external nutrient loading of today comes from agriculture.

Although phosphorus has been shown to be the primary nutrient leading to excessive algal growth in fresh water, several biotic factors – such as interactions between nutrients, phytoplankton, zooplankton and planktivorous fish – may play a decisive role in the occurrence and maintenance of large algal blooms.

The aim of this investigation was to study the changes in the fish community of Lake Ringsjön, especially the most dominant planktivores, and the state of the zooplankton community during the seventies. The fish fauna is dominated by cyprinids, especially roach, and there are relatively few predatory fish. During the seventies the mean size of roach decreased, and measurements of the zooplankton community indicated that the predation pressure on zooplankton had increased. The mean sizes of cladocerans such as *Daphnia* and *Bosmina*, which were selected for by the planktivorous fish, decreased; the size of the calanoid *Diaptomus*, which was not preyed upon by the dominating fish species, did not change. The growth of zooplankton-feeding stages of several fish species was retarded, which meant that the growth of young perch decreased, while older roach were mainly affected. In the prevailing situation, planktivorous roach can maintain a numerous population of small individuals, whereas the predatory perch is at a disadvantage, and predation on zooplankton is intense.

Introduction

During the 20th century Lake Ringsjön has developed from a mesotrophic to a eutrophic lake, and the phytoplankton community has changed from a rather diverse community to a monoculture of blue-green algae. Although a sewage plant was built in 1975, which reduced the phosphorus load by more than 95%, the eutrophication process has accelerated during the seventies. External nutrients from agriculture and internal loading from sediments are the most important sources of phosphorus and nitrogen today.

The trophic level of lakes is primarily a function of nutrient levels, but it is also influenced by biotic factors (Henrikson *et al.*, 1980). Interactions between nutrients, phytoplankton, zooplankton and fish may play a decisive role in the occurrence and maintenance of large algal blooms (Hrbáček *et al.*, 1961; Brooks, 1969; Andersson *et al.*,

58

1978). For example, unchecked nannoplankton growth during early summer prepares the ground for blue-green algae. The presence of efficient filter feeders such as large zooplankton species counteracts this process. The most efficient zooplankton grazers are in turn controlled by size-selective predation by fish and fish fry (Hrbáček, 1962; Brooks & Dodson, 1965; Nilsson & Pejler, 1973; Stenson, 1972, 1976; Nilssen, 1978; Vijverberg & van Densen, 1984).

The aim of the present investigation was to study changes in the fish community of Lake Ringsjön, in particular the most dominant planktivores, and the state of the zooplankton community between 1972 and 1982. The period

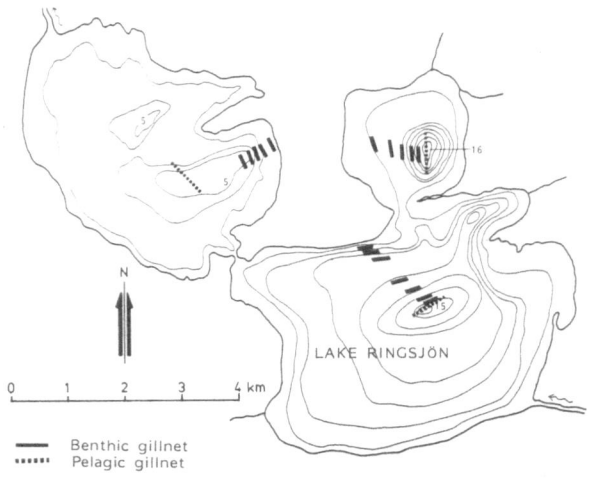

Fig. 1. Map of Lake Ringsjön.

Table 1. Proportion of zooplankton species in the diet of whitefish and in samples taken with a net of mesh size 75 μm.

	1972				1973			
	Whitefish stomachs Size class, cm			Net plankton %	Whitefish stomachs Size class, cm			Net plankton %
	10–15	15–20	20–30		10–15	15–20	20–30	
Daphnia spp.	16	25	38	17	27	25	26	26
Bosmina coregoni	–	–	–	0	25	41	32	17
Chydorus sphaericus	58	12	16	70	23	5	2	10
Leptodora kindtii	20	+	3	+	–	1	2	0
Eudiaptomus graciloides	+	–	+	2	–	+	–	25
Cyclops spp.	3	63	43	11	15	28	36	22
Miscellaneous	3	–	–	+	10	–	2	0
Total	100	100	100	100	100	100	100	100
Number of stomachs	19	2	23		3	8	30	
Catch	57	5	28		7	14	61	
	1981				1982			
Diaphanosoma brachyurum	–	–	–	–	–	5	7	11
Daphnia spp.	–	+	4	5	20	20	15	17
Bosmina coregoni	–	37	51	7	25	9	15	7
Chydorus sphaericus	–	50	38	83	10	1	2	24
Leptodora kindtii	–	1	1	0	23	63	29	1
Eudiaptomus graciloides	–	–	+	3	–	1	20	33
Cyclops spp.	–	6	5	2	22	1	10	7
Miscellaneous	–	6	–	0	–	–	2	0
Total		100	100	100	100	100	100	100
Number of stomachs	–	5	11		1	2	19	
Catch	0	12	22		2	3	62	

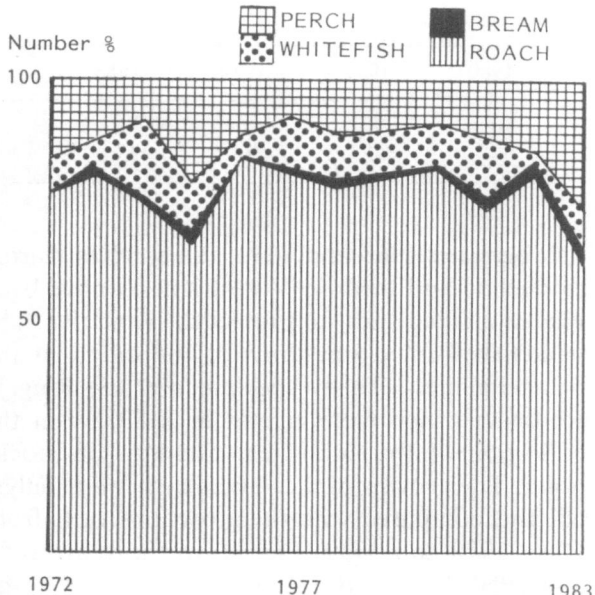

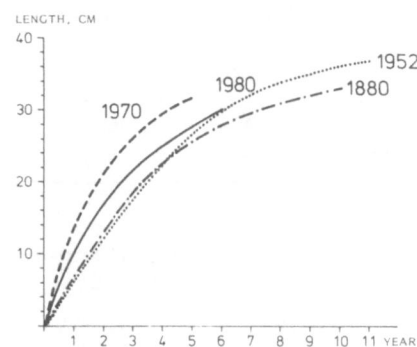

Fig. 3. Growth of whitefish, as length/ age relationship, during different periods.

Fig. 2. Species composition in the gillnet catches from the test-fishing programme.

was characterized by a large increase in the incidence of blooms of blue-green algae.

Lake Ringsjön is situated in the province of Scania (Skåne) in southern Sweden. The lake area is 41 km² and the mean depth 4.5 m. The mean transparency has decreased from 1 m in 1970 to 0.5 m in 1980 (Fig. 1).

A total of 8 fish species were recorded when test fishing with gillnets. Ranked in order of abundance they were: roach, perch, whitefish, bream, pike-perch, burbot, pike and rudd. The use of gillnets as a sampling method for fish has its limits. Eel are not caught in the gillnets and the proportion of bream in the fish community is probably underestimated (Hamrin, 1985). Lake Ringsjön has a dense eel population, but eel was not included in this investigation. According to commercial catches with fyke nets the bream population is dense and has increased during the last ten years. So the fish fauna of lake Ringsjön is dominated by cyprinids, which means that plankton-feeding and benthic-feeding fish species dominate the fish fauna. The piscivorous fish are relatively rare.

Material and methods

The test fishing was carried out in September each year. The fish were caught with gillnets of multiple mesh-size of 10, 12.5, 16.5, 22, 25, 30, 33, 38, 43, 50, 60 and 75 mm (Hammar & Filipsson, 1985), and fish from one or two years of age up to adult size were caught. The nets were set at different depths in the benthic and pelagic zones. Samples of scales and opercula for age analysis, and stomachs for feeding analysis, were collected from different size groups. Food organism identification was, if possible, carried out to the species level. The proportion of different food items was estimated using the per cent method. Zooplankton was collected with a plankton net of 75 μm mesh size. A minimum of 100 individuals of each species were measured. All individuals in one or two subsamples were measured.

Results

Roach, perch and whitefish dominated in the gillnet catches (Fig. 2); the study concentrates on describing the food and growth of these species.

Whitefish

Lake Ringsjön has a well-known whitefish stock that was first described by Valencienne in 1848.

Table 2. Mean weight of perch (g) in different years.

	1973	1974	1975	1976	1980	1981	1982	1983
Pelagic gillnet	39	46	76	70	30	28	8	11
Benthic gillnet	40	–	–	65	64	80	21	23

It is an obligate planktivore and according to the nomenclature of Svärdson (1979) the species is *Coregonus nilssoni* Valencienne. As the eutrophication process proceeds, the fish community generally changes so that cyprinid fish increase in abundance and coregonid fish decrease. After a peak in the commercial catches of white-

fish between 1968 and 1971, the catches started to decline and from 1974 restrictions have been placed on the fishery to protect the whitefish. The whitefish formed one-third of the catch in the pelagic gillnets. Thirty specimens of the Ringsjö whitefish which were caught in 1883 are in the collection of the zoological museum of Stockholm. The growth of these whitefish was analysed, and compared with the growth records from 1950, 1970 and 1980 (Fig. 3). In the years 1883 and 1950 the growth rate was similar, but in the early seventies the growth rate improved and became very rapid. During the seventies growth was retarded.

The diet of the whitefish reflected the zooplankton composition of the lake, as seen from the plankton samples. Some zooplankters, such as *Daphnia*, *Bosmina* and *Cyclops*, were slightly more

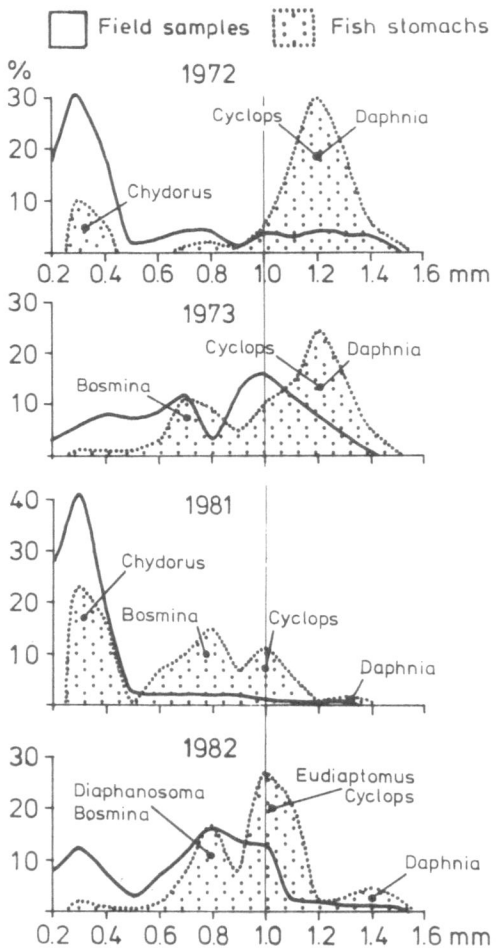

Fig. 4. Size distribution of zooplankton in plankton samples and whitefish stomachs from four years between 1972 and 1982.

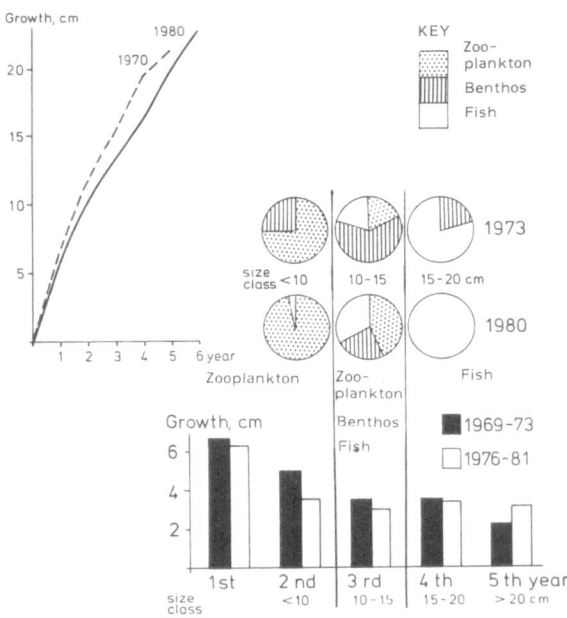

Fig. 5. Diet and growth in length of different age- and size-groups of perch.

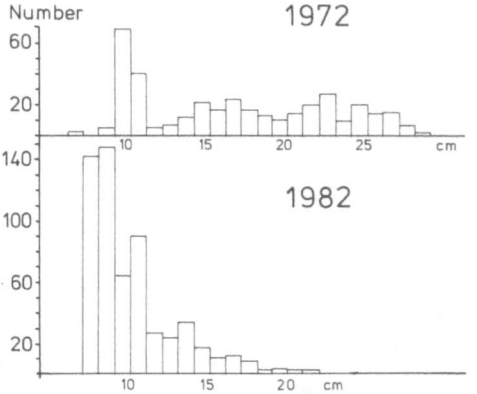

Fig. 6. The length/frequency distribution of roach in 1972 and 1982.

common in the stomachs than in the plankton net, whereas the calanoid *Eudiaptomus* was under-represented (Table 1). The whitefish seemed to prefer large slow-moving prey such as *Daphnia*. The calanoid *Eudiaptomus*, which moves in a jumpy manner, was only eaten in 1982, although it was available in all the other years of sampling. In 1981 the small species *Bosmina* and *Chydorus* dominated in the diet and the proportion of *Daphnia* was small. Measurements of the size of zooplankton in the gut contents showed that the mean size of whitefish prey decreased in the beginning of the eighties (Fig. 4), when there were few prey larger than 1 mm in size.

Perch

In 1920, perch was the second most dominant fish in the commercial catches, after roach. The yearly yield was around 8 tons. In the sixties the yield was around 3 tons and in the early eighties about 1 ton. In the test-fishing catches, perch larger than 20 cm decreased in abundance during the seventies and the mean weight declined in the eighties (Table 2).

The gut contents of different-sized perch show that zooplankton dominated in the diet up to the size of 10 cm. In the size group 10-15 cm zooplankton and benthic food were consumed, and most perch larger than 15 cm fed on fish (Fig. 5). Of the zooplankton, perch preferred large mobile prey like *Leptodora*, calanoid copepods and large species of *Daphnia*. In a laboratory experiment the small species *Chydorus* was ranked low by perch (Lessmark, 1983). However, *Chydorus* was fairly common in the stomachs from Lake Ringsjön, especially in 1980, which indicates a lack of more preferred species (Table 3). The small proportion of benthic food in the same year also indicates some limitation of this type of prey. In 1972 and 1973 the prey fish consisted of roach and perch. In 1980, 1981 and 1982 only perch was consumed.

The short-supply of larger zooplankton and

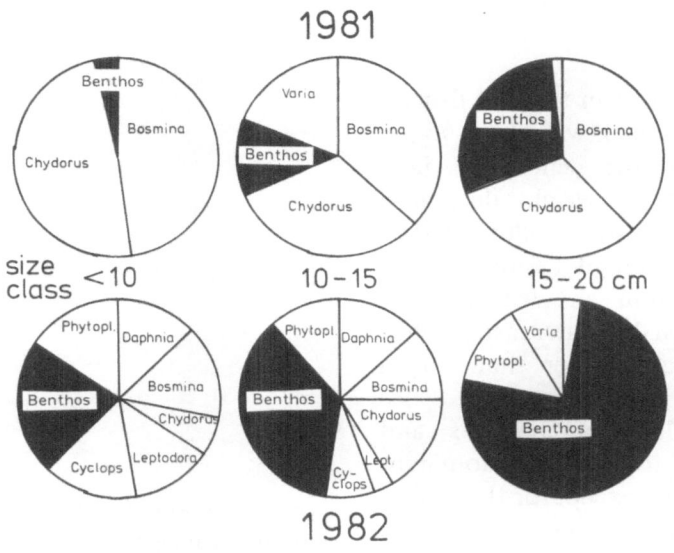

Fig. 7. Diet of different size of roach.

Table 3. Composition of the diet of perch and composition of zooplankton in samples taken with a net of mesh size 75 μm.

	1973			1980		
	Perch stomachs Size class, cm		Net plankton %	Perch stomachs Size class, cm		Net plankton %
	< 10	10–15		< 10	10–15	
Daphnia spp.	11	+	26	13	33	47
Bosmina coregoni	–	6	17	2	–	6
Chydorus sphaericus	2	–	10	25	–	27
Leptodora kindtii	–	9	0	40	8	0
Eudiaptomus graciloides	24	–	25	–	–	14
Cyclops spp.	38	3	22	18	3	6
Miscellaneous	–	–	0	1	+	0
Zooplankton total	75	18	100	99	44	100
Benthic food	25	62		1	23	
Fish	0	20		0	33	
Total	100	100		100	100	
Number of stomachs	5	21		16	24	
Catch	28	98		19	32	

benthic prey resulted in retarded growth for young perch in the early eighties, whereas the growth of piscivorous perch improved. However, the number of large perch had declined, so the mean growth of the whole population was lower in the eighties than in early seventies (Fig. 5).

Roach

Roach was the dominant fish in the test fishery. In 1972 the proportion of roach was about 75 per cent by number and 70 per cent by weight. However, the proportion by weight decreased during the seventies, which also was illustrated by a decrease in the mean weight (Table 4). The length-frequency distribution showed that the proportion of small roach had increased, while there were very few roach larger than 20 cm (Fig. 6).

Roach fed on zooplankton, benthos and vegetable food. Zooplankton was the dominant food item for all size groups except for the largest (Fig. 7). The zooplankton consisted of Cladocera and a small proportion of cyclopoids. This is in

agreement with laboratory experiments in which Lessmark (1983) found that the most preferred zooplankton species ranked: *Leptodora kindtii*, large *Daphnia* spp., *Bosmina*, *Chydorus* and then cyclopoid copepods. In the gut contents of roach, small species such as *Bosmina* and *Chydorus*

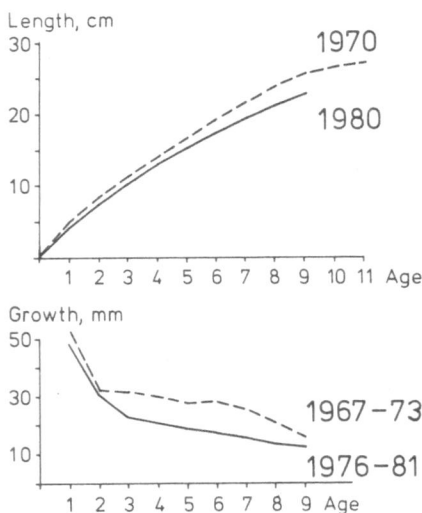

Fig. 8. Growth of roach, from two periods as length/ age relationship (above) and annual increment/age relationship (below).

Table 4. Yield from gillnet catches expressed as abundance and as weight in kg.

Species	Number							
	1972	1973	1976	1978	1980	1981	1982	1983
Roach	881	1318	1548	745	796	215	1149	228
Bream		10	3	8	15	6	19	11
Whitefish	100	91	90	97	79	41	74	34
Perch	190	208	214	112	88	28	239	82
Pike-perch	2	4	4	3	6	6	37	23
Pike		1	1					
Other fish	3	4		1		2		
	Weight							
Roach	41	39	24	10	13	3	21	4
Bream		2	2	1	2	+	1	1
Whitefish	5	8	11	7	6	3	7	3
Perch	10	8	14	5	4	2	4	2
Pike-perch	+	1	1	1	1	2	18	10
Pike		1	1					
Other fish	+	2		1		+		
Roach $\overline{\text{M}}$ g	46	29	16	14	17	13	14	17

dominated the diet in 1981, while *Daphnia*, which ranked number two in the test, was in the minority. This indicated a strong competition for zooplankton during this year.

The growth of roach decreased during the seventies. However, the change in growth was not the same for all size groups. The growth of large roach was retarded, whereas the growth of young roach did not change (Fig. 8).

Zooplankton

In 1883 Trybom visited Lake Ringsjön and gave a detailed report on its fish and zooplankton fauna (Trybom, 1893). He mentioned that there were many small invertebrates that were suitable for fish food, but that there were few species compared with other lakes. The most common species were *Daphnia longispina* and *Eudiaptomus graciloides*.

The sample of zooplankton taken in 1973 consisted of nearly the same species as in 1883, but the proportions of these species had changed

(Table 5). Two species *Holopedium gibberum* and *Bythotrephes longimanus* had disappeared. There were two species of *Daphnia*, *D. longispina* s.1. and *D. cucullata*, instead of one. *D. cucullata* was the most common. *Chydorus sphaericus* had

Table 5. Occurrence of zooplankton species in Lake Ringsjön.

Species	1883*	1973	1982
Diaphanosoma brachyurum			+ +
Holopedium gibberum	+		
Daphnia longispina	+ + +	+	+
Daphnia cucullata		+ +	+ +
Bosmina coregoni	+ +	+ +	+
Chydorus spaericus		+ +	+ + +
Bythotrephes longimanus	+ +		
Leptodora kindtii	+ +		+
Eudiaptomus graciloides	+ + +	+ + +	+ + +
Cyclops spp.	+	+ + +	+

* Trybom (1893)
+ + + most common (1883) > 20% (1973, 1982)
+ + common (1883) 10–20% (1973, 1982)
+ sparse (1883) < 10% (1973, 1982)

64

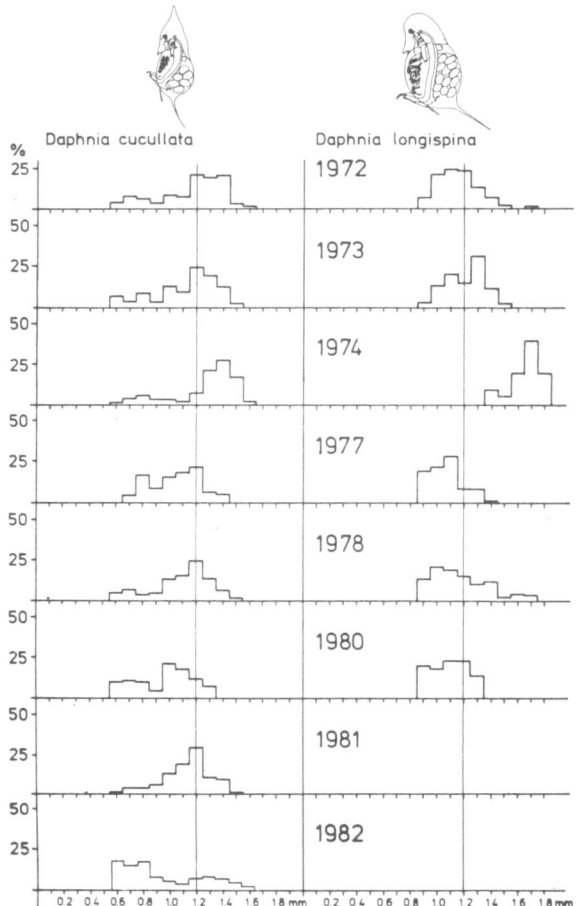

Fig. 9. Size distribution of *Daphnia cucullata* and *Daphnia longispina* between 1972 and 1982.

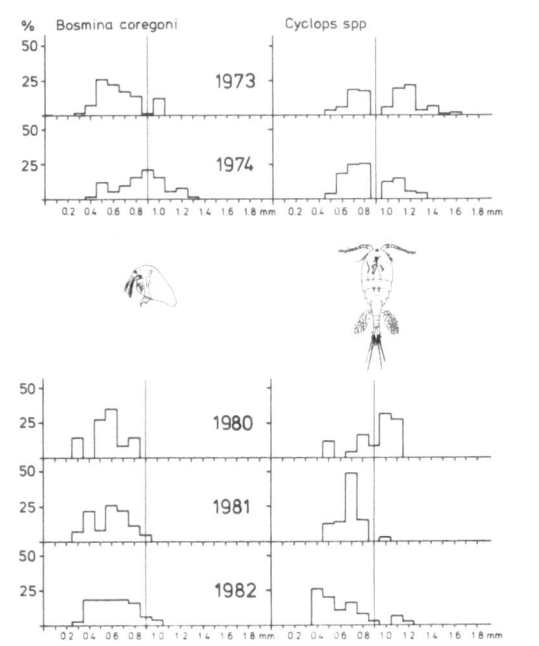

Fig. 10. Size distribution of *Bosmina coregoni* and *Cyclops* spp. between 1973 and 1982.

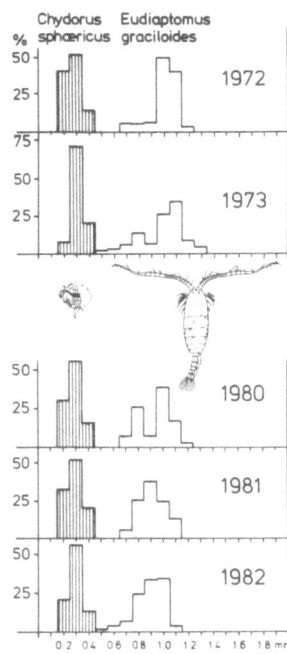

Fig. 11. Size distribution of *Chydorus sphaericus* and *Eudiaptomus graciloides* from 1972–1982.

spread from the littoral zone out into the pelagic zone, and constituted 10% of the zooplankton sample. In the seventies and early eighties there were further shifts in the balance of species. *Daphnia longispina* became very rare but a new species, *Diaphanosoma brachyurum*, was recorded in the pelagic zone in 1982.

Intensive predation by fish on zooplankton reduces the mean sizes of selected prey species. The size of different zooplankton species was measured. The two species of *Daphnia* decreased in size during the seventies and the numbers of individuals larger than 1.2 mm were diminished (Fig. 9). The mean sizes of *Bosmina coregoni* and *Cyclops* spp. were also reduced, and there were few individuals larger than 0.9 mm (Fig. 10). All these species-groups are common as fish food and

the Cladocera were highly selected by roach. *Cyclops* is selected by bream (Winfield *et al.*, 1983). The size of the calanoid *Eudiaptomus graciloides* varied less from year-to-year, although individuals more than 1 mm in size became rarer in 1981 and 1982. The size of the small *Chydorus sphaericus* was stable over the years (Fig. 11). *Eudiaptomus* was eaten by perch and whitefish, but not by roach. *Chydorus* was eaten by roach, bream, small whitefish and perch but was ranked low by the different fish species.

Discussion

At the beginning of the century roach and perch were the most common fish species in Lake Ringsjön. The stocks of pike and burbot were relatively small. Pike-perch was introduced in the twenties, but did not form a dense population. In 1970 the proportion of perch in the fish community had declined and except for eel, there were few predatory fish in the seventies.

Roach and perch are the most common fish species in many eutrophic Swedish lakes. Lessmark (1983) has shown that perch is the better adapted to the temperature regime in Swedish lakes. When there is a good supply of animal food of different sizes, perch has an advantage over roach. But if intensive predation on zooplankton results in a predominance of small zooplankton species, roach has an advantage. Roach is a more adept predator on small zooplankton. Moreover, when there is a shortage of animal food, roach can utilize vegetable food, although growth is retarded. This means that roach can maintain abundant but slow-growing populations, even if competition for zooplankton is intense (Persson, 1982; Lessmark, 1983). In this respect roach plays a key role in the process of eutrophication.

During the seventies the mean size of roach in Lake Ringsjön decreased and there were few individuals larger than 20 cm. The mean sizes of bream and perch also decreased. This means that a greater proportion of the fish stock was feeding on zooplankton. Measurements of zooplankton confirmed that the size-selective predation by fish on zooplankton increased during this period. The mean size of selected zooplankton species declined.

MacQueen *et al.* (1986) studied the impacts of bottom-up (producer controlled) and top-down (consumer controlled) forces on biomass and on different steps of the food web. They came to the conclusion that planktivores will have a negative impact on zooplankton biomass but the relationship between zooplankton biomass and phytoplankton is weak. In eutrophic lakes, however, large-bodied *Daphnia* are efficient enough to have a negative impact on phytoplankton biomass. In enclosure experiments, Hessen (1984) showed that a combination of dense roach populations and nutrients in excess gave rise to a strong increase in phytoplankton biomass. The biomass exceeded those in enclosures with only fish or only nutrients added.

The combination of the nutrient supply from agriculture, internal nutrient loading and the reduction of large-bodied *Daphnia* is probably the main reason for the great blooms of blue-green algae in Lake Ringsjön in the late seventies. The high abundance of small roach during the same period was the main reason for the greater predation on zooplankton. As there are few piscivorous fish in Lake Ringsjön, there is little regulation of small roach. Another factor which may have indirectly influenced the predation of roach on zooplankton is the large stock of eel. The eels feed mainly on benthos and competition for benthic food may have forced roach to feed on zooplankton. The increase in the bream population, as indicated by the commercial catches, may be important from the same point of view.

The greater competition for zooplankton was reflected in the retarded growth of zooplankton-feeding stages of roach, perch and whitefish. Large roach suffered most. The growth of roach is positively related to the presence of animal food and large zooplankton. When animal food is in short supply, intraspecific competition affects large roach most severely (Lessmark, 1983). The growth of young perch feeding on zooplankton and benthic food also declined. The lack of benthic food sometimes results in a bottleneck situation for perch. In Lake Ringsjön, fairly small individuals were able to turn to a piscivorous diet and it seemed as if some perch could proceed directly from a zooplankton diet to a fish diet. The pelagic habitat of Lake Ringsjön facilitates piscivory among perch (Persson, 1983). Stomach analyses showed that cannibalism increased in the eighties. This reduced the intraspecific competition among perch, and the growth of piscivorous perch improved (cf. Alm, 1946; Sumari, 1971). The intensified predation on zooplankton has, however, placed perch at a disadvantage, and the population has declined. The short supply of preferred zooplankton species has also affected the whitefish population, and the growth of whitefish has declined.

In the prevailing situation, in which there is a large population of small roach, the recovery of Lake Ringsjön will be delayed.

Acknowledgements

I am most grateful to Olof Filipsson for performing the test-fishing programme and for fruitful discussions throughout the study. I would also like to thank Gun Svensson and Gun Odén for valuable work in the laboratory, Monica Bergman for the skilful lay-out work on the illustrations and Eva Sers for preparing tables and typing the manuscript. Catherine Hill kindly improved my English.

References

Alm, G., 1946. Reasons for the occurrence of stunted fish population with special reference to the perch. Rep. Inst. Freshw. Res., Drottningholm 25. 146 p.

Andersson, G., H. Berggren, G. Cronberg & C. Gelin, 1978. Effects of planktivorous and benthivorous fish on organisms and water chemistry. Hydrobiologia 59: 9–15.

Brooks, J. L., 1969. Eutrophication and changes in the composition of the zooplankton. In Eutrophication: causes, consequences, correctives. Washington D.C.: 236–255.

Brooks, J. L. & S. I. Dodson, 1965. Predation, body size and composition of plankton. Science 150: 28–35.

Hammar, J. & O. Filipsson, 1985. Ecological testfishing with the Lundgren gillnets of multiple mesh size: the Drottningholm technique modified for Newfoundland Arctic char populations. Rep. Inst. Freshwat. Res., Drottningholm 62: 12–35.

Hamrin, S. F., 1985. The fish community and feeding resources in Lake Vombsjön 1983. Limnol. Inst., Univ. Lund. 61 p. (In Swedish.)

Henrikson, L., H. G. Nyman, H. G. Oscarson & J. A. E. Stenson, 1980. Trophic changes, without changes in the external nutrient loading. Hydrobiologia 68: 257–263.

Hessen, D. O., 1984. From a grazer food-chain to a detritus food-chain. Effects of manipulations with fish and nutrients in a eutrophic lake. In Interactions between trophic levels in freshwater. Nordic limnological symposium, Fagerfjell, Norway. 13 p. (In Norwegian with abstract in English.)

Hrbáček, J., 1962. Species composition and the amount of zooplankton in relation to the fish stock. Rozpr. Česk. Akad. Ved Rada Mat. Prir. Ved 72: 1–116.

Hrbáček, J., M. Dvořaková, V. Kořinek & L. Procházková, 1961. Demonstration of the effect of the fish stock on the species composition of zooplankton and the intensity of metabolism of the whole plankton association. Verh. Internat. Ver. Limnol. 14: 192–195.

Lessmark, O., 1983. Competition between perch (Perca fluviatilis) and roach (Rutilus rutilus) in south Swedish lakes. Inst. Limnol., Univ. Lund. Doctoral Dissertation. 172 p.

McQueen, D. J. & J. R. Post & E. L. Mills, 1986. Trophic relationships in freshwater pelagic systems. Can. J. Fish. aquat. Sci. 43: 1571–1581.

Nilssen, J. P., 1978. Eutrophication, minute algae and inefficient grazers. Mem. Ist. ital. Idrobiol. 36: 121–138.

Nilsson, N.-A. & B. Pejler, 1973. On the relation between fish fauna and zooplankton composition in north Swedish lakes. Rep. Inst. Freshw. Res., Drottningholm 53: 51–77.

Persson, L., 1982. Food consumption and competition in a roach (Rutilus rutilus) and a perch (Perca fluviatilis) population with special reference to intraspecific resource partitioning. Inst. Limnol, Univ. Lund. Doctoral Dissertation. 13 p.

Persson, L., 1983. Food consumption and competition between age classes in a perch Perca fluviatilis population in a shallow eutrophic lake. Oikos 40: 197–207.

Stenson, J. A. E., 1972. Fish predation effects on species composition of the zooplankton community in eight small forest lakes. Rep. Inst. Freshw. Res., Drottningholm 52: 132–148.

Stenson, J. A. E., 1976. Significance of predator influence on composition of Bosmina sp. populations. Limnol. Oceanogr. 21: 814–822.

Sumari, O., 1971. Structure of the perch populations in some ponds in Finland. Ann. Zool. Fenn. 8: 406–421.

Svärdson, G., 1979. Speciation of Scandinavian Coregonus. Rep. Inst. Freshw. Res., Drottningholm 57. 95 p.

Trybom, F., 1893. The fauna and flora and the fishery of Lake Ringsjön in the province of Malmöhus. Medd. Kongl. Lantbruksstyr. 4. 46 p. (In Swedish.)

Vijverberg, J & W. L. T. van Densen, 1984. The role of the fish in the foodweb of Tjeukemeer. Verh. int. Ver. Limnol. 22: 891–896.

Winfield, I. J., G. Peirson, M. Cryer & C. R. Townsend, 1983. The behavioural basis of prey selection by underyearling bream (Abramis brama (L.)) and roach (Rutilus rutilus (L.)). Freshwat. Biol. 13: 139–149.

Hydrobiologia **191**: 67–73, 1990.
P. Biró and J. F. Talling (eds), Trophic Relationships in Inland Waters.
© 1990 *Kluwer Academic Publishers.*

The dynamics of a population of roach (*Rutilus rutilus* (L.)) in a shallow lake: is there a 2-year cycle in recruitment?

Martin R. Perrow, Graeme Peirson & Colin R. Townsend
School of Biological Sciences, University of East Anglia, Norwich, NR4 7TJ, U.K.

Key words: population cycle, *Rutilus rutilus*, intraspecific competition

Abstract

Recruitment success of roach varied dramatically between 1978 and 1985 in Alderfen Broad, a small lake in eastern England. All size classes of roach feed to a significant extent upon zooplankton, but the underyearling fish have the greatest effects upon the abundance, species composition and mean size of zooplankton. During years of good recruitment (1979, 1981, 1983 and 1985) when the 0 + age group was abundant, they showed poor growth as a result of the depression of their prey populations. Older fish also tended to grow poorly in these years and may have been less fecund the following year. In years of poor recruitment (1980, 1982 and 1984), with the release of the depressive effect upon the zooplankton exerted by underyearling fish, the older size classes tended to grow well with higher fecundity the following season, giving rise to good recruitment of underyearling fish, even when the number of spawners was low. The evidence indicates that there is a 2-year cycle of roach recruitment in Alderfen and this will be described.

Introduction

The interactions between zooplankton and zooplanktivorous fish are of critical importance in determining not only microcrustacean abundance and species composition (Hrbacek *et al.*, 1961; Brooks & Dodson, 1965; Hall, 1970; Wells, 1970; Stenson, 1976; Hamrin, 1983; Mills & Forney, 1983) but also the diet, growth and survivorship of the fish (Kempe, 1962; Frank, 1970; Kuznetsov, 1972; Noble, 1975; Broughton *et al.*, 1977; Lemly & Dimmick, 1982). In particular, when the intensity of predation is sufficiently high the preferred prey classes become limiting and intraspecific competition, both within and between year classes, may be evident (Cryer *et al.*, 1986; Hamrin & Persson, 1986).

Alderfen Broad is a small (4.7 ha), shallow (mean depth 0.8 m, maximum depth 1.3 m) lake in eastern England (National grid reference TG 354196). A previous paper described how recruitment success of roach (*Rutilus rutilus* (L.)), the lake's principal predator of pelagic zooplankton, varied dramatically between 1979 and 1982, with large populations of underyearling fish in 1979 and 1981 and small populations in 1980 and 1982 (Cryer *et al.*, 1986; Townsend, 1988; Townsend & Perrow, 1988). It was postulated that a two-year cycle of recruitment occurred as a result of abundant underyearling fish depressing cladoceran densities which, in turn, led to poor growth of adult fish and low fecundity the following year. However, this hypothesis rested on only four years' data and required further substantiation. In the present paper, we provide 8 years' information on recruitment and growth of roach and reassess the original hypothesis.

In the years prior to 1979 Alderfen Broad was devoid of macrophytes, being generally turbid with dense phytoplankton populations. Early in 1979 a nutrient-rich input, derived from agricultural runoff and septic tank overflow, was diverted from the lake. This had consequences for water chemistry and phytoplankton populations (Moss *et al.*, 1986) and macrophytes returned from 1981 to 1984, being particularly dense in 1982 but declining again thereafter. The fish community also changed from one that was cyprinid-dominated to a cyprinid-percid mixture. However, the general pattern of abundance and year class structure of the roach population has not altered substantially.

Methods

Underyearling roach are notoriously difficult to sample quantitatively. We used as an index of recruitment of underyearling fish (hatched in June) the abundance of the year class when it achieves 1 + status in the year following hatching. This was assessed as catch per unit effort (CPUE) of 1 + fish taken by electrofishing (using a pulsed D.C. unit -Weiss, 1977 from a 3-m fibreglass dinghy) of the entire perimeter of the lake each October. At this time of year roach are concentrated around the margins and are conveniently taken (Peirson, 1986). Each year between 1 and 4 perimeters surveys were performed, and CPUE was expressed as mean number of 1 + fish per perimeter survey. Information on the relative abundance of older year classes was obtained in the same way.

The fork-length of each roach (including underyearlings) was measured and a sample of the population was aged by scale-reading with a microfiche viewer. Estimates of size for each class of roach in October 1978 (not measured at the time) were derived from known lengths in May 1979 (Peirson, 1986), back-calculated to allow for the small amount of growth that occurs during the winter months. Instantaneous growth rates (G), October to October, were estimated from

$$G = \frac{\mathrm{Log_e}\, L_2 - \mathrm{Log_e}\, L_1}{t}$$

Fig. 1. Index of underyearling recruitment of roach in Alderfen Broad during a period of 8 years. The index is taken as the average catch-per-unit-effort (per complete electrofishing of the lake perimeter) of yearling fish in October of the year following hatching.

where L_1 = initial length
         L_2 = final length
         t   = time lapsed, 1 year

Results

The index of underyearling recruitment shows convincing evidence of alternation of bad (even) and good (odd) years (Fig. 1) (the probability of this pattern occurring by chance is < 0.01, and the

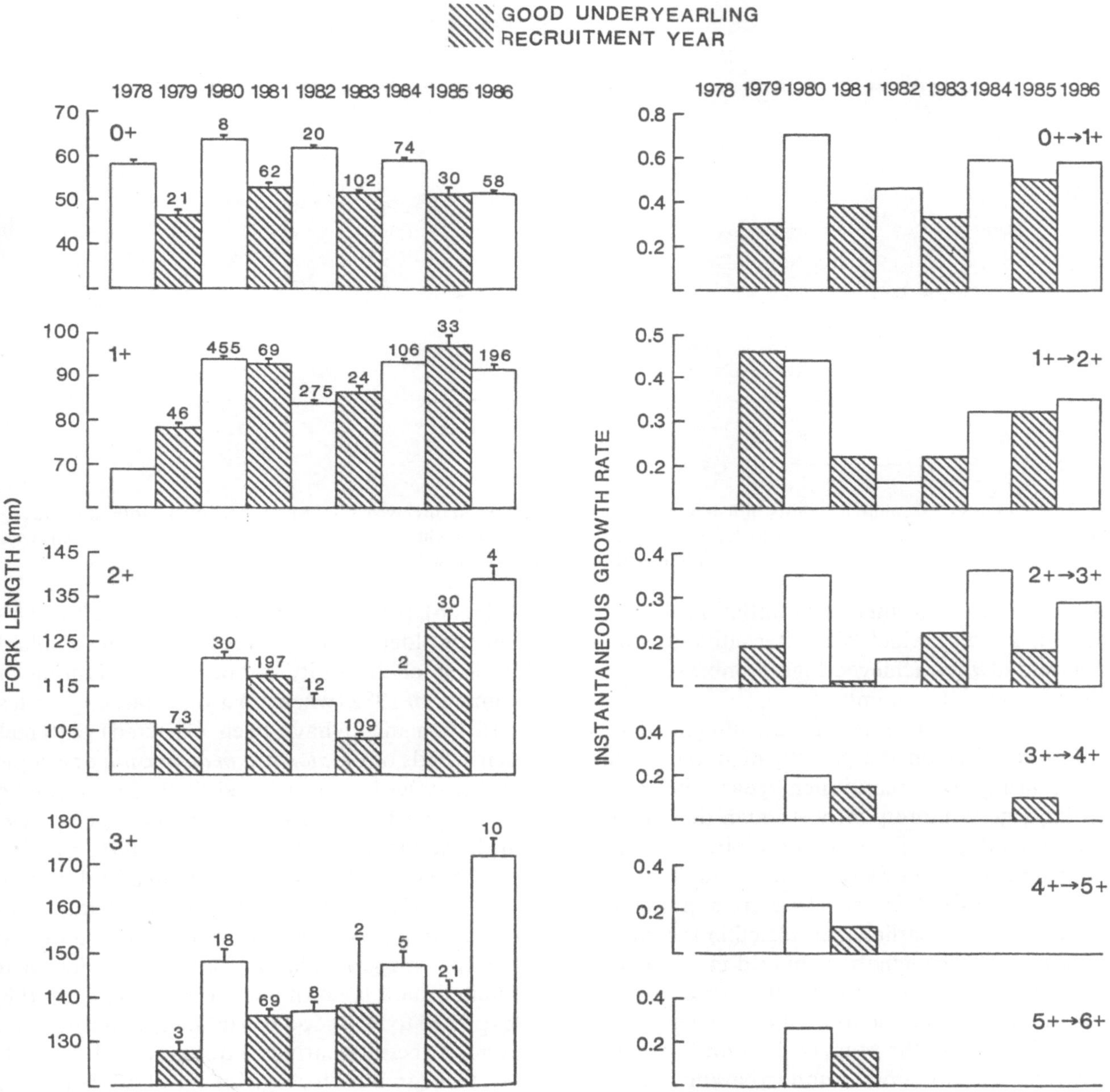

GOOD UNDERYEARLING
RECRUITMENT YEAR

Fig. 2. Size and growth of the different age classes of roach in Alderfen Broad taken in October surveys (a) mean fork length (± S.E.) achieved in October (b) instantaneous growth rate estimated from October to October.

70

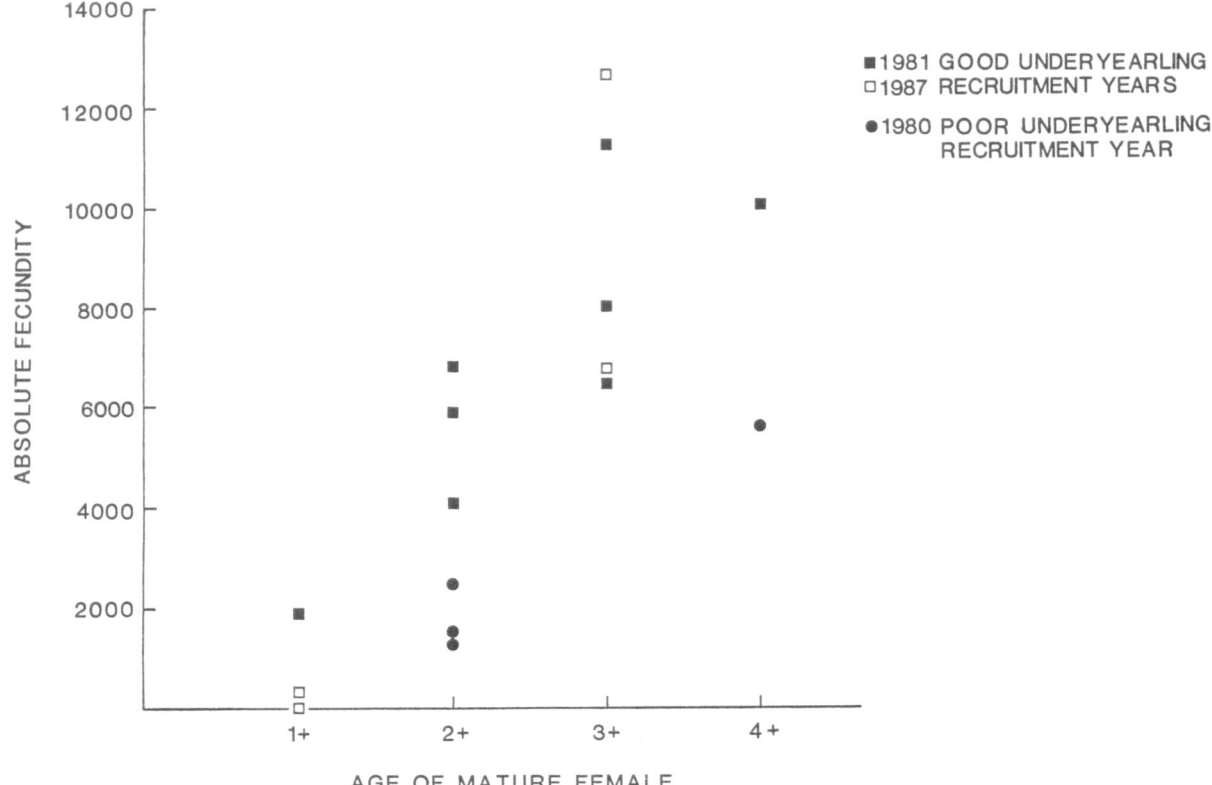

Fig. 3. Estimates of absolute fecundity of fish of a variety of age classes in three years. -1981, -1987, -1980. Note that 1981 and 1987 were years of good recruitment which followed years of good growth whereas 1980 was a year of poor recruitment which followed a year of poor growth.

pattern receives further substantiation from results for 2 + fish which also alternated in abundance and always achieved high numbers in years following high 1 + numbers.

The relative success of recruitment had a profound effect on the growth of underyearling fish. During good recruitment years the underyearlings grew more poorly through the summer and attained a smaller size in October than their counterparts in bad recruitment years (Fig. 2a).

To what extent is the alternating pattern of growth of underyearlings paralleled in the growth of older fish? The lengths achieved in october by 1 +, 2 + and 3 + do not show a clear pattern though there is an indication of consistently good growth in some of the years of poor underyearling recruitment and good underyearling growth (1980, 1984, 1986). When expressed as instantaneous growth rates a clearer pattern emerges

(Fig. 2b), particularly for the 0 + → 1 +, 2 + → 3 + and the older classes (where few data are available because of the rarity of specimens). During the summer of 1982 (when most year classes grew less well than might have been expected) extremely dense beds of *Ceratophyllum demersum* developed and may well have reduced feeding efficiency of roach, which have been shown to forage less well in dense plant structure (Peirson *et al.*, 1985).

The suggestion of poorer growth of older fish in years of good recruitment could have important consequences for the condition and fecundity of the spawning population in the following year. Limited data from mature female roach in 1987 (expected to be a good recruitment year) and 1981 (a good recruitment year) do indicate higher absolute fecundities for a given age class than roach in 1980 (a poor recruitment year) (Fig. 3).

Discussion

In general, annual variation in the recruitment of freshwater fish may be predominantly under climatic control (Mills & Mann, 1985). However, feeding conditions in the period before spawning may also be influential (Lyagina, 1972; Kuznetsov & Khalitov, 1978), and Hamrin & Persson (1986) have shown how recruitment in a population of vendace, *Coregonus albula*, alternates because of competition for crustacean zooplankton between underyearling recruits and the spawning population (almost entirely 1 + fish).

In the case of roach, most studies have pointed to irregular recruitment with a few very strong year classes, which have coincided with highly favourable climatic conditions, separated by poorly represented or non-existent year classes (e.g. Mann, 1973; Goldspink, 1978; Linfield, 1980). Why in Alderfen Broad should there be such a regular alternation of roach recruitment success? Unlike vendace the spawning population of roach normally consists of a large number of age classes, up to 12 + (Papageorgiou, 1979; Spivak *et al.*, 1979), many of which have quite different diets to that of underyearlings. However, in Alderfen Broad roach older than 3 + are very rare (probably mainly as a result of intensive predation by pike, *Esox lucius*) and the spawning population consists primarily of 2 + and some 3 + fish (Peirson, 1986). The diets of these fish overlap significantly with underyearlings, all taking predominantly planktonic crustaceans (Cryer *et al.*, 1986). Thus, competition can occur between recruits and fish old enough to spawn the following year, in a manner analogous to that described for vendace (Hamrin & Persson, 1986).

The alternation of recruitment success is well matched with differences in zooplankton biomass and composition during the period 1979 to 1982 (Cryer *et al.*, 1986). During years when fry were abundant (1979 and 1981) the summer zooplankton was dominated by cyclopoid copepods and rotifers, which are infrequently taken by roach because of efficient escape responses and small size, respectively. The preferred prey items of roach, planktonic cladocerans (Winfield *et al.*,

1983), tended to decline in abundance abruptly as they entered the diet of underyearlings during the summer. In contrast, when fry were scarce in 1980 and 1982 the zooplankton was dominated by cladocerans which maintained a high standing crop throughout the summer. The poor feeding conditions in 1979 and 1981 has a depressive effect on growth of underyearling fish in comparison to 1980 and 1982. This trend of poor growth when underyearling density is high was also show in 1983 and 1985 (and 1977 – see Peirson, 1986), and we propose that this is a result of intraspecific competition for a preferred prey resource which becomes limiting in late summer (all age classes of roach show hardly any growth in winter).

Given the high dietary overlap of roach up to 3 + we might expect the growth of older fish to be influenced by competition with underyearlings in good recruitment years and this is seen in the annual variation in instantaneous growth rates (Fig. 2b). Since there is generally a positive relationship between growth and fecundity (Mackay & Mann, 1969; Nikolskii, 1969; Papageorgiou, 1979), a depression of growth rate in older roach during years of good recruitment and poor growth of underyearlings would be expected to lead to lowered fecundity the following year. Limited data from 1980, 1981 and 1987 support this contention (Fig. 3). The bulk of Alderfen's spawning population consists of young fish (with even some 1 + female fish attaining maturity, at least in good recruitment years) and so depression of fecundity may be expected to have a profound effect with few eggs being laid. The resulting low numbers of underyearling fish would not significantly depress cladoceran populations, growth of underyearlings and older fish would be good and recruitment the following year may be expected to be high.

During the years 1979–1985, there were major ecological changes which could have interrupted the postulated two year cycle. The return of macrophytes in 1982 not only affected the zooplankton community by increasing the importance of non-planktonic Cladocera, but also greatly increased the densities of macro-invertebrates which may have provided an alternative source of prey for older fish, thereby potentially reducing competition between year classes. Underyearling roach have also been shown to be less efficient foragers in structurally complex environments compared to open water (Winfield, 1986), and this could account for low growth rates of young fish in 1982. The onset of macrophyte growth also favoured an increase in the number of perch, and as these are zooplanktivorous in their first few months of life and highly efficient predators of zooplankton among macrophytes (Winfield, 1986), they represent a major potential competitor for zooplankton. Even though the potential influence of such factors is great, the 2-year cycle of roach recruitment has gone on uninterrupted.

Acknowledgements

We wish to express our thanks to the Norfolk Naturalist's Trust for permission to work at Alderfen Broad, and to the Dean of the School of Biological Sciences for providing facilities. During the study, one of us (M.R.P.) was in receipt of an N.E.R.C. studentship, and one (G.P.) an S.E.R.C. research studentship. David Jordan, Martin Cryer, Geoff Cleveland, Rick Gowing, Andy Wood, Kay Schofield, John Mather and Shirley Wilkins assisted with field-work.

References

Brooks, J. L. & S. I. Dodson, 1965. Predation, body size, and composition of plankton. Science 15: 28–34.

Broughton, N. M., N. V. Jones & G. W. Lightfoot, 1977. The growth of 0-group roach in some Humberside waters. Proceedings of the eighth British Coarse Fish Conference, University of Liverpool: 53–60.

Cryer, M., G. Peirson & C. R. Townsend, 1986. Reciprocal interactions between roach Rutilus rutilus, and zooplankton in a small lake: Prey dynamics and fish growth and recruitment. Limnol. Oceanogr. 31: 1022–1038.

Frank, S., 1970. A contribution to the growth of young roach, Rutilus rutilus, with a discussion concerning the rate of growth. Vestnik. Čs. Spol. Zool. 34: 164–169.

Goldspink, C. R., 1978. Comparative observations on the growth rate and year class strength of roach, Rutilus rutilus, in two Cheshire lakes, England. J. Fish Biol. 12: 421–433.

Hall, D. J., 1970. Predator-prey relationships between yellow perch and Daphnia in a large temperate lake. Trans. am. Micros. Soc. 90: 106–107.

Hamrin, S. F., 1983. The food preference of vendace, Coregonus albula, in South Swedish forest lakes including the predation effect on zooplankton population. Hydrobiologia 101: 121–128.

Hamrin, S. F. & L. Persson, 1986. Asymmetrical competition between age classes as a factor causing population oscillations in an obligate planktivorous fish species. Oikos 47: 223–232.

Hrbáček, J., M. Dvořaková, V. Kořinek & L. Procházková, 1961. Demonstration of the effect of the fish stock on the species composition of zooplankton and the intensity of metabolism of whole plankton association. Verh. int. Ver. Limnol. 14: 192–195.

Kempe, O., 1962. The growth of roach, Rutilus rutilus, in some swedish lakes. Reports of the Institute of Freshwater Research, Drottningholm 44: 42–104.

Kuznetsov, V. A., 1972. The growth pattern of the larvae and young of some freshwater fish at different stages of development. J. Ichthyol. 12: 433–442.

Kuznetsov, V. A. & N. K. Khalitov, 1978. Alterations in the fecundity and egg quality of the roach, Rutilus rutilus, in connection with different feeding conditions. J. Ichthyol. 18: 63–70.

Lemly, A. D. & J. F. Dimmick, 1982. Growth of young-of-the-year and yearling centrarchids in relation to zooplankton in the littoral zone of lakes. Copeia 2: 305–321.

Linfield, R. S., 1980. Ecological changes in a lake fishery and their effects on a stunted roach, Rutilus rutilus, population. J. Fish. Biol. 16: 123–144.

Lyagina, T. N., 1972. The seasonal dynamics of the biological characteristics of the roach, Rutilus rutilus, under conditions of varying food availability. J. Ichthyol. 12: 210–226.

Mackay, I. & K. H. Mann, 1969. Fecundity of two cyprinid fishes in the River Thames, Reading, England. J. Fish. Res. Bd Can. 26: 2795–2805.

Mann, R. H. K., 1973. Observations on the age, growth, reproduction and food of the roach, Rutilus rutilus, in two rivers in southern England. J. Fish. Biol. 5: 707–736.

Mills, E. L. & J. L. Forney, 1983. Impact on Daphnia pulex of predation by young yellow perch in Oneida lane, New York. Trans. am. Fish. Soc. 112: 154–161.

Mills, C. A. & R. H. K. Mann, 1985. Environmentally induced fluctuations in year-class strength and their implications for management. J. Fish. Biol. 27 (suppl. A): 209–226.

Moss, B., H. Balls, K. Irvine & J. Stansfield, 1986. Restoration of two lowland lakes by isolation from nutrient-rich water sources with and without removal of sediment. J. appl. Ecol. 23: 391–414.

Nikolskii, G. V., 1969. Theory of fish population dynamics as the background for rational exploitation and management of fishery resources (Translated, J. E. S. Bradley). Oliver and Boyd.

Noble, R. L., 1975. Growth of young yellow perch, Perca flavescens, in relation to zooplankton populations. Trans. am. Fish. Soc. 104: 731–741.

Papageorgiou, N. K., 1979. The length-weight relationship, age, growth and reproduction of the roach, *Rutilus rutilus*, in Lake Volvi. J. Fish Biol. 14: 529–538.

Peirson, G., M. Cryer, I. J. Winfield & C. R. Townsend, 1985. The impact of reduced nutrient loading on the fish community of a small isolated lake, Alderfen Broad. Proceedings of the fourth British Freshwater Fisheries Conference, University of Liverpool: 167–175.

Peirson, G., 1986. The ecology of coarse fish in Alderfen Broad, with special reference to recruitment and early life history. Unpubl. PhD Thesis, University of East Anglia.

Spivak, E. G., G. N. Pinus & S. V. Sentishcheva, 1979. The age composition of the spawning population and the characteristics of the spawners size-age structure and fecundity of the roach, *Rutilus rutilus*, spawning in Kakhovka reservoir. J. Ichthyol. 19: 75–80.

Stenson, J. A. E., 1976. Significance of predator influence on composition of *Bosmina* spp. populations. Limnol. Oceanogr. 21: 814–822.

Townsend, C. R., 1988. Fish, fleas and phytoplankton. New Scientist 118: 67–70.

Townsend, C. R. & M. R. Perrow, 1988. Eutrophication can produce population cycles in roach, *Rutilus rutilus*, by two contrasting mechanisms. J. Fish. Biol. (in press)

Weiss, D. M., 1977. A high power pulse generator for electric fishing. J. Fish Biol. 4: 453–466.

Wells, L., 1970. Effects of alewife predation on zooplankton populations in Lake Michigan. Limnol. Oceanogr. 15: 556–565.

Winfield, I. J., Peirson, G., Cryer, M. & Townsend, C. R., 1983. The behavioural basis of prey selection by underyearling bream, *Abramis brama*, and roach, *Rutilus rutilus*. Freshwat. Biol. 13: 139–149.

Winfield, I. J., 1986. The influence of simulated aquatic macrophytes on the zooplankton consumption rate of juvenile roach, *Rutilus rutilus*, rudd, *Scardinius erythrophthalmus*, and perch, *Perca fluviatilis*. J. Fish. Biol. 29 (Supplement A): 37–48.

Hydrobiologia **191**: 75–85, 1990.
P. Biró and J. F. Talling (eds), Trophic Relationships in Inland Waters.
© 1990 *Kluwer Academic Publishers.*

Phosphorus eutrophication research in the lake district of south western Friesland, The Netherlands. Preliminary results of abiotic studies

H. J. W. J. Van Huet
Limnological Institute, Tjeukemeer Laboratory, De Akkers 47, 8536 VD Oosterzee, The Netherlands

Key words: eutrophication, phosphorus, hydrology, polder lake, modelling, loading, sediments

Abstract

The water quality of the lakes in south western Friesland is influenced by a rather complex hydrology. The purpose of the abiotic part of the eutrophication project, started in 1984 and focused on phosphorus, is to model hydrology and phosphorus dynamics, in order to compare scenarios for policy and management.

A brief survey is given of the preliminary results of the abiotic studies: hydrology, water quality, external loading from surrounding polders, sedimentary phosphorus and internal loading. The two largest lakes, Tjeukemeer and Slotermeer, are compared regarding these processes.

Introduction

The shallow (1–2 m) lakes – Tjeukemeer, Slotermeer, Brandemeer, Groote Brekken and Koevorder Meer – in the south western part of the province of Friesland are highly eutrophicated. Summer chlorophyll *a* concentrations are often above 150 μg l^{-1} while total P and total N concentrations are mostly above 0.2 mg l^{-1} and 2.0 mg l^{-1}, respectively (De Haan & Moed, 1984). The lakes (area 40 km^2) are part of the 'boezem', a system of interconnected lakes and canals (Fig. 1) with a total water surface area of 140 km^2. The boezem is used hydrologically for regulating the water table of Friesland, for flushing the boezem network or for water supply of the province of Groningen.

Generally during April-October there is inlet of water from IJsselmeer with chloride concentrations up to 300 mg l^{-1} and simultaneously inflow takes place into the surrounding polders. During October-April humic-rich water from the surrounding polders is pumped into the boezem system while the boezem water can be pumped into IJsselmeer or flows into the Waddenzee. This man-made regime is reflected in, for example, the chloride concentrations in Tjeukemeer: generally high concentrations in the summer period and low in the winter period. The humic-rich water in winter in this lakes leads to high optical density values in winter (De Haan & Moed, 1984). There is hardly any upward or downward seepage in the studied area, although local seepage may occur. The waste water of mainly housing in the agricultural area is dephosphorized at Lemmer and Sloten (capacities 20000 and 16500 inhabitant-equivalents, respectively). However, mainly at some places in the polders, sewage water may directly or indirectly reach the polder ditches. Also in polders manure may directly reach the ditches. For more details about the limnology, hydrology, nutrient concentrations, and algal periodicity, see Beattie *et al.* (1978), Leenen (1982), Moed & Hoogveld (1982), De Haan &

76

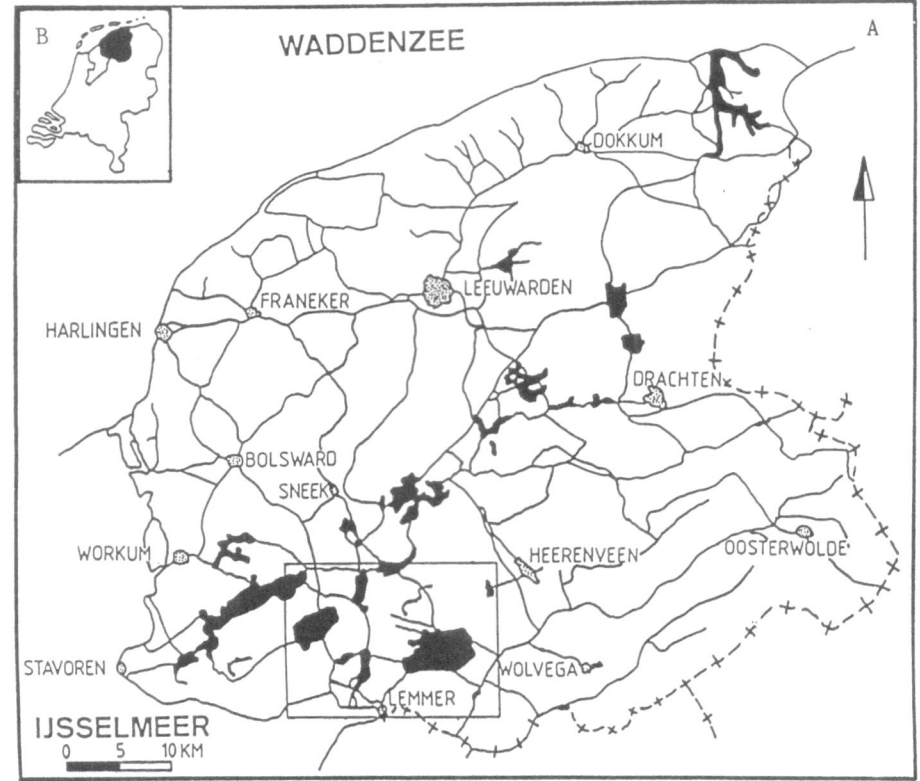

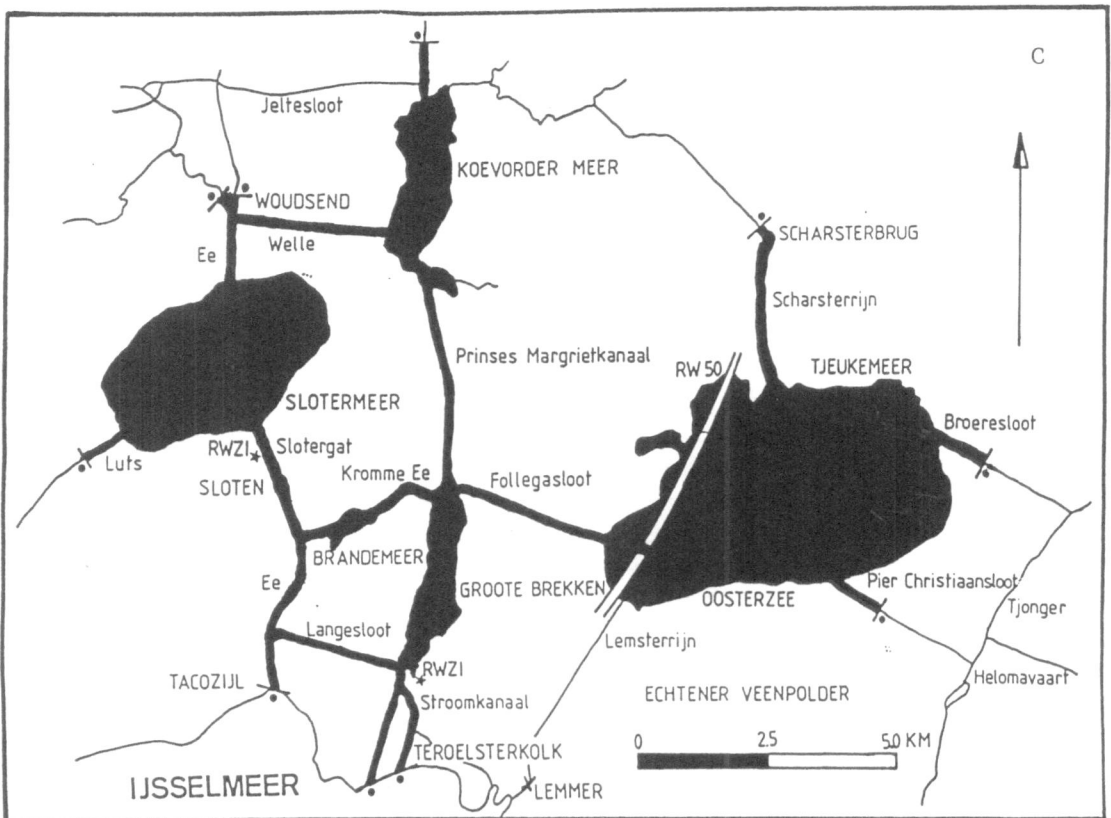

Fig. 1. (a) The Friesian 'boezem', a system of interconnected lakes. (b) Its position in the Netherlands. (c) The south western lake district. .–: System boundaries; RWZI: Dephosphorization Units (from Van Huet *et al.*, 1987).

Moed (1984), Claassen (1986) and Van Huet *et al.* (1987).

Thus although much research has been done on the trophic relationships especially in Tjeukemeer, up till the 80's hardly any investigations were made in modelling and detailed research of hydrology, external loading, sedimentary P and internal loading. In 1984 a eutrophication project started to study these processes, while the south western part of the boezem system was chosen as a study area because of relatively reasonable chances for restoration. The study of biotic and abiotic processes are the two main activities of research. Finally, the integrated results of these studies should lead to scenario simulations for policy and management.

Results and discussion

Measuring and modelling hydrology

In meteorologically normal years in summer near Lemmer there is an average inflow of IJsselmeer water of $260 \times 10^6 \, \text{m}^3$, while $120 \times 10^6 \, \text{m}^3$ of water is pumped out of the polders into the boezem system. In 1985, however, with much rainfall these values were $60 \times 10^6 \, \text{m}^3$ and $210 \times 10^6 \, \text{m}^3$, respectively. Especially Tjeukemeer received much polder water in this year; as much as 91% of the yearly inflow originated from surrounding polders, being 50% in normal years (Van Huet *et al.*, 1987).

Figure 2 shows the water quantities let in near Lemmer, Tacozijl and Stavoren and the quantities pumped into IJsselmeer near Lemmer and Stavoren during 1984–1986. It can be concluded that during these years there is mainly inflow near Lemmer and mainly pumping near Stavoren. In 1985 inflow of water was low. In 1986 water near Tacozijl was let in, influencing chloride concentrations in Slotermeer (see also Fig. 3a).

It appeared that generally water residence times in Tjeukemeer are about half of those in Slotermeer: 1–2 weeks to 3 months and 2 weeks to 6 months, respectively. Water velocity measurements were done during three charac-

teristic situations: when water of IJsselmeer was let in near Lemmer, when water was pumped into IJsselmeer near Lemmer and Stavoren and when a fairly hard north eastern wind was blowing. In the case of water inflow at Lemmer, 25–40% of the water was transported through the Follegasloot in direction Tjeukemeer, and 10–15% in direction Slotermeer. In the case of pumping boezem water into IJsselmeer near Lemmer and Stavoren, the Lemmer station pumped out hardly any water that originated from Slotermeer, while 30% originated from Tjeukemeer. Wind appeared to have a great influence on the water transport (Van Huet *et al.*, 1987).

Leenen (1982) used a chloride model to calculate the inflow and outflow of Tjeukemeer in 1971, resulting in a regression line between inflow to the lake and inflow near Lemmer and Stavoren. He concluded that an assumption of instantaneous mixing in the lake, often used in mathematical models, is only a mathematical tool and not realistic.

Brinkman *et al.* (1987) reported upon a water quantity and quality model for the whole boezem network. No wind data were used. The simulation results of these authors for nitrogen were quite satisfactory. Results for phosphorus were not as good, probably because data of internal loading were poor. Scenario calculations showed that flushing the whole boezem system with water of IJsselmeer seemed to have a positive effect by a shift from a dominancy of blue-green algae towards one of green algae, due to the shorter residence times. Flushing, however, will lead to higher chloride concentrations in summer. The authors concluded that this macro model could be a valuable tool for an integral water quality management.

Although in general water velocities are low ($0–10 \, \text{cm s}^{-1}$), an attempt was made to model hydrology of the south western system using a more detailed network. A node-link mathematical Chezy-model with wind terms (Orlob, 1972; Lijklema & Van Straten, 1977) was used. For the Friesian situation data on water levels at the system boundaries, the partial water balance and wind are needed. The simulation results describe

78

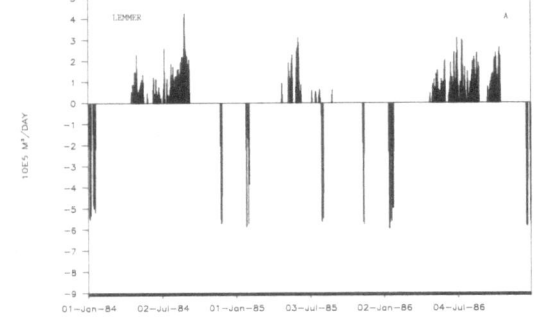

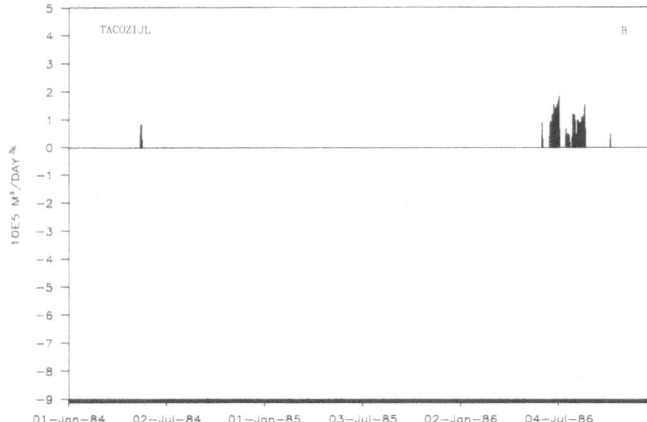

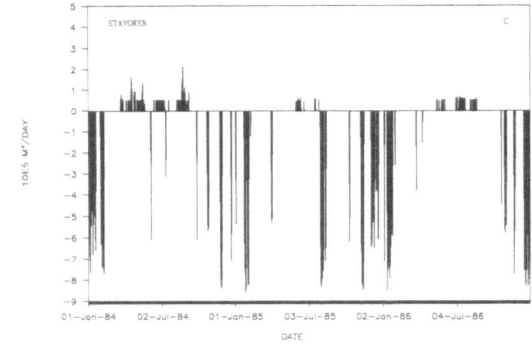

Fig. 2. Water quantities in 10^6 m^3 day^{-1} let in ($+$) or pumped into IJsselmeer ($-$) during 1984–1986; (a) Lemmer, (b) Tacozijl, (c) Stavoren.

the complete water balance of the system, water velocities (and thus discharges) in all canals and water levels at all nodes. Research is in progress, and results of simulation periods and sensitivity analysis will be reported elsewhere.

The differences in hydrology of Tjeukemeer and Slotermeer are reflected in the Cl$^-$ and total-P concentrations of the lake water. In Fig. 3 these concentrations are given for 1984–1986. The chloride sequence for Slotermeer (Fig. 3a) shows fewer fluctuations than that for Tjeukemeer (Figs. 3c, e). Chloride concentrations in Groote Brekken are highest because this lake is situated close to the inflow location (Fig. 3g).

Total-P concentrations in summer are mostly above 0.15 mg l^{-1} (Figs. 3b, d), which is the basic water quality standard of this component for the period April–September laid down by the Water Action Programme 1985–1989 of the Dutch Government. The winter total-P concentrations for Tjeukemeer are higher than for Slotermeer (Figs. 3b, d). In winter when polder water was flowing into the east of Tjeukemeer, average total-P concentrations were 0.28 mg l^{-1} for 1984/1985 and 0.33 mg l^{-1} for 1985/1986 and 1986/1987 (Fig. 3f). Near the inlet location at Lemmer, while water of IJsselmeer was flowing into the boezem system, the average total-P concentration during 1984–1985 was 0.18 mg l^{-1} (Fig. 3h). Thus total P concentrations in polder water are higher than in water of IJsselmeer. As Tjeukemeer is receiving more water of surrounding polders than Slotermeer (see also next paragraph), P loading in winter is highest in this lake.

Apart from the above Cl$^-$ fluctuations, the water regime is reflected in the water colour of Tjeukemeer: greenish in summer because of algal growth and brownish in winter because of humic polder water.

During 1984–1986 transparency values of 20–60 cm were measured. The values are a result of algal growth, humic compounds from polder water and/or resuspension by wind. In the same period oxygen concentrations of 5–14 mg l^{-1} were measured, while the pH ranged from 8.5 to 10.5.

Since 1984 the Echtener Veenpolder, a polder south of Tjeukemeer, was intensively studied (see Fig. 4). Daily discharges were registered and at

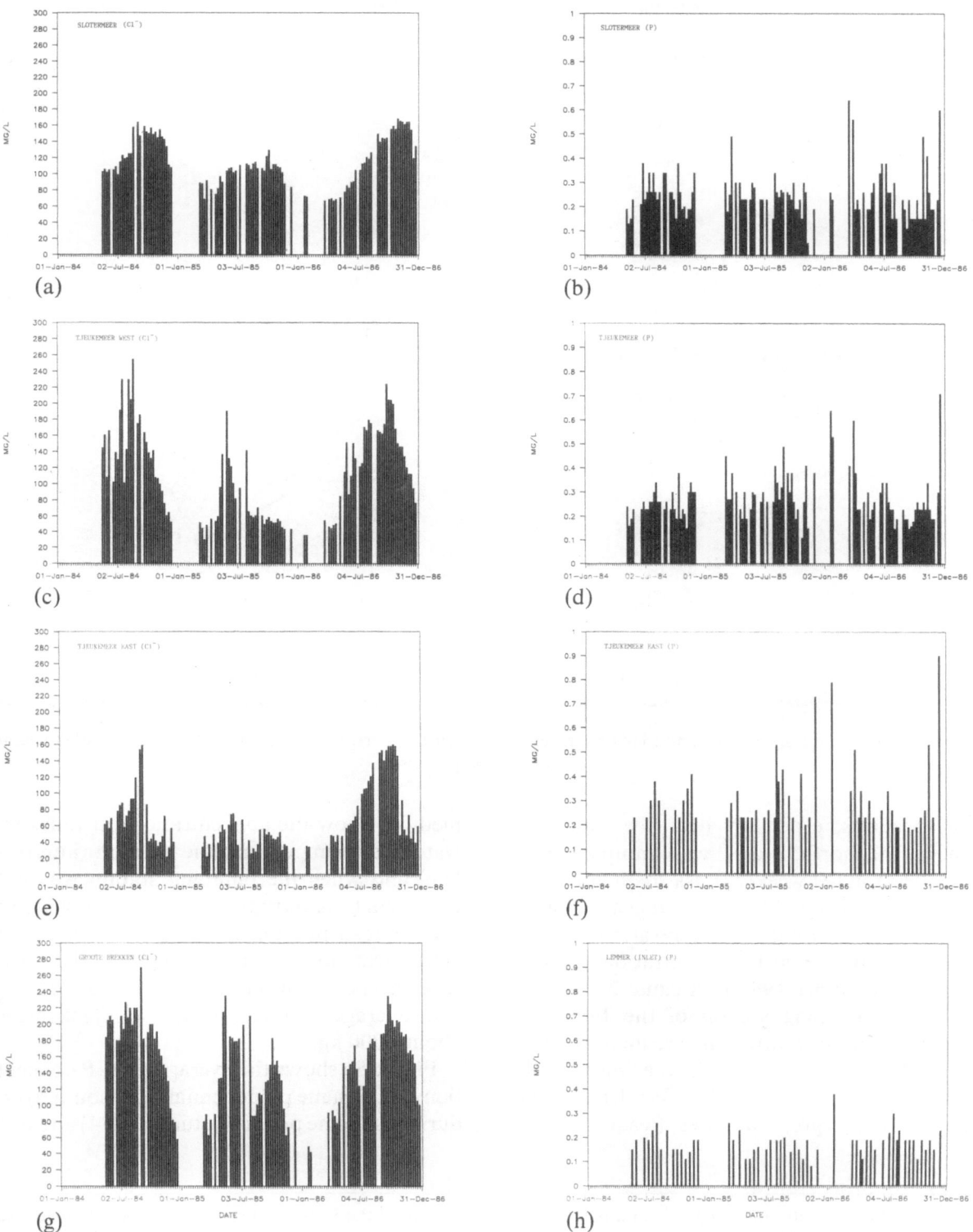

Fig. 3. Chloride (a, c, e, g) and total phosphorus (b,d,f,h) concentrations in mg l^{-1} during 1984–1986.

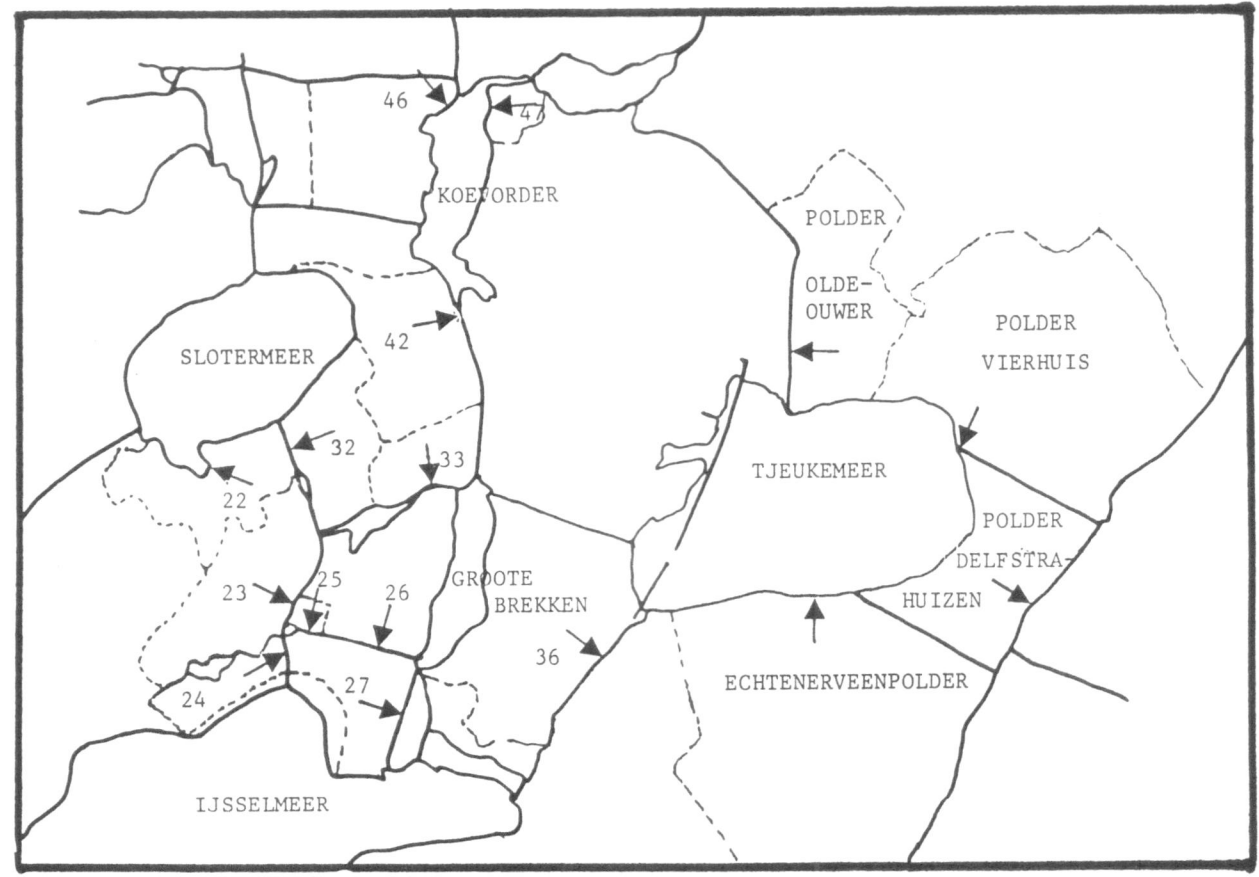

Fig. 4. The south western lake district and 15 of its surrounding sampled polders. The arrows mark the locations of the pumping stations of the polders.

several locations, especially near the pumping station in the north, water was sampled and analysed for total-P weekly. Less frequently discharges of 14 other polders were registered and water analysed for total-P concentrations.

Figure 5a shows total-P concentrations during September-December 1984 in a canal 30 metres south of the pumping station of the Echtener Veenpolder. The concentrations are increasing in time, which might be due to increasing rainfall causing run-off and/or increased P-release from sediments. The regression lines (least squares method) show the concentrations at times when water was pumped into the boezem (interrupted line) and when the pumping station was not in operation (uninterrupted line). Probably pumping caused resuspension resulting in higher total-P concentrations. The total-P load in the boezem from this polder during this period was 2150 kg P. The average standing stock of Tjeukemeer is about 8000 kg.

Figure 5b shows the average total-P concentrations in the same polder canal from south to north during the same period (autumn, 1984) and during

Fig. 5. (a) Total-P concentrations (mg l^{-1}) in a polder canal in the north of the Echtener Veenpolder about 30 meters ▷ of the pumping station. ■: Pumping station not in operation; □: Pumping station in operation. (b) Average total-P concentration (mg l^{-1}) at 5 locations from south to north (distance in metres) in the same canal. □: Autumn 1984; x: Spring 1986.

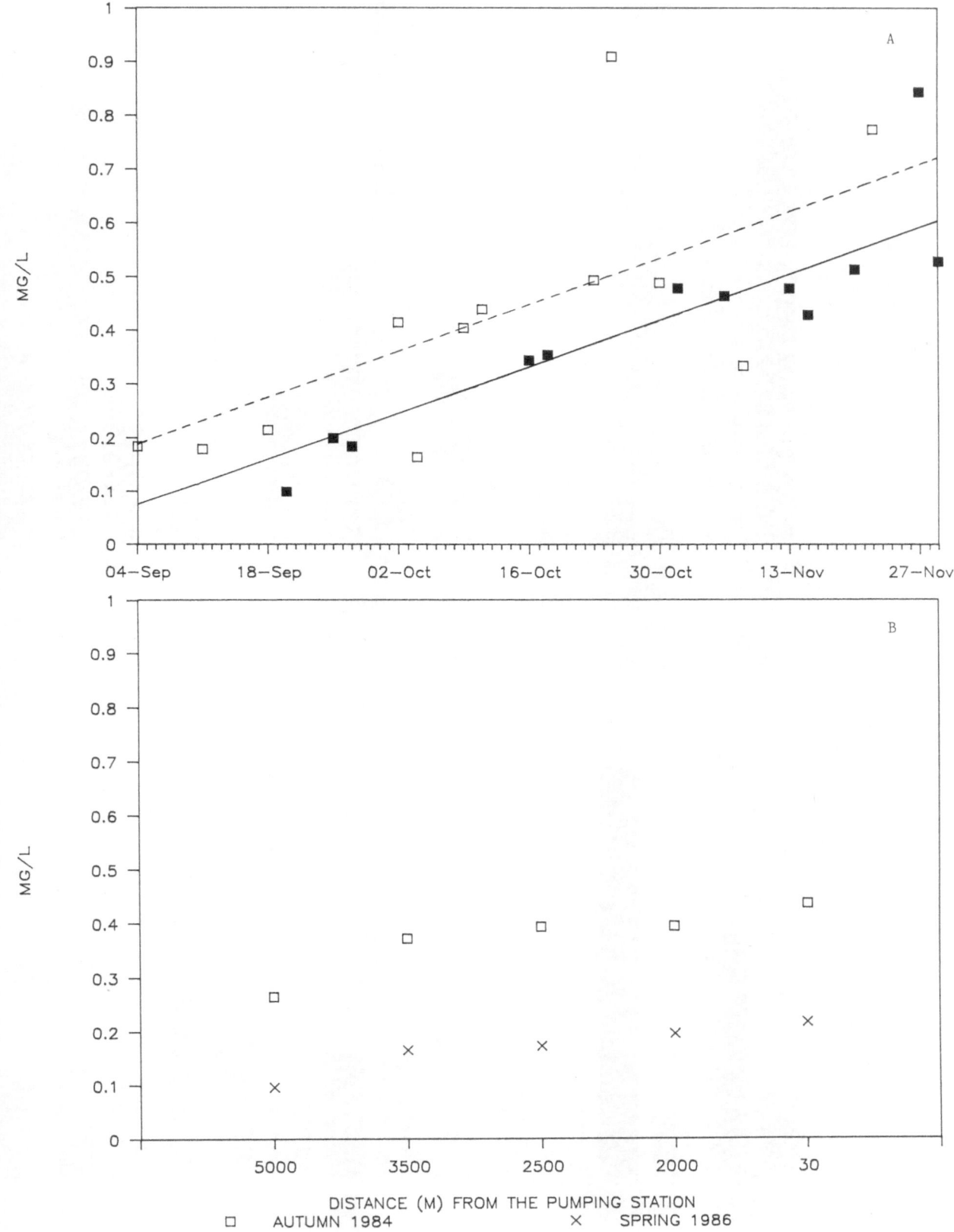

DISTANCE (M) FROM THE PUMPING STATION
☐ AUTUMN 1984 ✕ SPRING 1986

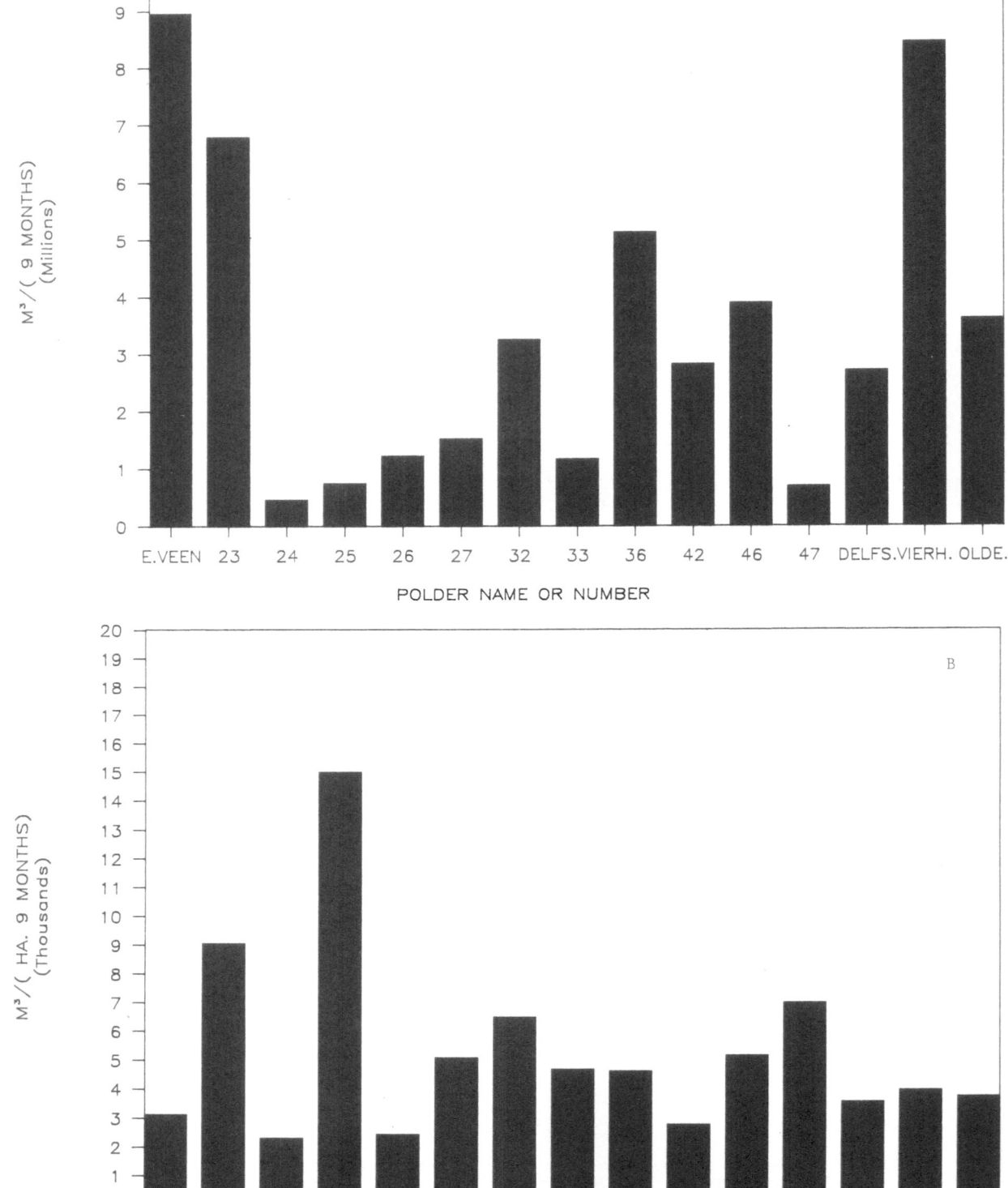

spring 1986. The canal transports water of the whole polder to the pumping station. Concentrations are increasing because from south to north the Echtener Veenpolder is receiving more and more water from agriculture and housing. The water in the south of the polder could be characterized as boezem water because in the south there is inflow of boezem water for flushing. Thus there is a P contribution from the polder.

Figure 6a shows the quantities of water of 14 other polders (see Fig. 4), which are pumped into the boezem system during April-December 1985, and Fig. 6b shows the quantities per hectare. Two important conclusions can be drawn: first, Tjeukemeer received much more polder water than for instance Slotermeer (Fig. 6a) and second, except

for polders 23 and 25 the relative quantities did not greatly differ (Fig. 6b). Apart from this Tjeukemeer is receiving polder water from the small river Tjonger. Thus the eastern part of the studied area is receiving more polder water than the western part. In 1985 the average total-P concentration of water of the Echtener Veenpolder which was pumped into Tjeukemeer was > 0.4 mg l^{-1}.

Sedimentary P and internal loading

Sampling of sediments, for studying horizontal and vertical distribution of phosphorus and for measuring and modelling internal loading, took

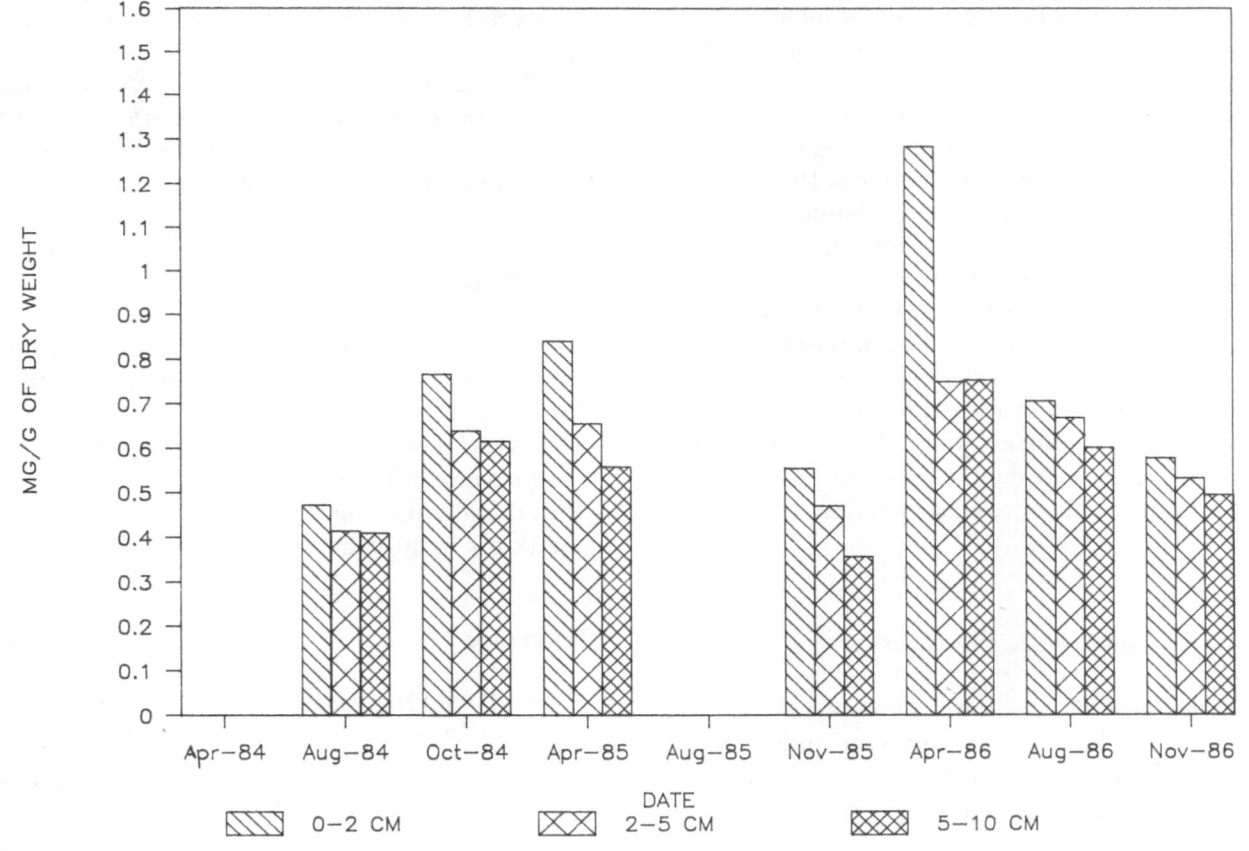

Fig. 7. Average total-P contents in sediments of all sampling stations in mg g^{-1} of dry weight.

◁ *Fig. 6.* (a) Water quantities of 15 polders during April-December 1985 pumped into the boezem system. (b) The same expressed per hectare.

place during 1984–1986 at 35 locations. The total-P concentrations in sediments varied between 0.1 and 4.0 mg P g^{-1} of dry weight, which was in agreement with other shallow eutrophicated Dutch lakes. There was a significant correlation with Kjeldahl nitrogen concentrations. Sediments of Tjeukemeer were mostly peaty, while sediments of Slotermeer were mostly sandy. Yet the P contents of these sediments did not differ much and were relatively low. Peaty sediments of Brandemeer showed relatively high P contents, just as did muddy sediments near or in shipping channels and sediments of inflow/outflow locations of the lakes. Probably at places where water velocities are relatively low a sedimentation of fine particles can take place, and these sediments have high P contents. Figure 7 shows average total-P contents of all stations for 7 sampling dates. P contents decrease with increasing sediment depth. At present it is not clear if there are also yearly fluctuations.

Extraction experiments according to the excretion scheme of Hieltjes & Lijklema (1980) showed that most P was iron- and aluminium-bound, except for sediments of Slotermeer in which mainly loosely bound P was measured. Release experiments with winter samples showed an increase flux after 1 week under laboratory conditions. Probably mineralisation due to activation of the microbial flora was an important process. In general P fluxes were less than 1 mg P m^{-2} d^{-1}. As more detailed research is in progress no further results of release experiments are given here.

Summary and concluding remarks

Although there are so far no complete water and phosphorus balances and thus no modelling results for policy and management, yet some important conclusions can be drawn:

1) The water of the studied area is eutrophic. Total-P concentrations in summer are mostly above 0.15 mg l^{-1}, which is the basic water quality standard of the Dutch Government.

2) Hydrology plays an important role in the studied area. Especially the water quality of Tjeukemeer is influenced by the summer and winter hydrology. Slotermeer appeared to be a relatively isolated lake.

3) Hydrology in the summer period can be roughly studied by monitoring Cl$^-$ concentrations at several locations.

4) In 1985 it was estimated that about 90% of the inflowing water into Tjeukemeer originated from polders.

5) Generally the total-P concentrations of polder water are higher than those of IJsselmeer water.

6) Total-P contents in sediments varied from 0.1 to 4.0 mg g^{-1} of dry weight. Most P was Fe- and Al-bound. There was a slight tendency for decreasing P content with increasing sediment depth.

7) Phosphorus release experiments with winter sediments showed that P release was less than 1 mg P m^{-2} d^{-1}. Probably the process was influenced by mineralisation.

Acknowledgements

The phosphorus eutrophication research project is supported by a grant of the Dutch Ministry of Housing, Planning and Environment and of the Department of Public Works and Environment of the Province of Friesland. Dr S. Parma is thanked for critically reading the manuscript and Ms M. Donk for improvement of the English text.

References

Beattie, D. M., H. L. Golterman & J. Vijverberg, 1978. An introduction to the limnology of the Friesian lakes. Hydrobiologia 58: 49–64.

Brinkman, J. J., P. S. Griffioen, S. Groot & F. L. Los, 1987. A water quantity and water quality model for the evaluation of water management strategies in the Netherlands -Application to the province of Friesland-. In: Beck, M. B., ed., Systems analysis in water quality management. Pergamon Press 1987.

Claassen, T. H. L., 1986. Eutrofiering en algengroei in het Friese boezemwater. H$_2$O (19), 12: 268–275 & 279.

De Haan, H. & J. R. Moed, 1984. Phosphorus, nitrogen and chlorophyll-a concentrations in a typical Dutch polder lake, Tjeukemeer, in relation to its water regime between 1968 and 1982. Wat. Sci. Tech. 17: 733–743.

Hieltjes, A. H. M. & L. Lijklema, 1980. Fractionation of inorganic phosphates in calcareous sediments. J. envir. Qual. 9: 405–407.

Leenen, J. D., 1982. Hydrology of Tjeukemeer. Hydrobiologia 95: 199–203.

Lijklema, L. & G. Van Straten, 1977. De waterbeweging in het plassengebied van N.W. Overijssel. H_2O (10), 16: 360–363.

Moed, J. R. & H. L. Hoogveld, 1982. The algal periodicity in Tjeukemeer during 1968–1978. Hydrobiologia 95: 223–234.

Orlob, G. T., 1972. Mathematical modeling of estuarial systems. Int. Symp. on modeling techniques in water resources systems. Ed. A.K. Biswas, Env. Canada, Ottawa.

Van Huet, H. J. W. J., H. De Haan & T. H. L. Claassen, 1987. Fosfaateutrofieringsonderzoek in het merengebied van zuid-west Friesland. H_2O (20), 6: 131–135.

Hydrobiologia **191**: 87–95, 1990.
P. Biró and J. F. Talling (eds), Trophic Relationships in Inland Waters.
© 1990 *Kluwer Academic Publishers.*

Phosphorus dynamics following restoration measures in the Loosdrecht Lakes (The Netherlands)

Louis Van Liere,[1] Ramesh D. Gulati,[1] Frederick G. Wortelboer[2] & Eddy H. R. R. Lammens[3]
[1] *Limnological Institute, Rijksstraatweg 6, 3631 AC The Netherlands*; [2] *National Institute of Public Health & Environmental Hygiene, P.O. Box 1, 3720 BA Bilthoven, The Netherlands*; [3] *Limnological Institute, De Akkers 47, 8536 VD, Oosterzee, The Netherlands*

Key words: lake restoration, phosphorus-dynamics, total-phosphorus, chlorophyll *a*

Abstract

External phosphorus loads to three shallow lakes in the Netherlands were reduced by eliminating waste-water discharge and by dephosphorization of the supply water, with which water level is controlled. Concentrations of total-phosphorus and chlorophyll *a* were significantly reduced during 1980–1986 in L. Breukeleveen, but not in L. Vuntus and L. Loosdrecht. In 1983–1986 the phosphorus flow through several trophic levels was determined. Changes over these years were not significant. External input to the lakes still contributes substantially to the phosphorus input. Release from the sediments also contributed to the cycling of the phosphorus. Excretion by large crustacean zooplankters was important in phosphorus recycling, and delivered 20–30% of the daily phytoplankton phosphorus demand. A similar contribution is expected from fish. If one wants recovery of the lakes to be accelerated, additional measures are needed.

Introduction

The Loosdrecht lakes system consists of a series of interconnected lakes (Fig. 1), which vary in area and are situated about 20 km south of Amsterdam. They are presently the subject of an intensive study of the workgroup Water Quality research Loosdrecht lakes (WQL) since 1983 (Loogman & Van Liere, 1986). Information on the morphometry of the lakes is given in Table 1. Within a few decades some parts of the Loosdrecht lakes (Lake Loosdrecht, Lake Breukeleveen and Lake Vuntus) changed from clear-water, charophyte-dominated ecosystems into highly turbid, phytoplankton-dominated

systems with high concentrations of seston and chlorophyll *a* (Table 2). The external phosphorus load to the lakes was about 2 g P m^{-2} year before measures to decrease this external load became effective. From 1970–1986 surrounding villages were provided with sewer systems, so that no untreated domestic waste water enters the lakes. Since 1984 the phosphorus-rich supply water from the River Vecht to the lakes, needed to maintain water level, has been replaced by water from Amsterdam-Rhine Canal, which is dephosphorized before entry into the lakes (Fig. 1). The external phosphorus load is at present 0.3 g P m^{-2} year^{-1} (Kal *et al.*, 1984).

Here the response of the ecosystem to the

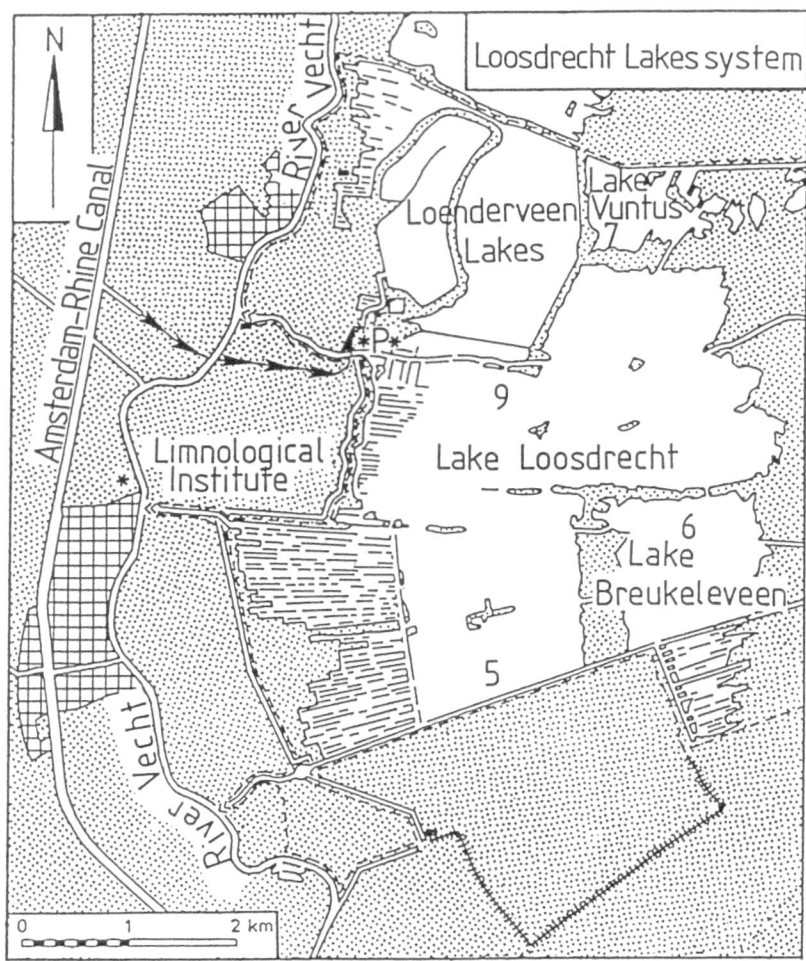

Fig. 1. Loosdrecht lakes area: (≪) – Locks that were used for shipping and/or inlet of water from River Vecht; (arrows) – pipeline from Amsterdam-Rhine Canal to Lake Loosdrecht; (*P*) – dephosphorization unit, capacity 17 $m^3 s^{-1}$; (5, 6, 7 and 9): sampling stations.

reduction in phosphorus load is presented on the basis of the chlorophyll *a*-total phosphorus relation and the phosphorus flow through various trophic levels. Possible explanations of the responses are discussed, and suggestions made for future management measures.

Methods

The various parameters were determined according to standard limnological methods, as described elsewhere (Gulati *et al.*, 1982; Van Liere *et al.*, 1986a, 1986b). The lakes were sampled fortnightly in 1983–1985, and monthly in 1986. Although the phosphorus flows within seston due to primary production and the grazing, assimilation and egestion of zooplankton were measured

Table 1 Morphometry of the Loosdrecht lakes.

	L. Loos-drecht	L. Breuke-leveen	L. Vuntus
Surface area (km²)	9.79	1.79	0.88
Mean depth (m)	1.85	1.45	1.45
Water residence time (yr)	0.6	0.4	0.6

Table 2 Indicators of the trophic status of Lake Loosdrecht.

	1955[a]	1983	1986
Mean zooplankton (> 150 μm) mass concenctration (mg l^{-1})	0.15	0.70	0.75
Mean seston concentration (< 150 μm, mg l^{-1})	0.50	12.2	28.5
Chlorophyll *a* concentration (summer average, μg l^{-1})	2–5	125	130
Total phosphorus concentration (summer average, mg l^{-1})	?	0.12	0.12

[a] After Geelen (1985).

by carbon-tracer methods, the data have been recalculated to phosphorus flows, using C : P ratios of seston (< 150 μm) and zooplankton (> 150 μm). Zooplankton (> 150 μm) excretion was directly measured, or calculated on the basis of a temperature-specific excretion rate (Den Oude & Gulati, 1989). Data on daily phosphorus losses of zooplankton (mortality) were calculated from the balance between assimilation rate and excretion rate. External inputs were calculated from hydrological data (B. F. M. Kal, pers. comm.) and data sets provided by the Municipal Waterworks Amsterdam. Loss factors were estimated by empirical modelling (Wortelboer, unpublished results). Mineralization rate is not included in the scheme; it has been measured only three times in 1986. Sinke (Van Liere *et al.*, 1986b) found the phosphorus-mineralization rate of sestonic matter to be the same as phosphorus uptake during primary production. This suggests that mineralization of seston occurs mainly in the water phase and the top-layer of the sediments (epipelon) that is in close contact with the aerobic water. Waterplants and fish have been excluded from the scheme (but see next section).

Results and discussion

Studying time courses of concentrations of both total-phosphorus and chlorophyll *a* is time-consuming and difficult, but can be simplified by using summer-averages (Fig. 2). In Lake

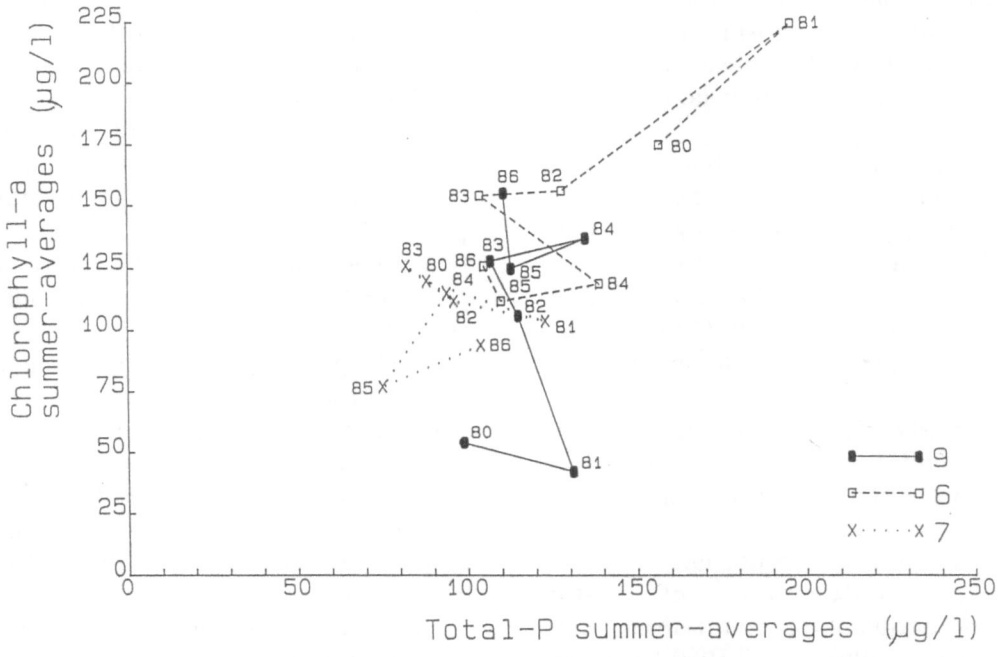

Fig. 2. Correlation diagram between 'summer averaged total phosphorus concentration' and 'summer averaged chlorophyll *a* concentration'. Concentrations are averaged over the months April–September. The values of station 5 were always close to those of station 9.

Loosdrecht an almost unchanged total phosphorus concentration was accompanied by a threefold increase in chlorophyll *a* concentration (1980–1983). During that timespan trichome-forming prokaryotes increased in concentration, from summer values of 15 000 trichomes per ml in 1981 to 160 000 trichomes per ml in 1983 (Boesewinkel *et al.*, 1984), dominated in summer by the prochlorophyte *Prochlorothrix hollandica* (Burger-Wiersma *et al.*, 1986). L. Breukeleveen showed a delayed reaction to the stopping of waste water discharge from the surrounding villages. Both the concentrations of total-phosphorus and chlorophyll *a* decreased markedly during 1980–1984. From that time the same irregular pattern as in L. Loosdrecht was observed. Multiple linear regression analysis revealed the trend towards lower concentrations of both chlorophyll *a* and total-phosphorus in L. Breukeleveen to be significant in the period 1980–1986. Trends in L. Loosdrecht and L. Vuntus were not significant over those years.

To study changes more closely, a descriptive model is presented to show the phosphorus flow through various trophic levels of the ecosystem (Fig. 3). Since values for phosphorus flows in L. Loosdrecht are similar to those observed in L. Breukeleveen, data of the latter lake has been left out of the Figure.

External phosphorus input was high in 1983 in L. Loosdrecht mainly because of the supply of water from R. Vecht. This has less influence on L. Vuntus, which is more indirectly affected by the supply water (Fig. 1). After the completion of water management measures in 1984 the external phosphorus input was reduced to much lower values.

The release of phosphorus from the sediments in the period under study has decreased in L. Loosdrecht, but it still contributes considerably to the total load. Differences in release rate between the various lakes are not large.

The average concentration of seston (< 150 μm) and zooplankton in the three years did not differ significantly (Gulati, 1990). They are in the range expected considering the meteorological variations. The same is true for primary production. Zooplankton and seston (< 150 μm) concentrations are higher in L. Loosdrecht than in L. Vuntus. The primary production rate is lower in L. Vuntus; presumably here the impact of phosphorus-limited algal growth is more important (see below), although this does not follow directly from data presented in Fig. 3. Zooplankton excretion contributes substantially to the daily phytoplankton phosphorus demand (20-30%).

The model has some inherent disadvantages. Since it averages over a time-period it does not consider the dynamics of the system. Still the model should reveal the general trends in the phosphorus flow through the food chain in the course of the process of recovery. The descriptive model needs to be supplemented by a mathematical model, which will enable some insight into the dynamics of the ecosystem, and to predict the course of changes expected during restoration of the lakes (Kouwenhoven & Aldenberg, 1986).

Summarizing, a three-year period after completing the measures to reduce the external phosphorus load is too short to draw definite conclusions about the effect of those measures. Therefore an extended study has been started and will continue for the next four years.

Reasons for the resilience of the ecosystem are:

(i) The external phosphorus input still contributes considerably to the phytoplankton phosphorus demand (Fig. 3). Discharge of superfluous polderwaters and precipitation have now become considerable in the phosphorus input since 1984 (Kal, 1986).

Flushing could be an additional measure to control the external load by removing phosphorus from the system (Engelen, 1982). Dephosphorized water is then supplied to the lakes, preferably in summertime when the total phosphorus concentration is high. After mixing with lake water the surplus is discharged again. Repeated supply/discharge will remove phosphorus from the system. Positive results through flushing have been reported (Hosper & Meyer, 1986).

(ii) Release from sediments is an important component of the phosphorus budget (Fig. 3).

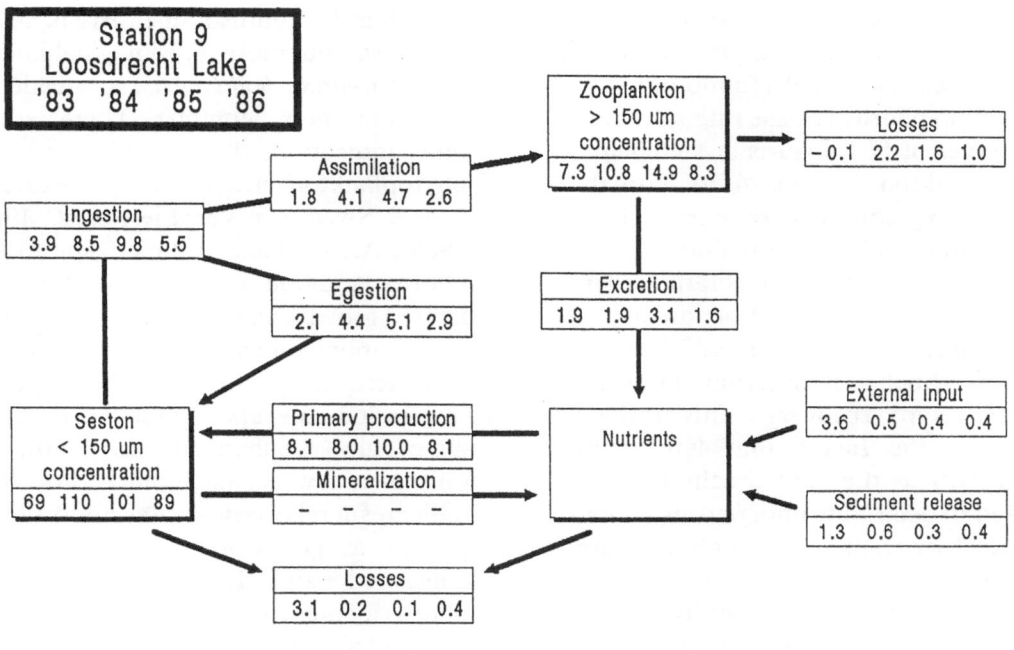

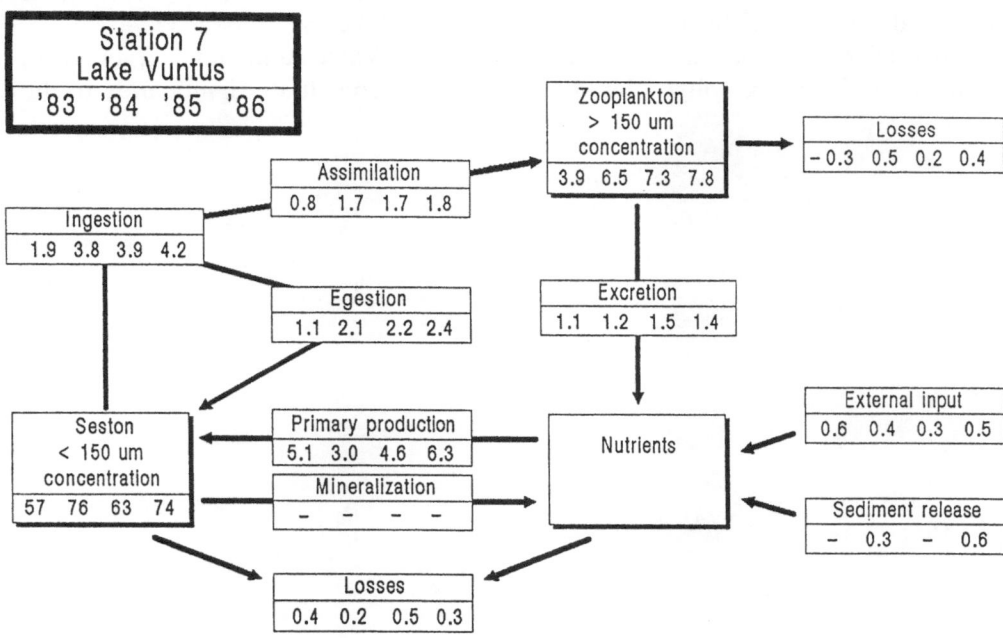

Fig. 3. Phosphorus flow diagram for L. Loosdrecht (station 9) and L. Vuntus (station 7). Since in the years of the study L. Breukeleveen did not differ much from L. Loosdrecht, data on it have been omitted. Data of the various phosphorus flows and concentrations are, from left to right, for 1983, 1984, 1985 and 1986, respectively. Phosphorus flow (squares) and concentration (shadowed squares) are averaged over the period May–September. Rates are expressed in μg P l^{-1} day^{-1}, concentrations in μg P l^{-1}. The effects of waterplants and fishes have been omitted, but see text.

92

Removal of sediments might be needed to reduce the phosphorus release (Björk, 1985), although Boers (Van Liere *et al.*, 1986b) found only a 30% reduction in phosphorus release rate after removing the top-layer of L. Loosdrecht sediments.

(iii) Phytoplankton limited in its growth by light-energy supply contains comparatively more phosphorus than phosphorus-limited phytoplankton; in the case of cyanobacteria under the conditions prevailing in the Loosdrecht lakes system the content can vary threefold (Zevenboom, 1986). Such an internal phosphorus storage may retard recovery. This raises the question about the factor that limits phytoplankton growth in the Loosdrecht lakes. The methods used to determine phosphorus limitation gave contradictory results. Although phosphorus uptake kinetics (Riegman & Mur, 1986) revealed that phosphorus limited phytoplankton growth in all of the lakes, one must take care, since the uptake rate of phytoplankton samples was compared with uptake rates of cultures of the cyanobacterium *Oscillatoria agardhii*, found sparsely in the lakes. These data have now been compared with uptake rates of *Prochlorothrix hollandica*. Phosphorus, then, is limiting phytoplankton growth in L. Vuntus and L. Breukeleveen; as for L. Loosdrecht more physiological information on *Prochlorothrix hollandica* is needed (Burger-Wiersma, pers. comm.). There are, however, other indications that phosphorus is not limiting phytoplankton growth in L. Loosdrecht (Rijkeboer & Sweerts in Van Liere *et al.*, 1986b; Gons, 1989). Also enhanced dark respiration on phosphorus addition is indicative for phosphorus-limited growth, since the energy required for both consumption and storage of phosphorus can only be generated by oxidation of glycogen. The dark respiration rate upon phosphorus addition was almost always enhanced in L. Vuntus (Fig. 4), but only on a few occasions in L. Loosdrecht. This indicates a relatively greater importance of phosphorus as growth-limiting factor in L. Vuntus compared with L. Loosdrecht. The average phosphorus content of the seston also supports this point (Table 3), although a contribution of resuspended matter (Gons, 1989) must not be overlooked. We expect a more important role of phosphorus as growth-limiting factor in the future in all of the lakes.

(iv) Waterplants have largely disappeared, and do not contribute significantly to the phosphorus

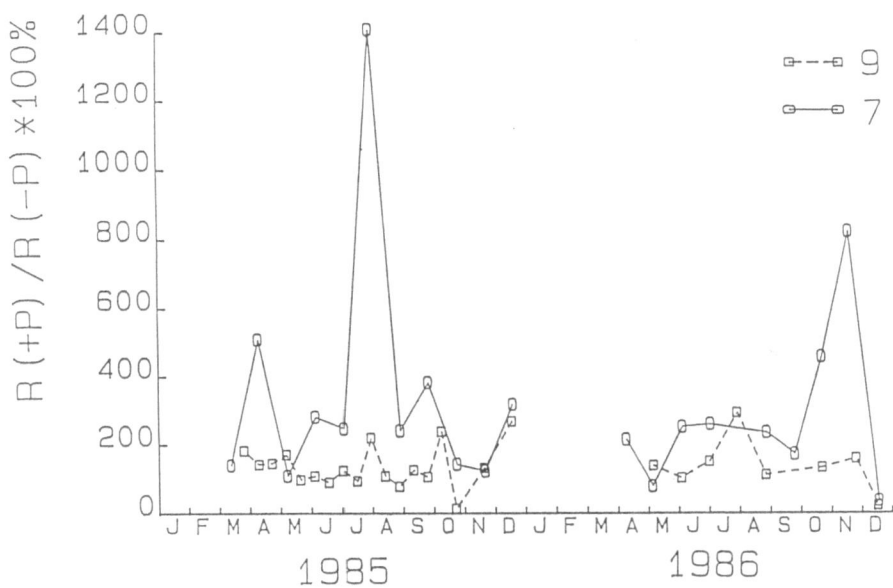

Fig. 4. Effect of phosphorus addition (13 μM) on dark respiration rates, in phytoplankton samples taken from L. Loosdrecht and L. Vuntus; relative differences are based on controls in which no phosphorus was added.

Table 3 Phosphorus content (% dry weight) of the seston (< 150 μm) averaged over the months May–September of L. Loosdrecht (averaged over stations 5 and 9), L. Vuntus and L. Breukeleveen.

	L. Loosdrecht	L. Breukeleveen	L. Vuntus
1983	0.36	0.24	0.23
1984	0.34	0.24	0.24
1985	0.33	0.24	0.27
1986	0.24	0.20	0.26

balance of the Loosdrecht lakes system at present (Malthus, 1990). However, they can become important after recolonization. First, reduced water circulation due to the presence of waterplants may induce anoxic conditions at the lake bottom, resulting in higher phosphorus release rates if compared with oxic conditions (Moss *et al.*, 1986). Second, waterplants are able to take up phosphorus from the interstitial water, and after decay it is available again. Since the latter process occurs mainly in autumn and winter direct impact might be low, although phosphorus adsorbed in winter by the sediments will be available next summer through release reactions.

(v) Excretion by the large zooplankton recycles a considerable amount of phosphorus to the phytoplankton (Fig. 3). The role of rotifers in phosphorus recycling has not been realized fully. From Gulati *et al.* (1987) it can be calculated that the influence of rotifers on phosphorus regeneration, through their extremely large number, can be in the range of 45–95% of the total zooplankton phosphorus-excretion (see also Gulati, 1990) After improvement of the water quality, the importance of rotifers may decrease, but may be offset by increase in larger crustaceans. Improvement in water quality presumably will be accompanied by lower numbers of trichome-forming phytoplankton species and higher numbers of green algae and diatoms, leading to higher grazing pressure of the large crustacean zooplankton, and total excretion may even be enhanced compared to the present situation.

(vi) Phosphorus storage in fish is high in the Loosdrecht lakes (Table 4), although higher

values have been reported (Kitchell *et al.*, 1975). The fish population in the Loosdrecht lakes rose to high levels (20–50 g m^{-2} wet weight) as a consequence of the higher concentrations of food organisms accompanying eutrophication. Some 60–70% of the fish population consists of bream (*Abramis brama*), especially bream of older year-classes (Van Densen *et al.*, 1986). We roughly estimate that an average fish population of 35 g m^{-2} wet weight can deliver 1 μg P l^{-1} d^{-1} to the system by metabolic excretion and decay. More attention to the quantitative aspects of phosphorus regeneration by fish is necessary.

The paramount importance, however, of the high number of large bream is not only their ability to control zooplankton abundance, but also their feeding behaviour on the bottom fauna; they disturb the sediments quite vigorously during feeding, which might have consequences for sediment phosphorus release.

Biomanipulation by stocking of the main predator pike-perch (*Stizostedion lucioperca*) could be promising. However, two aspects must be considered. First, before biomanipulation in the Loosdrecht lakes can become operative the abundance of trichomes needs reduction to a level that will not interfere with the growth of large crustacean zooplankters. Dawidowicz *et al.* (1988) observed poor reproduction of *Daphnia magna* in cultures containing filaments from L. Loosdrecht (150 μg l^{-1} chlorophyll *a*). Second, even large pike-perch cannot swallow large bream. The latter must be controlled in a different way for active fish management to be effective. Reducing the high number of large bream certainly

Table 4 Breakdown of phosphorus in different compartments in L. Loosdrecht, assuming 20–50 g m^{-2} wet weight of fish. % P based on data from Kitchell *et al.*, 1975.

	Concentration μg l^{-1}	% of total
Dissolved phosphorus	15	4–7
Seston (< 150 μm) phosphorus	95	29–46
Zoopl. (> 150 μm) phosphorus	10	3–5
Fish phosphorus	85–210	42–63

depress the recycling of phosphorus due to fish metabolism and decay.

Summarizing, the recovery of eutrophic lakes upon reduction of external phosphorus load may be retarded by various processes involved with the internal phosphorus load. If the external phosphorus load can be kept lower than the loss factors, restoration will eventually proceed. The year x, in which one can see waterplants growing on the bottom of the Loosdrecht lakes is, so far, unpredictable. If recovery needs to be accelerated, which is an anthropocentric question, additional measures at crucial moments are needed.

Acknowledgements

The authors are greatly indebted to the several members of the workgroup WQL that have determined the various parameters. Special thanks are due Ben Kal (Institute of Earth Sciences, Free University Amsterdam) and the Municipal Waterworks Amsterdam for the use of their dataset. Loes Breebaart and Klaas Siewertsen arranged the chemical and biological data. Onno van Tongeren performed the multiple regression analysis. The first author's work was supported by the Ministry of Science and Education and the Ministry of Housing, Physical Planning and Environment. The latter ministry also supported the third author.

References

Björk, S., 1985. Lake restoration techniques. In: Lakes pollution and recovery. Proc. Int. Congress European Water Pollution Control Association, Rome April 15–18, 1985; 202–212.

Boesewinkel-de Bruyn, P. J., L. Van Liere & B. Z. Salome, 1984. The Loosdrecht lakes restoration project: phytoplankton species composition. In: S. Parma & R. D. Gulati (eds), Institutes of the Royal Netherlands Academy of Arts and Sciences, Progress report 1983. Verh. Kon. Ned. Akad. Wet., Tweede Reeks 82: 32–33.

Burger-Wiersma, T., M. Veenhuis, H. Korthals, C. C. M. van der Wiel & L. R. Mur, 1986. A new prokaryote containing chlorophylls a and b. Nature 320: 262–264.

Dawidowicz, P., Z. M. Glicwicz & R. D. Gulati, 1988. Can Daphnia prevent a blue-green algal bloom in hypertrophic lakes? A laboratory test. Limnologica 19: 21–26.

Den Oude, P. J. & R. D. Gulati, 1989. Phosphorus and nitrogen excretion rates of zooplankton from the eutrophic Loosdrecht lakes, with notes on other P sources for phytoplankton requirements. Hydrobiologia 169: 379–390.

Engelen, G. B., 1982. Hydrology and watermanagement of the Loosdrecht lakes area in relation to water quality improvement by inlet of coagulated Amsterdam-Rhine Canal water (in Dutch). WQL-report 1982-3, 11 pp.

Engelen, G. B., 1986. Interactions of hydrological systems and eutrophication of Loosdrecht lakes. Hydrobiol. Bull., Amsterdam 20: 17–25.

Geelen, J. F. M., 1955. The plankton of the lakes. Report no. 38 (in Dutch), Amsterdam Municipal Waterworks.

Gons, H. J., 1988. Changes in sestonic matter in the Loosdrecht lakes in relation to production and resuspension. Hydrobiologia.

Gulati, R. D., K. Siewertsen & G. Postema, 1982. The zooplankton: its community structure, food and feeding and its role in the ecosystem of Lake Vechten. Hydrobiologia 95: 127–163.

Gulati, R. D., J. Rooth & J. Ejsmont-Karabin, 1987. A laboratory study of feeding and assimilation in Euchlanis dilatata lucksiana. Hydrobiologia 147: 289–296.

Gulati, R. D., 1990. Zooplankton structure in Loosdrecht lakes in relation to the trophic status and the recent restoration measures. Hydrobiologia 191: 173–188.

Hosper, H. & M-L. Meyer, 1986. Control of phosphorus loading and flushing as restoration methods for Lake Veluwe, The Netherlands. Hydrobiol. Bull. 20: 183–194.

Kal, B. F. M., G. B. Engelen & Th. E. Cappenberg, 1984. Loosdrecht lakes restoration project: Hydrology and physico-chemical characteristics of the lakes. Verh. int. Ver. Limnol. 22: 835–841.

Kal, B. F. M., 1986. Monthly mass balances for compartments of the Loosdrecht lakes: approach and preliminary results. Hydrobiol. Bull. 20: 27–39.

Kitchell, J. F., J. F. Koonce & P. S. Tennis, 1975. Phosphorus flux through fishes. Verh. int. Ver. Limnol. 19: 2178–2484.

Kouwenhoven, P. & T. Aldenberg, 1986. A first step in modelling plankton growth in the Loosdrecht lakes. Hydrobiol. Bull. 20: 135–145.

Loogman, J. G. & L. Van Liere, 1986. Restoration of shallow lake ecosystems with emphasis on Loosdrecht Lakes. Proceedings of the Water Quality research Loosdrecht Lakes symposium. Hydrobiol. Bull. 20: 3–259.

Malthus, T. J., 1990. An Assessment of the importance of emergent and floating-leaved macrophytes to trophic status in the Loosdrecht lakes. Hydrobiologia 191: 257–263.

Moss, B., H. Balls, K. Irvine & K. Stansfield, 1986. Restoration of two lowland lakes by isolation with and without removal of sediments. J. Appl. Ecol. 23: 391–414.

Riegman, R. & L. R. Mur, 1986. Phosphate uptake kinetics of the natural phytoplankton population from the Loosdrecht lakes. Limnol. Oceanogr. 31: 983–988.

Van Densen, W. L. T., C. Dijkers & R. Veerman, 1986. The fish community of the Loosdrecht lakes and the perspectives for biomanipulation. Hydrobiol. Bull. 20: 147–163.

Van Liere, L., L. Van Ballegooijen, W. A. De Kloet, K. Siewertsen, P. Kouwenhoven & T. Aldenberg, 1986a. Primary production in the various parts of the Loosdrecht lakes. Hydrobiol. Bull., Amsterdam 20: 77–85.

Van Liere, L., P. C. M. Boers, P. J. Den Oude, H. J. Gons, R. D. Gulati, M. Rijkeboer, A. Sinke, J-P. R. A. Sweerts, J. Bril & L. Postma, 1986b. Water Quality research Loosdrecht lakes, studying and modelling the impact of water management measures on the internal nutrient cycle. Final report ENV 839-NL (N) to the European Community, 155 pp.

Zevenboom, W., 1986. Ecophysiology of nutrient uptake, photosynthesis and growth. Can. Bull. Fish. aquat. Sci. 214: 391–422.

Hydrobiologia **191**: 97–103, 1990.
P. Biró and J. F. Talling (eds), Trophic Relationships in Inland Waters.
© 1990 *Kluwer Academic Publishers.*

Seasonal variability of N:P ratios in eutrophic lakes

J. Barica
Lakes Research Branch, National Water Research Institute, Burlington, Ontario, Canada, L7R 4A6

Key words: nutrient ratios, lakes, seasonal change, nitrogen fixation

Abstract

The ratios of different forms of nitrogen to phosphorus (particulate, total, total dissolved, and dissolved inorganic N : P) from four eutrophic lakes are presented to assess their variability during two consecutive growing seasons. Particulate and total N : P ratios showed the lowest amplitude of fluctuations, while the total inorganic N : P ratios showed the highest. All N : P forms demonstrated substantial variations and seasonal pattern over the growing period April–October. It was the short-lasting spring minima of the N : P ratios, ranging from less than 1 : 1 to 6 : 1, which triggered the onset of nitrogen-fixing cyanophyte blooms. Seasonal mean values were as high as 20 : 1 to 30 : 1 and misleading in assessing nitrogen limitation. The nitrogen fixation rapidly restored the N : P values to normal levels (TN : TP of 15 : 1 and over).

Introduction

The concept of nutrient limitation for plant growth and the role of nitrogen to phosphorus ratios has been known since the beginning of modern agriculture (Liebig, 1855; Leonardson & Ripl, 1980) and introduced to limnology at its early stage. The normal elemental composition of the main building blocks of plankton biomass, and generally accepted proportions of the key nutrients in the algal cells, were analyzed by Lund (1970) and Vollenweider (1985).

Forsberg (1979) presented a generalized diagram of trophic states in relation to N : P ratios; with low N : P ratios (< 10 : 1) being characteristic of nitrogen-limited eutrophic systems. Smith (1983) provided a summary of data from 17 lakes showing that a N : P ratio of 29 : 1 (as total N and P) is a borderline below which lakes favour development of N-fixing cyanophytes, whereas

higher ratios disfavour it. Schindler (1977), Leonardson & Ripl (1980) and Barica *et al.* (1980) demonstrated experimentally that manipulation of the N : P ratios by addition of N (as nitrate or ammonia, or both) leads to substantial changes in phytoplankton composition, with low ratios favouring the cyanophytes, and the higher ratios favouring chlorophytes or non-fixing blue-greens.

However, there has been no uniformity in the forms used to calculate N : P ratios in the literature. They vary from cellular ratios, considered in Vollenweider (1985), to total, total dissolved and total inorganic N : P in the water, the latter used in experimental enrichment studies. Average data for a growing season (Smith, 1983) or seasonal averages (late winter: Niemi, 1979; winter and summer: Allan & Kenney, 1978) were mostly used, assuming negligible variation of the ratios within the growing season.

This paper compares different forms of expression of the N : P ratios and demonstrates significant variability within a growing season, re-analyzing experimental data from four eutrophic lakes in Western Canada (Barica, 1975).

Study area and methods

Definition of terms and methodology

The various forms of nitrogen and phosphorus and their ratios presented in this paper are defined as follows:

Particulate nitrogen (Part. N): analyzed on a portion of a GF/C filter using Carlo Erba elemental analyzer for C and N;

Particulate phosphorus (Part. P.): analyzed by ignition and acid digestion on a portion of the GF/C filter.

Part. N : Part. P = ratio of the concentrations by weight.

Dissolved inorganic nitrogen (DIN); The sum of ammonia, nitrite and nitrate nitrogen ($\mu g \, l^{-1}$); all measured in filtrate, determined colorimetrically using phenol-hypochlorite method for ammonia and sulphanilamide-naphthylethylene diamine method for nitrite. Nitrate was determined using automated procedures (Cd reduction).

Dissolved inorganic phosphorus (DIP) = soluble reactive phosphorus (SRP), determined in a filtrate using the acid molybdate-ascorbic acid method.

DIN : DIP = ratio of the concentrations by weight

Total dissolved nitrogen (TDN) and total dissolved phosphorus (TDP): analyzed as nitrate and SRP respectively on part of a filtrate irradiated with UV radiation.

TDN = DIN + dissolved organic N
TDP = SRP + dissolved organic P
TDN : TDP = ratio of the concentrations by weight

Total nitrogen (TN) = Part. N + TDN
Total phosphorus (TP) = Part. P + TDP

Selected study-lakes

Four small prairie lakes of varying levels of eutrophication and algal biomass were selected as being characteristic of a total of about 100 lakes studied during 1971–1983 near Erickson, southwestern Manitoba, in central Canada (Barica, 1975). The lakes were monitored at short intervals over a period of several years covering all seasons, but particulate C, N and P values were obtained only during 1971–1974 and later discontinued. The study period for this paper is therefore limited to those years. The lakes are presented according to the advancement of natural eutrophication in the following order:

Lake 885

A true hypereutrophic lake, with heavy blooms of *Aphanizomenon flos-aquae* each year, and their regular massive collapses and resulting fish kills between July and August, with additional partial collapses of phytoplankton and after. These include preceding blooms of chrysophytes (*Chrysochromulina parva*) in May. Maximum chlorophyll *a* levels for the study period were 211–283 $\mu g \, l^{-1}$.

Lake 154

A hypereutrophic lake also, but collapses of *Aphanizomenon* blooms do not occur regularly each year. Partial collapses of *Microcystis* and *Merismopedia* blooms in combination with diatoms (*Nitzschia*, *Synedra* and *Cyclotella*) were recorded (Kling, 1975). Maximum chlorophyll *a* levels for the study period were 141–280 $\mu g \, l^{-1}$.

Lake 302

A eutrophic lake, without total collapses of algal blooms. The predominant algae were species of *Microcystis*, *Anabaena*, *Cryptomonas*, *Rhodomonas*, and spring blooms of diatoms. The blooms die-off gradually, without causing a total oxygen depletion. Maximum chlorophyll *a* levels for the study period were 108–110 $\mu g \, l^{-1}$.

LAKE 885

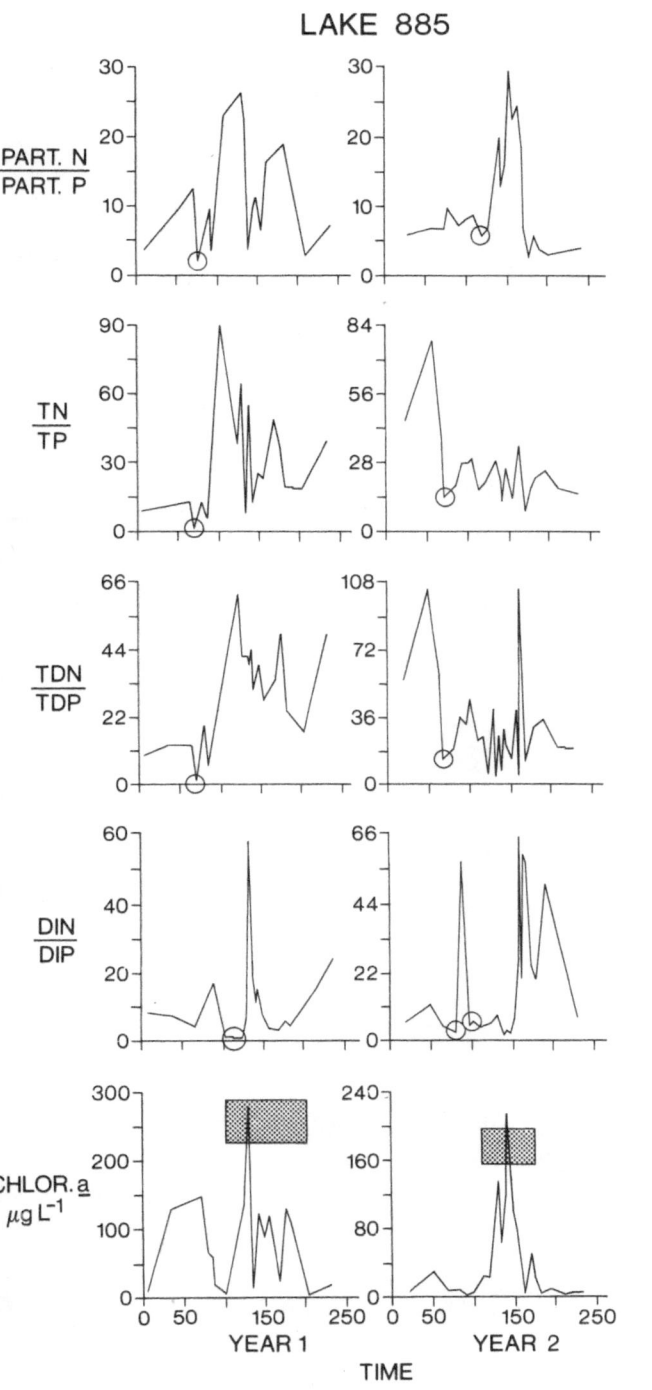

LAKE 154

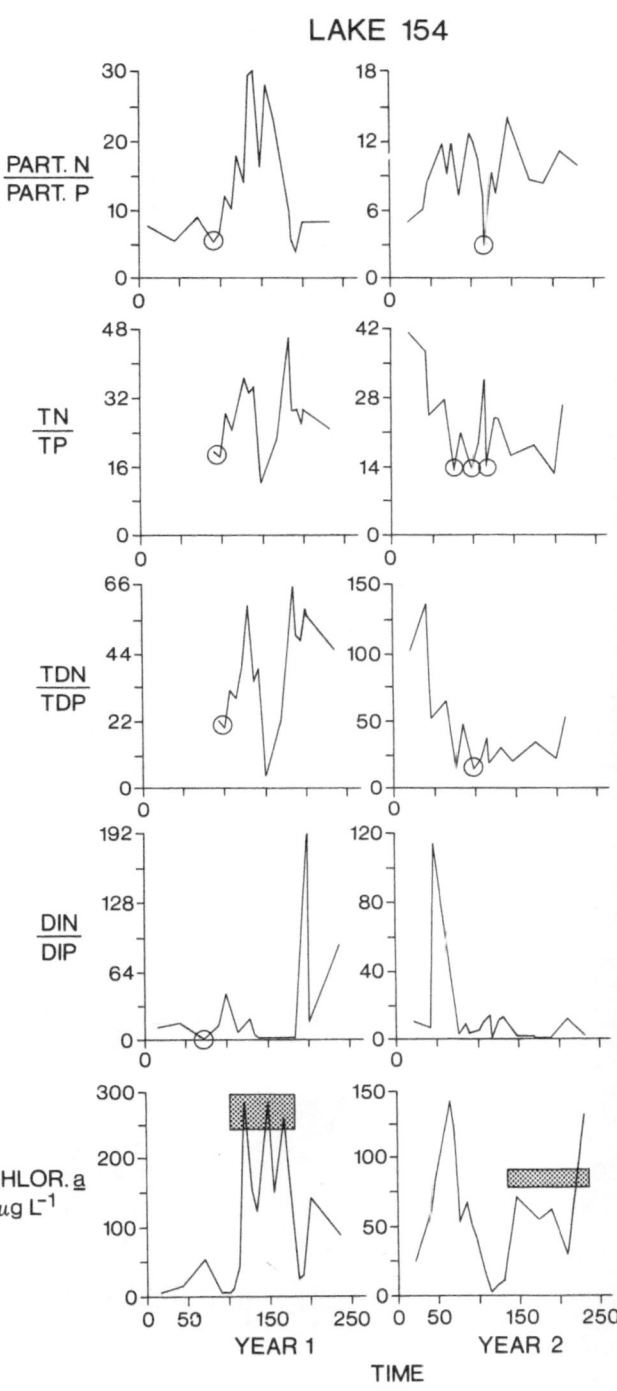

Fig. 1. Lake 885: variations of different forms of N : P ratios and chlorophyll *a* during two consecutive growing seasons. Shaded areas indicate cyanophyte bloom periods; circles show N : P ratio minima preceding them. Time in Julian days.

Fig. 2. Lake 154: variations of different forms of N : P ratios and chlorophyll *a* during two consecutive growing seasons. Shaded areas indicate cyanophyte bloom periods; circles show N : P ratio minima preceding them. Time in Julian days.

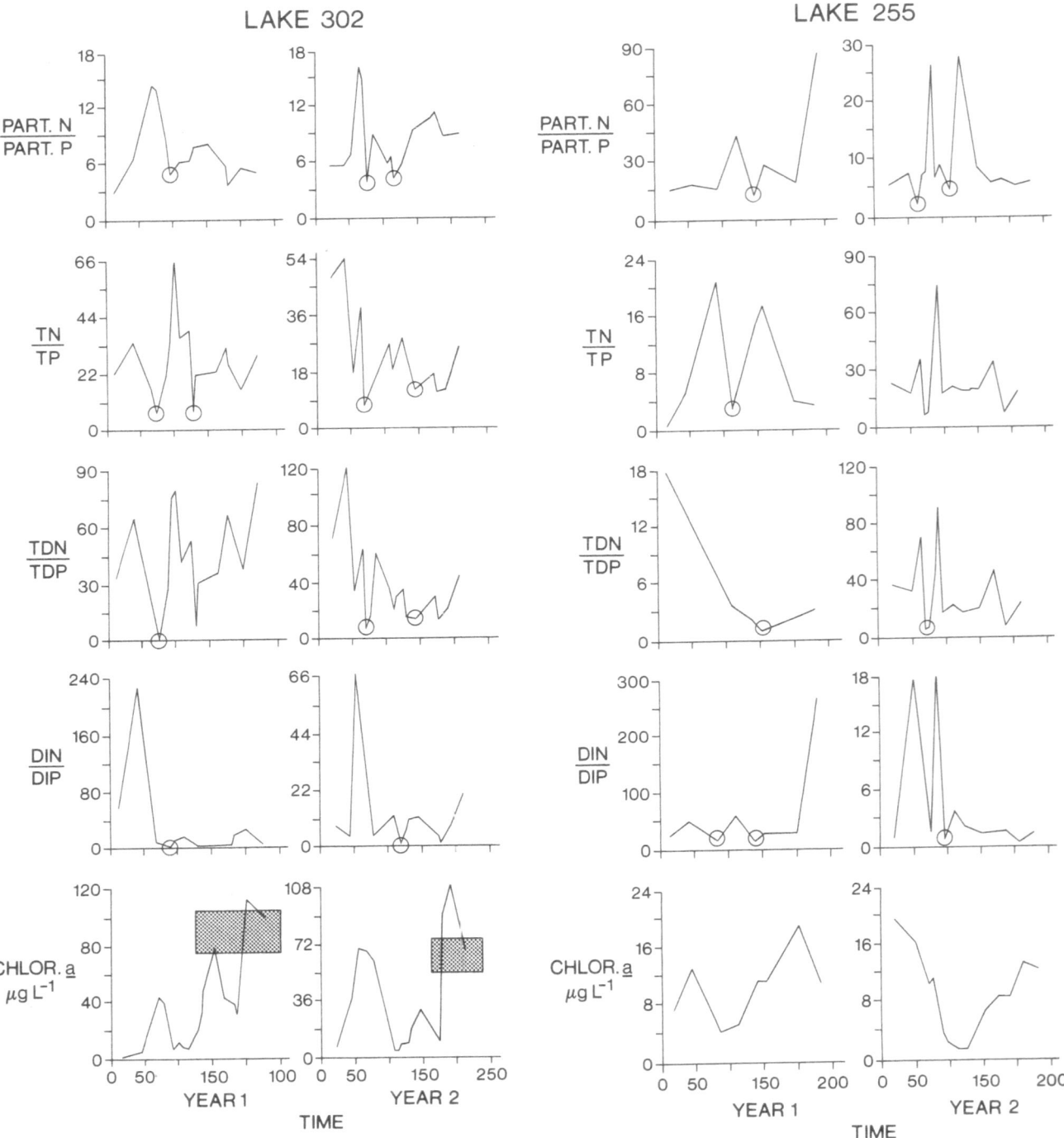

Fig. 3. Lake 302: variations of different forms of N : P ratios and chlorophyll *a* during two consecutive growing seasons. Shaded areas indicate cyanophyte bloom periods; circles show N : P ratio minima preceding them. Time in Julian days.

Fig. 4. Lake 225: variations of different forms of N : P ratios and chlorophyll *a* during two consecutive growing seasons. Circles show N : P ratio minima. Time in Julian days.

Table 1 N : P forms and chlorophyll *a* in experimental lakes.

Parameter	Maximum	Minimum	Mean	Std. Dev.	Coeff. of variation
Lake 885					
Part N/Part P					
Year 1	26.5	1.8	10.7	7.3	68.0
Year 2	30.2	3.0	12.0	8.0	66.7
TN/TP					
Year 1	89.9	1.2	27.2	22.3	82.1
Year 2	76.8	7.6	24.4	13.9	57.1
TDN/TDP					
Year 1	61.8	0.4	28.3	16.9	59.9
Year 2	102.7	2.0	32.6	28.1	86.3
DIN/DIP					
Year 1	57.9	0.6	11.4	13.2	114.9
Year 2	64.5	1.3	17.9	21.0	117.2
Chl.-*a* (μg l^{-1})					
Year 1	283.0	7.4	81.0	69.0	85.2
Year 2	211.0	2.0	47.7	58.8	123.3
Lake 154					
Part N/Part P					
Year 1	30.0	3.6	12.8	8.4	65.7
Year 2	14.0	2.8	7.6	2.7	30.7
TN/TP					
Year 1	45.7	11.7	27.3	7.9	29.2
Year 2	41.2	12.2	22.7	8.6	38.1
TDN/TDP					
Year 1	64.3	3.6	38.6	16.9	43.9
Year 2	137.0	14.2	42.5	33.9	79.7
DIN/DIP					
Year 1	192.5	1.0	32.1	51.5	160.2
Year 2	113.7	0.2	12.9	27.3	211.2
Chl.-*a* (μg l^{-1})					
Year 1	280.0	4.0	100.7	93.4	92.7
Year 2	141.0	3.0	54.1	41.2	76.1
Lake 302					
Part N/PartP					
Year 1	14.3	2.8	7.1	3.2	45.7
Year 2	16.3	3.9	8.3	3.5	42.7
TN/TP					
Year 1	64.7	6.6	26.7	14.1	53.0
Year 2	53.8	7.4	23.3	13.8	59.0
TDN/TDP					
Year 1	82.0	0.7	44.0	25.4	57.7
Year 2	120.3	6.5	36.1	28.6	79.0
DIN/DIP					
Year 1	227.8	0.2	27.0	59.8	221.0
Year 2	66.0	0.3	10.5	16.0	152.3
Chl.-*a* (μg l^{-1})					
Year 1	110.0	1.9	35.8	32.4	90.5
Year 2	108.5	3.4	40.3	33.6	1132.0

Table 1 (*Continued*)

Parameter	Maximum	Minimum	Mean	Std. Dev.	Coeff. of variation
Lake 255					
Part N/Part P					
Year 1	87.0	12.4	29.7	25.1	84.7
Year 2	29.0	2.7	9.9	7.6	76.7
TN/TP					
Year 1	20.9	0.4	8.6	7.6	87.9
Year 2	76.8	7.2	26.5	18.6	70.3
TDN/TDP					
Year 1	17.6	0.7	5.3	6.9	129.5
Year 2	93.0	7.1	33.5	24.6	73.4
DIN/DIP					
Year 1	263.7	12.1	7094.4	84.2	142.9
Year 2	18.6	0.6	4.5	6.6	146.1
Chl.-*a* (μg l^{-1})					
Year 1	19.0	4.0	10.1	4.8	47.6
Year 2	20.0	2.0	9.7	5.6	57.8

Lake 255

The least eutrophic lake of the four, with low summerkill risk and no visible algal blooms (maximum chlorophyll *a* levels for the study period were 19–20 μg l^{-1}). Phytoplankton was composed of Chrysophyceae, diatoms and cryptomonads.

Results and discussion

Variations of N : P ratios during a growing season

Time-series of the various forms of N : P ratios and chlorophyll *a* during two consecutive growing seasons (April–October of 1972 and 1973) for the four study lakes are presented in Figures 1–4. The low N : P ratios (< 5 : 1), referred to by Allan & Kenney (1979), occur for only a short period of time (a few weeks) usually in the spring or fall. The spring minima in lakes 885 and 154 preceded development of *Aphanizomenon flos-aquae* blooms and may be assumed to trigger the onset of N-fixing species. Then, the lake system seems to adjust its N : P balance to come close to 'normal' ratios, which then persist for the duration of the bloom, and respond to its physiological state

(note some decreases in ratio during the bloom collapse period). Seasonal mean values of N : P ratios, often used in the literature, are therefore misleading. It is the short-lasting minima of the N : P ratios which characterize the nutrient limitation status of a lake. Considering only the seasonal mean values, all four study lakes – which are distinctly eutrophic to hypereutrophic – would fall below these categories if using the boundaries of either Forsberg (1979) or Smith (1983) (seasonal mean values of the TN : TP, TDN : TDP, and DIN : DIP ratios were > 20 : 1 in most cases). Only Part. N : Part. P ratios show values expected for eutrophic lakes (< 10 : 1). At the same time, most of the N : P minimum values in the study-lakes were in the range of < 1 : 1 to 6 : 1.

Differences between the various forms of N : P ratios

Statistical evaluation of the experimental data, particularly for their minimum, maximum and mean values, standard deviations and coefficients of variability (Neville & Kennedy, 1964), are presented in Table 1. Particulate and total N : P ratios appear to be the least fluctuating forms, and are closest to the true cellular ratios. TDN : TDP ratios fluctuate more, and the most extreme fluctuations are found in DIN : DIP values. This is understandable, since the extreme fluctuations of water quality are characteristic of hypereutrophic lakes as a result of nutrient oscillations and frequent pulses of regenerated nutrients associated with collapse of the *Aphanizomenon flos-aqua* blooms (Barica, 1974; Barica *et al.*, 1980). DIN : DIP ratios may be useful for N or P enrichment experiments, but their validity for an overall characterization of lakes is questionable. Particulate N : P ratios are more elaborate to obtain, so total N : P ratios appear to be the most practicable. The cellular component in TN : TP ratios provides a stabilizing factor, as the response to instantaneous changes in chemistry of a water column is buffered by physiological processes in the algal cells and creates a considerable time lag. On the other hand, the DIN : DIP ratios reflect the momentary situation in the lake and therefore fluctuate erratically. Therefore, they are not a reliable indicator of the nutrient limitation.

The highest amplitudes of fluctuations in all N : P forms, and the highest coefficients of variation, were found in Lake 255 which does not show nitrogen-fixing blooms.

Effect of nitrogen fixation and build-up

Onset of nitrogen-fixing species provides the necessary supply of nitrogen needed to compensate for its deficit. Brownlee & Murphy (1983) measured the total input of fixed nitrogen to Lake 885 during one month of fully developed *Aphanizomenon* bloom to be 2.73 kg ha^{-1}. However, the build-up of the nitrogen in the system was not permanent; on the die-off of the *Aphanizomenon* bloom, the N : P ratios quickly fell back to spring low levels. This phenomenon was also observed during N-addition experiments (Barica *et al.*, 1980) where the accumulated nitrogen was soon removed from the system, presumably through denitrification. The eutrophic lakes cannot store the added nitrogen for an extended period and cannot carry it over to the next growing season.

Acknowledgements

I wish to acknowledge with thanks computer-programming work of Mike Fellowes; technical assistance in data processing of Jim Gibson and Cheriene Vieira; and critical comments on the manuscript by Hugh Dobson, Colin Gray and R.A. Vollenweider.

References

Allan, R. J. & B. C. Kennedy, 1978. Rehabilitation of eutrophic prairie lakes in Canada. Verh. int. Ver. Limnol. 20: 214–224.

Barica, J., 1974. Some observations on internal recycling, regeneration and oscillation of dissolved nitrogen and phosphorus in shallow self-contained lakes. Arch. Hydrobiol. 73: 334–360.

Barica, J., 1975. Geochemistry and nutrient regime of saline eutrophic lakes in the Erickson-Elphinstone district of southwestern Manitoba. Fish. Mar. Serv. Can., Tech. Rep. 511: 82 p.

Barica, J., H. Kling & J. Gibson, 1980. Experimental manipulation of algal bloom composition by nitrogen addition. Can. J. Fish. aquat. Sci. 37: 1175–1183.

Brownlee, B. G. & T. P. Murphy, 1983. Nitrogen fixation and phosphorus turnover in a hypertrophic prairie lake. Can. J. Fish. aquat. Sci. 40: 1853–1860.

Forsberg, C., 1979. Die physiologischen Grundlagen der Gewässer-Eutrophierung. Wasser Abwasser Forsch. 12, 2: 40–45.

Kling, H., 1975. Phytoplankton successions and species distribution in prairie ponds of the Erickson – Elphinstone district, South-western Manitoba. Fish. Mar. Serv. Can., Tech. Rep. 512: 31 p.

Leonardson, L. & W. Ripl, 1980. Control of undesirable algae and induction of algal successions in hypertrophic lake ecosystems. In: J. Barica & L. R. Mur (eds.), Hypertrophic ecosystems. Developments in Hydrobiology 2, 57–65.

Lund, J. W. G., 1970. Primary production. Proc. Soc. Water Treatment Exam. 19, 332–358.

Neville, A. M. & J. B. Kennedy, 1964. Basic statistical methods for engineers and scientists. International Textbook Comp., Scranton, Pennsylvania. 470 pp.

Niemi, A., 1979. Blue-green algal blooms and N : P ratio in the Baltic Sea. Acta Bot. Fennica 110: 57–61.

Schindler, D. W., 1977. Evolution of phosphorus limitation in lakes. Science 195: 260–262.

Smith, V. H., 1983. Low nitrogen to phosphorus ratios favor dominance by blue-green algae in lake phytoplankton. Science 221: 669–671.

Vollenweider, R. A., 1985. Elemental and biochemical composition of plankton biomass; some comments and explanations. Arch. Hydrobiol. 105: 11–29.

Hydrobiologia **191**: 105–110, 1990.
P. Bíró and J. F. Talling (eds), Trophic Relationships in Inland Waters.
© 1990 *Kluwer Academic Publishers.*

Trophic response to phosphorus in acidic and non-acidic lakes in Nova Scotia, Canada

Joseph J. Kerekes,[1] Anthony C. Blouin[2] & Stephen T. Beauchamp[1]
[1] *Canadian Wildlife Service, Environment Canada, c/o Biology Department, Dalhousie University, Halifax, Nova Scotia, Canada, B3M 2X9*; [2] *Water Resources Division, Department of Environment and Lands, Newfoundland and Labrador, Post Office Box 8700, St. John's, Newfoundland, A1B 4J6*

Key words: phosphorus, trophic response, phytoplankton, acid lakes

Abstract

Twenty lakes (oligotrophic or eutrophic) with a wide range of acidity (pH 3.5 to 7.6) show a typical trophic response to total phosphorus with respect to algal biomass (OECD relationship), irrespective of their acidity. Zooplankton abundance is also related to total phosphorus, except for an outlier lake which is very acidic and eutrophic. This lake, however, has an abundant benthic and pelagic insect fauna and shows an overall 'normal' trophic response to phosphorus. In three lakes where planktonic primary production at light optimum (P_{max}) was measured, it was highest in the most acid lake (pH 4.4) which has the largest total phosphorus concentration.

Introduction

In response to concern regarding the possible environmental damage caused by the atmospheric deposition of acidifying substances, there is a concentrated scientific effort to study and understand this potential problem (i.e. Drablos & Tollan, eds., 1980; Martin, 1986). Nova Scotia has an abundance of dilute, poorly buffered, acidified waters which are well suited for such study, along with other waters which are also acidic because of drainage from organic soils and disturbed pyritic rocks (Kerekes *et al.*, 1986). Studies conducted since 1979 include the characterization of precipitation and the water quality and hydrology of selected lakes and their biota. Biological studies have included investigation of the trophic status of the lakes and the effect of total phosphorus and

pH on trophic response. This paper reports on the apparent trophic response of lake zooplankton and planktonic primary production to phosphorus and acidity.

Methods

Shallow, unstratified lakes were sampled near the surface, whereas stratified lakes were sampled at three depths (surface, mid-depth and near-bottom) using a 2-l Van Dorn bottle. Phytoplankton was also sampled with the same Van Dorn bottle at three depths, and zooplankton with a 32-l Schindler-Patalas plankton trap fitted with a 35-micron mesh net. All sampling was conducted at the deepest point of each lake.

For phytoplankton 350 ml of lake water were

treated with 150 ml of Transeau's preservative (6 parts water, 3 parts 100 percent ethanol, 1 part 40 percent formaldehyde). Samples were counted using the Utermöhl sedimentation technique with a Zeiss inverted microscope at a magnification of 265 ×. Replicate 25-ml subsamples were settled and at least two transects of each settling chamber were enumerated. Phytoplankters were counted as the number of cells of each species and were recorded as cells per liter (no large colonial species were encountered). For the common species, 20 individuals were measured for three linear dimensions (length, width, height) and cell volume was calculated. Effective spherical diameters (ESD) were calculated for the species' volumes by approximating shape to a similar geometrical figure.

Zooplankton was enumerated using a Wild dissecting microscope at a magnification of 50–60 ×. Two or three subsamples were counted.

Chlorophyll *a* was determined by a fluorometric method, and total phosphorus concentration was measured by the persulfate oxidation method (Kerekes, 1975).

pH was determined from samples stored in air-free polyethylene bottles at 25 C with a Radiometer pH meter 29 calibrated with standard buffer solutions.

Planktonic primary production was measured, using a modification of the ^{14}C-technique of Steemann-Nielsen, in a light incubator as described by Fee (1973, 1980).

Sampling period

Most lakes were sampled (for zooplankton, chlorophyll *a* and algal biomass) five times, at approximately six week intervals between May and November 1983. Lakes in the Halifax-Sackville area (Lakes 11 to 14, Fig. 1) were sampled on four occasions and Layton's Lake (Lake 15) was sampled twice.

Beaverskin, Pebbleloggitch and Kejimkujik Lakes were normally sampled for planktonic primary production twice each month during the growing season, and once a month while ice-covered during the period May 1979 to April 1981.

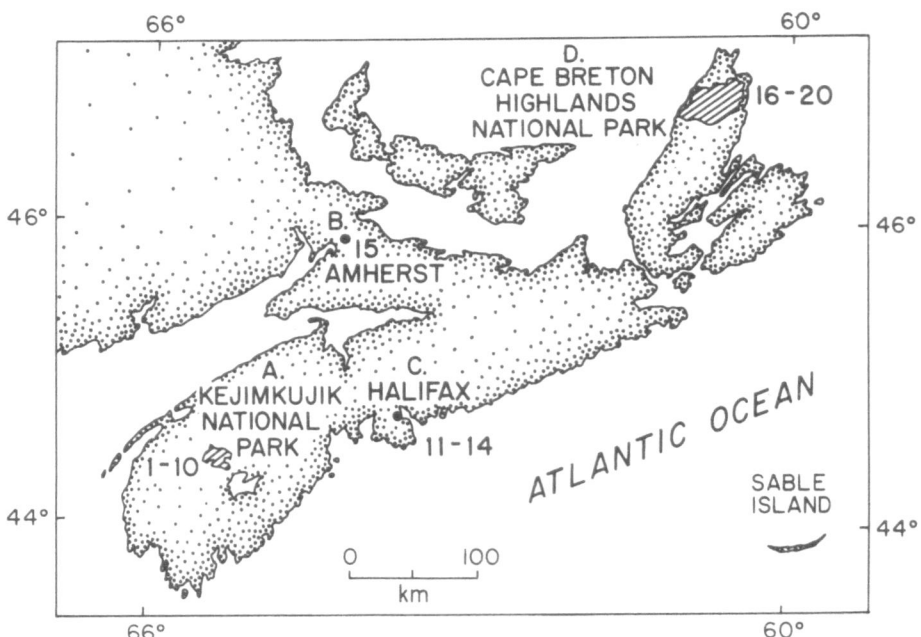

Fig. 1. The location of study lakes in Nova Scotia sampled between May and November 1983. The lakes' names, with identification numbers and notes, are given in Table 1.

The study-lakes

The twenty lakes were selected from across Nova Scotia to provide a range of pH and nutrient conditions (Fig. 1, Table 1). Ten lakes are located in Kejimkujik National Park (Lakes 1–10), five in Cape Breton Highlands National Park (Lakes 16–20), four in the Halifax-Sackville area (Lakes 11–14) and one in the Amherst Point National Wildlife Area (Lake 15). Background morphometric, water quality and limnological data for these lakes may be found in Kerekes (1975, 1983), Kerekes *et al.* (1982, 1984, 1986), Howell & Kerekes (1982), Clifford (1984), and MAPC (1972).

Kejimkujik, Grafton and McGinty Lakes lie on a sandstone-shale conglomerate bedrock originating in the late Devonian/early Carboniferous period (Roland, 1982). Big Dam East and Big Dam West Lakes occur north of Kejimkujik Lake

on a border between the above bedrock and a shale-limestone area originating in the Cambrian period. Pebbleloggitch, Beaverskin, Peskawa, Mountain and Puzzle Lakes are on or near the border of a Carboniferous-Devonian granite/- diorite bedrock and a Cambrian shale-limestone bedrock southwest of Kejimkujik Lake. The lakes of the Halifax area (Kearney, Little Springfield, Drain & Lacey Mill) are situated in the Southern Upland Coastal geological region on a border zone between sandstone-shale conglomerate bedrock (Devonian/Carboniferous) and shale-limestone bedrock (Cambrian). Freshwater, Cann's and Warren lakes are located on Carboniferous granite/diorite bedrock; Freshwater Lake is on the coast, while Cann's and Warren Lakes are at higher altitude. French Lake is near the maximum altitude area of the plateau (430 m) in an area of gneiss/schist bedrock of the Helikian period. Presqu'ile Lake is a coastal lake on an

Table 1. Physical and chemical data for 20 study-lakes (Mean values taken over all dates and depths).

Lake	Code	pH	TP	Zoopl.	Algae	Cell. Vol.	Chl. *a*	Colour	Al	$\bar{z}$	Surface area
1) Big Dam East	A	6.2	7.7	74.5	1630	1.55	0.77	7.0	0.09	2.3	45.5
2) Big Dam West	B	5.0	10.0	103	1200	1.50	1.39	108.0	0.27	2.5	105.0
3) Kejimkujik	J	5.0	10.1	86	774	1.64	1.00	80.3	0.16	4.4	24400.0
4) Grafton	H	5.9	12.9	84.6	1830	1.08	1.19	30.7	0.07	2.8	270.0
5) McGinty	M	6.2	19.5	290	768	3.31	2.46	79.0	0.09	1.4	4.4
6) Puzzle	R	5.2	8.4	117	1730	0.55	0.92	17.0	0.04	2.7	33.7
7) Mountain	N	4.9	7.1	53.7	2180	0.81	0.55	9.5	0.08	4.3	136.0
8) Beaverskin	C	5.4	8.9	215	27400	1.30	1.01	4.0	0.03	2.2	39.5
9) Pebbleloggitch	O	4.4	10.6	198	364	0.74	1.27	118.0	0.21	1.4	33.4
10) Peskawa	P	4.8	10.8	100	396	0.74	0.80	64.0	0.22	3.2	388.0
11) Kearney	I	6.1	8.4	21.2	849	0.21	0.39	5.4	0.13	9.2	63.9
12) L. Springfield	S	3.5	6.1	26	665	0.85	0.49	3.8	3.62	4.0	13.7
13) Drain	E	4.2	38.5	133	5060	15.5	10.3	10.8	0.84	0.6	16.3
14) Lacey Mill	K	4.6	6.0	72.9	543	0.70	0.23	3.8	0.35	1.5	16.0
15) Laytons	L	7.5	45.7	1540	2810	10.1	3.09	6.3	0.10	2.1	11.3
16) Freshwater	G	6.8	9.0	155	624	2.13	1.11	4.3	0.02	6.5	42.2
17) Canns	D	6.2	7.0	117	1490	2.13	0.60	6.7	0.11	2.0	10.4
18) Warren	T	5.7	10.7	36.1	160	0.20	0.85	74.7	0.23	15.9	89.8
19) French	F	5.5	10.0	96.6	589	1.48	0.46	43.3	0.14	1.0	7.0
20) Presqu'ile	Q	7.6	14.7	335	5080	1.87	2.76	5.0	0.03	2.1	4.4

Key: pH – pH (units); TP – Total phosphorus (mg m^{-3}); Zoopl. – Total zooplankton (10^3 indiv. m^{-3}); Algae – Algal abundance (10^3 indiv. m^{-3}); Cell Vol. – Algal cell volume (mm^3) l^{-1}); Chlor *a*. – Chlorophyll *a* (mg m^{-3}); Colour – Hazen units; Al – Aluminum (mg l^{-1}); $\bar{z}$ – Mean depth (m); Area – surface area (hectares); Code – Identifying code used on graphs ‖ signifies two coincident observations on Figures 3, 4, and 5.

108

early Carboniferous sandstone-shale conglomerate bedrock. Layton's Lake, near Amherst in northern Nova Scotia, is in a sandstone-shale-coal formation of the late Carboniferous period.

The majority of the lakes (nos 3, 4, 11 to 14, 17 to 19) in Kejimkujik and Cape Breton Highlands National Parks are pristine and free from direct human influence. The catchments of four lakes, Kejimkujik, Grafton, Freshwater and Presqu'ile, receive a small amount of road salt during the winter. Kejimkujik, Grafton and Freshwater Lakes have some rural housing on their catchments (Kerekes, 1983; Kerekes et al., 1982). The eight colored lakes (nos 2 to 5, 9, 10, 18, 19 color > 30 Hazen units) receive organic-rich drainage from peat bogs. These lakes have a somewhat larger concentration of total phosphorus, and their pH is as much as one unit lower than that of clearwater lakes lying on a similar geological formation (Kerekes et al., 1982, 1986).

Kearney Lake has some housing on its catchment and a paved road runs along its length. Little Springfield Lake has been influenced by construction that exposed pyritic bedrock, causing very acidic drainage (water pH 3.5) and large aluminum concentrations (3.6 mg l^{-1}). Little Springfield Lake drains into Drain Lake, which also receives sewage from a trailer park (Kerekes et al., 1984). Lacey Mill also receives some acidic drainage from disturbed pyritic rock. Layton's Lake is meromictic and subject to internal loading of phosphorus (Howell & Kerekes, 1982).

The study-lakes vary greatly in size from 4.4 ha to 24.4 km^2 (Table 1). The mean depths range from 0.6 to 15.9 m. The majority of the lakes are shallow ($\bar{z} < 3.0$ m) and they do not stratify permanently during summer.

Results and discussion

The limnological data are summarized in Table 1. The lakes range in pH from 3.5 to 7.6. The majority of lakes are oligotrophic, one (Presqu'ile) is mesotrophic and two (Drain, Layton's) are eutrophic.

There is an insignificant correlation between pH and total phosphorus ($r^2 = 0.08$), chlorophyll a ($r^2 = 0.00$), algal cell volume ($r^2 = 0.00$), and zooplankton abundance ($r^2 = 0.30$). The data of the relationship for mean chlorophyll a and mean total phosphorus concentration (Fig. 2) plot within the 95% confidence interval of single values on the log-log plot developed for these variables in the 'OECD lakes' by Vollenweider & Kerekes (1981a, b). This suggests that these lakes show a typical trophic response with respect to chlorophyll a, regardless of their acidity. The very acidic (pH 4.2) and eutrophic Drain Lake plots above the OECD regression line, but within the 95% confidence interval of single values; this is a typical response of a lake that receives a large, biologically highly available, anthropogenic nutrient load (Kerekes et al., 1984). The other eutrophic lake (Layton's, pH 7.5) plots well below the OECD line which is typical for a shallow lake that supports dense, extensive growth of macrophytes (Howell & Kerekes, 1987). There is also a strong relationship between algal cell volume and

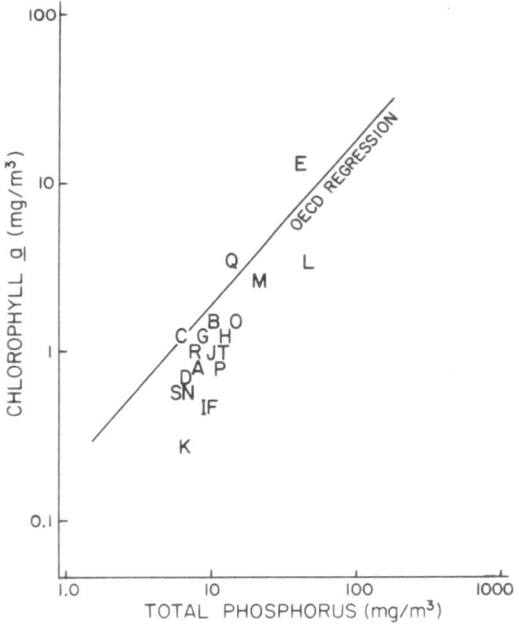

Fig. 2. Mean growing-season concentration of chlorophyll a versus mean in lake concentration of total phosphorus in relation to the OECD regression line (Vollenweider & Kerekes, 1981a, b). Capital letters representing lakes are identified in Table 1.

chlorophyll a (r^2 = 0.84, Fig. 3) and between total phosphorus and algal cell volume (r^2 = 0.83, Fig. 4).

Planktonic primary production at light optimum (P_{max}) in the growing season is 50.3, 60.0 and 66.8 mg C m^{-3} day^{-1} in Beaverskin, Kejimkujik and Pebbleloggitch Lakes respectively. It is largest in the most acidic lake (Pebbleloggitch, pH 4.4) which has the highest total phosphorus concentration among the three lakes where primary production was measured. The least acidic (pH 5.4), Beaverskin Lake with the lowest total phosphorus concentration, has the lowest P_{max} value among the three lakes. The mean P_{max} values in the three lakes are not significantly different, but they suggest a positive response to phosphorus regardless of acidity. This is consistent with the findings of Kerekes (1974) in five oligotrophic lakes (pH 6.4 to 7.2), where P_{max} was significantly correlated with total phosphorus concentration.

The above relationships (total phosphorus *vs* chlorophyll a, algal cell volume, P_{max} suggest that pH does not modify the relationship of primary production and total phosphorus. This does

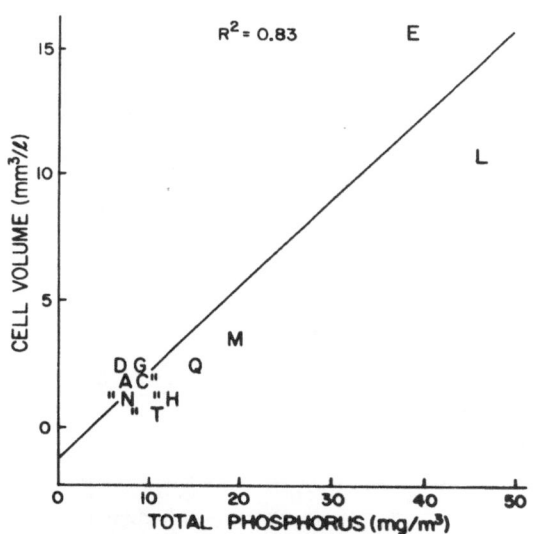

Fig. 4. Relationship between total phosphorus concentration and algal cell volume in the study-lakes.

not mean that acidity does not influence individual species or groups of species (i.e. Findlay & Kasian, 1986; Harvey & McArdle, 1986; Malley & Chang, 1986; Mills & Schindler, 1986). Kerekes & Freedman (1989) suggested that overall biological production in acidic lakes may remain high if the nutrient supply remains high. However they noted that acid-tolerant species were present in the acidic lakes.

There is also a strong positive correlation between total phosphorus and zooplankton numbers, with the eutrophic Drain Lake being an obvious outlier exhibiting a relatively sparse zooplankton (Fig. 5). This probably is not a consequence of the lake's high acidity (pH 4.2) and total aluminum concentration (0.84 mg l^{-1}), since the very acidic (pH 3.5), oligotrophic Little Springfield Lake has an even higher aluminum concentration (3.6 mg l^{-1}) and supports an apparently 'normal' zooplankton abundance with respect to total phosphorus (Table 1, Fig. 5). Other biological studies of Drain Lake indicate that it has a rich benthic fauna and an abundance of pelagic insects (Clifford, 1984; Kerekes *et al.*, 1984; Schell, 1987). Drain Lake is also noted for its dense population of breeding waterfowl (> 0.5 brood ha^{-1}) which depend on aquatic insects for food. There is no clear explanation for the smaller

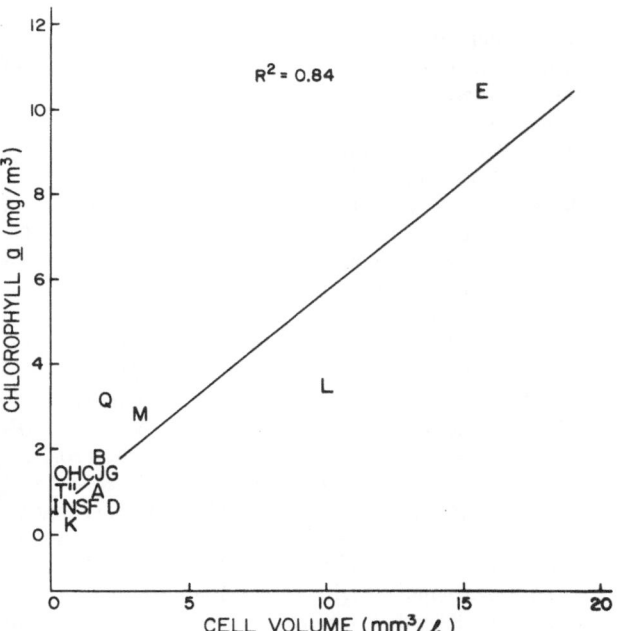

Fig. 3. Relationship between algal cell volume and chlorophyll a concentration in the study-lakes.

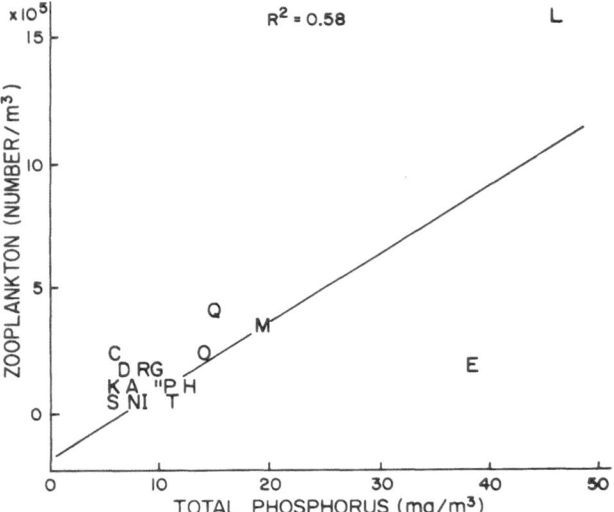

Fig. 5. Relationship between total phosphorus concentration and zooplankton number in the study-lakes.

zooplankton abundance. A possible explanation could be a negative interaction between the abundant insect population and zooplankton in this essentially fishless lake. With the exception of the zooplankton abundance, Drain Lake along with the other study lakes exhibits an 'average' quantitative trophic response to nutrients. Thus, it can be considered that the twenty lakes investigated in Nova Scotia show an 'average' trophic response to total phosphorus, and that acidity does not affect the overall relationship.

References

Clifford, P. J., 1984. Effects of acidification and eutrophication on benthic invertebrates in three Nova Scotia Lakes. B.Sc. thesis. Biology Department, Dalhousie University, Halifax, Nova Scotia. 38 pp.

Fee, E. J., 1973. A numerical model for determining integral primary production and its application to Lake Michigan. J. Fish. Res. Bd Can. 30: 1447–1468.

Fee, E. J., 1980. Important factors for estimating annual phytoplankton production in the Experimental Lakes Area. Can. J. Fish. aquat. Sci. 37: 513–522.

Findlay, D. L. & S. E. M. Kasian, 1986. Phytoplankton community responses to acidification of lake 223. Experimental Lakes Area. Northwestern Ontario. Wat. Air. Soil. Pollut. 30: 719–726.

Harvey, H. H. & J. M. McArdle, 1986. Composition of the benthos in relation to pH in the LaCloche Lakes. Wat. Air. Soil. Pollut. 30: 529–536.

Howell, G. D. & J. J. Kerekes, 1982. Ectogenic meromixis at Layton's Lake, Nova Scotia, Canada. J. Freshwat. Ecol. 1: 483–493.

Howell, G. D. & J. J. Kerekes, 1987. Primary production of two small lakes in Atlantic Canada. Proc. Nova Scotia Inst. Sci. 37: 71–88.

Kerekes, J. J., 1974. Limnological conditions in five oligotrophic lakes in Terra Nova National Park, Newfoundland. J. Fish. Res. Bd Can. 31: 555–583.

Kerekes, J. J., 1975. Phosphorus supply in undisturbed lakes in Kejimkujik National Park, Nova Scotia, Canada. Verh. int. Ver. Limnol. 19: 3490–357.

Kerekes, J. J., 1983. Predicting trophic response to phosphorus addition in Kejimkujik National Park, Nova Scotia, Canada. Proc. Nova Scotia Inst. Sci. 33: 7–18.

Kerekes, J. J., S. Beauchamp, R. Tordon & T. Pollock, 1986. Sources of sulphate and acidity in wetlands and lakes in Nova Scotia. Wat. Air. Soil. Pollut. 31: 207–214.

Kerekes, J. and B. Freedman, 1989. Characteristics of three acidic lakes in Kejimkujik National Park, Nova Scotia, Canada, Arch. Environ. Contam. Toxicol. 18: 183–200.

Kerekes, J. J., B. Freedman, G. Howell & P. Clifford, 1984. Comparison of the characteristics of an acidic eutrophic, and an acidic oligotrophic lake near Halifax, Nova Scotia. Wat. Pollut. Res. J. Can. 19: 1–10.

Kerekes, J. J., G. Howell, S. Beauchamp & T. Pollock, 1982. The response of three lake basins to acid precipitation in central Nova Scotia between June 1979 and May 1980. Int. Revue ges. Hydrobiol. 67: 679–694.

Malley, D. F. & P. S. S. Chang, 1986. Increase in the abundance of cladocera at pH 5.1 in experimentally acidified lake 223, Experimental Lakes Area, Ontario. Wat. Air. Soil. Pollut. 309: 629–638.

Martin, H. C., ed., 1986. Acid precipitation. Proceedings of the International Symposium of Acidic Precipitation, Muskoka, Ontario, Sept. 15–20, 1985. Reprinted from Water, Air and Soil Poll. Vol. 30. D. Reidel Publishing Co. Dordrecht/Boston. 1053 p. and 1118 p.

Mills, K. H. & D. W. Schindler, 1986. Biological indicators of lake acidification. Wat. Air. Soil. Pollut. 30: 779–789.

MAPC, 1972. Water quality survey for selected metropolitan area lakes. Metropolitan Area Planning Committee. Halifax, Nova Scotia.

Roland, A. E., 1982. Geological background and physiography of Nova Scotia. Nova Scotia Inst. Sci., Halifax, N. S.

Schell, V., 1987. Preliminary survey of benthic macroinvertebrates in selected lakes of the Kejimkujik Area in: Workshop Proc. Kejimkujik Calibrated Catchments Program. LRTAP Liaison Office. Downsview, Ontario 109–115.

Vollenweider, R. A. & J. J. Kerekes, 1981a. OECD cooperative program on monitoring of inland waters (eutrophication control). Synthesis Report. OECD Secretariat. Environmental Directorate. Paris, France. 289 p.

Vollenweider, R. A. & J. J. Kerekes, 1981b. Background and summary results of the OECD Cooperative Programme on Eutrophication. p. 25–36. In: Restoration of lakes and inland waters. Int. Symp. on Inland Waters and Lake Restoration, Sept. 8–12, 1980. Portland, Maine. EPA, Washington, D.C. EPA 440/5-81-110.

Hydrobiologia **191**: 111–122, 1990.
P. Biró and J. F. Talling (eds), Trophic Relationships in Inland Waters.
© *1990 Kluwer Academic Publishers.*

Trophic interactions among heterotrophic microplankton, nanoplankton, and bacteria in Lake Constance

Thomas Weisse
Limnological Institute, University of Konstanz, P.O. Box 5560, D-7750 Konstanz, F.R.G.

Key words: microbial loop, HNF, ciliates, dilution experiments, Lake Constance

Abstract

A considerable portion of the pelagic energy flow in Lake Constance (FRG) is channelled through a highly dynamic microbial food web. In-situ experiments using the lake water dilution technique according to Landry & Hasset (1982) revealed that grazing by heterotrophic nanoflagellates (HNF) smaller than 10 μm is the major loss factor of bacterial production. An average flagellate ingests 10 to 100 bacteria per hour. Nano- and micro-ciliates have been identified as the main predators of HNF. If no other food is used between 3 and 40 HNF are consumed per ciliate and hour. Other protozoans and small metazoans such as rotifers are of minor importance in controlling HNF population dynamics.

Clearance rates varied between 0.2 and 122.8 nl HNF^{-1} h^{-1} and between 0.2 and 53.6 μl ciliate^{-1} h^{-1}, respectively.

Ingestion and clearance rates measured for HNF and ciliates are in good agreement with results obtained by other investigators from different aquatic environments and from laboratory cultures. Both the abundance of all three major microheterotrophic categories – bacteria, HNF, and ciliates – and the grazing pressure within the microbial loop show pronounced seasonal variations.

Introduction

It has now widely been accepted that a highly dynamic 'microbial loop' (Azam *et al.*, 1983) consisting of pelagic bacteria, autotrophic pico- and nanoplankton, heterotrophic nanoflagellates, and microciliates is an integral part of the planktonic food web (for reviews, see Pomeroy, 1974, 1984; Williams, 1981; Porter *et al.*, 1985; E.B. Sherr *et al.*, 1986).

A number of recent investigations from different marine areas has shown that heterotrophic nano-flagellates (HNF) are the major consumers of free-living pelagic bacteria (summarized by Sherr & Sherr, 1984). Small aloricate ciliates (E. B.

Sherr *et al.*, 1986; Sherr *et al.*, 1989) and chloro-plast-containing chrysomonads (Bird & Kalff, 1986, 1987; Estep *et al.*, 1986) have also been implicated as potentially significant bacterial grazers. The fate of HNF and other protozoan production, however, is virtually unknown. Thus, the issue whether the microbial loop forms a 'link' or a 'sink' of carbon and energy flow along the planktonic food web cannot be answered at present (E. B. Sherr *et al.*, 1986).

In freshwater habitats our present knowledge of the microbial loop is extremely scarce. Fresh-water species of HNF and microciliates have been cultured on a bacterial diet in the laboratory (for review see Sherr & Sherr, 1984). There is also some evidence that HNF grazing may control the population dynamics of freshwater pelagic bacteria (Güde, 1986; Sanders & Porter, 1986; Bloem & Bär-Gilissen, 1989; Bloem et al., 1989). However, studies in situ on grazing of freshwater HNF are still lacking. The same holds true for the grazing impact of freshwater ciliates on bacteria and HNF.

This paper presents results of experiments in situ on growth and grazing of pelagic bacteria, HNF, and microciliates in Lake Constance (FRG). The trophic relationships among the principal members of the microbial loop will be discussed and compared with results obtained in similar marine studies.

The present investigation is part of an extended study on the carbon flow in Lake Constance within the 'Sonderforschungsbereich 248'.

Material and methods

Integral samples covering the entire water column from 0 to 6 m were taken with a 4-liter volume, 2-meter long water sampler (developed by the Limnological Institute Constance) at the deepest site of 'Überlinger See', the north western bay of Lake Constance. Samples were taken on 23 occasions between 15 September 1986 and 11 August 1987 at weekly to biweekly intervals. In January and February 1987 no sampling was conducted for technical reasons. To estimate growth rates and grazing pressure within the microbial community, experiments were conducted in situ using a dilution technique and analysis bags (modified after Landry & Hasset, 1982). Natural samples were pre-filtered through 200-μm-meshed gauze to exclude larger zooplankton, poured into a pre-rinsed 30-liter volume plastic tub and carefully mixed. Half of the water was sterilized by filtering through membrane filters of 0.2 μm pore size. The remaining water containing natural microplankton assemblages was then combined with the filtered water in ratios 1:0, 1:1, and 1:2. Each dilution mixture was dispensed in triplicate into 1-liter volume dialysis sacs (Spetrapor 2, 12 000 MW cutoff). The sacs were individually

placed into coarse synthetic fibre mesh bags, tied into a plastic frame, and the whole device connected to a surface buoy so that the dialysis sacs were suspended 3–3.5 m below the water surface. In situ incubation lasted for 24 hours. The synthetic netting reduced light intensity (measured as photons) by less than 4% of the ambient light. The time lag between sampling and the beginning of the experiments was about half an hour. At the beginning and end of experiments a subsample of 100 ml from each dialysis sac was fixed with Lugol's iodine solution and another subsample of about 25 ml with formalin (final concentration 1.5%). Bacteria and HNF concentrations were counted in formalin fixed samples using epifluorescence microscopy and DAPI staining according to Porter & Feig (1980). Depending on varying cell concentrations between 2 and 10 ml of these samples were filtered through a 0.2 μm nuclepore filters for epifluorescence counting. Ciliates were counted in 50 ml Lugol-fixed subsamples under an inverted microscope using the Utermöhl (1958) technique.

Growth and grazing coefficients were calculated according to Landry & Hasset (1982) assuming exponential population increase. Net growth rate μ is measured in each dilution mixture from the change in population density during the incubation period (t):

$$\mu = \ln \frac{N_t \cdot 1}{N_0 t} \qquad (1)$$

with N_0 and N_t as the initial and final cell concentrations at the beginning and end of incubation, respectively.

As the grazing impact is continuously reduced with increasing dilution of natural lake water, net growth rates are linearly related to the dilution factor. In a dilution series linear regression analysis between the dilution factor and the measured net growth rate yields the gross growth rate (k) as the Y-axis intercept and the grazing coefficient (g) as the negative slope of this relationship. Bacterial cell counts were converted to biomass using mean cell volumes for each season (winter, spring, clear-water phase, summer, and autumn) measured by Simon (1987). Bacterial cell volumes in the upper 6 m varied between 0.039 μm^3 in the clear-water phase (June) and 0.057 μm^3 in mid-summer. HNF were classified by their longest linear dimension into three size categories: <2 μm, 2–5 μm, and 5–10 μm. Larger HNF than 10 μm in length and ciliates were found occasionally, but occurred in numbers too low to be quantified. For each HNF size class a mean volume was calculated assuming cell shapes as rotational ellipsoids with circular cross-sections.

Ingestion and clearance rates of HNF and ciliates were calculated according to Davis & Sieburth (1984):

Ingestion rate

$$I \text{ (Bact. HNF}^{-1}\text{h}^{-1}) = \frac{g \cdot N_{\text{Bact.}}}{N_{\text{HNF}}} \qquad (2)$$

with the grazing coefficient g (in units h^{-1}) and the average concentrations of HNF (N_{HNF}) and bacteria ($N_{\text{Bact.}}$) during the experiments, and

$$\text{Clearance rate } C \text{ (nl HNF}^{-1}\text{h}^{-1}) = \frac{I}{N_{\text{Bact.}}} \qquad (3)$$

Results

Near-surface water temperatures and water transparency measured as Secchi disc readings during the period of investigation are indicated in Fig. 1. Decreasing water temperatures from September through December 1986 and increasing vertical mixing led to a deterioration of mean light conditions for phytoplankton in the upper 6 m of the water column. With decreasing phytoplankton biomass water transparency continuously increased until mid-December. However, in February 1987, when no experiments were conducted, transparency was even higher (13.7 m on 23 February). The increase in water temperature at the onset of stratification and the concomitant decrease in visibility in mid-April mark the beginning of the spring phytoplankton bloom. The phytoplankton biomass peak was reached on 27 April, when the average chlorophyll concentration in the upper 5 m was 75.5 μg chl.a l^{-1}.

The corresponding phytoplankton biovolume was 4283 mm^3 m^{-3} (Braunwarth & Tilzer, unpubl. results). The clear-water phase characterized by low phytoplankton biomass is obvious in Fig. 1 from the end of June through mid-July, when the transparency was more than 5 m. Yet, probably due to the unusually bad weather conditions the clear-water phase was less expressed than in former years when transparency increased from 1 to 10 m within a few days (Lampert, 1978; Geller, 1980).

Cell numbers of pelagic bacteria and heterotrophic nanoflagellates (HNF) are given in Fig. 2. Mean bacterial concentrations in the upper 6 m of the water column varied between 0.60 $\times$ 10^6 cells ml^{-1} (20 October 1986) and 6.52 $\times$ 10^6 cells ml^{-1} (29 June 1987). HNF abundance was lowest in April (0.54 $\times$ 10^3 cells ml^{-1} on 6 April 1987) and highest in early summer (8.14 $\times$ 10^3 cells ml^{-1} on 1 June 1987). From September through December 1986 bacteria and HNF populations varied inversely. In spring 1987 bacterial concentrations closely followed phytoplankton biomass, whereas HNF numbers started to increase with a time lag of about two weeks. A second peak of bacterial abundance was reached in early July after a previous decline of HNF abundance. This pattern of a typical predator-prey relationship is more obvious if one considers biomass instead of cell numbers (Fig. 3). Through most of the year HNF biomass is a smaller fraction, about 10 and 20%, of bacterial biomass. In late spring and early summer, however, the biovolume of HNF reaches that of free-living bacteria. The grazing pressure on pelagic bacteria is shown in Fig. 4. In autumn 1986, decreasing grazing pressure coincided with an increase of bacterial cell numbers and biomass (Figs. 2, 3). In 1987, the changes in bacterial population grazing rates roughly corresponded with varying bacterial biomass. The low grazing rates measured in June coincided with decreasing HNF populations. HNF seem to be the major bacterivores in Lake Constance, since the ingestion rate of HNF closely paralleled grazing pressure on bacteria (Fig. 5). Here and in the following calculation of HNF clearance rates I assumed that HNF were the sole

114

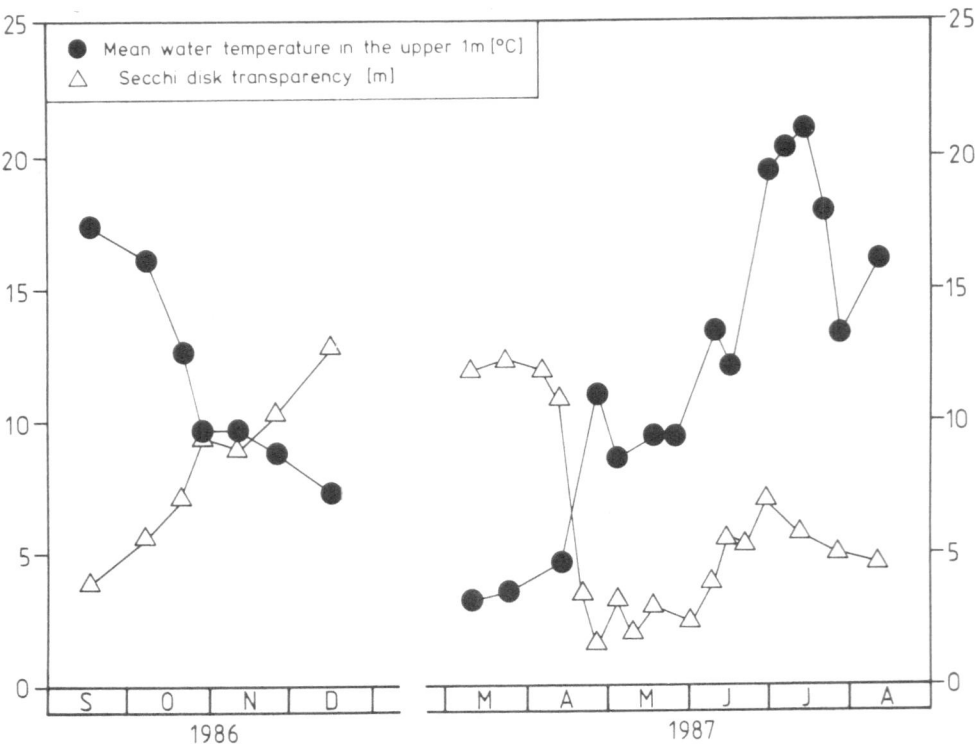

Fig. 1. Near-surface water temperature and water transparency at the sampling location.

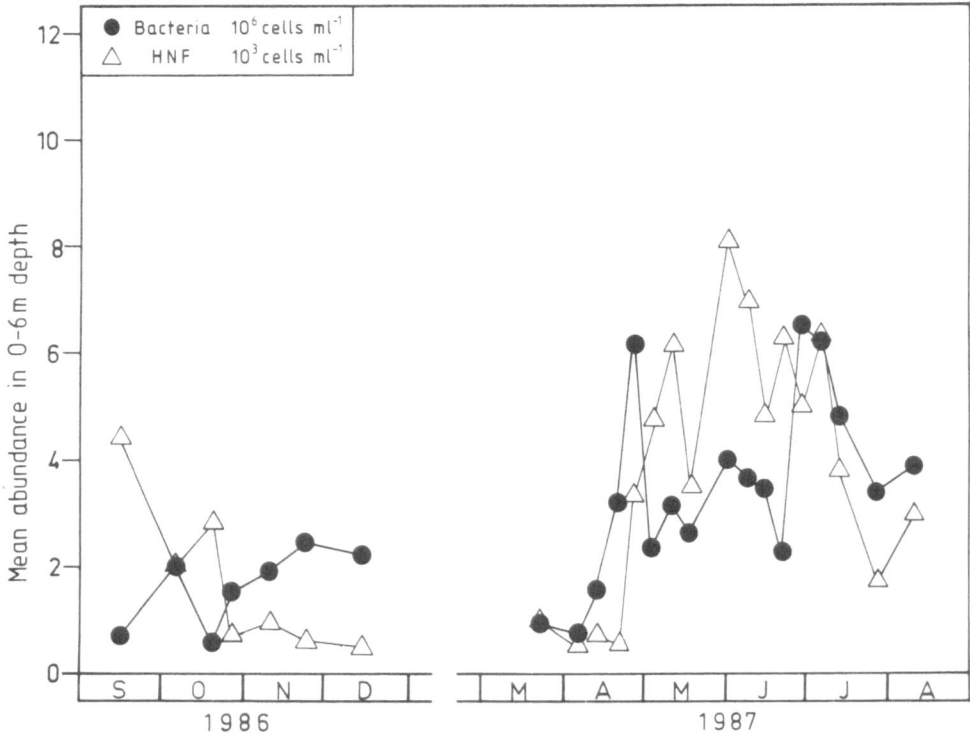

Fig. 2. Mean concentrations of free-living bacteria and heterotrophic nanoflagellates (HNF) in the upper 6 m of the water column.

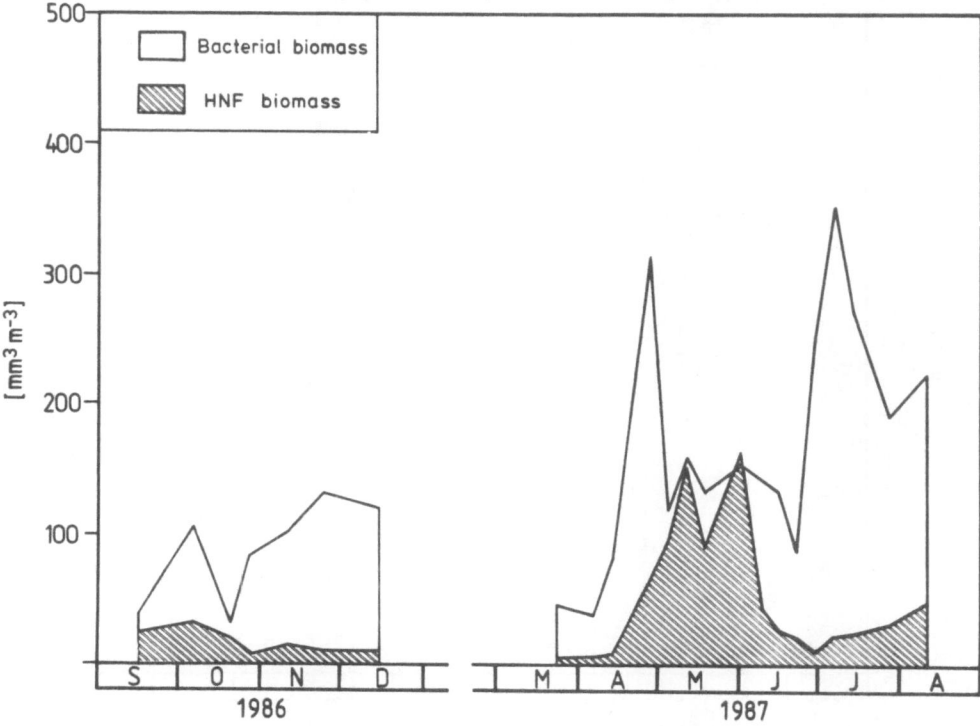

Fig. 3. Mean biomass (given as biovolume) of the pelagic bacteria and of heterotrophic nanoflagellates (HNF) in the upper 6 m of the water column.

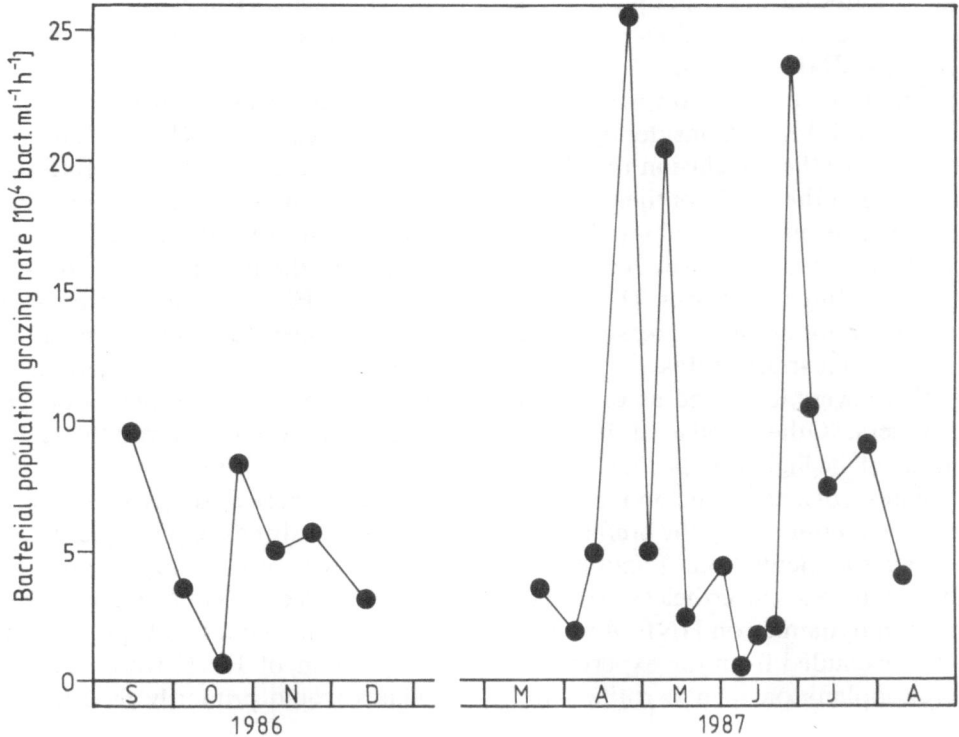

Fig. 4. Bacterial population grazing rates obtained from experiments in situ.

116

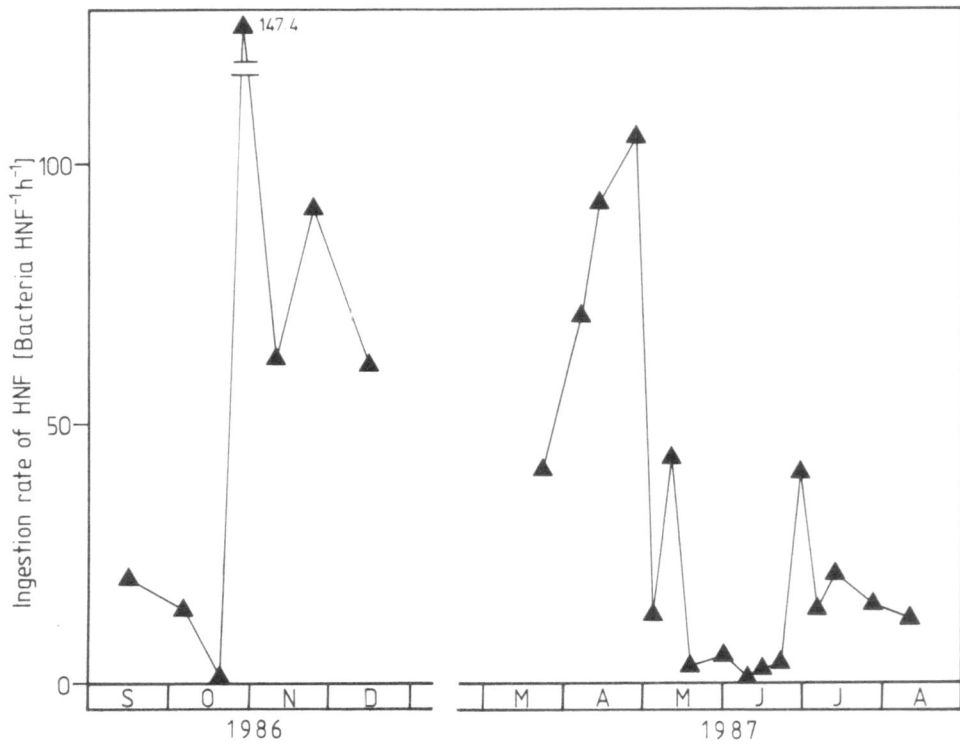

Fig. 5. Ingestion rates of heterotrophic nanoflagellates (HNF) calculated from experiments.

bacterial predators during the experiments. The results must therefore be regarded as maximum estimates. Typically HNF ingest between 10 and 100 bacteria per hour. Apparently low values were obtained in May and June. Considering HNF clearance rates (Fig. 6) the conclusion that HNF become food limited at the height of their population standing stock is even more striking. In June clearance rates were almost zero. The recovery of HNF populations in July and August (Figs. 2, 3) was paralleled by a moderate increase in per capita ingestion and clearance rates.

So far only HNF were identified as the major bacterial consumers. Other potential bacterial feeders are either of negligible importance as in the case of ciliates (discussed below) or were excluded from the experiments by the prefiltration of the water used for incubations (cladocerans and copepods). Ciliates nevertheless strongly affected population dynamics on HNF. As larger zooplankton was excluded from the experiments and other microzooplankton such as rotifers and

developmental stages of crustacea was of minor importance, the grazing pressure on HNF (Fig. 7) has to be primarily caused by ciliate feeding. The first grazing peak in April obviously prevented the rapid increase in HNF standing stocks parallel to that of bacteria, because HNF production had started to increase simultaneously with bacterial production. The abundance of nano- and microciliates in the upper 6 m of the water column is given in Fig. 8. It is remarkable that ciliates reached their first spring peak simultaneously with phytoplankton and bacteria, but two weeks earlier than HNF. The decline of ciliate concentrations from the end of April through mid-May was the period of maximum increase of HNF biomass. These findings were confirmed during a detailed study of the phytoplankton spring bloom and the response of the microbial loop conducted in 1988 (Weisse *et al.*, submitted). Yet, later in summer the course of population changes of ciliates and of HNF were positively related. If ciliates would primarily feed upon HNF during

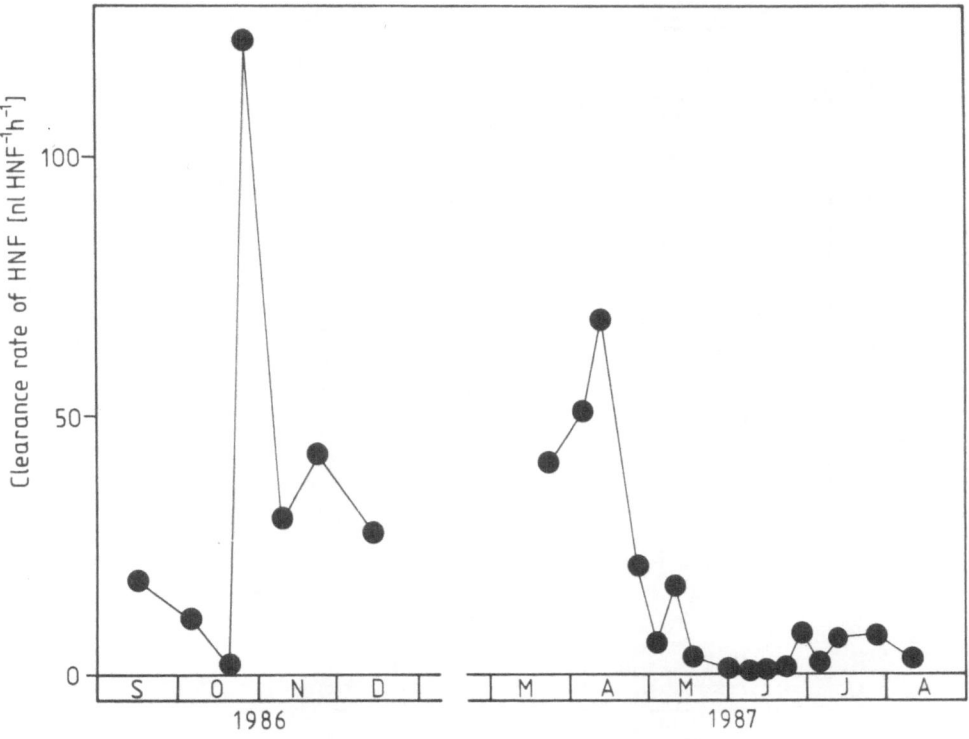

Fig. 6. Clearance rates of heterotrophic nanoflagellates (HNF) calculated from experiments.

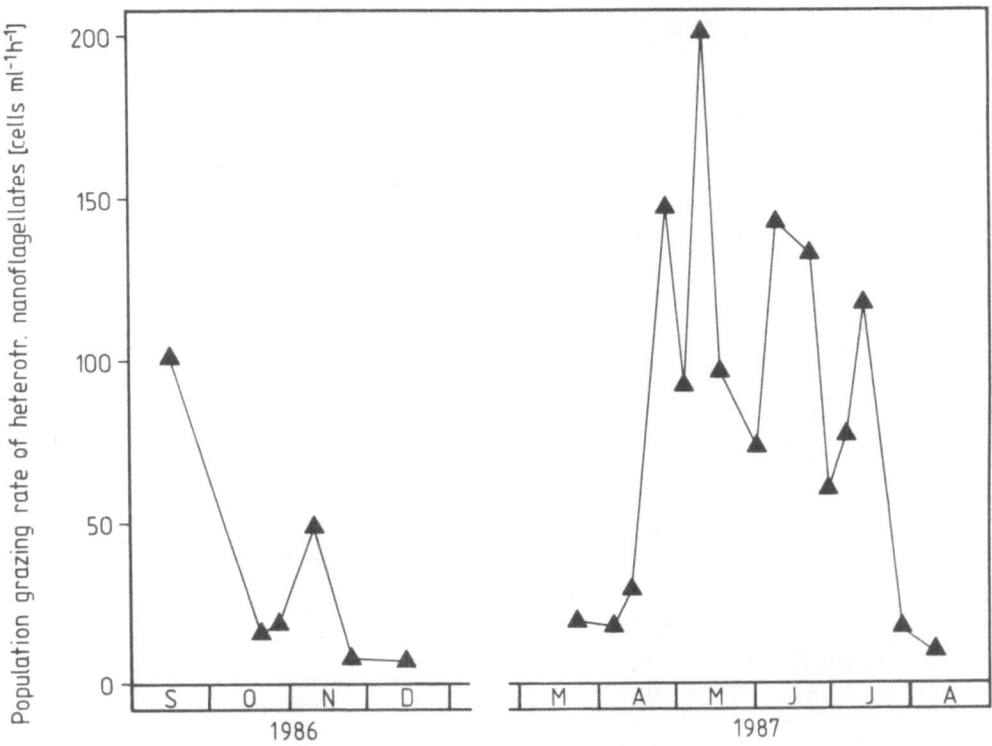

Fig. 7. Grazing pressure on heterotrophic nanoflagellates (HNF) measured from in situ experiments.

118

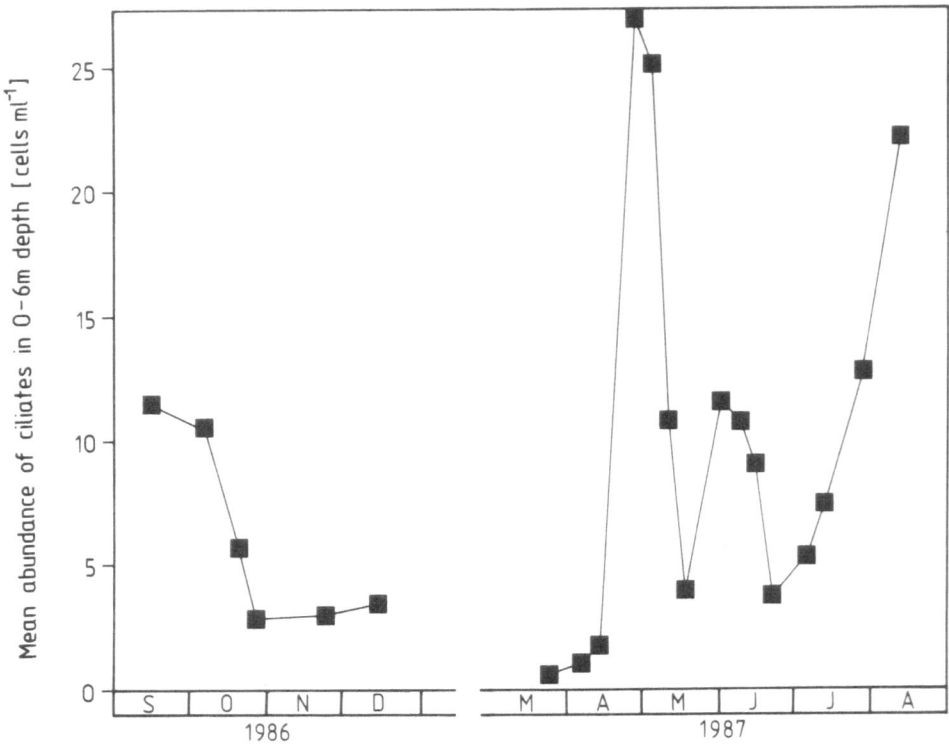

Fig. 8. Mean abundance of nano- and micro-ciliates in the upper 6 m of the water column.

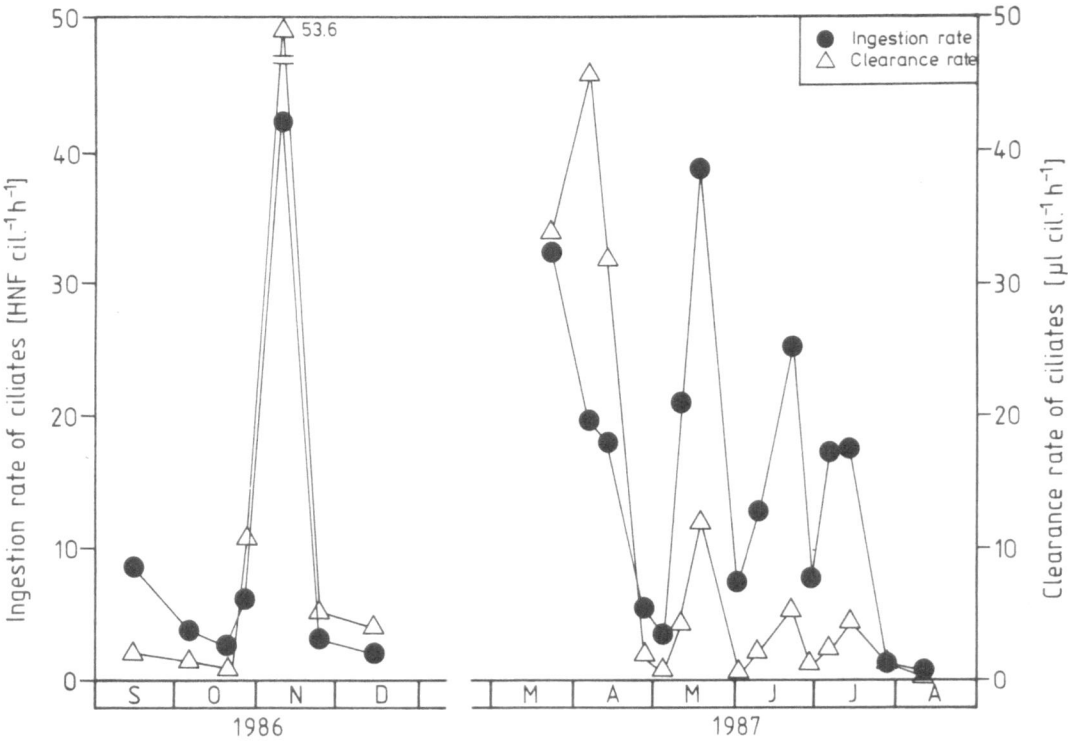

Fig. 9. Ingestion and clearance rates of ciliates calculated from in situ experiments.

this period HNF biomass should have collapsed instead of slowly recovering. Ingestion and clearance rates of ciliates reported in Fig. 9 may clarify the trophic relationships between HNF and ciliates. If HNF were the sole food item of ciliates those would have exerted heavy grazing pressure on HNF at the onset of the spring phytoplankton bloom when the build-up of high standing stocks of HNF was retarded. Thereafter, rapidly decreasing grazing pressure through ciliates enabled the mentioned increase in HNF cell number and biomass. From Fig. 9 it is also apparent that ciliates would have starved in July and August if no other food sources than HNF were available. The increasing ciliate populations after the clearwater phase suggest that ciliates switched to a phytoplankton diet. The decreasing ciliate grazing pressure was accompanied by a shift in ciliate species composition from spring through summer (Müller, 1989). As a consequence ciliate grazing pressure on HNF was reduced thus allowing a recovery of HNF populations. This interpretation is in good agreement with measured grazing rates on HNF (cf. Fig. 7). As it is not known to what extent phytoplankton was grazed by ciliates during the spring bloom, the calculated clearance and ingestion rates reported in Fig. 9 have to be regarded as minimum values.

Discussion

Results presented in this paper were obtained from experiments *in situ* using a lake water dilution technique. This technique has been first proposed by Landry & Hasset (1982), who measured the grazing impact of marine microzooplankton on phytoplankton assemblages in coastal waters off Washington, USA. Since then it has been used several times in marine studies (Landry *et al.*, 1984; Campbell & Carpenter, 1986; Verity, 1986a, b). Three basic assumptions are involved: (1) linear relationship between the dilution factor and grazing, (2) constant gross growth independent from the dilution factor, and (3) exponential cell growth. The first two assumptions are the most critical ones. Feeding is only linearly related

to the probability of encounter between predator and prey as long as neither a lower threshold concentration is transgressed nor substrate concentration is saturating. In the first case feeding ceases and calculated mean ingestion rates are underestimated; in the second case further increase in substrate concentration is not followed by increased ingestion rates. Thus, if two dilution mixtures are in the range of substrate saturation, no grazing effect is measurable at all or calculated grazing rates are gross underestimations. Consequently, the dilution mixtures have to be chosen in a way that prey concentration is neither too high nor too low. The optimal dilution ratios were tested in preliminary experiments conducted in August and September 1986 (Weisse, unpubl.). In these experiments dilutions beyond a ratio of 1 : 3 natural to filtered lake water sometimes revealed no further increase in net growth rates of bacteria compared to less diluted mixtures. Thus, the existence of a lower feeding threshold for HNF grazing on bacteria at a bacterial concentration of about 0.8×10^6 cell ml^{-1} is possible in Lake Constance in summer. Fenchel (1982) reported a feeding threshold at concentrations of 1×10^6 cell ml^{-1}. Yet, the feeding threshold itself might depend on concentration. In oligotrophic waters of the Red Sea where near-surface bacterial concentrations are in the range of $0.5-0.9 \times 10^6$ cells ml^{-1} I found some indication for the existence of a feeding threshold when bacterial abundance fell below 0.6×10^6 cells ml^{-1} (Weisse, 1989). Seasonal variation of the level of the feeding threshold might explain why grazing rates were indeed linearly related to the dilution factor even in March and early April, when bacterial concentrations were considerably below 1×10^6 cells ml^{-1} in the most diluted dialysis sacs. The second assumption, that gross growth is constant in all dilution mixtures, is best accomplished using dialysis sacs and in-situ incubation as substrates can cross the dialysis membrane (Verity, 1986a). Furthermore, the natural light climate was not strongly altered by the experimental set-up used in the present study. Dialysis sacs have meanwhile been used successfully in several studies on growth and grazing rates of natural pico-, nano-, and micro-plankton assemblages (Landry & Hasset, 1982; Landry et al., 1984; Verity, 1986a, b).

Ingestion and clearance rates reported in this study are well within the range of published data. Feeding of HNF measured elsewhere with different methods and various species usually yielded ingestion rates between 10 and 80 bacteria consumed $HNF^{-1} h^{-1}$ and clearance rates between 0.2 and 79 nl $HNF^{-1} h^{-1}$ (reviewed by B. F. Sherr et al., 1986). In a study conducted parallel to the present in 'Obersee', the main basin of Lake Constance, Jürgens & Güde (1990) measured ingestion rates ranging from 4–34 bacteria $HNF^{-1} h^{-1}$ depending on the method used. Therefore, the conclusion seems to be justified that HNF in Lake Constance can thrive solely on a bacterial diet. This interpretation is further supported by the fact that HNF production (in terms of biovolume) amounts to about 15% of bacterial production on the annual average. If we assume a gross growth efficiency of 30% (Fenchel, 1982; Sherr et al., 1983; Sherr & Sherr, 1984; Landry et al., 1984) HNF would consume about 50% of the annual bacterial production in Lake Constance. However, in terms of carbon biomass the cropping of bacterial production by grazing flagellates would be somewhat lower as the weight-specific carbon content of bacteria is obviously higher than that of protozoa (Simon & Azam, 1988). Similar consumption values have been found in comparable marine studies. Sherr et al. (1984) calculated that HNF in Georgia estuaries and offshore waters grazed 30 to 50% of daily bacterial production. Landry et al. (1984) reported that although HNF were the dominant bacterivores in shallow waters off Hawaii maintenance of relatively stable bacterial concentrations could not be attributed solely to grazing by HNF. Thus, although grazing by HNF is the main loss factor of bacterial production in Lake Constance as has already been assumed from indirect evidence by Simon (1987) and Güde (1986) other loss processes such as consumption by larger zooplankton, sedimentation, or autolytic cell disintegration are together of equal importance as HNF grazing. As stated by Güde (1986, 1988) crustacean zooplankters are not important grazers of bacteria in Lake Constance. The two dominant cladoceran species in Lake Constance, Daphnia galeata and D. hyalina have been classified as low efficiency bacterial feeders (Geller & Müller, 1981; Gophen & Geller, 1984; Güde, 1988). In the trophogenic zone ciliates are also of minor importance as bacterial consumers. The predominant ciliate species occurring in the upper 20 m of the water column are known as non-bacterivores or low-efficiency bacteria feeders. In the deeper water the fraction of high-efficiency bacteria feeders among the ciliate community might be higher (Weisse & Müller, 1989; Müller et al., 1990). As larger zooplankton was excluded from the dilution experiments, ciliates were the top predators within the microbial plankton assemblage. In previous studies it has been shown experimentally that ciliates feed upon flagellates (Sheldon et al., 1986; Verity & Villareal, 1986 and references therein). On the assumption that ciliates ingested no other food than HNF clearance rates between 0.2 and 53.6 μl $ciliate^{-1} h^{-1}$ were calculated. This agrees well with previous estimates. Ciliates clearance rates reported in literature generally vary between 0.1 and 85 μl $ciliate^{-1} h^{-1}$ (Spittler, 1973; Heinbokel, 1987a, b; Fenchel, 1980; Capriulo, 1982; Rivier et al., 1985; Verity, 1985). The measured community grazing rate on HNF amounted to 1600 $HNF ml^{-1} d^{-1}$. The calculated HNF production rate is in the same order of magnitude (Weisse, unpublished). Therefore, it is concluded that ciliates largely control HNF population dynamics in Lake Constance throughout most of the year. This might not be true in summer, when grazing pressure on HNF declined although ciliate populations increased, because ciliates used other food during that time.

Grazing and clearance rates of HNF and ciliates have been calculated on simplified assumptions. Other food sources such as autotrophic picoplankton, which occurs in typical concentrations of 10^3 to 10^5 cell ml^{-1} (Weisse, 1988; Weisse & Schweizer, 1990), nanophytoplankton, detritus, and attached bacteria have to be considered as potential food for HNF and ciliates.

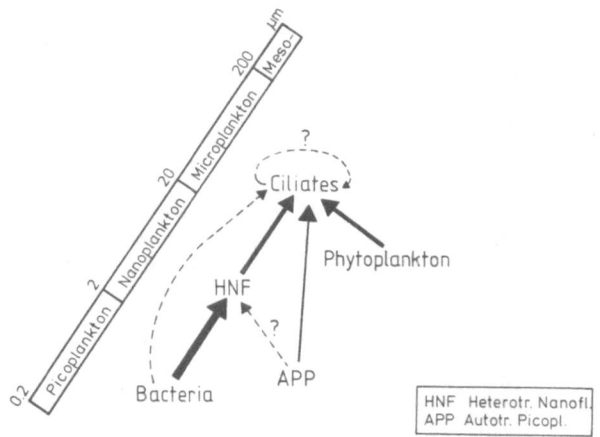

The microbial loop in Lake Constance

Fig. 10. The microbial loop in Lake Constance. Known fluxes of organic matter are indicated by solid arrows, hypothetical fluxes by dashed lines. The size range of the main categories is shown at the left margin.

Several ciliate species have been found in Lake Constance which are well known as carnivores. Some are large forms > 200 μm, but others belong to the microplankton (Weisse & Müller, 1989). It is therefore assumed that group-cannibalism also occurs among the ciliate community. However, no estimates can be made up to date as to what extent ciliates feed upon ciliates. The same holds true for the HNF assemblage.

Our present knowledge of the microbial loop in Lake Constance is summarized in Fig. 10. Within the microbial loop a linear food chain consisting of bacteria-HNF-ciliates is a potential major pathway transferring bacterial production to higher levels of the planktonic food web. However, it remains a goal for future research whether HNF and ciliate production represent a significant pathway for organic carbon transfer from bacteria to larger zooplankton, or conversely, whether the microbial loop is a sink and dead end of the energy flow within the planktonic system.

Acknowledgements

The study was supported by the Deutsche Forschungsgemeinschaft within the Sonderforschungsbereich 248 'Stoffhaushalt des Bodensees'. The technical assistance of G. Baldringer is grateful acknowledged. W. Geller and H. Müller provided helpful comments on the manuscript.

References

Azam, F., T. Fenchel, J. G. Field, J. S. Gray, L. A. Meyer-Reil & F. Thingstad, 1983. The ecological role of water-column microbes in the sea. Mar. Ecol. Prog. Ser. 10: 257–263.

Bird, D. F. & J. Kalff, 1986. Bacterial grazing by planktonic algae. Science 231: 493–495.

Bird, D. F. & J. Kalff, 1987. Algal phagotrophy: regulating factors and importance relative to photosynthesis in *Dinobryon* (Chrysophyceae). Limnol. Oceanogr. 32: 277–284.

Bloem, J. & M.-J. B. Bär-Gilissen, 1989. Bacterial activity and protozoan grazing potential in a stratified lake. Limnol. Oceanogr. 34: 297–309.

Bloem, J., F. M. Ellenbroek, M.-J. B. Bär-Gilissen & T. E. Cappenberg, 1989. Protozoan grazing and bacterial production in stratified Lake Vechten estimated with fluorescently labeled bacteria and by thymidine incorporation. Appl. Environ. Microbiol. 55: 1787–1795.

Campbell, L. & E. J. Carpenter, 1986. Estimating the grazing pressure of heterotrophic nanoplankton on *Synechococcus* spp. using the sea water dilution and selective inhibitor techniques. Mar. Ecol. Prog. Ser. 33: 121–129.

Capriulo, G. M., 1982. Feeding of field collected tintinnid micro-zooplankton on natural food. Mar. Biol. 71: 73–86.

Davis, P. G. & J. McN. Sieburth, 1984. Estuarine and oceanic microflagellate predation of actively growing bacteria: estimation by frequency of dividing-divided bacteria. Mar. Ecol. Prog. Ser. 19: 237–246.

Estep, K. W., P. G. Davis, M. D. Keller & J. McN. Sieburth, 1986. How important are algal nanoflagellates in bacterivory? Limnol. Oceanogr. 31: 646–650.

Fenchel, T., 1980. Suspension feeding in ciliated Protozoa: feeding rates and their ecological significance. Microb. Ecol. 6: 13–25.

Fenchel, T., 1982. Ecology of heterotrophic microflagellates. II. Bioenergetics and growth. Mar. Ecol. Prog. Ser. 8: 225–231.

Geller, W., 1980. Stabile Zeitmuster in der Planktonsuccession des Bodensees (Überlinger See). Verh. Ges. Ökol. 8: 373–387.

Geller, W. & H. Müller, 1981. The filtration apparatus of Cladocera: Filter mesh-sizes and their implications on food selectivity. Oecologia (Berl.) 49: 316–321.

Gophen, M. & W. Geller, 1984. Filter mesh size and food particle uptake by *Daphnia*. Oecologia (Berl.) 64: 408–412.

Güde, H., 1986. Loss processes influencing growth of planktonic bacterial populations in Lake Constance. J. Plankton Res. 8: 795–810.

122

Güde, H., 1988. Direct and indirect influences of crustacean zooplankton on bacterioplankton in Lake Constance. Hydrobiologia 159: 63–73.

Heinbokel, J. F., 1978a. Studies on the functional role of tintinnids in the southern California Bight. I. Grazing and growth rates in laboratory cultures. Mar. Biol. 47: 177–189.

Heinbokel, J. F., 1978b. Studies on the functional role of tintinnids in the southern California Bight. II. Grazing rates of field populations. Mar. Biol. 47: 191–197.

Jürgens, K. & H. Güde, 1990. Seasonal changes in the grazing impact of phagotrophic flagellates on bacteria in Lake Constance. Marine Microbial Food Webs: (in press).

Lampert, W., 1978. Climatic conditions and planktonic interactions as factors controlling the regular succession of spring algal blooms and extremely clear water in Lake Constance. Verh. int. Ver. Limnol. 20: 969–974.

Landry, M. R. & R. P. Hasset, 1982. Estimating the grazing impact of marine micro-zooplankton. Mar. Biol. 67: 283–288.

Landry, M. R., L. W. Haas & V. L. Fagerness, 1984. Dynamics of microbial plankton communities: experiments in Kaneohe Bay, Hawaii. Mar. Ecol. Prog. Ser. 16: 127–133.

Müller, H., 1989. The relative importance of different ciliate taxa in the pelagic food web of Lake Constance. Microb. Ecol. 18: (in press).

Müller, H., W. Geller & A. Schöne, 1990. Pelagic ciliates in Lake Constance: Comparison of epilimnion and hypolimnion. Verh. int. Ver. Limnol. 24: (in press).

Pomeroy, L. R., 1974. The ocean's food web. A changing paradigm. Bio Science 24: 499–504.

Pomeroy, L. R., 1984. Significance of microorganisms in carbon and energy flow in marine ecosystems. In M. J. Klug & C. A. Reddy (eds.), Current perspectives in microbial ecology. American Society for Microbiology, Washington, D.C.: 405–411.

Porter, K. G. & Y. S. Feig, 1980. The use of DAPI for identifying and counting aquatic microflora. Limnol. Oceanogr. 25: 943–948.

Porter, K. G., E. B. Sherr, B. F. Sherr, M. Pace & R. W. Sanders, 1985. Protozoa in planktonic food webs. J. Protozool. 32: 409–415.

Rivier, A., D. C. Brownlee, R. W. Sheldon & F. Rassoulzadegan, 1985. Growth of microzooplankton: a comparative study of bacterivorous zooflagellates and ciliates. Mar. Microb. Food Webs 1: 51–60.

Sanders, R. W. & K. G. Porter, 1986. Use of metabolic inhibitors to estimate protozooplankton grazing and bacterial production in a monomictic lake with an anaerobic hypolimnion. Appl. envir. Microbiol. 52: 101–107.

Sheldon, R. W., P. Nival & F. Rassoulzadegan, 1986. An experimental investigation of a flagellate-ciliate-copepod food chain with some observations relevant to the linear biomass hypothesis. Limnol. Oceanogr. 31: 184–188.

Sherr, B. F. & E. B. Sherr, 1984. Role of heterotrophic protozoa in carbon and energy flow in aquatic ecosystems. In M. J. Klug & C. A. Reddy (eds.), Current perspectives in microbial ecology. American Society for Microbiology, Washington, D.C.: 412–423.

Sherr, B. F., E. B. Sherr & T. Berman, 1983. Grazing, growth, and ammonium excretion rates of a heterotrophic microflagellate fed with four species of bacteria. Appl. envir. Microbiol. 45: 1196–1201.

Sherr, B. F., E. B. Sherr & S. Y. Newell, 1984. Abundance and productivity of heterotrophic nanoplankton in Georgia coastal waters. J. Plankton Res. 6: 195–202.

Sherr, B. F., E. B. Sherr, T. L. Andrew, R. D. Fallon & S. Y. Newell, 1986. Trophic interactions between heterotrophic Protozoa and bacterioplankton in estuarine water analyzed with selective metabolic inhibitors. Mar. Ecol. Prog. Ser. 32: 169–179.

Sherr, E. B., F. Rassoulzadegan & B. F. Sherr, 1989. Bacterivory by pelagic choreotrichous ciliates in coastal waters of the NW Mediterranean Sea. Mar. Ecol. Prog. Ser. 55: 235–240.

Sherr, E. B., B. F. Sherr & G.-A. Paffenhöfer, 1986. Phagotrophic Protozoa as food for metazoans: a 'missing' trophic link in marine pelagic food webs? Mar. Microb. Food Webs 1: 61–80.

Simon, M., 1987. Biomass and production of small and large free-living and attached bacteria in Lake Constance. Limnol. Oceanogr. 32: 591–607.

Simon, M. & F. Azam, 1988. Protein content and protein synthesis rates of planktonic marine bacteria. Mar. Ecol. Prog. Ser. 48: (in press).

Spittler, P., 1973. Feeding experiments with tintinnids. Oikos (Suppl.) 15: 128–132.

Utermöhl, H., 1958. Zur Vervollkommnung der quantitativen Phytoplankton-Methodik. Mitt. int. Ver. Limnol. 9: 1–38.

Verity, P. G., 1985. Grazing, respiration, excretion, and growth rates of tintinnids. Limnol. Oceanogr. 30: 1268–1282.

Verity, P. G., 1986a. Grazing of phototrophic nanoplankton by microzooplankton in Narragansett Bay. Mar. Ecol. Prog. Ser. 29: 105–115.

Verity, P. G., 1986b. Growth rates of natural tintinnid populations in Narragansett Bay. Mar. Ecol. Prog. Ser. 29: 117–126.

Verity, P. G. & T. A. Villareal, 1986. The relative food value of diatoms, dinoflagellates, flagellates, and cyanobacteria for tintinnid ciliates. Arch. Protistenkd. 131: 71–84.

Weisse, T., 1988. Dynamics of autotrophic picoplankton in Lake Constance. J. Plankton Res. 10: 1179–1188.

Weisse, T., 1989. The microbial loop in the Red Sea: dynamics of pelagic bacteria and heterotrophic nanoflagellates. Mar. Ecol. Prog. Ser. 55: 241–250.

Weisse, T. & Müller, H., 1989. Significance of heterotrophic nanoflagellates and ciliates in large lakes: evidence from Lake Constance. In M. M. Tilzer & C. Serruya (eds.), Functional and structural properties of large lakes. Science Tech. Publ., Madison, WI: (in press).

Weisse, T. & A. Schweizer, 1990. Seasonal and interannual variation of autotrophic picoplankton in a large prealpine lake (Lake Constance). Ver. int. Ver. Limnol. 24: (in press).

Williams, P. J. LeB., 1981. Incorporation of microheterotrophic processes into the classical paradigm of the planktonic food web. Kieler Meeresforsch., Sonderh. 5: 11–28.

Hydrobiologia **191**: 123–128, 1990.
P. Biró and J. F. Talling (eds), Trophic Relationships in Inland Waters.
© 1990 *Kluwer Academic Publishers.*

Influence on phytoplankton biomass in lakes of different trophy by phosphorus in lake water and its regeneration by zooplankton

Jolanta Ejsmont-Karabin[1] & Irena Spodniewska[2]
[1] *Polish Academy of Sciences, Institute of Ecology, Hydrobiological Station, ul. Leśna 13, 11-730 Mikolajki, Poland*; [2] *ul. Darwina 18 m 118, 03-488 Warsaw, Poland*

Key words: lakes, eutrophication, phosphorus regeneration, phosphorus concentration, zooplankton, phytoplankton

Abstract

In 49 unpolluted lakes of north-eastern Poland the biomass of algae in summer is significantly related to the concentration of total phosphorus and to the rate of phosphorus regeneration by zooplankton. Using a model with equations describing these relationships, the biomass of blue-green algae and other phytoplankton groups was predicted for 14 polluted lakes. A good approximation of actual values was obtained only for the biomass of blue-green algae calculated from the estimated rate of P regeneration by zooplankton in these lakes. It is hypothesized that more-or-less edible algae of other classes did not show dependence on the rate of input of regenerated P because their biomass was heavily reduced by grazing of zooplankton.

Introduction

Investigations of the eutrophication process in lakes have mainly involved the relationship between the total amount of phosphorus (as a limiting factor) and the quantity of phytoplankton as biomass or chlorophyll *a* concentration. However, most of the total phosphorus is contained in phosphorus compounds not directly available to phytoplankton, especially during the period of summer stratification. As frequently noted (e.g. Vollenweider, 1968; Nicholls & Dillon, 1978) a clear relationship between total-P concentration and algal density (frequently expressed as chlorophyll *a*) thus cannot be of a direct character. The total phosphorus content in lake water is a potential source of phosphate available to phyto-

plankton only if the organic matter containing phosphorus is mineralized. Taking into account that bacteria should be considered as consumers rather than regenerators of phosphorus in the epilimnion (Uehlinger, 1986), the excretion of P by zooplankton can be considered as the main mechanism of phosphorus regeneration, in addition to phosphatase action.

Consequently, a positive relationship can be expected between the rate of P uptake by phytoplankton, determining its standing crop, and the rate of P regeneration by zooplankton.

The purpose of this paper is to document the relationship between the biomass of phytoplankton and the rate of P regeneration by zooplankton. Because of their distinct ecology and special importance to eutrophication processes, blue-

124

green algae are here analyzed separately. Thus two groups of phytoplankton are distinguished: blue-green algae and other algae.

Material and methods

The data used were collected in 1976–1979 from lakes of north-eastern Poland, of which 49 are clean and 14 contaminated with domestic sewage. Samples were taken on one occasion during the period of summer stratification about early July from the deepest place in each lake, at 1-m intervals over the epilimnion or from surface to bottom in lakes without thermal stratification.

The biomass of algae was calculated from the numbers of the different species and the volume of their colonies, trichomes and cells. The results

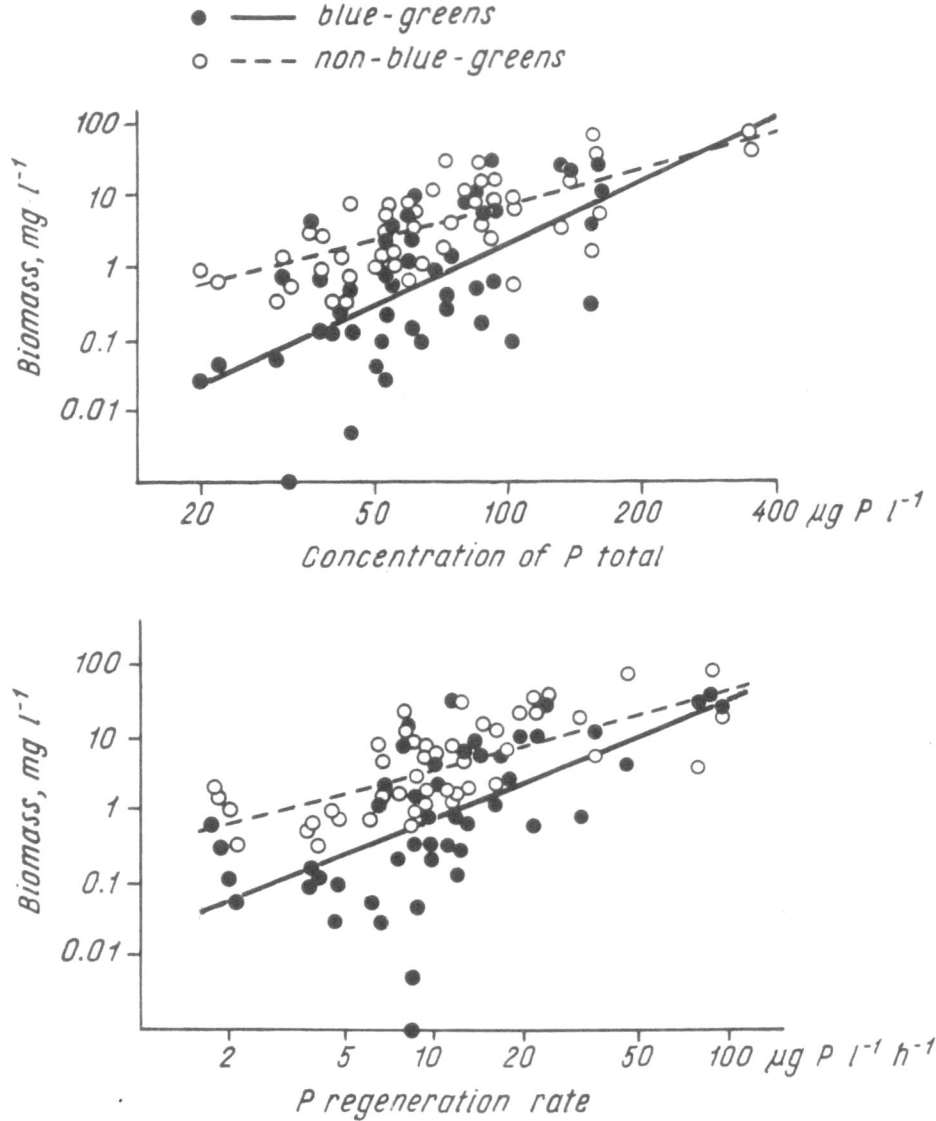

Fig. 1. Relationship between the concentration of total-P in lake water and biomass of blue-green algae and other algae (upper panel) and the rate of P regeneration by zooplankton and the biomass of blue-green algae and of other algae (lower panel) in the epilimnion of 49 unpolluted lakes during the period of summer stagnation.

have already been published by Spodniewska (1983, 1986), where a more detailed description of the method used for phytoplankton estimates can be found.

The rate of phosphorus regeneration has been assessed by using regression equations describing the effects of individual weights of animal and ambient temperature on the rate of P excretion (Ejsmont-Karabin, 1984). The data used in the calculations included densities as well as biomass of all rotifer and crustacean species. The methods of material collection and examination have been described by Karabin (1985a).

Most data on the total-P concentration in lake water are taken from Kajak & Zdanowski (1983). This phosphorus was determined by Zdanowski (1983) in samples taken from these lakes concurrently with samples of phytoplankton and zooplankton.

Results

An analysis of numerous lakes with differing nutrient content (Ejsmont-Karabin, 1983) shows that the rate of phosphorus regeneration by

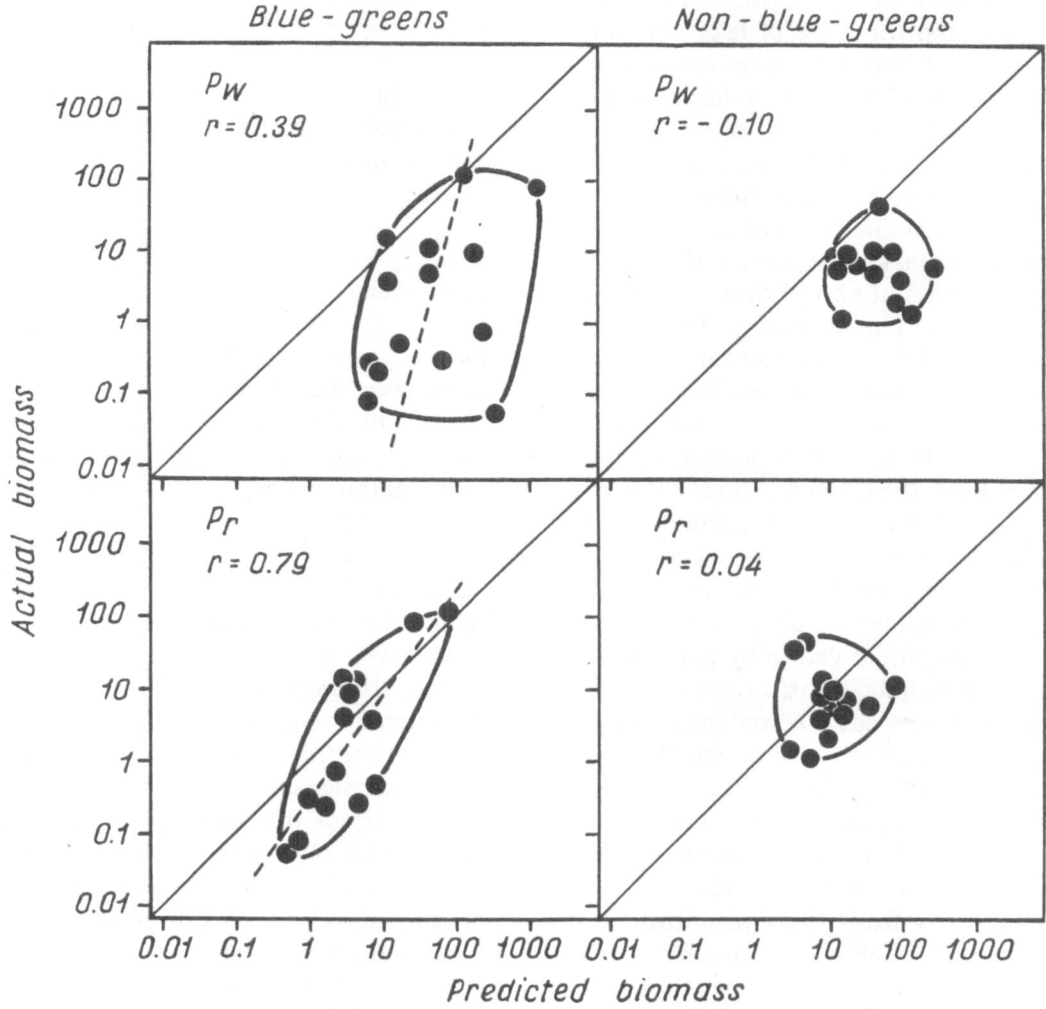

Fig. 2. Comparison of predicted and measured biomass of blue-green algae (left 2 panels) and predicted and measured biomass of the other phytoplankton groups (right 2 panels) in 14 lakes with point sources of contamination. P_w – biomass of algae predicted from the concentration of total-P in lake water. P_r – biomass of algae predicted from the rate of P regeneration

zooplankton is increasing with increasing P content so that the turnover time of phosphorus in the epilimnion is maintained at an almost constant level, although the pool of total-P is variable. The increases in both P concentration and regeneration rate in the unpolluted lakes are correlated with biomass of both blue-green algae and the other phytoplankton (Fig. 1). There are highly significant power relationships (Table 1).

At the same time, a strong correlation can be expected among all the factors analyzed when nutrient content is changing. The question as to which of the P-sources determines standing crop of the phytoplankton can be answered by analyzing this relationship for a set of lakes in which changes in nutrient concentrations are not correlated with changes in the rates of their regeneration. This situation is observed in heavily contaminated lakes (Ejsmont-Karabin, 1983). The zooplankton in such lakes, being limited by factors resulting from contamination and by other non-trophic environmental factors, does not efficiently mineralize the excess of phosphorus entering the lakes by the inflows. The taxonomic composition of zooplankton in these lakes is usually variable, and typically does not correspond with their trophic status (Karabin, 1985a). For these lakes the abundance of the two phytoplankton groups was predicted from the concentrations of total-P in the water and the P regeneration rate by zooplankton using the regressions calculated for unpolluted lakes as a model (Table 1).

The calculated values of algal biomass were compared with empirical values by correlational analysis after logarithmic transformation (Fig. 2). For a complete fit between the predicted biomass and measured biomass, the points should be on the marked diagonal. It can be seen that the biomass of blue-green algae predicted from the P regeneration rate (P_r) provides a good approximation of the empirical values. However, the equation calculated from the concentration of total-P in water (P_w) yields too high a biomass of blue-green algae.

The biomasses of the other phytoplankton groups predicted from the model largely depart from the empirical values found for these polluted

Table 1 Regression equations describing the relation of the biomass of blue-green algae and of other groups of phytoplankton to the concentration of total-P in lake water, and to the rate of phosphorus regeneration by zooplankton in 49 unpolluted lakes in the period of summer stratification.

Y	X	Total-P in lake water $\mu g \, l^{-1}$	Phosphorus regeneration rate $\mu g P \, l^{-1} \, d^{-1}$
Biomass of blue-green algae mg wet wt. l^{-1}		$Y = 6.45 \times 10^{-6} \, X^{2.78}$ $r = 0.60$ $p < 0.001$	$Y = 0.0162 \, X^{1.63}$ $r = 0.57$ $p < 0.001$
Biomass of the other groups of phytoplankton mg wet wt. l^{-1}		$Y = 0.00435 \, X^{1.60}$ $r = 0.66$ $p < 0.001$	$Y = 0.300 \, X^{1.05}$ $r = 0.69$ $p < 0.001$

lakes. The equation describing the effect of total-P concentration in water predicts biomass values that are almost always several times the empirical values.

Discussion

The results of this analysis seem to support the hypothesis that P uptake by algae is directly affected by the rate of supply of available phosphorus forms, mostly by the P regenerated by zooplankton rather than depending on the concentration of total-P. This is inferred from the fact that the biomass of the two phytoplankton groups calculated from total-P concentration was much higher than their actual biomass.

However, another phenomenon noted in this study is striking: the rate of P regeneration seems to determine only the abundance of blue-green algae. From the point of view of the functioning of a phytoplankton community, the hypothesis that a part of this community depends on the input of phosphorus regenerated by zooplankton, whereas the rest is not, does not seem to be acceptable. Why do not non-blue-green algae exhibit such a relationship?

Unlike other groups of phytoplankton, especially nanoplankton, blue-green algae are relatively little-consumed by zooplankton. In addition, the importance of species capable of feeding on blue-green algae (e.g. *Daphnia hyalina*: Knisely & Geller, 1986) is decreasing with rising nutrient content in the lakes (Karabin, 1985a), thus also with growing biomass of blue-green algae. The increase in the importance of blue-green algae with increasing nutrient content accounts for an increasing pressure of zooplankton (the abundance of which is also increasing) on non-blue-green species of algae, the available food. This means that the increase in the biomass of algae resulting from the input of regenerated nutrients is offset by the consumption of these algae by zooplankton. This hypothesis is, however, inconsistent with the results obtained by Reinertsen *et al.* (1986), who found that the addition of herbivorous zooplankton led to an increase in population of the green alga *Staurastrum*, whereas such an increase did not occur in the blue-green alga *Anabaena*. According to the authors the more homogeneous release of phosphorus by the zooplankton would allow *Staurastrum* to outcompete *Anabaena*. The importance of the mode of nutrient supply upon the outcome of competition for an algal resource has also been emphasized by Sakshaug & Olsen (1986). However, the latter describe variation of the total phytoplankton community, which can consist of both edible and non-edible species, and also show differential responses to environmental factors.

The present interpretation is that the phytoplankton as a whole reaches a standing crop corresponding to the input of mineral phosphorus forms characteristic of each lake. However, only the abundance of algae is determined by the input of available nutrients, whereas the species composition of this community is determined by a number of factors both biotic and abiotic.

How is it possible that phosphorus regenerated by zooplankton, and originating from only a small fraction of algae, can cover the requirements of the total phytoplankton community? This paradox can be explained by analyzing the trophic groups of zooplankton and their role in phosphorus regeneration. With increasing nutrient content in the lakes, the proportion of the species living on detritus and bacteria is increasing (Karabin, 1985b).

To sum up, these analyses suggest that the rate of phosphorus regeneration by zooplankton can determine the abundance of algae. Blue-green algae, which are lightly consumed by zooplankton, can almost reach a maximum abundance allowed by the input of regenerated nutrients, so that their densities can be determined mostly by the rate of phosphorus regeneration. The other forms of phytoplankton, although also dependent on the input of mineral phosphorus excreted by zooplankton, do not exhibit such a dependence, being heavily reduced by zooplankton grazing.

In addition to zooplankton grazing (this paper; Porter, 1977), light-limitation has also been proposed as an important factor influencing algal community structure (Smith, 1986). Blue-green algae possess several strategies by which they can outcompete other species under conditions of low light availability (Zevenboom & Mur, 1984). Zevenboom (1982) suggested that low light availability favors the growth of *Oscillatoria agardhii* over other species of algae. Blue-green assemblages of the lakes under study were dominated by *O. agardhii*, so that the role of light-limitation may also be important in these lakes. It is likely that light and zooplankton grazing both act to regulate the phytoplankton species composition, whereas their densities are determined mostly by the rate of nutrient regeneration.

Acknowledgement

The authors are deeply indebted to Dr Ramesh D. Gulati for valuable comments on the manuscript.

References

Ejsmont-Karabin, J., 1983. Ecological characteristics of lakes in north-eastern Poland versus their trophic gradient. VIII. Role of nutrient regeneration by planktonic rotifers and crustaceans in 42 lakes. Ekol. pol. 31: 411–427.
Ejsmont-Karabin, J., 1984. Phosphorus and nitrogen excretion by lake zooplankton (rotifers and crustaceans) in relationship to individual body weights of the animals, ambient temperature and presence or absence of food. Ekol. pol. 32: 3–42.

128

Kajak, Z. & B. Zdanowski, 1983. Ecological characteristics of lakes in north-eastern Poland versus their trophic gradient. I. General characteristics of 42 lakes and their phosphorus load. Study objectives and scope. Ekol. pol. 31: 239–256.

Karabin, A., 1985a. Pelagic zooplankton (Rotatoria + Crustacea) variation in the process of lake eutrophication. I. Structural and quantitative features. Ekol. pol. 33: 567–616.

Karabin, A., 1985b. Pelagic zooplankton (Rotatoria + Crustacea) variation in the process of lake eutrophication. II. Modifying effect of biotic agents. Ekol. pol. 33: 617–644.

Knisely, K. & W. Geller, 1986. Selective feeding of four zooplankton species on natural lake phytoplankton. Oecologia 69: 86–94.

Nicholls, K. H. & P. J. Dillon, 1978. An evaluation of phosphorus-chlorophyll phytoplankton relationship for lakes. Int. Revue ges. Hydrobiol. 62: 141–154.

Porter, K., 1977. The plant-animal interface in freshwater ecosystems. Am. Sci. 65: 159–170.

Reinertsen, H., A. Jansen, A. Langeland & Y. Olsen, 1986. Algal competition for phosphorus: the influence of zooplankton and fish. Can. J. Fish. aquat. Sci. 43: 1135–1141.

Sakshaug, E. & Y. Olsen, 1986. Nutrient status of phytoplankton blooms in Norwegian waters and algal strategies for nutrient competition. Can. J. Fish. aquat. Sci. 43: 389–396.

Smith, V. H., 1986. Light and nutrient effects on the relative biomass of blue-green algae in lake phytoplankton. Can. J. Fish. aquat. Sci. 43: 148–153.

Spodniewska, I., 1983. Ecological characteristics of lakes in north-eastern Poland versus their trophic gradient. VI. The phytoplankton of 43 lakes. Ekol. pol. 31: 353–381.

Spodniewska, I., 1986. Planktonic blue-green algae of lakes in north-eastern Poland. Ekol. pol. 34: 151–183.

Uehlinger, U., 1986. Bacteria and phosphorus regeneration in lakes. An experimental study. Hydrobiologia 135: 197–206.

Vollenweider, R. A., 1968. Scientific fundamentals of the eutrophication of lakes and flowing waters. OECD, Paris: 182 + 34 + 61.

Zdanowski, B., 1983. Ecological characteristics of lakes in north-eastern Poland versus their trophic gradient. III. Chemistry of the water in 41 lakes. Ekol. pol. 31: 287–308.

Zevenboom, W., 1982. N_2-fixing Cyanobacteria: Why they do not become dominant in shallow hypertrophic lakes. Hydrobiol. Bull., Amsterdam 16: 289–290.

Zevenboom, W. & L. R. Mur, 1984. Growth and photosynthetic response of the cyanobacterium Microcystis aeruginosa in relation to photoperiodicity and irradiance. Arch. Microbiol. 139: 232–239.

Hydrobiologia **191**: 129–138, 1990.
P. Biró and J. F. Talling (eds), Trophic Relationships in Inland Waters.
© 1990 *Kluwer Academic Publishers.*

Algal growth and loss rates in Lake Loosdrecht: first evaluation of the roles of light and wind on a basis of steady state kinetics

Herman J. Gons & Machteld Rijkeboer
Limnological Institute, Vijverhof Laboratory, Rijksstraatweg 6, 3631 AC Nieuwersluis, The Netherlands

Key words: Loosdrecht lakes, eutrophication, underwater light, resuspension, wind effect, *Prochlorothrix hollandica*

Abstract

Lake Loosdrecht (The Netherlands) is shallow, highly eutrophic and subject to frequent wind-induced resuspension of settled algae and detritus. The summer phytoplankton consists of filamentous prokaryotic species. Chlorophyll *a* levels are rather stable over the summer at a concentration of *ca.* 160 mg m^{-3}; losses due to grazing and sinking are small. Epipelic chlorophyll *a* concentrations range from 0 to 250, but *ca.* 50 mg m^{-2} is typical. *In situ* rates of change of chlorophyll *a* in the water column were related to specific growth rates predicted by a model for light-limited growth. In the model, incident light is partitioned among algae, tripton and background colour, to determine the light available for algal growth and cell maintenance. Model coefficients were derived primarily from laboratory studies of the growth of *Prochlorothrix hollandica*, an abundant species in the lake in summer. Presuming constant rates of loss due to grazing and sinking, for summers 1985 and 1986 some 56% of the variation in the chlorophyll *a* in the lake water was explained by change in light conditions alone and 77% by light and wind-driven resuspension of epipelic chlorophyll *a* together. These factors had little influence on the phytoplankton biomass in 1983 and 1984; other environmental conditions, e.g. phosphorus availability, may have been important. Also, the laboratory-derived growth kinetics of *P. hollandica* may not have been equally suitable for modelling in the four summers.

Introduction

A comprehensive research programme is being conducted on the effects of reduction of external phosphorus loading on the water quality of the Loosdrecht lakes system (Van Liere, 1986a). Three of the major lakes are shallow and highly eutrophic; in the recent summers Secchi disc depth did not exceed 0.4 m and chlorophyll *a* concentration ranged up to *ca.* 300 mg m^{-3}.

In the present study, temporal changes of chlorophyll *a* in the largest of these lakes, L.

Loosdrecht, will be considered with reference to the roles of light and wind energy in the dynamics of the phytoplankton. Some morphometric and hydrographic details on the lake are given below.

The lake used to be clear until the nineteen-fifties; only storms might turn the water turbid for some days. The very high light attenuation nowadays is due to particulate matter. Although the correlation between light attenuation and chlorophyll *a* is weak, because the detritus fraction of the seston is large and varying (Gons, 1987), it is evident that most of the present concentration of particles is the outcome of enhanced algal production due to eutrophication. However, it is difficult to quantify the formation of biomass using primary production measurements (Van Liere *et al.*, 1986), as well as the processes detracting from the particulate matter in the lake (Gons *et al.*, 1986a; Van Liere *et al.*, 1990). Due to the shallowness of the lake and its exposure to wind, frequent but irregular resuspension (Gons *et al.*, 1986b) may significantly influence these processes.

The species composition of the phytoplankton (Boesewinkel-de Bruyn *et al.*, 1984, 1986) has appeared to be quite stable. Filamentous 'blue-green' algae represent almost 100% of the algal biomass from May until the year-end. It has been found recently that the most abundant species is *Prochlorothrix hollandica*, which is prokaryotic like the blue-green algae, but with respect to pigmentation it relates to green algae (Burger-Wiersma *et al.*, 1986). The species used to be recognized as '*Oscillatoria limnetica*' or classified among the unidentified filaments, and it is not yet clear whether it was equally important in all summers of the study period.

In qualitative terms, the persistent composition of the summer phytoplankton is allowed by an environment imposing low rates of loss upon the predominant algae. Thus high population densities can be maintained at growth rates that are low compared to the physiological potential (Gons *et al.*, 1986a).

The question as to whether light or a nutrient, such as phosphorus, limits the growth of the phytoplankton has not been answered satisfactorily. Obviously, the nature of the limitation is important in the system's response to the decreased external phosphorus loading.

The occurrence of light limitation can be anticipated from the ratio between depth of the mixed layer (1.9 m) and Secchi disc depth (0.35 m in summer). This 'optical depth' puts L. Loosdrecht almost precisely on the regression line for 'nutrient-saturated' water bodies given by Wofsy (1983).

Experimental results have been contradictory. From phosphate uptake kinetics of the lake water, Riegman & Mur (1986) concluded that phosphorus was growth-limiting during the greater part of the summer 1983, and that light limitation applied to short periods of the year only. They obtained similar results for 1984 (R. Riegman, pers. comm.), but chlorophyll-specific uptake was lower than in 1983.

Studies of the influence of phosphate loading on seston dynamics in summers 1985 and 1986 showed effect on phosphate uptake; however, neither change in chlorophyll *a* nor in sestonic dry weight depended on the supply rate of the nutrient. It was surmised that light had been growth-limiting (Sweerts *et al.*, 1986; Van Liere, 1986b).

A system like L. Loosdrecht, showing only gradual temporal changes in species composition, biomass and physico-chemical environment, may be considered as oscillating near 'steady-state'. The oscillations are due to either changes in the availability of light or a nutrient for the algae, or in the loss factors, i.e. grazing and sinking; of course, temperature fluctuations may also be of influence.

Steady state models have proven to be very useful for understanding phytoplankton growth in light-limiting conditions (e.g. Bannister, 1979; Wofsy, 1983), though among systems the parameters may differ markedly (Atlas & Bannister, 1980; Kirk, 1983). For L. Loosdrecht, we calibrated a simple model on the growth of *P. hollandica* in light-limited laboratory systems, and determined the specific light attenuation coefficients necessary for application to the field situation. Chlorophyll *a* levels dependent on growth rate and the partial light attenuation by detrital particles – i.e. tripton – have been considered by Rijkeboer & Gons (1990).

Here, we compare model predictions of growth rate with rates of change in measured concentrations of chlorophyll *a* ('apparent' growth rates). In theory, the relation between these two growth rates yields the overall loss rates of the algal community, and allows for an estimate of the counterbalancing effect of resuspension in the loss.

The lake

The Loosdrecht lakes area (Fig. 1) is situated 20 km south-east of Amsterdam, The Netherlands. The lakes were created as a result of peat dredging, enhanced by wind and wave erosion (Van Liere, 1986a). The part of the area called Kievitsbuurt still reflects this development. L. Loosdrecht (area = 9.8 km^2; mean depth = 1.9 m) formerly consisted of five lakes, now partly separated by islands. The various lake parts are quite uniform in depth.

For L. Loosdrecht, L. Breukeleveen and L. Vuntus, the catchment area totals 38 km^2, and the water renewal time is *ca.* 1 year. Seepage losses to low-lying polders exceed seepage gains by far (Kal *et al.*, 1984). During dry periods in summer the losses due to seepage and evaporation necessitate external supply for maintaining the legally required minimum water table in these lakes, for

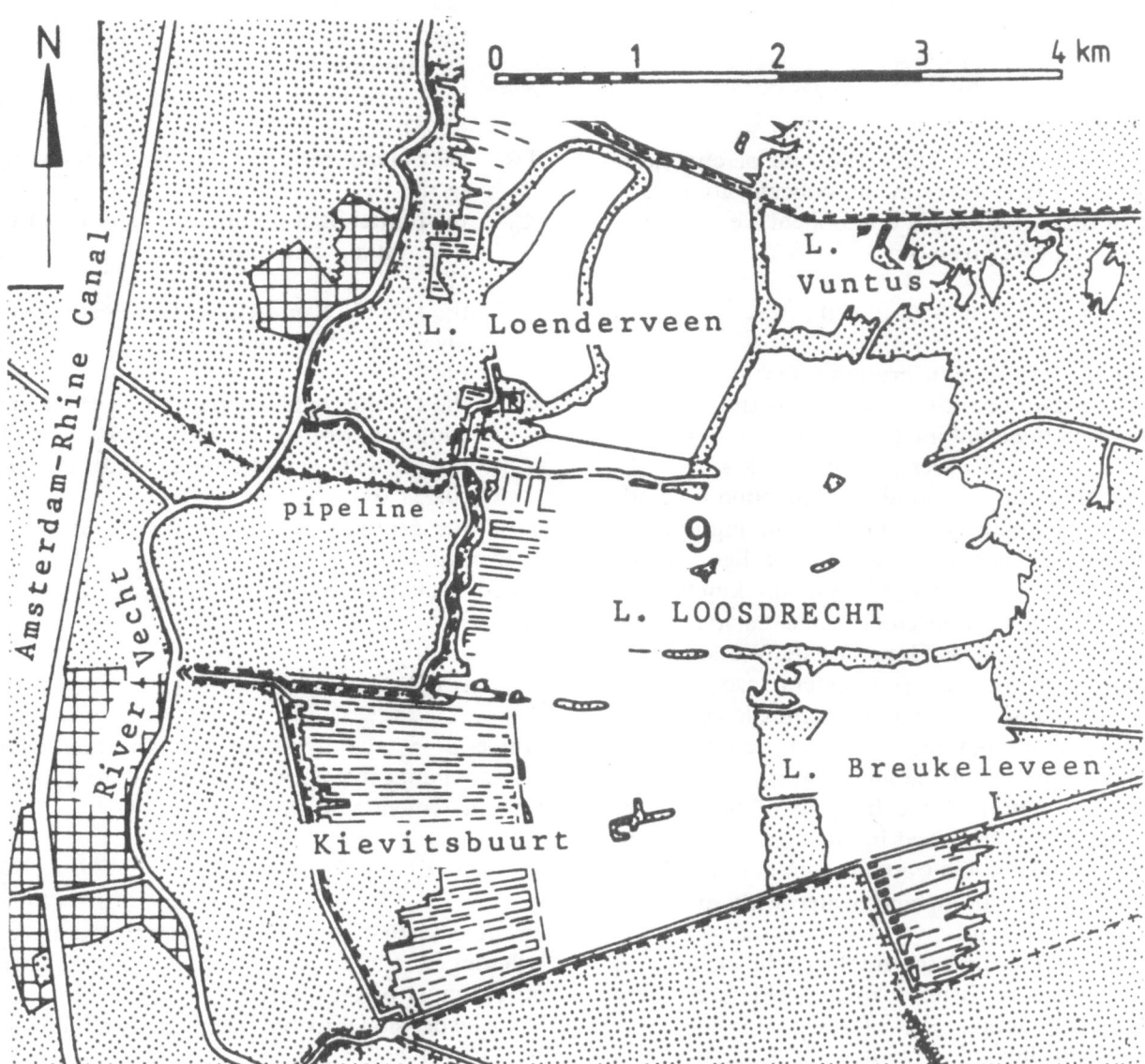

Fig. 1. Map of the Loosdrecht lakes area showing the sampling station 9.

132

which water from the polluted River Vecht has been used.

In 1984 the water loss was replaced by inflow from the Amsterdam-Rhine Canal, after removal of more than 90% of the total phosphorus by coagulation with $FeCl_3$. Presumably sanitation of the sewage disposal in the area has been equally important in the reduction of the external phosphorus loading from *ca.* 1.1 to 0.3 g P m^{-2} year^{-1}. A partial phosphorus balance for L. Loosdrecht is not yet available; see Van Liere *et al.* (1990) for more details.

Theory

In idealized steady state, the rate of change in population density or in the concentration X of any of the population's components, e.g. chlorophyll *a*, can be given as:

$$dX/dt = \mu \cdot X - D \cdot X = 0 \qquad (1)$$

where μ is the specific growth rate (time^{-1}), and D is the dilution rate (time^{-1}); in this study D represents the combined loss rates due to grazing and sinking, and other possible causes.

The value of μ of an algal population depends on the availability of the growth-limiting nutrient or light. We explore a model for light-limited growth on the basis of growth kinetics of *P. hollandica* in laboratory scale enclosures at 20 °C (Rijkeboer & Gons, 1990). We stress the points that several parameters are species-dependent, and are valid only for the lower half of the organism's potential range of growth rate.

The principles of the model are:
1. determination of the fraction of underwater light that is absorbed by the algae, dependent on the partial light attenuation coefficients of the algae, tripton, and water with dissolved substances, and
2. relating μ to the light uptake of the algae using the concepts 'true' growth yield and maintenance energy.

For light-limited growth in L. Loosdrecht in summer, μ depends on 3 variables only; the

coefficients have been derived in Rijkeboer & Gons (1990). For algal biomass expressed in chlorophyll *a* units:

$$\mu = \frac{Q \cdot p \cdot Y}{(CHL \cdot k_c + TR \cdot k_{tr}/k_c + K_B/k_c) \cdot z} - \mu_e \qquad (2)$$

where Q is the insolation (kJ m^{-2} day^{-1}),
p a constant for conversion of the insolation to subsurface photosynthetically available radiation (0.38),
Y the 'true' growth yield on light for biomass as chlorophyll *a* (0.044 mg kJ^{-1}),
CHL the concentration of chlorophyll *a* (mg m^{-3}),
TR the concentration of tripton as dry weight (g m^{-3}),
K_B the partial light attenuation coefficient due to water and dissolved compounds (1.2 m^{-1}),
k_c the specific light attenuation coefficient of algae in chlorophyll *a* units (0.011 m^2 mg^{-1}),
k_{tr} the specific light attenuation coefficient of tripton in dry weight units (0.23 m^2 g^{-1}),
μ_e the specific maintenance rate constant (0.031 day^{-1}), and
z the depth of the mixed layer (1.85 m).

In a natural system, growth will seldomly be exactly balanced by losses, hence $dX/dt <> 0$, and

$$\mu - D = \mu' \qquad (3)$$

where μ' is the apparent growth rate, which relates change from concentration X_0 to X_t over a time increment t according to:

$$(\ln X_t - \ln X_0)/t = \mu' \qquad (4)$$

In the case that there is an external source contributing to the concentration X in the system, eq. (3) must be adjusted.

Resuspension may represent such an input. Let the average concentration of a component in the layer of resuspendable particles be R (g m^{-2}) and its concentration in water be X (g m^{-3}). We consider that during a time interval t resuspension at a fraction F_r of the lake area has completely mixed the layer into the water column. Then the input of the component into the water column during the interval has been:

$$F_r \cdot R/z \text{ (g m}^{-3}) \qquad (5)$$

and the input per unit of the component already present in water, and per unit may be defined as the specific rate of change due to resuspension:

$$\mu_r = F_r \cdot R/(X \cdot z \cdot t) \qquad (6)$$

The role of resuspension in changes in concentration of a component in the water column may now be evaluated; adjustment of eq. (3) gives:

$$\mu - (D - \mu_r) = \mu', \text{ or } \mu - D = \mu' - \mu_r \quad (7)$$

Materials and methods

Lake water and epipelon samples from station 9 (Fig. 1) were collected as described by Van Liere et al. (1986a) and Gons et al. (1986b), respectively. From 1983 until 1985 sampling was biweekly, but in 1986 its frequency was once every 4 weeks. The samples were prefiltered over 150 μm mesh.

Sestonic dry weight and chlorophyll were determined after filtration on Whatman glass fibre filters with 0.7 μm effective pore diameter; for epipelic chlorophyll centrifugation was applied. Chlorophyll a was measured and corrected for phaeopigment according to Moed & Hallegraeff (1978).

Insolation and wind data have been provided by the Royal Netherlands Meteorological Institute for the station in De Bilt, which is situated at about 11 km from the southern border of the lake.

Resuspension due to wind-driven surface waves has been computed by applying simple wave theory (Smith, 1979; Carper & Bachmann, 1984; Gons et al., 1986b). Resuspension should occur once half the wavelength exceeds the water depth. The percentage lake area subject to resuspension was estimated for every day by computing the wavelength : depth ratios for 37 grid points on the lake, from the maximum of the average hourly windspeed and the wind direction in that hour.

Results

Besides the derived coefficients (eq. 2), model predictions of the specific growth rate in the field require data on insolation and the dry weight and chlorophyll a of seston (Fig. 2).

The changes in chlorophyll a from May until September only are discussed, because

1. P. hollandica was abundant not earlier than from May onward;
2. the water temperature should not deviate too much from the experimental temperature 20 °C, since both pigmentation and endogenous metabolism of the cells are temperature-dependent (Harris, 1978; Kirk, 1983), and thus the values of Y, k_c and μ_e. In May the water is generally colder than 15°, and it may reach ca. 25° in July or August; after mid-September its temperature almost invariably drops below 15 °C.

The tripton concentration was estimated according to:

$$TR = DW - 0.048 \cdot CHL \qquad (8)$$

where DW is the dry weight concentration of seston (g m^{-3}), and 0.048 the dry weight (g) associated with chlorophyll a (mg) in light-limited P. hollandica (Rijkeboer & Gons, 1990). Derived in this way, the share of tripton in the sestonic dry weight generally varied between 65 and 80%. For 1984 similar values were obtained through conversion from biovolume measurements (Gons, 1987). K_{TR}, i.e. the estimated partial attenuation coefficient of tripton, varied between 2 m^{-1} in June and September 1983 to 13 m^{-1} in August 1986.

Differences in physical conditions and in the time course of chlorophyll a between the summers will be discussed later.

134

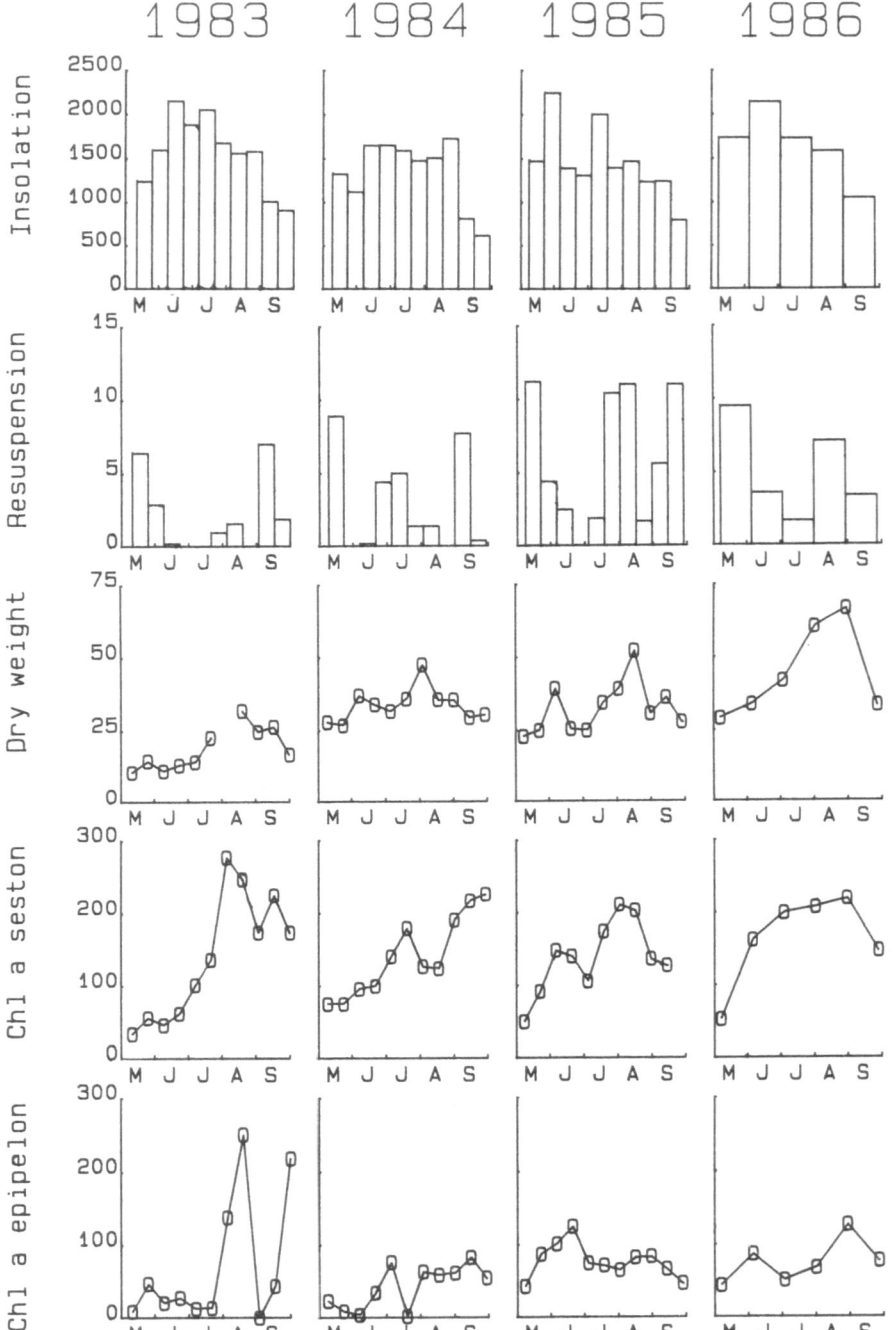

Fig. 2. Daily insolation (kJ m^{-2} day^{-1}), resuspension frequency (% lake area day^{-1}), and concentrations of sestonic dry weight (g m^{-3}), sestonic chlorophyll *a* (mg m^{-3}) and epipelic chlorophyll *a* (mg m^{-2}) of L. Loosdrecht during summers 1983-1986.

First, we consider the relationship between the apparent growth rates (eq. 4) on a chlorophyll a basis with the specific growth rates (eq. 2) that should be expressed when light was growth-limiting. The specific growth rates were computed per sample interval, i.e. generally 2 weeks, but 4 weeks in 1986, on the basis of the averaged chlorophyll a and dry weight concentrations for the interval. For each year, the start was the interval preceeding the first sampling day in June, and the end the second half of September.

The predicted μ values (Fig. 3) spanned the complete physiological range of *P. hollandica*: the organism's μ_{max} at 20° is *ca.* 0.5 day^{-1} (T. Burger-Wiersma, pers. comm.). However, there was a notable lack of mid-range values, and the high rates were predicted for 1983 only. Average values of μ in the 4 summers ranged from 0.12 in 1984 and 1986 to 0.29 day^{-1} in 1983.

The pooled apparent growth rates correlated positively with μ, but linear fit according to eq. (3) was weak (Table 1). A reasonable fit was obtained only for the summers 1985 and 1986.

Hypothetically, part of the scatter in the data (Fig. 3) has been due to resuspension. It has been established that in L. Loosdrecht *ca.* 25% of the storage of chlorophyll a is epipelic, and that resuspension frequently affects large parts of the lake area (Fig. 2; see also Gons *et al.*, 1986ab).

The values of μ_r (eq. 6) were of the same order of magnitude as μ', but, of course, never negative. Great differences existed in average μ_r: in 1983

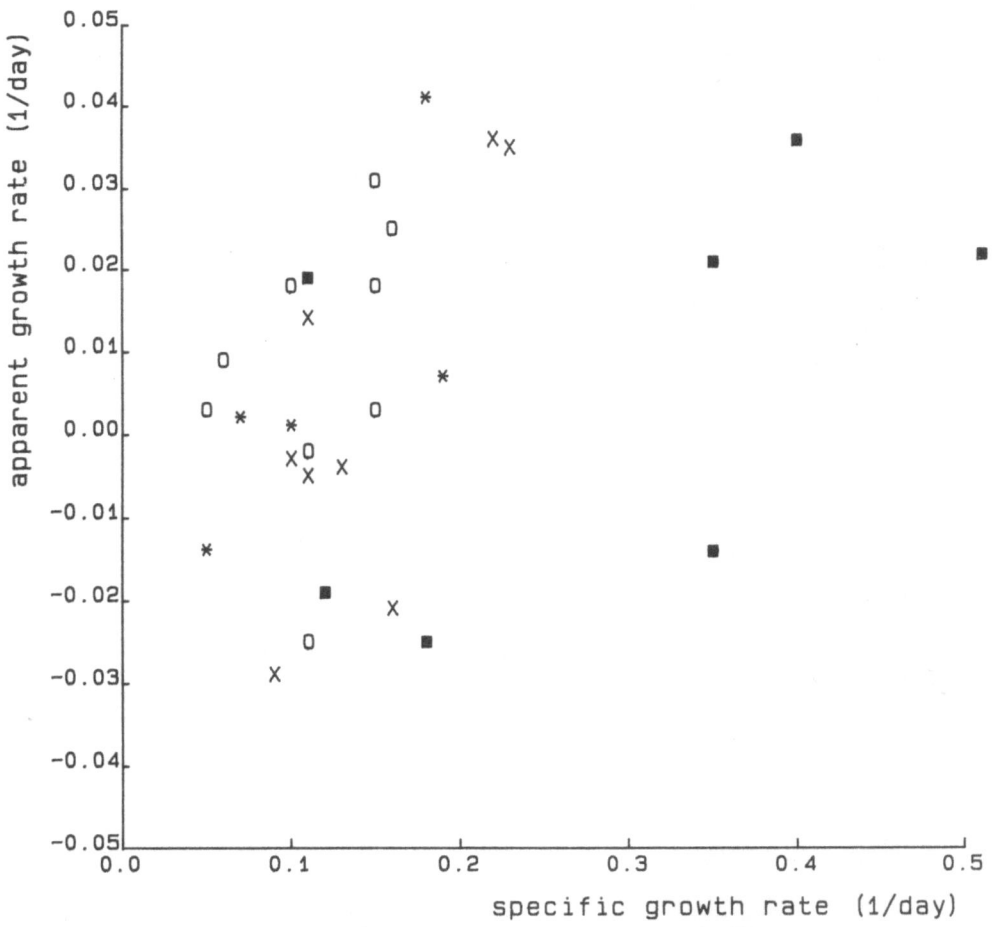

Fig. 3. Relationship between apparent growth rate (1983: solid squares; 1984: circles; 1985: crosses; 1986: asterisks) and predicted specific growth rate.

Table 1. Relationship between apparent growth rate and specific growth rate (μ) in L. Loosdrecht during summers 1983–1986, before (μ') and after ($\mu' - \mu_r$) correction of the apparent rate for resuspension.

Year(s)	N	μ' vs $\cdot \mu$	r^2	$(\mu' - \mu_r)$ vs $\cdot \mu$	r^2
1983–1986	29	$0.07 \cdot \mu - 0.005$	0.14	$0.09 \cdot \mu - 0.016$	0.21
1983	7	$0.08 \cdot \mu - 0.018$	0.28	$0.09 \cdot \mu - 0.024$	0.28
1984	9	$0.16 \cdot \mu - 0.009$	0.14	$0.17 \cdot \mu - 0.015$	0.18
1985	8	$0.33 \cdot \mu - 0.045$	0.56	$0.36 \cdot \mu - 0.060$	0.77
1986	5	$0.24 \cdot \mu - 0.021$	0.56	$0.19 \cdot \mu - 0.028$	0.77

this was 0.003, but in 1986 0.014 day^{-1}. Maximally, μ_r reached *ca.* 20% of the μ value predicted for the sampling interval.

Taking into account the proposed reduction in deposition loss (eq. 7), the correlation coefficients of linear regression (Table 1) improved, except for 1983. The effect of the correction made by subtracting μ_r from μ' on the data distribution is shown for 1985 (Fig. 4).

So, for summers 1985 and 1986, but not for 1983 and 1984, much of the variation in μ', and thus in the chlorophyll *a* concentration, has been explained by the combined effects of light and wind; however, the slope in the relation between $(\mu' - \mu_r)$ and μ (Table 1) deviated considerably from unity, as should apply if D was constant (eq. 7). The values of *D*, i.e. the combined loss rate due to grazing and sinking, are represented by the intercepts (Table 1) and were only 0.06 and 0.03 day^{-1} in 1985 and 1986, respectively.

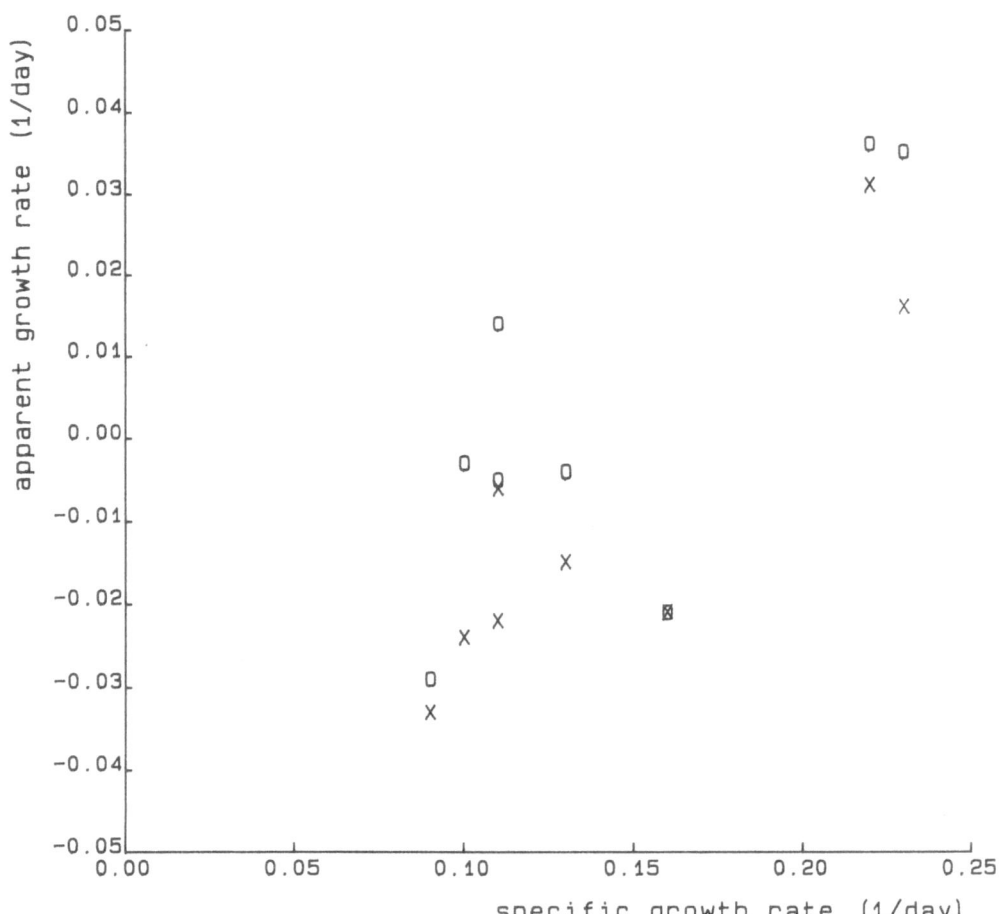

Fig. 4. Apparent growth rate in relation to predicted specific growth rate in 1985 before (circles) and after (crosses) subtraction of the rate of change due to resuspension.

Discussion

Strictly speaking, eq. (2) only applies when:

1. growth proceeds light-limited at approximately steady state, whereby the growth rates do not exceed $^1/_2\,\mu_{max}$;
2. the temperature does not deviate much from 20 °C, see above;
3. the phytoplankton biomass consists largely of *P. hollandica*.

These conditions may have been violated more or less severely; see Introduction with respect to the last.

In 1983, the underwater light at relatively low concentrations of chlorophyll *a* and dry weight, and favourable insolation (Fig. 2), in principle allowed for high growth rates until July. However, the predicted growth rates were not reflected in the values of μ' in this period (Fig. 2). Although theoretically (eq. 3) enhanced loss rate, say by grazing, may have been the reason, it is more likely that growth was not light-limited. Riegman & Mur (1986) concluded that phosphorus limitation applied to the greater part of summer 1983. Remarkably though, they found a low phosphate uptake capacity and supposed that light had been limiting during June and part of July.

For summer 1984, only low μ values were predicted. As in 1983, the correlation between μ' and μ was weak (Table 1). This summer was characterized by even but low insolation (Fig. 1), and marked fluctuation in water temperature; particularly in July the water was relatively cold. Also, as mentioned earlier, phosphate uptake experiments indicated growth to the phosphorus- rather than light-limited.

The chlorophyll *a* maximum in 1984 occurred at the end of September, when the water temperature was only 13 °C, and the insolation had been less than in any of the other years at this time. Two circumstances still allow the assumption that light had been limiting: the steepest increase in the chlorophyll *a* concentration coincided with the greatest insolation, while the further increase may have been caused by the high resuspension frequency in the first part of September (Fig. 1).

Also, because the insolation in September was very low compared to the summer mean, the value of μ may have been underestimated as a consequence of enhanced growth yield at low irradiance. Incident light-dependent change in the efficiency of conversion of light energy into biomass has been reported by Van Liere & Mur (1979) and Gons & Mur (1980).

Summers 1985 and 1986 had respectively, low and variable, and 'normal' insolation, and the average frequency of resuspension was similarly high. It has not been established why the sestonic dry weight concentration in 1986 was higher than in the earlier years; anyway, this has been the reason why the predicted growth rates remained low compared with 1983, when insolation was similar.

For 1985 and 1986 it now seems indubitable that light had been growth-limiting, even more so since laboratory results demonstrated that phosphate supply had no effect on the dynamics of chlorophyll *a* (Sweerts *et al.*, 1986; Van Liere, 1986b).

Considering the boundary conditions for the applicability of the model, the linear fits (Table 1) for the last two summers have not been disappointing, but the slopes were lower than those predicted by eq. (7); hence the loss rates may also have been underestimated. Conversely, temporal change in loss rate may been a reason for the lower slope. Generally (results not shown), the low μ values were predicted for intervals during which temperature was some 5° lower than 20 °C: consequently, the grazing component in the loss may have been lowered too.

Out of several possibilities, only one further explanation is given here for deviation from linearity according to eq. (7), namely that the internal phosphorus loading of L. Loosdrecht (Boers, 1986; Van Liere, 1986b; Van Liere *et al.*, 1990) is 'saturating' growth under average conditions of insolation and vertical extinction, but unable to support more production – the product of biomass and μ - when the light availability for algae improves. In this view, Fig. 3 simply represents a saturation curve, and part of the conflicting results of the past years may be reconciled.

138

Acknowledgements

Part of this study was financed by the Ministry of Housing, Physical Planning and Environment. Loes Breebaart, Ronny van Keulen and Heb Roon are thanked for analyses, computations, and assistance in the field.

References

Atlas, D. & T. T. Bannister, 1980. Dependence of mean spectral extinction coefficient of phytoplankton on depth, water color, and species. Limnol. Oceanogr. 25: 157–159.

Bannister, T. T., 1979. Quantitative description of steady state, nutrient-saturated algal growth, including adaptation. Limnol. Oceanogr. 24: 76–96.

Boers, P. C. M., 1986. Studying the phosphorus release from the Loosdrecht Lakes sediments, using a continuous flow system. Hydrobiol. Bull. 20: 51–60.

Boesewinkel-de Bruyn, P. J., 1984. Kwantitatief fytoplanktononderzoek in de Loosdrechtse Plassen in 1983. W.Q.L.-report, Limnological Institute, Nieuwersluis, The Netherlands.

Boesewinkel-de Bruyn, P. J., 1986. Kwantitatief fytoplanktononderzoek in de Loosdrechtse Plassen in 1984. W.Q.L.-report, Limnological Institute, Nieuwersluis, The Netherlands.

Burger-Wiersma, T., M. Veenhuis, H. Korthals, C. C. M. Van der Wiel & L. R. Mur, 1986. A new prokaryote containing chlorophylls a and b. Nature 320: 262–264.

Carper, G. L. & R. W. Bachmann, 1984. Wind resuspension of sediments in a prairie lake. Can. J. Fish. aquat. Sci. 41: 1763–1767.

Gons, H. J., 1987. De relatie tussen doorzicht en slib in de Loosdrechtse Plassen in verband met de zwemwaternorm. W.Q.L.-report, Limnological Institute, Nieuwersluis, The Netherlands.

Gons, H. J. & L. R. Mur, 1980. Energy requirements for growth and maintenance of Scenedesmus protuberans Fritsch in light-limited continuous cultures. Arch. Microbiol. 125: 9–17.

Gons, H. J., R. D. Gulati & L. Van Liere, 1986a. The eutrophic Loosdrecht Lakes: current ecological research and restoration perspectives. Hydrobiol. Bull. 20: 61–75.

Gons, H. J., R. Veeningen & R. Van Keulen, 1986b. Effects of wind on a shallow lake ecosystem: redistribution of particles in the Loosdrecht Lakes. Hydrobiol. Bull. 20: 109–120.

Harris, G. P., 1978. Photosynthesis, productivity and growth: the physiological ecology of phytoplankton. Arch. Hydrobiol. Beih. 10: 1–171.

Kal, B. F. M., G. B. Engelen & Th. E. Cappenberg, 1984. Loosdrecht Lakes restoration project: hydrology and physico-chemical characteristics of the lakes. Verh. int. Ver. Limnol. 22: 835–841.

Kirk, J. T. O., 1983. Light and photosynthesis in aquatic ecosystems. Cambridge University Press, Cambridge, 401 pp.

Moed, J. R. & G. M. Hallegraeff, 1978. Some problems in the estimation of chlorophyll-a and phaeopigment from pre- and post-acidification spectrophotometric measurements. Int. Revue ges. Hydrobiol. 63: 787–800.

Riegman, R. & L. R. Mur, 1986. Phosphate uptake kinetics of the natural phytoplankton population from the Loosdrecht lakes. Limnol. Oceanogr. 31: 983–988.

Rijkeboer, M. & H. J. Gons, 1990. Light-limited algal growth in Lake Loosdrecht: steady state studies in laboratory scale enclosures. Hydrobiologia 191: 241–248.

Smith, I. R., 1979. Hydraulic conditions in isothermal lakes. Freshwat. Biol. 9: 119–145.

Sweerts, J-P. R. A., H. J. Gons & M. Rijkeboer, 1986. Phosphate uptake capacity of summer phytoplankton of the Loosdrecht Lakes in relation to phosphorus loading and irradiance. Hydrobiol. Bull. 20: 101–107.

Van Liere, L., 1986a. Loosdrecht Lakes, origin, eutrophication, restoration and research programme. Hydrobiol. Bull. 20: 9–15.

Van Liere, L. (ed.), 1986b. Water quality research Loosdrecht Lakes; studying and modelling the impact of water management measures on the internal nutrient cycle. W.Q.L.-report, Limnological Institute, Nieuwersluis, The Netherlands.

Van Liere, L., L. Van Ballegooijen, W. A. de Kloet, K. Siewertsen, P. Kouwenhoven & T. Aldenberg, 1986. Primary production in the various parts of the Loosdrecht Lakes. Hydrobiol. Bull. 20: 77–85.

Van Liere, L. & L. R. Mur, 1979. Growth kinetics of Oscillatoria agardhii Gomont in continuous culture, limited in its growth by the light-energy supply. J. gen. Microbiol. 115: 153–160.

Van Liere, L., R. D. Gulati, F. G. Wortelboer & E. H. R. R. Lammens, 1990. Phosphorus dynamics following restoration measures in the Loosdrecht lakes. Hydrobiologia 191: 87–95.

Wofsy, S. C., 1983. A simple model to predict extinction coefficients and phytoplankton biomass in eutrophic waters. Limnol. Oceanogr. 28: 1144–1155.

Hydrobiologia **191**: 139–148, 1990.
P. Biró and J. F. Talling (eds), Trophic Relationships in Inland Waters.
© 1990 *Kluwer Academic Publishers.*

Interactions between sediment and water in a shallow and hypertrophic lake: a study on phytoplankton collapses in Lake Søbygård, Denmark

Martin Søndergaard, Erik Jeppesen, Peter Kristensen & Ole Sortkjær
National Environmental Research Institute, Lysbrogade 52, 8600 Silkeborg, Denmark

Key words: phytoplankton collapses, hypertrophic, nitrogen, phosphorus, sedimentation

Abstract

Short-term changes in phytoplankton and zooplankton biomass have occurred 1–3 times every summer for the past 5 years in the shallow and hypertrophic Lake Søbygård, Denmark. These changes markedly affected lake water characteristics as well as the sediment/water interaction. Thus during a collapse of the phytoplankton biomass in 1985, lasting for about 2 weeks, the lake water became almost anoxic, followed by rapid increase in nitrogen and phosphorus at rates of 100–400 mg N m^{-2} day^{-1} and 100–200 mg P m^{-1} day^{-1}. Average external loading during this period was about 350 mg N m^{-2} day^{-1} and 5 mg P m^{-2} day^{-1}, respectively.

Due to high phytoplankton biomass and subsequently a high sedimentation and recycling of nutrients, gross release rates of phosphorus and nitrogen were several times higher than net release rates. The net summer sediment release of phosphorus was usually about 40 mg P m^{-2} day^{-1}, corresponding to a 2–3 fold increase in the net phosphorus release during the collapse. The nitrogen and phosphorus increase during the collapse is considered to be due primarily to a decreased sedimentation because of low algal biomass. The nutrient interactions between sediment and lake water during phytoplankton collapse, therefore, were changed from being dominated by both a large input and a large sedimentation of nutrients to a dominance of only a large input. Nitrogen was derived from both the inlet and sediment, whereas phosphorus was preferentially derived from the sediment. Different temperature levels may be a main reason for the different release rates from year to year.

Introduction

Reduced nutrient loading is not always sufficient to reduce eutrophication, since nutrients accumulated in the sediment can be released and provide significant internal loading (Boström *et al.*, 1982; Jacoby *et al.*, 1982). In particular phosphorus released from the sediment can severely prevent or delay the recovery of lakes (e.g. Bengtsson *et al.*, 1975; Peterson, 1982). In shallow lakes, where the input from the sediment is generally higher than in deeper lakes, this internal phosphorus loading sometimes can be a major contribution to the overall loading of the lake (Holdren & Armstrong, 1980; Lennox, 1984).

Traditionally, internal phosphorus loading has been regarded as a function of chemical and biological interactions, either within the sediment or between sediment and water. Less attention has been paid to how the biological structure in the lake water affects the internal loading. In shallow, small and hypertrophic lakes changes in the biological structure can be very rapid, and more or less dramatic collapses in phytoplankton biomass have been reported from many lakes (Barica, 1975; Fott et al., 1980). These changes affect the chemical structure in the lake water, but it is an open question to what extent these collapses can affect the internal loading.

This paper describes collapses in phytoplankton biomass and changes in nutrient cycling that take place every summer in the shallow and hypertrophic Lake Søbygård, Denmark. The aim was to determine the impact of these changes on the interaction between sediment and water, especially concerning nitrogen and phosphorus dynamics. A related paper (Jeppesen et al., 1989) discusses the biological interactions and factors controlling the collapses. The collapse observed in 1985 is used as a basis, supported by observations from 1986 and 1987.

Study area

Lake Søbygård is a small (0.39 km^2) and shallow (max. depth 1.9 m, mean depth 1.0 m) lake situated in the central part of Jutland, Denmark (9° 48′ E, 56° 15′ N) (see also Søndergaard et al., 1987). Water residence time is 15 to 30 days, and one main inlet contributes 70–80% of the total water input. The remaining water comes from springs around the lake.

For several decades Lake Søbygård received large quantities of poorly treated municipal wastewater. Since 1976, however, the wastewater has been biological treated, and in 1982 a chemical treatment was introduced to reduce the input of phosphorus. The external phosphorus loading thereby decreased from about 30 to 5 g P m^{-2} year^{-1}. Correspondingly, phosphorus retention in the lake changed from highly positive to negative. Therefore, the internal phosphorus loading today accounts for about 2/3 of the total phosphorus loading. Measurements have revealed a large pool of phosphorus in the sediment, high enough to support an internal phosphorus loading for many years (Søndergaard et al., 1987; Sønder-

gaard, 1988). Phosphorus concentrations in the upper 15 cm of the sediment vary from 6 to 12 mg P g dw^{-1}. A major part of this phosphorus is bound to iron (extractable with NH_4Cl or $NaOH$).

Due to the high nutrient loading the lake is hypertrophic. Primary production is up to 1500 g C m^{-2} year^{-1}, and Secchi transparency < 0.5 m most of the summer. Phytoplankton is dominated by diatoms during spring maxima, and by coccal green algae in summer and autumn. Cryptophytes are occasionally of importance (see Jeppesen et al., 1989). The lake water is well-mixed, without any vertical or horizontal stratification. Because of its shallowness, wind action often induces resuspension of sediment which occasionally causes a considerable increase in lake water concentrations of nutrients and suspended matter (Jensen & Kristensen, 1986; Søndergaard et al., 1987).

Rapid changes in phytoplankton biomass have not only been recorded in 1985 to 1987 as described in this paper, but also in the summers 1983 and 1984 (Andreasen et al., 1984; Jeppesen et al., 1989).

Methods

Concentrations of total phosphorus (tot-P), soluble reactive phosphorus (PO_4-P), total nitrogen (tot-N), and nitrite + nitrate (NO_{2+3}-N) have been monitored intensively during 1985–1987 in the main inlet, lake centre and in the outlet. Sampling frequency was 1–3 times a week. In periods with rapid changes samples were taken every day in the lake and every 4 hours in the outlet. Inlet and outlet water was collected by automatic sampler emptied every 1–4 days. The outlet concentrations were generally not different from those in the lake.

Orthophosphate was determined by the ascorbic acid-molybdenum blue method according to Murphy & Riley (1972), and total phosphorus as orthophosphate after persulphate digestion in an autoclave at 200 kPa for 30 min according to Koroleff (1970). Ammonium (NH_4-N) was determined by the phenol hypochlorite method according to Dansk Standard (1975) modified from Solórzano (1969), and nitrite + nitrate spectrophotometrically after reduction according to Dansk Standard (1975) and Dahl (1974). Usually nitrite concentrations are negligible compared to nitrate concentrations. Total nitrogen was measured as NO_{2+3}-N after potassium persulphate digestion according to Dansk Standard (1975) and Solorzano & Sharp (1980). Oxygen was measured continuously at 0.1 m depth using an oxygen probe. The probe was calibrated automatically once a day (Sortkjær & Jeppesen, 1987). Chlorophyll *a* was determined spectrophotometrically after extraction with ethanol according to Dansk Standard (1975) and Holm-Hansen & Riemann (1978).

Results

As described by Jeppesen *et al.* (1989), the collapses in phytoplankton biomass did not follow the same pattern from year to year. Furthermore, the collapses could be divided into different steps including a pre-clear-water period. The following results, however, focuse mainly on changes in nutrient concentrations during and shortly after a phytoplankton collapse.

In 1985 the variation during summer in the chemical and biological parameters of Lake Søbygård was especially characterized by one event. From the beginning of July to 20 July phytoplankton biomass, measured as chlorophyll *a*, decreased rapidly from about 800 to $< 50\ \mu g$ chl-a 1^{-1}. The decrease occurred in two steps, and most rapidly from 15 July, reaching a minimum of 12 μg chl-a 1^{-1} on 23 July (Fig. 1). This clear-water period lasted for 3–5 days, after which the phytoplankton biomass returned to the original level.

Due to low primary production, but probably sustained respiration at the sediment/water interface, oxygen concentrations in the lake water decreased from a hypersaturated level of about 20 mg O_2 1^{-1} to 1 mg O_2 1^{-1} at the end of the clear-water period (Fig. 1). Lake water tempera-

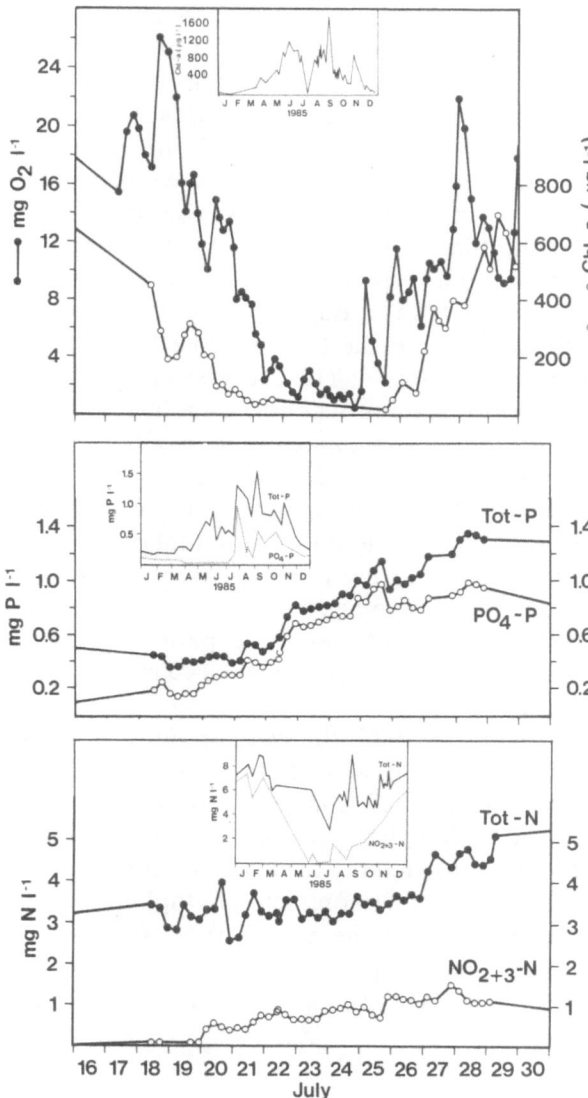

Fig. 1. Phytoplankton collapse from 18 to 30 July 1985. Upper: Chlorophyll *a* and oxygen concentrations in the outlet. Middle: Total N and NO_{2+3}-N in the outlet. Lower: Total phosphorus and PO_4-P in the outlet.

ture during the clear-water period was between 17 and 19 °C.

During the clear-water period both total nitrogen and nitrate increased markedly (Fig. 1). The increase in nitrate started on 20 July when the concentration of chlorophyll decreased to

$< 50\ \mu g$ chl-a l^{-1}, and continued until 28 July when chlorophyll reached a level of 300–400 μg chl-a l^{-1} again. Apart from minor fluctuations, the increase rate of nitrate in this period was consistently about 130 mg N m^{-2} day^{-1}. External nitrogen loading during the collapse was about 350 mg N m^{-2} day^{-1} (Table 1).

At the time of the phytoplankton collapse the external input of phosphorus varied between 3.9 and 5.3 mg P m^{-2} day^{-1}, while the output of phosphorus from the lake increased from 10 to 53 mg P m^{-2} day^{-1}. This increase in net internal phosphorus loading was reflected in rapidly increasing lake-water concentrations of both orthophosphate and total phosphorus (Fig. 1). From 20 July to 28 July, orthophosphate thus increased from 0.24 to 0.96 mg P l^{-1}, and total phosphorus from 0.37 to 1.33 mg P l^{-1}. Corrected for input and output, the average net release of phosphorus from the sediment for an 8-day period during the phytoplankton collapse was 145 mg P m^{-2} day^{-1}.

Contrary to 1985, the development of phytoplankton in 1986 was characterized by several periods with rapid changes and relatively low biomass, although the phytoplankton biomass in 1986 did not reach the same low level as in 1985. The first and most complete phytoplankton collapse appeared at the end of June, followed by others in late August, September and October, respectively (Fig. 2). At the collapse in June

chlorophyll a decreased from about 800 μg chl-a l^{-1} on 10 June to about 50 μg chl-a l^{-1} from 28 June to 1 July. Minimum biomass was reached on 29 June with 38 μg chl-a l^{-1}. The collapses that appeared later in the year were less marked.

Oxygen was not measured continuously during the collapse in June, but measurements indicate that oxygen concentrations may have been well below 5 mg O$_2$ l^{-1}. Ammonium increased at a rate of 65 mg NH$_4$-N m^{-2} day^{-1}. Lake water temperature during the collapse was from 22 to 24 °C.

Total nitrogen and nitrate did not increase to the same degree as during the collapse in 1985. Nitrate increased at a rate of 30 mg NO$_3$-N m^{-2} day^{-1} (Fig. 2). There are only few measurements of total nitrogen in this period, but at the end of the collapse total nitrogen began to increase concurrently with the increase in phytoplankton biomass. The increase-rate was about 150 mg N m^{-2} day^{-1}.

Concurrently with the decrease in phytoplankton biomass, phosphate concentrations increased from mid-June. The increase in total phosphorus did not start significantly until 28 June, when phytoplankton biomass had reached a level of 50 μg chl-a l^{-1}. The increase rate of both total phosphorus and phosphate was consistently about 0.20 mg P l^{-1} day^{-1} (Fig. 2). Corrected for input and output from the lake, average net release of phosphorus from the sediment in this

Table 1. Average P and N concentrations in main inlet and external P and N loadings during selected periods.

	tot-P		PO$_4$-P		tot-N		NO$_{2+3}$-N	
	mg l^{-1}	mg m^{-2} day^{-1}	mg l^{-1}	mg m^{-2} day^{-1}	mg l^{-1}	mg m^{-2} day^{-1}	mg l^{-1}	mg m^{-2} day^{-1}
1985	0.30	10.7	0.030	1.07	10.5	500	8.5	400
15–30 July	0.25	4.6	0.020	0.37	10.0	350	8.0	330
1986	0.25	12.6	0.020	1.0	9.5	440	8.5	400
24 June–4 July	0.20	4.8	0.015	0.38	10.0	250	9.5	220
1987	–	–	–	–	–	–	–	–
25 May–18 June	0.20	–	0.020	–	11.0	–	10.0	–

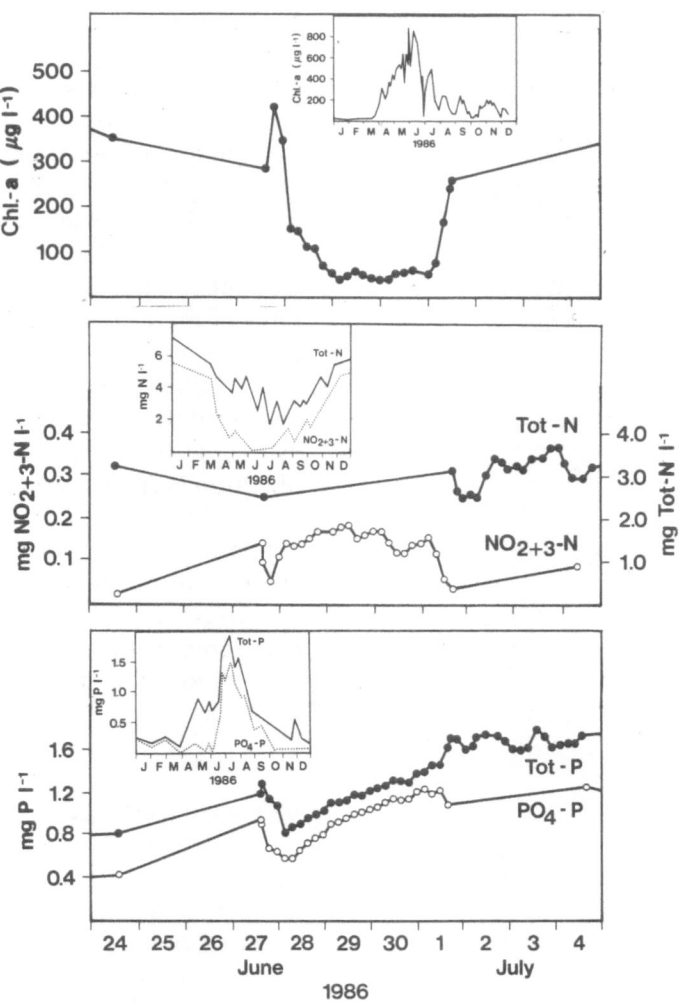

Fig. 2. Phytoplankton collapse from 24 June to 4 July 1986. Upper: Chlorophyll *a* concentrations in the outlet. Middle: Total N and NO_{2+3}-N in the outlet. Lower: Total phosphorus and PO_4-P in the outlet.

period was 240 mg P m^{-2} day^{-1}. Total phosphorus continued to increase for some time after the collapse, and remained high until mid-July.

In the first part of summer 1987 the seasonal development in phytoplankton was dominated by one major but relatively long-lasting collapse. This year it occurred already in early June, when chlorophyll *a* decreased from 430 μg chl-*a* l^{-1} on 20 May to <25 μg chl-*a* l^{-1} from 2 to 14 June (Fig. 3). Minimum concentrations of chlorophyll *a* were reached on 3 June with 3 μg chl-*a* l^{-1}, and on 11 June with only 1 μg chl-*a* l^{-1}. Lake water temperature during the collapse was from 13 to 15 °C.

Nitrate did not reach the same low concentrations before the collapse as in the previous years, probably because of the generally lower phytoplankton biomass in 1987. At the beginning of the clear-water period in late May concentrations were about 1.1 mg NO_3-N l^{-1}, but decreased to about 3.1 on 18 June (Fig. 3), corresponding to an increase rate of 90 mg NO_{2+3}-N m^{-2} day^{-1}. Total nitrogen increased at approximately the same rate. The ammonium increase rate was 75 mg NH_4-N m^{-2} day^{-1}.

Total phosphorus concentrations in lake water decreased during May until 3 June, after which the concentrations increased at a rate of 50 mg

144

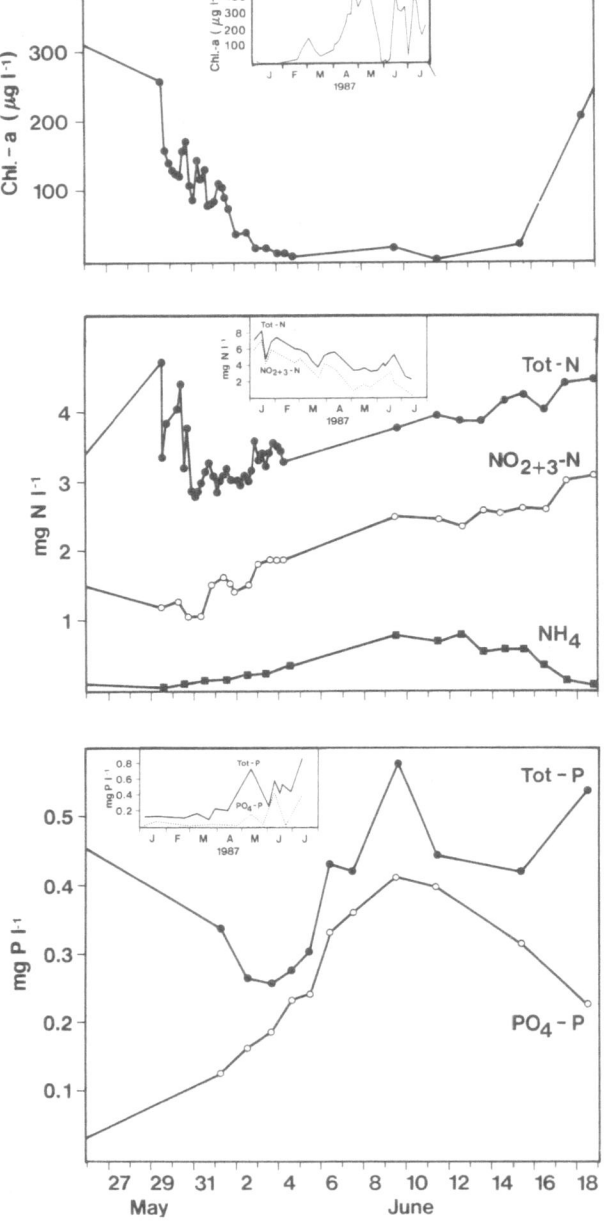

Fig. 3. Phytoplankton collapse from 26 May to 18 June 1987. Upper: Chlorophyll *a* concentrations in the outlet. Middle: Total N, NO_{2+3}-N and NH_4-N in the lake. Lower: Total P and PO_4-P in the lake.

tot-P m^{-2} day^{-1} (Fig. 3). Phosphate concentrations started to increase from 25 May, but most rapidly from 3 June at a rate of 40 mg PO_4-P m^{-2} day^{-1}.

Discussion

Marked seasonal variations in nutrient concentrations are not unusual in eutrophic lakes (Raaphorst & Brinkman, 1984; Yosida, 1982; Otsuki *et al.*, 1981). However, because of the very high nutrient level in both water and sediment, the variations in Lake Søbygård are large compared with most other lakes. This is also reflected in the rates of increase or decrease which take place in both the biological and chemical structure.

It is remarkable that the changes in phosphorus and nitrogen concentrations usually start almost exactly at or just prior to the date when phytoplankton biomass reaches its minimum – typically when chlorophyll *a* has decreased to < 100 μg chl-a l^{-1}. This supports the interpretation that the changes in nutrient concentration are related to the decreasing phytoplankton biomass. Average net increases of nitrogen and phosphorus during the periods of phytoplankton collapse are given in Table 2.

Increase of nitrogen:

Most of the nitrogen input to the lake, mainly in the form of nitrate, is either denitrified or incorporated as organic nitrogen in the phytoplankton. Denitrification may contribute significantly to the total removal of nitrogen in eutrophic lakes (Andersen, 1974; Kasper, 1985). Additionally, the very high primary production and nitrogen uptake by phytoplankton in Lake Søbygård, together with the shallow water, causes a high sedimentation of organic matter and thereby a removal of nitrogen from lake water to sediment. Secondarily, this sedimentation creates a potentially high consumption of oxygen and nitrate at the sediment water interface.

By use of sedimentation traps, the sedimentation of organic matter in Lake Søbygård, prior to the phytoplankton collapse in 1985, has been estimated as 4–6 g C m^{-2} day^{-1} (Kristensen & Jensen, 1987), and on average 800 mg N m^{-2} day^{-1}. Thus the external input of about 350 mg N m^{-2} day^{-1}, besides the outlet of organic nitro-

Table 2. Lake water temperature, average net increase of nitrogen and average net release of phosphorus during the collapses; in $mg\ m^{-2}\ day^{-1}$.

Period	Temp. °C	NO_{2+3}-N	tot-N	PO_4-P	tot-P
1985 15–30 July	17–19	130	130–400	145	145
1986 24 June–4 July	22–24	0–30	(150)	240	240
1987 25 May–15 June	13–15	90	90	40	50

gen (average 120 mg N m^{-2} day^{-1}), is usually compensated by high rates of both denitrification and sedimentation. Depending on the denitrification rate, the recycling and release of nitrogen from the sediment must according to this mass balance be of the order of 600–800 mg N m^{-2} day^{-1}.

During the phytoplankton collapse, however, phytoplankton biomass and therefore sedimentation is low, and the nitrogen concentration in the lake water will increase due to continued input of nitrogen from both inlet and sediment. The increase of nitrogen in 1985 during the collapse was mainly due to an increase of nitrate. The increase rate of 130 mg N m^{-2} day^{-1} must be due to the input of nitrate from the inlet which is not further taken up by phytoplankton, and/or a nitrification of ammonium released from the sediment. Unfortunately, we have no measurements of ammonium in this period, but according to the sequence in following years there must have been a marked increase of ammonium concentration during the collapse. The start of the increase of nitrate about 20 July corresponds very well to the time when phytoplankton and thereby sedimentation reached very low levels. Problems concerning release rates of nitrogen from the sediment, nitrification and denitrification activities during the collapse are, however, not yet completely elucidated.

The enhanced increase rate of total nitrogen at the end of the collapse in 1985 may partly be explained by the development of anoxic conditions in the beginning of the clear-water period.

Anoxia may shift a major part of the respiration at the sediment surface from using oxygen to using nitrate. Correspondingly, denitrification is increased. At the end of the collapse, however, when the lake water becomes oxic, part of the respiration will shift back to using oxygen, and thereby decrease the denitrification. Moreover, the increasing phytoplankton biomass at the end of the collapse with increased demand for nitrogen will tend to diminish the importance of denitrification.

The pattern of nitrogen increase in 1985 was supported by the development in both 1986 and in 1987, although the increase in those years was not as marked. Changes in temperature (Kamp-Nielsen, 1974), different phytoplankton minima and changed external nitrogen input (Table 1) may be reasons for the differences in nitrogen increase rates from year to year.

Increase of phosphorus:

The internal phosphorus loading in Lake Søbygård is usually restricted to the period from April to December. The net release of phosphorus in connection with phytoplankton collapses, and afterwards when phosphate is accumulated in the lake water, is therefore important for the overall internal phosphorus loading.

The increase in phosphorus concentration during phytoplankton collapses was very pronounced over all three years. The increase in PO_4-P usually starts before phytoplankton has

146

reached its minimum, whereas the increase in total phosphorus does not start until the phytoplankton biomass is close to zero. This pattern reflects first a release of phosphate from decomposing phytoplankton combined with decreased phytoplankton phosphate uptake, and second an increase of total phosphorus due to decreased sedimentation of phosphorus. The same mechanism as proposed for the increase in nitrogen is thus valid also for phosphorus, although the increase in total phosphorus is solely caused by an internal loading.

The daily sedimentation of phosphorus prior to the phytoplankton collapses in 1985 has been calculated as 125 mg P m^{-2} (Kristensen & Jensen, 1987). During the collapse input from the inlet and output through the outlet have been calculated as respectively 4 and 17 mg P m^{-2} day^{-1}. If the gross release rate of phosphorus during the collapse is unchanged, and water concentration is corrected for in- and output, this means that in the absence of sedimentation of organic phosphorus the increase of phosphorus concentration in the lake water would be 130 μg P l^{-1} day^{-1}. This increase rate fits well the one actually observed. Therefore, during phytoplankton collapse the net release or increase of phosphorus approaches the gross release rate. Accordingly, the occurrence of phytoplankton collapses provides an excellent opportunity to study gross release rates of phosphorus in a natural system.

The net release rates or approximated gross release rates recorded in 1985 to 1987 during collapses varied from 50 to 240 mg P m^{-2} day^{-1}. These are very high rates compared to values from other lakes (e.g. Stevens & Gibson, 1977; Boström et al., 1982), but are supported both by the measured sedimentation rates (Kristensen & Jensen, 1987) and by experiments with undisturbed sediment cores in the laboratory (Søndergaard, 1987; Søndergaard, 1989). The reason for the very high release rates is probably the unusually high phosphorus concentrations in the sediment, which create a potentially high release rate. Environmental conditions, such as a high decomposition rate due to a large sedimenta-

tion, frequent occurrence of resuspension, and a high proportion of iron-bound phosphorus sensitive to changes in redox potential and pH, are regarded as major factors causing the release (Søndergaard, 1988; Søndergaard, 1989).

The large variations in the release rates from year to year (Table 2) might partly be caused by the different temperature levels. Temperature can have a major effect on the release rate of phosphorus (Kamp-Nielsen, 1975; Lee, 1977; Kelderman & Van der Repe, 1982; Psenner, 1984). Above 15 °C Kamp-Nielsen (1975) found that an increase in temperature by 5 °C would lead to a doubling or more in the release rate of phosphorus. This strong temperature-dependence is comparable to Lake Søbygård in the temperature range 14-18 °C, although not 18–22 °C where the increase in release rate is less pronounced. The temperature-dependent release of phosphorus in Lake Søbygård signifies that a relatively large proportion of the net primary production in this shallow lake is decomposed in the sediment, and not in the lake water. This decomposition is very temperature-dependent, and if the release of phosphorus, as proposed by several authors (Lee, 1977; Psenner, 1984), is strongly affected by microbial and benthic activity, this might explain the temperature-dependent release. The temperature-dependence has been supported by experiments with undisturbed sediment cores (Søndergaard, 1987; Søndergaard, 1989). In these experiments the average release rate was determined to 20–30 mg P m^{-2} day^{-1} at 10 °C dependent on, e.g., pH, and to 130 mg P m^{-2} day^{-1} at 20 °C.

Besides the decreased sedimentation of phosphorus during the collapses, the low oxygen concentrations could promote a redox dependent release of phosphorus (Boström et al., 1982). However, there are no indications that gross phosphorus release is enhanced during phytoplankton collapse. The increased net release seems solely to be a matter of reduced sedimentation.

To summarize, rapid changes of phytoplankton biomass occur more or less dramatically every summer in the shallow and hypertrophic Lake Søbygård. These changes lead to considerable changes also in the chemical structure mainly because of reduced sedimentation of organic nitrogen and phosphorus.

Nitrogen concentrations increase during the phytoplankton collapse because of decreased sedimentation of organic N but continuing high input of N from the inlet and sediment. The recycling of nitrogen between sediment and water is very important for the overall nitrogen pathway.

Usually gross release of phosphorus is several times greater than the net release, only counterbalanced by a large sedimentation. During a collapse, however, phosphorus sedimentation is reduced to near zero, and net release of phosphorus will be close to the gross release rate. The release rate of phosphorus seems to be temperature-dependent.

Thus lake water nitrogen and phosphorus concentrations during phytoplankton collapse are changed from being dominated by both a large sedimentation (output) and a large input (N from the inlet and sediment, and P from the sediment) to being dominated by only a large input of nutrients.

Acknowledgement

This study was financially supported by the National Agency of Environmental Protection. The technical staff of National Environmental Research Institute, Silkeborg, are gratefully acknowledged for assistance.

References

Andersen, J. M., 1974. Nitrogen and phosphorus budgets and the role of sediment in six shallow Danish lakes. Arch. Hydrobiol. 74: 259–277.

Andreasen, K., M. Søndergaard & H. H. Schierup, 1984. En karakteristik af forureningstilstanden i Søbygård Sø – samt en undersøgelse af forskellige restaureringsmetoders anvendelighed til en begrænsning af den interne belastning. Publication No. 7 from The Botanical Institute, University of Århus, Denmark: 164 pp (in Danish).

Barica, J., 1975. Collapses of algal blooms in prairie pothole lakes: their mechanism and ecological impact. Verh. int. Ver. Limnol. 19: 606–615.

Bengtsson, L., S. Fleischer, G. Lindmark & W. Ripl, 1975. Lake Trummen restoration project I. Water and sediment chemistry. Verh. int. Ver. Limnol. 19: 1080–1087.

Boström, B., M. Jansen & C. Forsberg, 1982. Phosphorus release from sediments. Arch. Hydrobiol. Beih. Ergebn. Limnol. 18: 5–59.

Dahl, I., 1974. Intercalibrations of methods for chemical analysis of water. V. Results from intercalibrations of methods for determining nitrate and total nitrogen. Vatten 30: 180–186.

Dansk Standard, 1975. Vandundersøgelse. Dansk Standardiseringsråd, København (in Danish).

Fott, J., L. Pechar & M. Prazaková, 1980. Fish as a factor controlling water quality in ponds. In J. Barica & L. R. Mur (eds.): Hypertrophic ecosystems. Developments in Hydrobiology 2: 255–261.

Holdren, G. C. & D. Armstrong, 1980. Factors affecting phosphorus release from intact sediment cores. Envir. Sci. Technol. 14: 79–86.

Holm-Hansen, D. & B. Riemann, 1978. Chlorophyll a determination: improvements in methods. Oikos 30: 438–447.

Jacoby, J. M., D. D. Lynch, E. B. Welch & M. A. Perkins, 1982. Internal phosphorus loading in a shallow eutrophic lake. Wat. Res. 16: 911–919.

Jensen, P. & P. Kristensen, 1986. Sedimentation and resuspension measured by use of sedimentation traps. In: Physical processes in Lakes (ed. J. Virta), NHP-Report 16: 108–112.

Jeppesen, E., M. Søndergaard, O. Sortkjær, E. Mortensen & P. Kristensen, 1989. Interactions between phytoplankton, zooplankton and fish in a shallow, hypertrophic lake: a study on phytoplankton collapses in Lake Søbygård, Denmark. Hydrobiologia/Developments in Hydrobiology (this volume).

Kamp-Nielsen, L., 1974. Mud-water exchange of phosphate and other ions in undisturbed sediment cores and factors affecting the exchange rates. Arch. Hydrobiol. 73: 218–237.

Kamp-Nielsen, L., 1975. A kinetic approach to the aerobic sediment-water exchange of phosphorus in Lake Esrom. Ecol. Modelling 1: 153–160.

Kasper, H. F., 1985. The denitrification capacity of sediment from a hypertrophic lake. Freshwat. Biol. 15: 449–453.

Kelderman, P. & A. M. Van der Repe, 1982. Temperature dependence of sediment-water exchange in Lake Grevelingen, SW Netherlands. Hydrobiologia 92: 489–490.

Koroleff, F., 1970. Determination of total phosphorus in natural waters by means of persulphate oxidation. An Interlab. Report No. 3. Cons. int. Explor. Mer.

Kristensen, P. & P. Jensen, 1987. Sedimentation og resuspension i Søbygaard Sø. M.sc. thesis. Publ. No. 64 from The Freshwater Laboratory, The National Agency of Environmental Protection, Denmark, in cooperation with Botanical Institute, University of Århus, Denmark: 176 pp (in Danish).

148

Lee, G. F., 1977. Significance of oxic vs. anoxic conditions for Lake Mendota phosphorus release. In: Golterman, H. L. (ed.) Interactions between sediments and freshwater, Dr. W. Junk, B. V., The Hague: 294–306.

Lennox, L. J., 1984. Sediment-water exchange in Lough Ennell with particular reference to phosphorus. Wat. Res. 18: 1483–1485.

Murphy, J. & J. P. Riley, 1972. A modified single solution method for the determination of phosphate in natural waters. Analyt. Chim. Acta 27: 21–26.

Otsuki, A., K. Seiichi & K. Takkayoshi, 1981. Seasonal changes of the total phosphorus standing crop in a highly eutrophic lake: the importance of internal loading for shallow lake restoration.

Peterson, S. A., 1982. Lake restoration by sediment removal. Wat. Res. Bull. 18: 423–435.

Psenner, R., 1984. Phosphorus release patterns from sediments of a meromictic mesotrophic lake (Piberger See, Austria). Verh. int. Ver. Limnol. 22: 219–228.

Raaphorst, W. & A. G. Brinkman, 1984. The calculation of transport coefficients for phosphate and calcium fluxes across the sediment-water interface, from experiments with undisturbed sediment cores. Wat. Sci. Technol. 17: 941–951.

Solórzano, L., 1969. Determination of ammonia in natural waters by the phenolhypochlorite method. Limnol. Oceanogr. 14: 799–801.

Solórzano, L. & J. H. Sharp, 1980. Determination of total dissolved nitrogen in natural waters. Limnol. Oceanogr. 25: 751–754.

Sortkjær, O. & E. Jeppesen, 1987. Hyppige feltmålinger øger forståelsen af lavvandede søers udvikling. In Annual Report 1986 from The Freshwater Laboratory, The National Agency of Environmental Protection: 21–28 (in Danish).

Stevens, R. J. & C. E. Gibson, 1977. Sediment release of phosphorus in Lough Neagh, Northern Ireland. In: Golterman, H. L. (ed.) Interactions between sediment and freshwater. Dr. W. Junk. B. V., The Hague: 343–347.

Søndergaard, M., 1987. Fosfordynamikken i lavvandede søer med udgangspunkt i Søbygård Sø. Report from The National Agency of Environmental Protection, The Freshwater Laboratory, Silkeborg, Denmark: 142 pp (in Danish).

Søndergaard, M., E. Jeppesen & O. Sortkjær, 1987. Lake Søbygård: a shallow lake in recovery after a reduction in phosphorus loading. GeoJournal 14: 381–384.

Søndergaard, M., 1988. Seasonal variations in the loosely sorbed phosphorus fraction of the sediment of a shallow and hypertrophic lake. Envir. Geol. Wat. Sci. 11: 115–121.

Søndergaard, M., 1989. Phosphorus release from a hypertrophic lake sediment: experiments with intact sediment cores in a continuous flow system. Arch. Hydrobiol. 116: 45–59.

Yosida, T., 1982. On summer peak of nutrient concentrations in lake water. Hydrobiologia 92: 571–578.

Hydrobiologia **191**: 149–164, 1990.
P. Biró and J. F. Talling (eds), Trophic Relationships in Inland Waters.
© 1990 *Kluwer Academic Publishers.*

Interactions between phytoplankton, zooplankton and fish in a shallow, hypertrophic lake: a study of phytoplankton collapses in Lake Søbygård, Denmark

Erik Jeppesen, Martin Søndergaard, Ole Sortkjær, Erik Mortensen & Peter Kristensen
National Environmental Research Institute, Lysbrogade 52, DK-8600 Silkeborg, Denmark

Key words: phytoplankton collapses, hypertrophic lake, high pH, phytoplankton, zooplankton, fish

Abstract

Since 1983 severe phytoplankton collapses have occurred 1-4 times every summer in the shallow and hypertrophic Lake Søbygård, which is recovering after a ten-fold decrease of the external phosphorus loading in 1982. In July 1985, for example, chlorophyll *a* changed from 650 μg l^{-1} to about 12 μg l^{-1} within 3-5 days. Simultaneously, oxygen concentration dropped from 20-25 mg O$_2$ l^{-1} to less than 1 mg O$_2$ l^{-1}, and pH decreased from 10.7 to 8.9. Less than 10 days later the phytoplankton biomass had fully recovered. During all phytoplankton collapses the density of filter-feeding zooplankton increased markedly, and a clear-water period followed. Due to marked changes in age structure of the fish stock, different zooplankton species were responsible for the density increase in different years, and consequently different collapse patterns and frequencies were observed.

The sudden increase in density of filter-feeding zooplankton from a generally low summer level to extremely high levels during algae collapses, which occurred three times from July 1984 to June 1986, could neither be explained by changes in regulation from below (food) nor from above (predation). The density increase was found after a period with high N/P ratios in phytoplankton or nitrate depletion in the lake. During that period phytoplankton biomass, primary production and thus pH decreased, the latter from 10.8-11.0 to 10.5. We hypothesize that direct or indirect effects of high pH are important in controlling the filter-feeding zooplankton in this hypertrophic lake. Secondarily, this situation affects the trophic interactions in the lake water and the net internal loading of nutrients. Consequently, not only a high content of planktivorous fish but also a high pH may promote uncoupling of the grazing food-web in highly eutrophic shallow lakes, and thereby enhance eutrophication.

A tentative model is presented for the occurrence of collapses, and their pattern in hypertrophic lakes with various fish densities.

Introduction

Marked shifts in species composition of phyto-plankton and zooplankton occur every summer in most lakes due to changes of environmental variables or interactions within the biological community. Hypertrophic lakes are characterized by dominance of a few species and a significant lack of feed-back mechanisms. Therefore, a marked shift in the populations of a species commonly results in great oscillations in the whole lake-water ecosystem, ranging from periods with dominance of autotrophic organisms and low transparency to periods with mainly heterotrophs and clear-water conditions (Uhlmann, 1971; Barica, 1975; Fott et al., 1980; Benndorf et al., 1984; Jeppesen et al., in press).

In most oligotrophic and mesotrophic lakes, species richness is higher and the food web more complex. Consequently, a collapse of a population of one species due to unfavourable conditions is, at least during summer, often compensated by an increase in abundance of one or more species belonging to the same trophic level (e.g. Stewart & Wetzel, 1986).

Nevertheless, the biologically buffered oligotrophic or mesotrophic lake periodically also suffers from lack of timing between autotrophy and heterotrophy, which results in similar but less pronounced changes in the ecosystem. The clear-water period in early summer in many lakes (Lampert et al., 1986; Stewart & Wetzel, 1986; Sommer et al., 1986) is an example of lack of timing.

The simplicity of the ecological network and the generally higher process rates in the hypertrophic lake make such lakes more suitable for studies of interactions within the biological community and of inter-relationships between the biological community and its chemical and physical environment. The hypertrophic lake is thus a full-scale laboratory experiment in the field, from which knowledge can be obtained in order to interpret results, including those more biologically complex lake systems.

We have studied the hypertrophic and highly oscillating shallow Lake Søbygård for 10 years, but most intensively after a reduction of its loading of organic matter and phosphorus. Here we analyse both the causes of phytoplankton collapses, which occur 1-4 times every summer, and the consequences of the collapses for the dynamics of other trophic levels.

The lake has been followed intensively since 1984. The most comprehensive sampling pro-gramme which included continuous measurements of oxygen and pH was, however, run during the single collapse in 1985. This event is described in detail, while the collapses in 1984, 1986 and 1987 are discussed more briefly.

In a related paper (Søndergaard et al., 1989) we discuss the impact of these shifts on the chemical environment and the internal loading.

Study-area

The calcareous Lake Søbygård is situated in central Jutland, Denmark (9° 48′ 35″ E, 56° 15′ 20″ N) (Fig. 1). It is 0.38 km^2 in area and shallow, with a mean depth of 1.0 m and maximum of 1.9 m. The hydraulic retention time is short, on average 15-20 days on a yearly basis and 21–30 days in summer. The lake is surrounded by decidous and coniferous forest, except to the west where it is exposed to prevailing wind. Emergent and floating-leaved macrophytes are only sparsely developed, and submerged vegetation is totally missing.

In the 1960's and 1970's the lake received large amounts of only mechanically treated sewage water from the sewage plant in the town of Hammel. In 1976 a biological treatment plant was established.

Marked changes in phosphorus loading have occurred during the last ten years. Until 1982, when the sewage plant was extended with a chemical step to remove phosphorus, the phosphorus loading of the lake was extremely high (28-33 g P m^{-2} yr^{-2}); since then it has been reduced to a much lower but still relatively high level of 4-7 g P m^{-2} yr^{-1}.

Corresponding to the changes in loading, the concentration of phosphorus in the main inlet has changed from 1-4 mg P l^{-1} before 1982 to 0.15-0.25 mg P l^{-1} since 1983. This decrease was, however, not reflected in an equivalent lower phosphorus concentration in the lake because of high phosphorus release from the sediment.

A brief description of the lake and loading history is given in Jeppesen et al. (1985) and Søndergaard et al. (1987), and a more comprehensive description by Jeppesen et al. (in press).

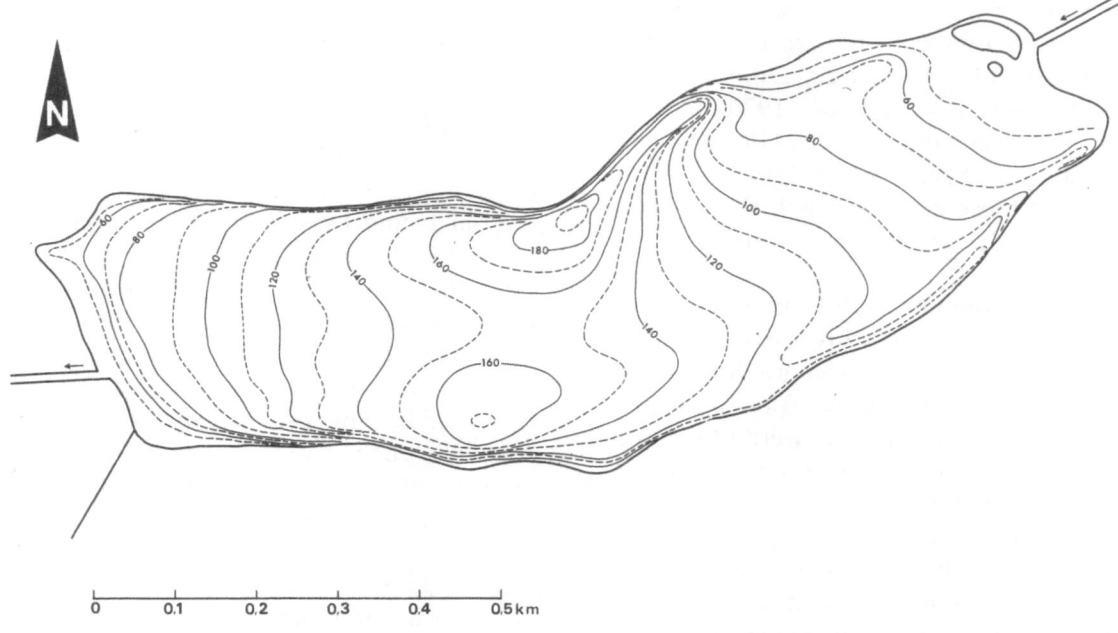

Fig. 1. Map of Lake Søbygård, with depth-contours in cm.

Methods

Phytoplankton:

Samples of 3.3 l were taken with a Patalas sampler from depths of 0.5 m and 1.2 m, then pooled, whereafter a 100-ml subsample was fixed with 1 ml lugol. Phytoplankton was quantified by counting a fixed water volume in an inverted microscope. Usually at least 100 individuals of each of the 2-3 dominating species were counted, and fewer of the other species. The maximum statistical error of the total number of algae was generally within 20-30% (S.E.).

Phytoplankton biomass expressed as bio-volume was estimated by fitting the species or subspecies at each sampling date to simple geometrical figures.

Zooplankton:

Samples of 3.3 l were taken with a Patalas sampler from depths of 0.5 m and 1.5 m, then pooled, whereafter 6.6 l were filtered through a mesh net with a pore size of 50 μm (1984) or 20 μm (1985-87), and finally fixed with 1 ml lugol solution and in 100 ml tap water. Triplicate samples were collected randomly in the pelagial at each sampling date.

During most periods with peak densities of zooplankton, 2-3 samples have been counted; otherwise only a single sample was counted. Various counting procedures have been used. Usually the samples were pre-filtered on a 140 μm net, and all animals retained on the filter were counted. In periods with high densities, however, animals were enumerated on subsamples.

Zooplankton < 140 μm was counted in an inverted microscope. After filtration on a 20 μm net, the animals retained were transferred into fifteen 2.9-ml chambers and either counted totally or partly using counting strips, at 40-100 fold magnification, depending on the size of the zooplankton. In periods with a high content of resuspended matter, it was necessary to dilute the samples 5-10 fold before counting, and zooplankton was then enumerated on sub-samples. Between 2 and 100% of each sample was counted on each sampling date.

Rotifer and cladoceran biomass were estimated from length measurements by use of published length-weight equations (rotifers: Dumont *et al.*, 1975; cladocerans: Bottrell *et al.*, 1976).

Length measurements were made on each sampling date on at least 20 rotifers and 20-50 cladocerans of these species which then contributed significantly to the total zooplankton biomass.

Biomass of cyclopoid nauplii was set to 0.2 μg d.w. indiv.$^{-1}$ according to Culver *et al.* (1985).

For estimation of zooplankton grazing rate on algae, the *in vitro* principle of Bjørnsen *et al.* (1986) was used. Water samples were filtered on a 20 μm mesh net to remove grazing animals. ^{14}C-marked bicarbonate (NaH ^{14}CO$_3$, 20 μCi per ampoule) was added to a subsample to a final concentration of 20-30 μCi per liter, and before the experiment incubated in light either in the laboratory for 18 hours at *in situ* temperature (until October 1985), or *in situ* for 2-3 hours (from October 1985). Prior to the grazing experiments the labelled phytoplankton was refiltered through a 20 μm mesh net.

The experiments were run on mid-lake composite samples (60-70 l) from depths of 0.5 and 1.5 m. Some 200 ml of ^{14}C-labelled phytoplankton were added to six 800-ml subsamples. Triplicate samples were incubated *in situ* for 20-60 min, depending on the temperature, while the remaining triplicate samples were used as blanks. Incubations were stopped by filtering the water from the experimental bottles through a 140 μm and then a 20 μm mesh net (diameter 25 mm). Zooplankters remained on the net were washed eight times in lake water by rapid back-filtration. The nets were drained on filter paper, transferred to glass vials, and assayed on a LKB-WALLAC 1210 liquid scintillation counter after dissolution in a 10 ml Ready-solv Hp/b (Beckmann) liquid for at least 24 h. Quenching was determined by the external standard-channels-ratio-method.

The ^{14}C-activity of phytoplankton before and after incubation was measured on 10-ml aliquots of the samples; filtered *in situ* on cellulose nitrate filters (0.45 μm), and in the laboratory transferred to glass vials and assayed as described above.

Mean ^{14}C-activity during the incubation was used as the tracer activity.

Chemical variables:

Oxygen was measured either manually at each sampling date by use of the modified Winkler-technique (Limnologisk Metodik, 1977), or over periods automatically every 0.5 h by use of oxygen probes (pHOX system). The pH was either measured manually at each sampling date or over periods automatically (Great Lakes Instruments) every 0.5 h. The probes were calibrated automatically once a day by pumping air-bubbled, oxygen-saturated and borax-buffered water (pH = 9.18) through the probe chambers. The probe signals were recorded on microchips. For more details, see Jeppesen *et al.* (1985) and Sortkjær & Jeppesen (1987). For other chemical variables, see Søndergaard *et al.* (1989).

Results

1985

The collapse in 1985 occurred in two steps (Figs. 2, 3). First, the algae biomass (largely *Scenedesmus* spp.) (Fig. 2) decreased during a period of 20 days, from 1200 μg chl.-*a* l^{-1} to a lower but still high level of 650 μg l^{-1} (Fig. 5). Then, a fast decline to 12 μg l^{-1} occurred within 3-5 days (Figs. 3, 5), followed by a rapid recovery after 3-4 days later. The second step of the collapse was accompanied by marked changes in the oxygen concentration from a highly super-saturated level of 20-25 mg O$_2$ l^{-1} to 1 mg O$_2$ l^{-1} within 7 days, after which a fast recovery was observed simultaneously with the reappearance of phytoplankton. The pH was reduced from 10.7 to 8.9 before it slowly increased again.

There was a tendency to a less pronounced change in the maximum photosynthetic capacity, P_{max}, than in chl.-*a* (Fig. 5). P_{max} thus altered from 1.45 to 1.25 (mg C l^{-1} h^{-1}) from 11 to 18 July, while chl.-*a* in the same period changed from 650 μg l^{-1} to 350 μg l^{-1}.

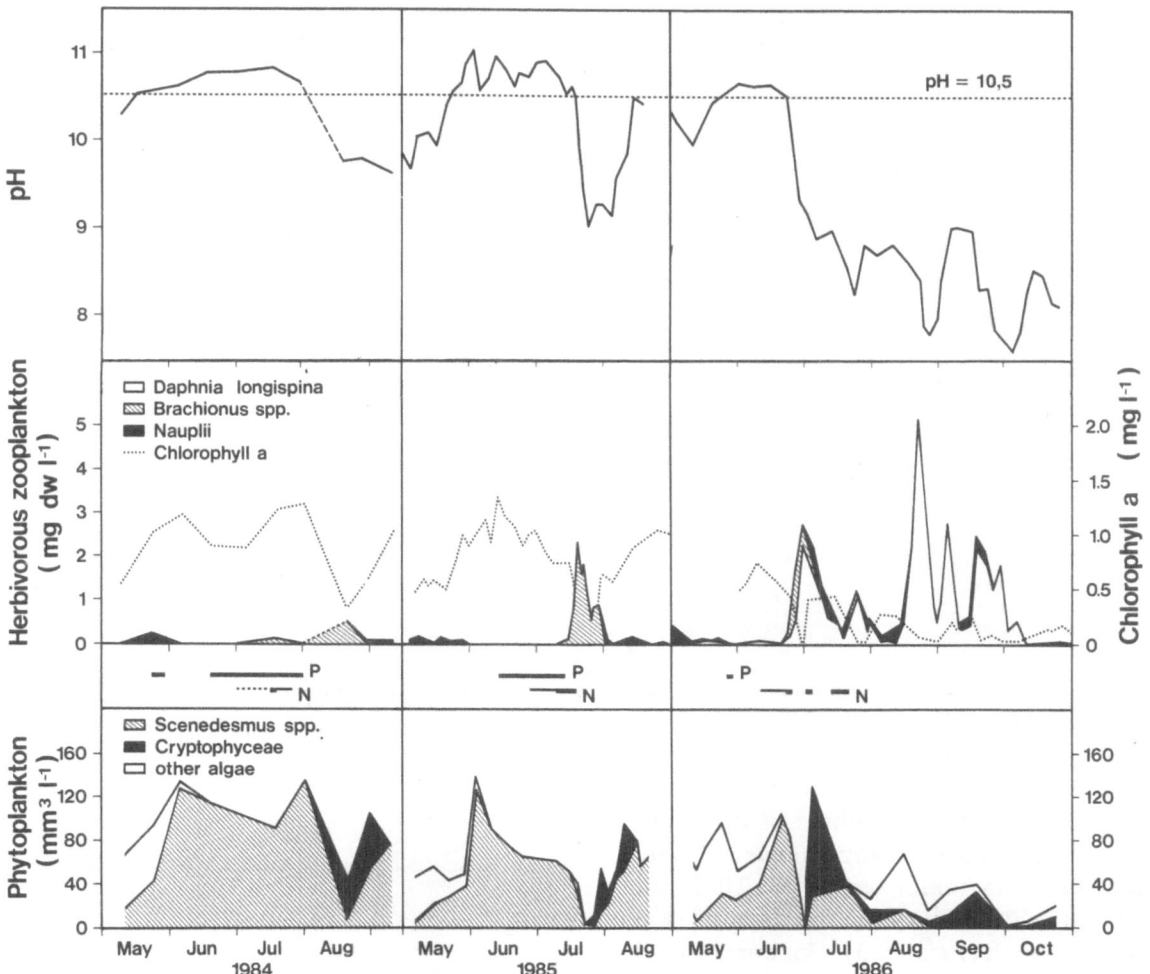

Fig. 2. Seasonal and year-to-year variation in pH, biomass of herbivorous zooplankton, and biovolume of various groups of phytoplankton. In between the latter two further incidences are shown; periods with low nitrogen input from the inlet (closed thin bars), and nitrate depletion ($< 2\ \mu g\ N\ l^{-1}$) in the lake (closed thick bars), and periods with orthophosphate concentration in the lake below $10\ \mu g\ P\ l^{-1}$.

During the first step in the collapse, pH was reduced from a mid-morning level of 10.7–10.8 to 10.5. Herbivorous zooplankton density was extremely low prior to the collapse, and remained low during its first step (Figs. 2 and 5).

In spite of the high level of total phosphorus and total nitrogen, the concentration of inorganic phosphorus was low ($< 20\ \mu g\ P\ l^{-1}$), and nitrate was fully depleted ($< 2\ \mu g\ N\ l^{-1}$) towards the end of the first step in the collapse (Fig. 2 and Søndergaard *et al.*, 1989).

The nitrogen content in the particulate organic

matter (mainly phytoplankton) ranged from 6.7 to 8.6% of organic d.w. from May to the clear-water period in July, and showed no trends during that period (Fig. 4). However, the N/P-ratio in the particulate organic matter showed considerable variations, from 6.2 in May to peak values of 10 and later 9.6 during the first step in the collapse (Fig. 4). In between those peaks the N/P-ratio declined to 8, concurrently with a tendency to a temporary increase in chl.-*a*.

The rapid decline in phytoplankton biomass in the second phase of the collapse followed the

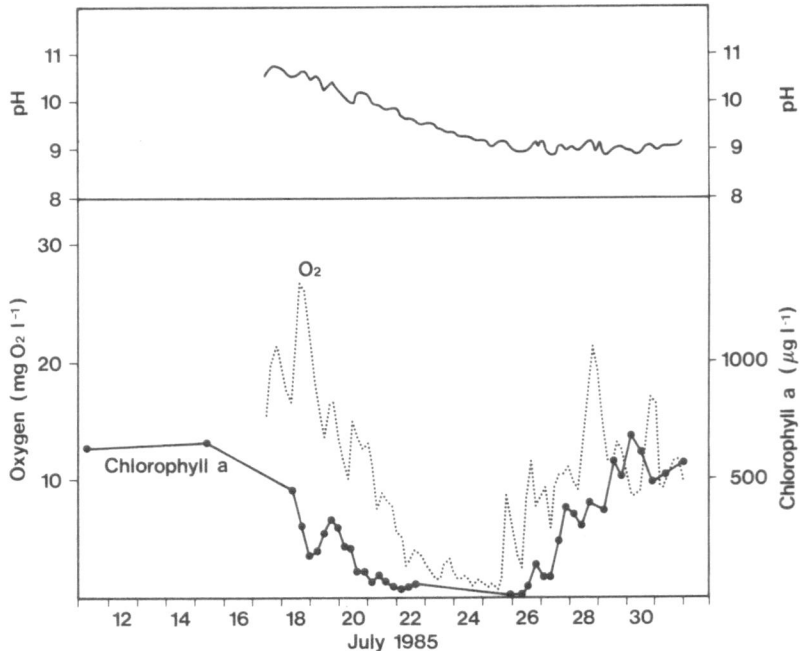

Fig. 3. Diel variations in pH and concentrations of oxygen and chlorophyll *a* during the phytoplankton collapse in 1985.

nitrate depletion (Fig. 2), and was accompanied by an abrupt increase in numbers of the rotifers *Brachionus calyciflorus* and *Brachionus urceolaris*, which reached a maximum density of 17 000 indiv. l^{-1} or 2.8 mg d.w. l^{-1} within 10 days after the initiation of growth (Figs. 2, 5). Concurrently, the zooplankton clearance rate on phytoplankton increased from near-zero values of 23 ml l^{-1} h^{-1} (Table 1).

The rotifer peak was followed by a rapid decrease in density and formation of mictic resting eggs (Fig. 5). Simultaneously, due to growth of especially *Cryptomonas spp.* and *Scenedesmus spp.* (Fig. 2), the phytoplankton biomass increased to 750 μg chl.-*a* l^{-1} within 14 days.

1984

In 1984 the sampling was run only bi-weekly. A collapse in phytoplankton was observed in August, accompanied by a temporary increase in the density of *Brachionus calyciflorus* (Fig. 2). Sampling was made towards the end of the col-

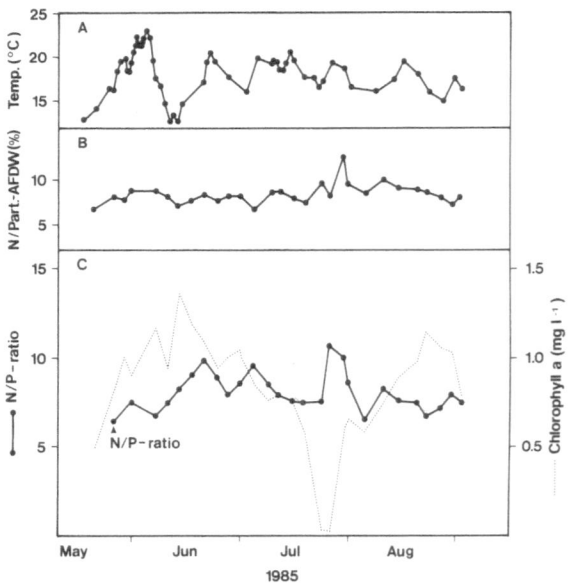

Fig. 4. Changes in A) temperature, nitrogen content in organic mater (AFDW) (mainly algae), B) N/P ratio in organic matter (AFDW), and C) chlorophyll *a* concentration in lake water during summer 1985.

Table 1. Mid-morning level of estimated zooplankton clearance rates of phytoplankton in Lake Søbygård (ml 1^{-1} h^{-1}).

Date	Zooplankton size groups			
	$> 140\ \mu m$	20–$140\ \mu m$	total ($> 20\ \mu m$)	
18 June 85	0	0	0	before collapse
19 July 85	0.7	10.3	10.9	during collapse
23 July 85	2.7	20.0	22.7	during collapse
8 Aug. 85	0	1.0	1.0	after collapse
24 June 86	0.2	0.2	0.2	before collapse
21 Aug. 86	0	21.7	21.7	during collapse
29 Aug. 86	13.7	5.9	19.6	during collapse

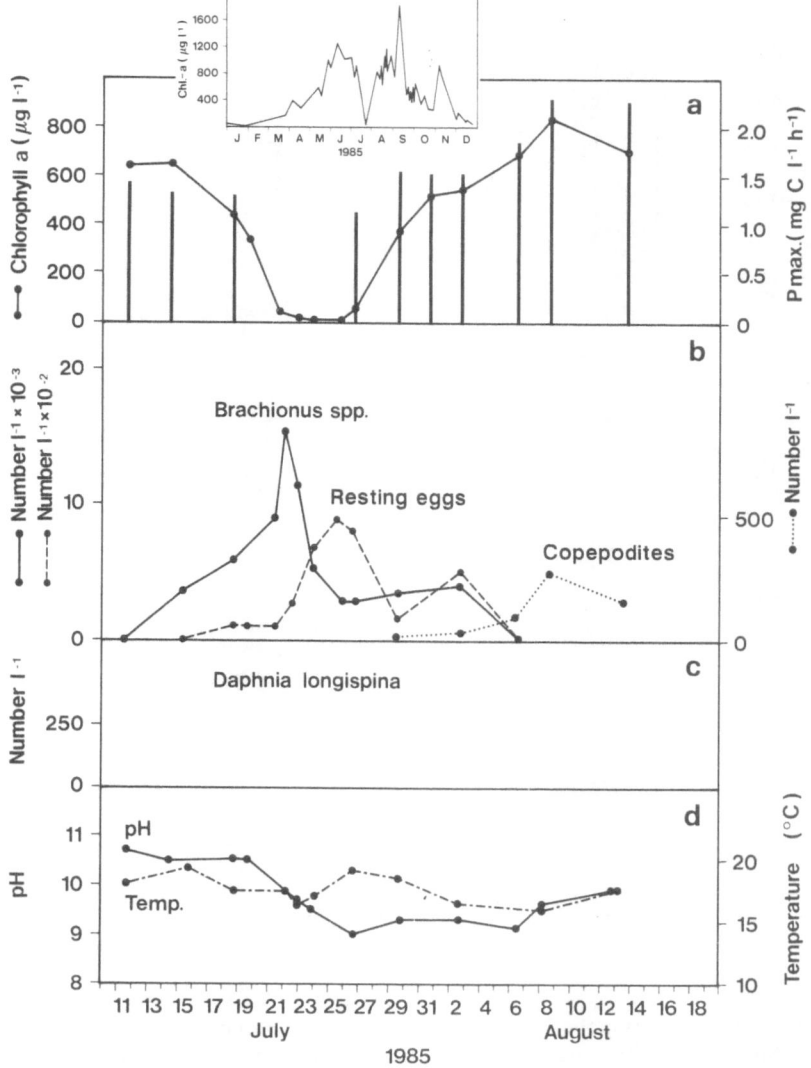

Fig. 5. Changes in selected variables during the collapse in 1985: a) chlorophyll *a* (full line) and maximum photosynthetic capacity, P_{max} (full bars). Small figure: Seasonal variation in chlorophyll *a* concentration. b) numbers of *Brachionus urceolaris* and *Brachionus calyciflorus* (full line), numbers of resting eggs (broken line), and numbers of copepodites (mainly of *Cyclops vicinus*) (dotted line). c) numbers of *Daphnia longispina* above 1 indiv. 1^{-1}. d) mid-morning level of pH and temperature.

lapse, at which the rotifer population was in the decreasing phase as indicated by high numbers of resting eggs. The rotifer density had probably been much higher earlier in the collapse.

This year cryptophytes and *Scenedesmus spp.* also dominated phytoplankton communities in post-collapse period of re-growth. A small de-

crease in phytoplankton biomass and a peak of *Brachionus urceolaris* were also observed in July (Fig. 2), but the *Scenedesmus* community recovered simultaneously with an increase in the nitrate input to the lake, and consequently an increase in the nitrate concentration in the lake (Fig. 2).

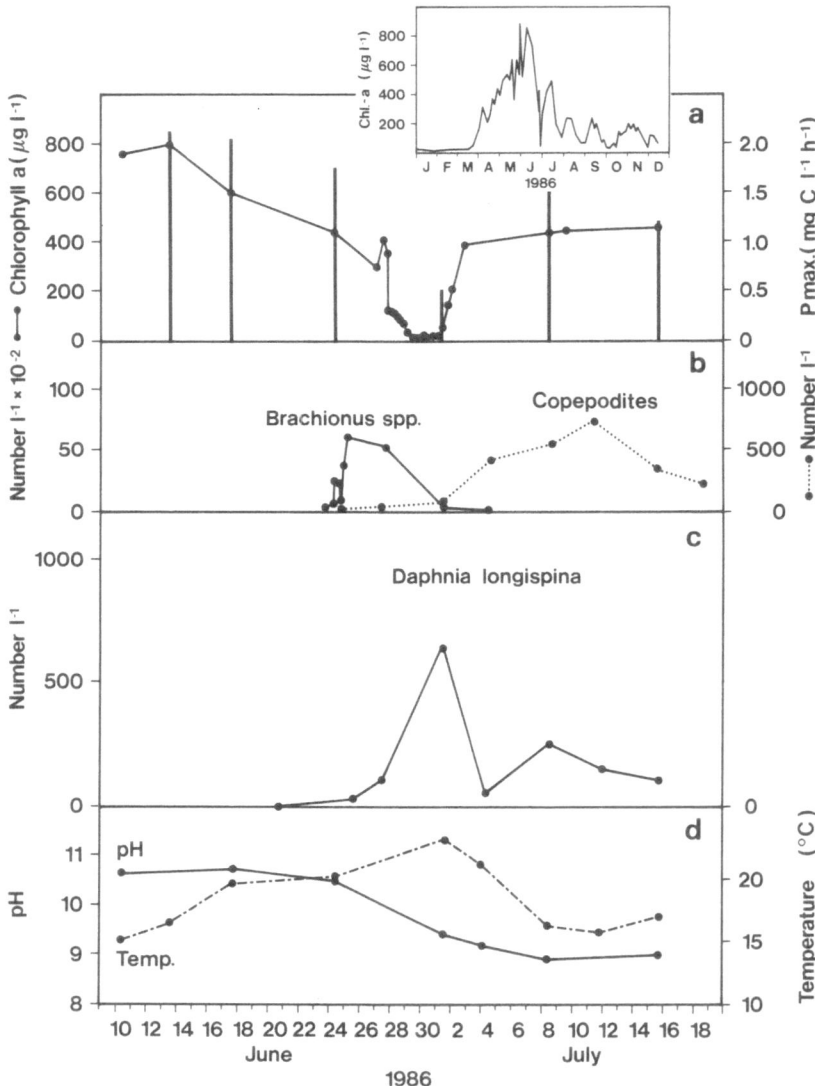

Fig. 6. Changes in selected variables during the first collapse in 1986: a) chlorophyll *a* (full line) and maximum photosynthetic capacity, P_{max} (full bars). Small figure: Seasonal variation in chlorophyll *a* concentration. b) numbers of *Brachionus urceolaris* and *Brachionus calyciflorus* (full line), and numbers of copepodites (mainly of *Cyclops vicinus*) (dotted lines). c) numbers of *Daphnia longispina*. d) mid-morning level of pH and temperature.

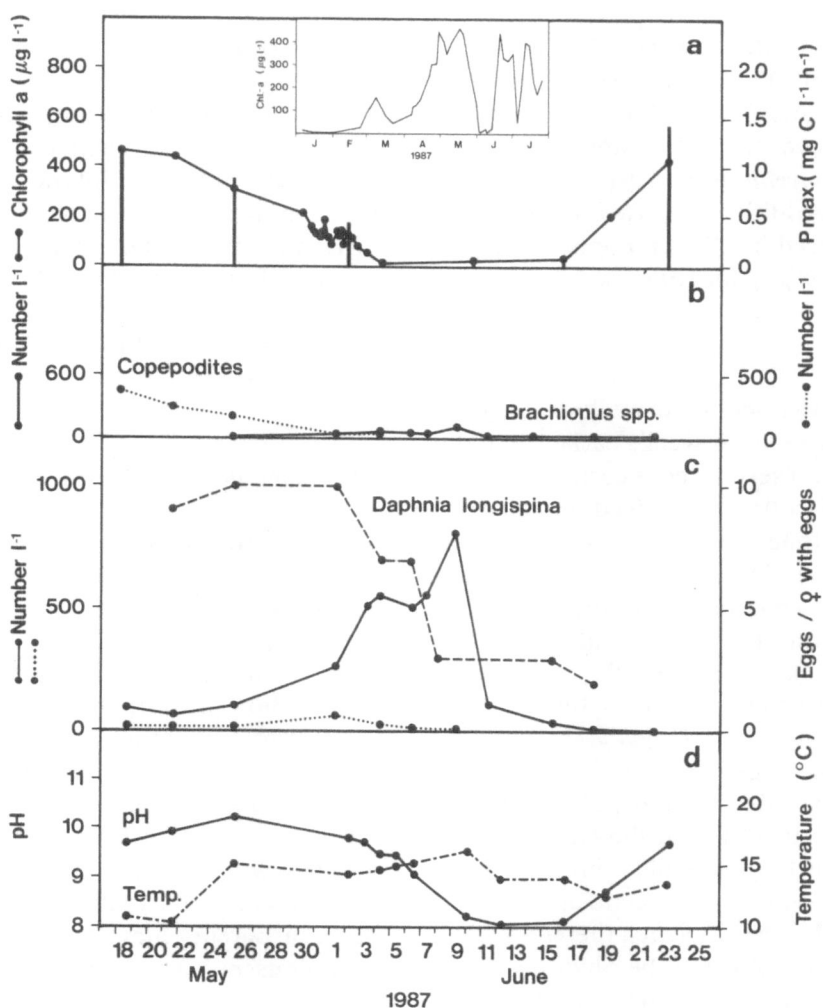

Fig. 7. Changes in selected variables during the first collapse in 1987: a) Chlorophyll *a* (full line) and maximum photosynthetic capacity, P_{max} (full bars). Small figure: Seasonal variation in chlorophyll *a* concentration. b) numbers of *Brachionus urceolaris* and *Brachionus calyciflorus* and numbers of copepodites (mainly of *Cyclops vicinus*). c) numbers of *Daphnia longispina* (full line), numbers of eggs per female with eggs (broken line), and numbers (left Y-axis) of female with eggs (dotted line) of *D. longispina*. d) mid-morning level of pH and temperature.

1986

In 1986 four events of phytoplankton collapses were observed. The first collapse occurred three weeks earlier than in 1985, but followed a remarkable similar pattern (Figs. 2, 6). It involved initial decline in chl.-*a* accompanied by a tendency to an increase in the P_{max}/chl.-*a* ratio, a decline in pH from 10.6-10.7 to less than 10.5, and nitrate

depletion ($< 2 \mu g$ N l^{-1}). Then a steep decrease followed in connection with an increase in the density of rotifers (mainly *Brachionus calyciflorus*), but in 1986 additionally succeeded by a marked increase in the abundance of *Daphnia longispina* and a simultaneously reduction in rotifers (Fig. 6). Probably due to the relatively high water temperature (Fig. 6) and high irradiance (not shown), a fast recovery of phytoplankton (mainly cryptophyceae) was then observed (Fig. 2). In 1986 the chl.-*a* level after the collapse only reached half of the pre-collapse level before a new decline occurred, simultaneously with a rapid increase in the density of *B. calyciflorus* and especially *D. longispina* (Fig. 2). Additionally, two further collapses were observed in 1986 in connection with new peaks of *D. longispina* (Fig. 2).

1987

Up to July there was one major collapse in 1987 (May) which occurred four weeks earlier than the first one in 1986, and seven weeks earlier than that in 1985. The 1987-collapse followed a somewhat different pattern than the collapses in the two preceding years (Fig. 7). Due to unusual climatic conditions in early summer, with only half of the mean irradiance for the period and low water temperature (Fig. 7), chl.-*a* level, P_{max}, and consequently pH were lower prior to the collapse than in the preceding years (Figs. 5–7).

Hence, chl.-*a* was then 450 μg l^{-1} in 1987, but 800-1200 μg l^{-1} in 1985-86, and pH 9.8-10.2 in 1987 compared to 10.7-10.8 in the two previous years. The nutrient level was also different prior to the collapse. Neither phosphate nor nitrate were depleted, and these nutrients, therefore, cannot have been responsible for the phytoplankton collapse in this year. However, the collapse occurred simultaneously with an increase in the density of *Daphnia longispina* from near-zero numbers to 800 indiv. l^{-1} (Figs. 2, 7), and a temperature increase from 10 °C to 14-15 °C (Fig. 7).

During the collapse, the number of egg-bearing females and the percentage of females with eggs were reduced markedly (Fig. 7), and simultaneously the number of eggs per female with egg decreased from 9-10 to 2-3 (Fig. 7). Subsequently, a steep decline in the density of *D. longispina* was observed.

In accordance with the lower water temperature and the expected lower re-growth rate of phytoplankton in 1987 compared to 1985 and 1986 (Figs. 5, 6), the clear-water period lasted longer than in the preceding years, being 13 days in 1987 compared to 3-5 days in 1985-86.

Discussion

In spite of large differences from year to year and within years, the collapses in phytoplankton generally seem to follow a common pattern. Hence, when the pre-collapse level of chl.-*a* and pH were above 700-800 μg l^{-1} and 10.5, respectively (July 1985, May 1986) the collapses in phytoplankton seem to occur in two steps, otherwise in one step (June-Sept. 1986, and May-July 1987) (Fig. 2).

First step in the collapse:

The first step in the two-step collapses may be due to one or more of the following factors: reduced primary production, increased grazing by zooplankton, infections by viral pathogens, bacteria or fungi, increased sedimentation rate, or exhaustion of nutrients.

The first hypothesis can probably be rejected, since the maximum potential production per unit phytoplankton volume or chl.-*a* generally showed tendency to increase during the first step in all events with a two-step collapse (Figs. 2, 5, 6). This might be due to the improved light conditions as transparency increased. The production per m^2 was consequently altered less than phytoplankton biomass during that period. The primary production was, nevertheless, either not transformed to growth, or the phytoplankton produced was eliminated due to an increase in one or another of the loss factors.

Zooplankton grazing is responsible for short-lasting clear-water periods in many lakes (e.g. Sommer *et al.*, 1986). However due to the low density of herbivorous zooplankton grazing on phytoplankton was insignificant during the first step of the collapse in 1985 and the first collapse in 1986 (Table 1 and Fig. 2).

Infections by various pathogens can affect sedimentation velocity of phytoplankton (e.g. Jewson *et al.*, 1981). During the first step in collapses mentioned, no evidence of infection of the dominating phytoplankton group, *Scenedesmus* was found. Neither encysted zoospores, sporangia nor hyphae were observed, and we did not find any increase in the numbers of dead cells during that period.

Nevertheless, by means of sediment traps and by mass balance calculations, Kristensen & Jensen (1987) did find a steady increase in the specific loss rate by sedimentation in Lake Søbygård 1985. This rose from about 2-3% d^{-1} in the exponential growth phase of *Scenedesmus* (May), to 14-17% d^{-1} in the stationary phase (June), with a further increase to 25% d^{-1} towards the end of the first step in the collapse. Large differences in the specific loss rate by sedimentation between phytoplankton in the exponential growth phase and in nutrient-limited stationary phase have been observed, both in laboratory investigations (e.g. Titman & Kilham, 1976) in lakes or *in situ* enclosures (e.g. Jassby & Goldmann, 1974; Reynolds & Wiseman, 1982). Thus Titman & Kilham (1976) found a 3.3-fold higher specific loss rate by sedimentation in the nutrient-limited stationary phase than in the exponential growth phase in laboratory experiments with *Scenedesmus quadricauda*, which is an abundant *Scenedesmus* species in Lake Søbygård. The observed increase in specific loss rates by sedimentation might, therefore, indicate nutrient limitation. This is supported by the fact that orthophosphate concentration was low, and nitrate was depleted towards the end of the first step in the collapse in 1985 (Fig. 2). Although the nitrogen content in particulate organic matter (mainly phytoplankton) showed only minor changes from May to the clear-water period in 1985, marked changes in the N/P-ratio and consequently in the P-content occurred during this period. Maximum N/P-ratios were obtained after a period of relatively low water temperature (Fig. 4), when the net phosphorus release from the sediment was relatively low (Søndergaard *et al.*, 1989). Since the release of phosphate from the sediment was an important contribution to the overall phosphate pool, which can be used for phytoplankton growth in summer (Søndergaard *et al.*, 1989), a reduced sediment release might explain the observed increase in the N/P-ratio in the phytoplankton. Since N/P-ratios near or

above 10 may indicate phosphorus limitation for *Scenedesmus* (Kunikane *et al.*, 1984), lack of phosphorus might have been a regulating factor for the decline in phytoplankton biomass in 1985. Conversely, the first two collapses in 1986 occurred when nitrate was depleted ($<2\ \mu$g N l^{-1}) but orthophosphate was above 25 μg P l^{-1}. Nitrate was also depleted during the end of the first step in the collapse in 1985 (Fig. 2 and Søndergaard *et al.*, 1989). Therefore, we suggest that the first step in the two-step collapses was due to exhaustion of nutrient, probably either phosphate or nitrate, or both.

Second step and single-step collapses:

The second and fastest step in the collapses in 1984 to June 1986 and the single step collapse since June 1986 occurred simultaneously with a marked increase in numbers of filter-feeding zooplankton, which changed from very low values to extremely high peak densities (Fig. 2). Accordingly, high grazing rates were obtained (Table 1).

The relative importance of the different filter-feeders during the collapse changed from year to year. The rotifers *Brachionus spp.* dominated totally in 1984 and 1985, whereas a peak of *Daphnia longispina* succeeded the rotifer maximum during the first collapse in 1986, after which the importance of the rotifers steadily declined. This structural shift from small rotifers to larger crustaceans may be explained by changes in the age-structure of the fish population, which is totally dominated by roach (*Rutilus rutilus*) and rudd (*Scardinius erythropthalmus*). Probably due to direct or indirect effects of high pH (e.g. ammonia), all planktivorous fish species did not spawn in 1984–86 (Timmermann, 1987). Consequently their younger year-classes, which are believed to have a more severe impact on structure and density of zooplankton in shallow eutrophic lakes than older age-classes (Cryer *et al.*, 1986), have gradually become less abundant since 1984. Concurrently, as expected, the predation pressure on the larger zooplankton has been reduced portionally, and we observed the well-described shift towards larger zooplankton (e.g. Hrbáček, 1962).

The increase in abundance of *D. longispina* in summer from 1984 until 1986 could, however, also have been self-strengthened by an increase in the production of resting eggs in autumn, according to a successive increase in numbers of mictic reproducing females. Conversely, some years earlier in summer 1983, both *D. longispina* and *Daphnia magna* were in a short time able to develop high densities from sediment isolated within 'in situ' enclosures without fish (Andreasen et al., 1984). The population growth, therefore, seems not to be seriously limited by the quantity of resting eggs in Lake Søbygård.

pH-effect:

Although changes in fish stock and perhaps to a minor extent a delay in recolonization might explain the year-to-year changes in the zooplankton structure during the collapses, it still remains striking that filter-feeding zooplankton in summer was abundant only during the phytoplankton collapses in 1984 and 1985 (Fig. 2). This was despite the absence of predaceous zooplankton (mainly *Cyclops vicinus*) from the pelagial during a long period in mid-summer. Insufficient amounts of food or poor food-quality could be a possible explanation for the absence of the filter-feeders. However, several authors have shown that various species of *Scenedesmus* are suitable as food for different species of *Daphnia*, including *D. longispina* (e.g. Infante, 1973; Geller, 1975; Horn, 1981; Muck & Lampert, 1984; Bloem & Vijverberg, 1984). Additionally, Lyche (1984) found that one of the two most abundant species of *Scenedesmus* in Lake Søbygård, *S. quadricauda*, is ingested and also assimilated with a high efficiency by *D. longispina*. Furthermore, Gilbert & Starkweather (1978) have shown that *Brachionus calyciflorus* is able to feed on the other abundant *Scenedesmus* species in Lake Søbygård, *S. acuminatus*, although it seems to select for smaller particles by use of various rejection mechanisms (Starkweather, 1978).

Therefore, neither food-quality nor insufficient amounts of food (see Fig. 2) can explain the delayed initiation of growth of *D. longispina* in 1986 compared to 1987, and the only short-term occurrence of *Brachionus* spp. in 1984-85.

However, high pH has been shown to influence survival and reproduction of zooplankton. Bogatova (1962), Walter (1969) and Ivanova (1969) all found an upper pH limit of 10.5-11.5 for survival of cladocerans, and from experiments in fish ponds O'Brien & deNoyelles (1972) suggested that high pH (above 10.6) was responsible for a temporary disappearance of *Ceriodaphnia reticulata* from the ponds. Additionally, Hessen & Nielssen (1985) found reduced egg production at high pH values, and also concluded that a change in pH from 9.5 to 10.5 was the main reason for the disappearance of cladocerans (among others *D. longispina*) and most rotifers in fishless enclosures in Lake Gjersjøen. Mitchell & Joubert (1986) obtained highest capacity of population increase and highest longevity of *B. calyciflorus* between pH 8.5 and 9.5, and lowest values of both parameters between 10 and 10.5 in a series of batch experiments run at pH intervals of 7.5 to 10.5. *B. calyciflorus* thus prefers moderately high pH, but is severely affected when pH is above 10.

Therefore, it seems reasonable to suggest that direct or indirect effects of high pH are responsible for low density of filter-feeders in Lake Søbygård prior to the collapses in 1984 and 1985, and to the first collapse in 1986. This hypothesis is further supported by the fact that *D. longispina* in 1987, when the spring pH was relatively low (below 10.2), was abundant already in late May compared to the beginning of July in 1986. In 1987 the first peak in *D. longispina* of 800 indice 1^{-1} occurred 15 days after a marked increase in water temperature from 10 to 14-15 °C. Hence, temperature seems to be a more important regulating factor in this year.

In summary, high pH thus may suppress reproduction and survival of filter-feeding zooplankton, and consequently their abundance in hypertrophic lakes. High pH may, therefore, affect the grazing food-web in a similar way as high numbers of planktivorous fish. Hence, in the short-term, high pH may increase the effect of the fish on the trophic structure in the lake, leading to an enhanced eutrophication in terms of increased chl.-*a* and decreased Secchi transparency. Conversely, on a long time scale (more than, say, two years), high pH may result in a temporary reduction of the predation pressure on filter-feeders owing to the negative effect of high pH on fish spawning. This leads to a high density of filter-feeding zooplankton and high grazing pressure on phytoplankton. Consequently, phytoplankton biomass and production are reduced, and pH then lowered, which again positively and in a self-amplifying manner affects the reproduction and survival of the filter-feeders. Concurrently, the chl.-*a* level decreases, and Secchi transparency increases.

Zooplankton dynamics during and after the collapse:

During the second step of the two-step collapses, in which pH dropped to below 10.5, and during the single-step collapses, the filter-feeding zooplankton showed a somewhat similar response pattern in 1984-87. First, an almost exponential increase in the population was observed. Second, when the phytoplankton biomass was almost exhausted, the numbers of cladoceran females with eggs as well as the numbers of eggs per female declined (1986 and 1987), and rotifers formed mictic resting eggs (1984 and 1985). Third, due to the low egg production in the cladocerans and the formation of only resting eggs by the rotifers, the population decreased abruptly. The second and the third statements indicate food-limitation, at least for the cladocerans. This initially affects the egg production, when the available food concentration reaches a lower threshold level of about 0.2 mg C l^{-1} (Lampert & Schober, 1978), and concurrently a decrease in the population density follows.

A similar pattern was observed by Lampert *et al.* (1986) during clear-water conditions in the Schönsee.

The formation of mictic eggs in *B. calyciflorus* has been shown to be induced by crowding (Gilbert & Starkweather, 1978), but it is also possibly an adaptive reaction to low food concentrations.

The rotifers reacted differently in 1986, as no mictic eggs were produced. The decline in population occurred simultaneously with a density increase of both *D. longispina* and copepodites of mainly *Cyclops vicinus* (Fig. 6). The rotifer decline might, therefore, be due to a competitive exclusion by the cladocerans, as shown by e.g. Gilbert & Starkweather (1978); or to predation by the copepodites and adult cyclopoids, which has been shown to have positive electivity for the genus *Brachionus* (Brandl & Fernando, 1978); or both factors could be involved. The total disappearance of rotifers in early August 1985 (Fig. 5) and in late August 1984 (Fig. 2), in spite of re-development of phytoplankton and the still only moderately high pH, could also be due to the increased density of copepodites and adult cyclopoids.

Hence, while high pH is suggested to be the major controlling factor for the pre-collapse densities of filter-feeding zooplankton in 1985 to June 1986, food and possibly also predation by carnivorous zooplankton were important regulators during and immediately after the collapses mentioned.

The post-collapse response pattern in the lake varied from year to year. In 1984 and 1985, when rotifers were the most abundant filter-feeders during the collapse, phytoplankton biomass and pH reached to the pre-collapse level shortly after the collapse, and the regulating role of pH was suggestively re-obtained. However, in 1986 and 1987, when *D. longispina* was more important, phytoplankton biomass and pH did not reach the pre-collapse level before a new collapse was encountered in connection with new peaks in zooplankton density (Fig. 2). The role of pH as a strong regulator of the abundance of filter-feeders was thus restricted to the period before the first collapse in 1986, and due to unusual weather conditions pH was less important in 1987 even in the pre-collapse period.

Clear-water periods and collapse pattern:

Clear-water periods occur in many eutrophic lakes, although the duration and the frequency vary considerably (Barica, 1975; Lampert & Schober, 1978; Fott *et al.*, 1980; Lampert *et al.*, 1986; Sommer *et al.*, 1986). In shallow hypertrophic lakes the composition of the fish stock, including density and age composition of planktivorous fish, seems to play the major role for the pattern and length of the clear-water phase. In fish-free conditions, either due to winter kill under ice, summer kill, or intensive fish harvesting, a high spring peak in phytoplankton is often followed by a long-lasting clear-water period (Barica, 1975; Fott *et al.*, 1980; Lynch, 1980; Andersson & Cronberg, 1984).

The clear-water period is sustained by large cladocerans (as *D. magna* and *D. pulex*) which frequently occur in high densities, and is often accompanied by dominance of the cyanobacterium, *Aphanizomenon flos-aquae* (Lynch, 1980; Andersson & Cronberg, 1984; Benndorf *et al.*, 1984). This pattern has also been found in Lake Søbygård in the seventies (Holm & Tuxen-Petersen, 1975; Andersen *et al.*, 1979; Jeppesen *et al.*, 1985) in connection with fish kill in summer.

Conversely, when the fish stock recovers, and the hypertrophic lakes explodes with young yearclasses of planktivorous fish, or alternatively when the lake is manually stocked with high densities of planktivorous fish, the phytoplankton community often changes to small chlorococcal greens (Fott *et al.*, 1980; Benndorf *et al.*, 1984; Jeppesen *et al.*, in press). In that case no or only a few short-term collapses, and consequently clear-water periods, may occur (Fig. 2). This is due to the fact that growth and survival of filter-feeding zooplankton, and then grazing on phytoplankton, is suppressed by both predation by planktivorous fish and by direct or indirect effects of elevated pH, the latter due to a more-or-less unlimited growth of phytoplankton. The collapses occur only when pH is reduced below about 10.5, e.g. due to a low nutrient or cloudy weather mediated reduction in primary production. High-pH tolerant rotifers (e.g. *Brachionus*

calyciflorus, *Brachionus urceolaris*) dominate the zooplankton community during the collapse.

However, when predation pressure of planktivorous fish on zooplankton is moderate, here due to lack of fish spawning for more than two years, phytoplankton is often dominated by cryptomonads (Fott *et al.*, 1980; Shapiro & Wright, 1984; Reinartsen & Olsen, 1984) which can compensate for grazing by high growth rates (Fott *et al.*, 1980), or large chlorococcal greens or gelatinous algae (Benndorf *et al.*, 1984; Jeppesen *et al.*, in press). In that case several collapses and short-lasting clear-water periods may occur during the summer in the hypertrophic lake, primarily due to grazing by intermediate-sized cladocerans (e.g. *D. longispina*). The oscillations may be due to alternations in 1) food conditions for zooplankton (e.g. below or above a threshold for egg production), 2) predation pressures by predatory zooplankton and fish (e.g. lack of timing in growth of prey and predator), 3) pH, or 4) climatic conditions.

Acknowledgements

This study was financially supported by the National Agency of Environmental Protection. The staff of National Environmental Research Institute are gratefully acknowledged for assistance. Special thanks are due to Jane Stougaard, Birte Laustsen and Ursula Gustavsen for careful quantification of phytoplankton and zooplankton, and to Hanne Rossen for typing the manuscript.

References

Andersen, J. M., J. Jensen, Aa. Kristensen & K. Kristensen, 1979. Undersøgelse af forureningstilstanden i Søbygård sø 1978. Report from Århus County: 29 pp (in Danish).

Andersson, G. & G. Cronberg, 1984. *Aphanizomenon flos-aqua* and *Daphnia*. An interesting plankton association in hypertrophic waters. In Bosheim & Nicholls (eds.), Interaksjoner mellom trofiske nivåer i ferskvann. Norsk Limnologförening, Blindern, Oslo, Norway: 63–76.

Andreasen, K., M. Søndergaard & H. H. Schierup, 1984. En

karakteristik af forureningstilstanden i Søbygård sø – samt en undersøgelse af forskellige restaureringsmetoders anvendelighed til en begrænsning af den interne belasting. Publ. No. 7 from The Botanical Institute, University of Århus, Denmark: 164 pp (in Danish).

Barica, J., 1975. Collapses of algal blooms in prarie pothole lakes: their mechanism and ecological impact. Verh. int. Ver. Limnol. 19: 606–615.

Benndorf, J., H. Kneschke, K. Kossatz & E. Penz, 1984. Manipulation of the pelagical food web by stocking with predaceous fishes. Int. Revue ges. Hydrobiol. 69: 407–428.

Bjørnsen, P. K., J. B. Larsen, O. Geertz-Hansen & M. Olsen, 1986. A field technique for the determination of zooplankton grazing on natural bacterioplankton. Freshwat. Biol. 16: 245–253.

Bloem, J. & J. Vijverberg, 1984. Some observations on diet and food selection of *Daphnia hyalina* (Cladocera) in an eutrophic lake. Hydrobiol. Bull. Amsterdam 18: 39–45.

Bogatova, I. B., 1962. Lethal ranges of oxygen content, temperature and pH for some representatives of the family Chydoridae. Zool. Zh. 41: 58–62 (in Russian).

Bottrell, H. H., A. Duncan, Z. M. Gliwicz, E. Grygierek, A. Herzig, A. Hillbricht-Ilkowska, H. Kurasawa, P. Larsson & T. Weglenska, 1976. A review of some problems in zooplankton production studies. Norw. J. Zool. 24: 419–456.

Brandl, Z. & C. H. Fernando, 1978. Prey selection by the cyclopoid copepods *Mesocyclops edax* and *Cyclops vicinus*. Verh. int. Ver. Limnol. 20: 2505–2510.

Cryer, M., G. Peirson & C. R. Townsend, 1986. Reciprocal interactions between roach, *Rutilus rutilus*, and zooplankton in a small lake: prey dynamics and fish growth and recruitment. Limnol. Oceanogr. 31: 1022–1038.

Culver, D. A., M. M. Bourcherle, D. J. Bean & J. W. Fletcher, 1985. Biomass of freshwater Crustacean zooplankton from length-weight regressions. Ca. J. Fish. aquat. Sci. 42: 1380–1390.

Dumont, H. J., I. Van De Velde & S. Dumont, 1975. The dry weight estimate of biomass in a selection of Cladocera, Copepoda and Rotifera from plankton, periphyton and benthos of continental waters. Oecologia (Berl.) 19: 75–97.

Fott, J., L. Pechar & M. Prazaková, 1980. Fish as a factor controlling water quality in ponds. In J. Barica & L. R. Mur (eds.): Hypertrophic ecosystems. Developments in Hydrobiology 2: 255–261.

Geller, W., 1975. Die Nährungsaufnahme von *Daphnia pulex* in Abhängigkeit von der Futterkonzentration, der Temperatur, der Körpergrösse und dem Hungerzustand der Tiere. Arch. Hydrobiol/Suppl. 48: 47–107.

Gilbert, J. L. & P. L. Starkweather, 1978. Feeding in the rotifer *Brachionus calyciflorus* III. Direct observations on the effect of food type, food density, change in food type and starvation on the incidence of pseudotrochal screening. Verh. int. Ver. Limnol. 20: 2382–2388.

Hessen, D. O. & J. P. Nielssen, 1985. Factors controlling rotifer abundancies in a Norwegian eutrophic lake: an experimental study. Ann. Limnol. 21: 97–105.

Holm, T. F. & F. Tuxen-Petersen, 1975. Gudenåen 1974. Rep. from Lab. Physical Geography, Geological Institute, University of Århus, Denmark: 108 pp (in Danish).

Horn, W., 1981. Phytoplankton losses due to zooplankton grazing in a drinking water reservoir. Int. Revue ges. Hydrobiol. 69: 781–817.

Hrbáček, J., 1962. Species composition and the amount of zooplankton in relation to fish stock. Rozpravy Československe Akademie Ved. 72,10: 114 pp.

Infante, A., 1973. Untersuchungen über die Ausnutzarkeit verschiedener Algen durch das Zooplankton. Arch. Hydrobiol./Suppl. 42: 340–405.

Ivanova, M. B., 1969. The influence of active water reaction on filtering rate of Cladocera. Pol. Arch. Hydrobiol. 16: 115–124.

Jassby, A. D. & C. R. Goldmann, 1974. Loss rates from a lake phytoplankton community. Limnol. Oceanogr. 19: 618–624.

Jeppesen, E., O. Sortkjær, Aa. Rebsdorf, M. Søndergaard, P. Jensen, P. Kristensen, K. Andreasen, H. H. Schierup & J. M. Andersen, 1985. Recovery of a shallow lake. Reprint from a poster session ('Phosphorus in freshwater ecosystems', Uppsala, Sweden, 1985). Publ. No. 33 from the Freshwater Laboratory, National Agency of Environmental Protection: 10 pp.

Jeppesen, E., M. Søndergaard, E. Mortensen, J. P. Jensen, P. Kristensen, B. Riemann, H. J. Jensen, J. P. Müller, O. Sortkjær, K. Christoffersen, S. Bosselmann & E. Dall (in press). Fish Manipulation as a lake restoration tool in shallow, eutrophic lakes: analysis of three case-studies and data from 300 Danish lakes. Proc. Int. Conf. 'Biomanipulation, Tool for Watermanagement', Developments in Hydrobiology.

Jewson, D. H., B. H. Rippey & W. K. Gilmore, 1981. Loss rates from sedimentation, parasitism and grazing during the growth, nutrient limitation, and dormancy of diatom crop. Limnol. Oceanogr. 26: 1045–1056.

Kristensen, P. & P Jensen, 1987. Sedimentation og resuspension i Søbygård sø. M. sc. thesis. Publ. No. 64 from the Freshwater Laboratory, The National Agency of Environmental Protection, Silkeborg, Denmark, in cooperation with Botanical Institute, University of Århus, Denmark: 176 pp (in Danish).

Kunikane, S., M. Kaneko & R. Maehare, 1984. Growth and nutrient uptake of green algae *Scenedesmus dimorphus* under a wide range of nitrogen/phosphorus ratios. 1. Experimental study. Wat. Res. 18: 1299–1311.

Lampert, W. & U. Schober, 1978. Das regelmässige Auftreten von Frühjahrs-Algenmaximum und 'Klarwasserstadium' im Bodensee als Folge klimatischen Bedingungen und Wechselwirkungen zwischen Phyto- und Zooplankton. Arch. Hydrobiol. 82: 364–386.

Lampert, W., W. Fleckner, H. Rai & B. E. Taylor, 1986. Phytoplankton control by grazing zooplankton: a study on the spring clearwater phase. Limnol. Oceanogr. 31: 478–490.

Limnologisk Metodik, 1977. Freshwater Laboratory, University of Copenhagen and Akademisk Forlag: 163 pp (in Danish).

Lyche, A., 1984. Experimentelle studier av beitbarhet og assimilerbarhed av 6 planktonalger for *Daphnia longispina*. In S. Bosheim and Nicholls (eds.), Interaksjoner mellom trofiske nivåer i ferskvann. Norsk Limnologförening, Blindern, Oslo, Norway: 49–62 (in Norwidish).

Lynch, M., 1980. *Aphanizomenon* blooms: Alternate control and cultivation by *Daphnia pulex*. Am. Soc. Limnol. Oceanogr. Spec. Symp. 3: 299–304. Univ. Press New England, Hanover.

Mitchell, S. A. & J. H. B. Joubert, 1986. Effect of elevated pH on the survival and reproduction of *Brachionus calyciflorus*. Aquaculture 55: 215–220.

Muck, P. & W. Lampert, 1984. An experimental study on the importance of food conditions for the relative abundance of calanoid copepods and cladocerans. 1. Comparative feeding studies with *Eudiaptomus gracilis* and *Daphnia longispina*. Arch. Hydrobiol. Suppl. 66: 157–169.

O'Brien, W. J. & F. deNoyelles, 1972. Photosynthetically elevated pH as a factor in zooplankton mortality in nutrient enriched ponds. Ecology 53: 605–614.

Reinertsen, H. & Y. Olsen, 1984. Effects of fish elimination on the phytoplankton community of a eutrophic lake. Verh. int. Ver. Limnol. 22: 649–657.

Reynolds, C. S. & S. W. Wiseman, 1982. Sinking losses of phytoplankton in closed limnetic systems. J. Plankton Res. 5: 203–234.

Shapiro, J. & D. I. Wright, 1984. Lake restoration by biomanipulation: Round Lake, Minnesota, the first two years. Freshwat. Biol. 14: 371–383.

Sommer, U., Z. M. Gliwicz, W. Lampert & A. Duncan, 1986. The PEG-model of seasonal succession of plankton in freshwater. Arch. Hydrobiol. 106: 433–471.

Sortkjær, O. & E. Jeppesen, 1987. Hyppige feltmålinger øger forståelsen af lavvandede søers udvikling. In Annual Report 1986 from the Freshwater Laboratory, The National Agency of Environmental Protection: 21–28 (in Danish).

Starkweather, P. L., 1978. Behavioral determinants of diet quantity and diet quality in *Brachionus calyciflorus*. In Special symposium. Amer. Soc. of Limnol. Oceanogr.: 151–157.

Stewart, J. & R. G. Wetzel, 1986. Cryptophytes and other microflagellates as couplers in planktonic community dynamics. Arch. Hydrobiol. 106: 1–19.

Søndergaard, M., E. Jeppesen & O. Sortkjær, 1987. Lake Søbygård: a shallow lake in recovery after a reduction in phosphorus loading. GeoJournal 14: 381–384.

Søndergaard, M., E. Jeppesen, P. Kristensen & O. Sortkjær, 1990. Interactions between sediment and water in a shallow and hypertrophic lake: a study on phytoplankton collapses in Lake Søbygård, Denmark. Hydrobiologia 191: 139–148.

Timmermann, M., 1987. Manglende reproduktion af skalle (*Rutilus rutilus*) (L.)) og rudskalle (*Scardinius erythrophthalmus* (L.)) og skalle- (*R. rutilus* (L.)), brasen- (*Abramis brama* (L.)) of aborre- (*Perca fluviatilis* L.) yngelens densitet og vækst i Væng Sø. Publ. No. 85 from the Freshwater Laboratory, The Agency of Environmental Protection, Silkeborg, Denmark: 81 pp (in Danish).

Titman, D. & P. Kilham, 1976. Sinking in freshwater phytoplankton: some ecological implications of cell nutrient status and physical mixing processes. Limnol. Oceanogr. 21: 409–417.

Uhlmann, D., 1971. Influence of dilution, sinking and grazing rate on phytoplankton populations of hyperfertilized ponds and micro-ecosystems. Mitt. int. Ver. Limnol. 19: 100–124.

Walter, B., 1969. Interrelations of Cladocera and algae. Ph.D. thesis. Westfield Coll., Univ. London.

Hydrobiologia **191**: 165–171, 1990.
P. Biró and J. F. Talling (eds), Trophic Relationships in Inland Waters.
© 1990 *Kluwer Academic Publishers.*

Phytoplankton and zooplankton (Cladocera, Copepoda) relationship in the eutrophicated River Danube (Danubialia Hungarica, CXI)

Anna Bothár & Keve T. Kiss
Hungarian Danube Research Station of the Hungarian Academy of Sciences, H-2131 Göd, Hungary

Key words: River Danube, phytoplankton, zooplankton, primary productivity, secondary productivity, ecological efficiency

Abstract

The seasonal variation in primary production, individual numbers, and biomass of phyto- and zooplankton was studied in the River Danube in 1981. The secondary production of two dominant zooplankton species (*Bosmina longirostris* and *Acanthocyclops robustus*) was also estimated. In the growing season (April-Sept.) individual numbers dry weights and chlorophyll *a* contents of phytoplankton ranged between $30\text{-}90 \times 10^6$ individuals, l^{-1}, $3\text{-}12$ mg l^{-1}, and $50\text{-}170$ μg l^{-1}, respectively. Species of Thalassiosiraceae (Bacillariophyta) dominated in the phytoplankton with a subdominance of Chlorococcales in summer. Individual numbers and dry weights of crustacean zooplankton ranged between 1400-6500 individuals m^{-3}, and 1.2-12 mg m^{-3}, respectively. The daily mean gross primary production was 970 mg C m^{-3} d^{-1}, and the net production was 660 mg C m^{-3} d^{-1}. *Acanthocyclops robustus* populations produced 0.2 mg C m^{-3} d^{-1} as an average, and *Bosmina longirostris* populations 0.07 mg C m^{-3} d^{-1}. The 'ecological efficiency' between phytoplankton and crustacean zooplankton was 0.03%.

Introduction

Investigations on phytoplankton and zooplankton (crustaceans) have been carried out weekly for 20 years at the Hungarian Danube Research Station. In the seventies pronounced changes were recorded in the trophic status of the Danube: both the averages and the maxima of algal biomass increased, reaching even ten times the former values. From that time changes in species composition of planktonic crustaceans, and the increase of their abundance, also indicate the gradual eutrophication of the River Danube.

The main purpose of this paper is to characterize the seasonal variations in primary production, individual numbers, and biomass of phyto- and zooplankton. The secondary production of two dominant zooplankton species (*Bosmina longirostris* and *Acanthocyclops robustus*) was also estimated.

Materials and methods, study area

The sampling site lies in the middle reaches of the Hungarian Danube section at Göd, above Budapest, longitudinal distance 1669 km of the river. Samples were taken weekly from the flow in 1981.

Net and gross primary productivity were estimated by Dvihally *et al.* (1982) with the dark and light bottle method (see detailed description of this method in Kothé, 1981).

The chlorophyll *a* concentration was measured by photometer after filtration and methanol-extraction (Felföldy, 1980). Phytoplankton counts were made by the Utermöhl technique, biomass was estimated from the cell dimensions (Kristiansen, 1971), dry weights and organic carbon contents were estimated from the proportions by weight commonly used: wet weight = 100%, dry weight = 20%, organic C = 10%. We also calculated primary productivity from these data on the basis of Gutelmacher's work (1986).

Zooplankton samples were taken using a centrifugal self-priming pump. Some 200 l of water from the surface and 200 l from near the bottom were filtered through a plankton net (mesh size: 75 μm). The filtrates were condensed to a volume of 150 ml and preserved in 4% formalin. All crustacean specimens in the samples were determined and counted. The numbers of individuals in the samples taken from the two layers were added up and referred to a cubic meter.

For estimation of the production of *Bosmina longirostris* the recruitment time was used (for detailed description see Bothár, 1986). The calculation of the production of *Acanthocyclops robustus* was carried out by the technique of growth increment summation as suggested by Rigler & Downing (1984). The embryonic and post-embryonic development times were determined in the laboratory under natural food conditions (Bothár, 1987).

Results

In the growing season, from April till October, water discharge conditions were favourable for phytoplankton and zooplankton development. After a big flood in April the water discharge was smaller than the mean low-water discharge till the end of July. After a short flood at the end of July, the water discharge was again smaller than the

mean low-water discharge, with little fluctuations till the middle of October (Fig. 1).

Daily averages of gross primary productivity ranged from 750-1650 mg C m^{-3} d^{-1} and those of net primary productivity from 530-1060 mg C m^{-3} d^{-1} (Table 1). During flood-waves, primary productivity could be measured only in the surface layers, whereas the maximum values were observed during low-water discharge, when the phytoplankton numbers were also high (Dvihally *et al.*, 1982).

The initial phytoplankton peak appeared in April. It was followed by a permanent bloom with small declines till the end of June when the greatest phytoplankton peak occurred. There fol-

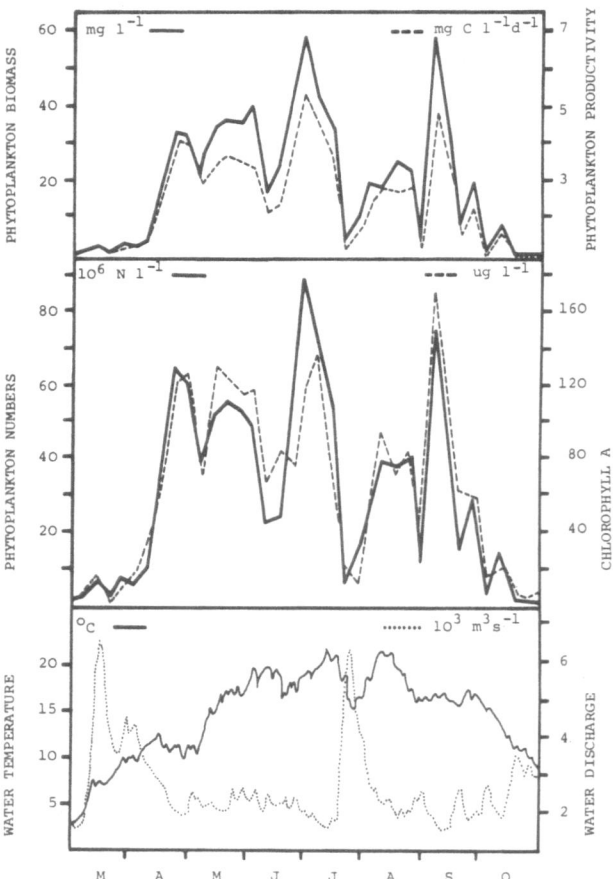

Fig. 1. Water temperature and water discharge of the River Danube during 1981; also chlorophyll *a* content, abundance, biomass and calculated productivity values of phytoplankton.

Table 1. Daily averages of primary and secondary productivity data referring to unit volume, unit surface and to the water volume discharged daily (V_d) through the cross-section of the River Danube. P/B values refer to the growing season.

		April	May	June	July	Aug.	Sept.	P/B
Phytoplankton	$mg\ C\ m^{-3}$	1910	3180	3240	3210	2590	2680	
biomass	$mg\ C\ m^{-2}$	9090	12240	12800	14350	10490	9300	
	$10^5\ kg\ C\ V_d^{-1}$	4.97	6.19	6.48	8.17	5.39	4.56	
Zooplankton	$mg\ CCm^{-3}$	0.85	5.80	3.47	1.60	1.95	1.61	
biomass	$mg\ C\ m^{-2}$	4.04	22.33	13.71	7.15	7.90	5.59	
	$kg\ C\ V_d^{-1}$	221.60	1129.50	694.90	407.40	406.20	274.40	
Gross	$mg\ C\ m^{-3}\ d^{-1}$	750	1650	1340	1620	1250	1220	
prim. prod.	$mg\ C\ m^{-2}\ d^{-1}$	3570	6350	5290	7240	5060	4230	
	$10^5\ kg\ C\ V_d^{-1}\ d^{-1}$	1.95	3.21	2.68	4.12	2.60	2.08	
Net	$mg\ C\ m^{-3}\ d^{-1}$	530	870	970	1060	940	910	
prim. prod.	$mg\ C\ m^{-2}\ d^{-1}$	2520	3350	3830	4740	3810	3160	
	$10^5\ kg\ C\ V_d^{-1}\ d^{-1}$	1.38	1.69	1.94	2.69	1.95	1.55	
Calculated	$mg\ C\ m^{-3}\ d^{-1}$	1750	2520	2080	2360	1810	1710	
prim. prod.	$mg\ C\ m^{-2}\ d^{-1}$	8330	9700	8210	10550	7330	5930	133
	$10^5\ kg\ C\ V_d^{-1}\ d^{-1}$	4.56	4.91	4.16	6.00	3.77	2.91	
Bosmina	$mg\ C\ m^{-3}\ d^{-1}$		0.03	0.17	0.01	0.10	0.03	
longirostris	$mg\ C\ m^{-2}\ d^{-1}$		0.12	0.67	0.04	0.40	0.10	46
production	$kg\ C\ V_d^{-1}\ d^{-1}$		5.80	34.00	2.50	20.80	5.10	
Acanthocyclops	$mg\ C\ m^{-3}\ d^{-1}$		0.18	0.23	0.19	0.21	0.16	
robustus	$mg\ C\ m^{-2}\ d^{-1}$		0.69	0.91	0.85	0.85	0.55	20
production	$kg\ C\ V_d^{-1}\ d^{-1}$		34.40	46.10	48.40	43.70	27.30	

lowed a radical decrease in phytoplankton numbers at the end of July, but by the low-water period in August and September the phytoplankton bloom showed two more peaks (Fig. 1).

In spring the phytoplankton featured a strong dominance (nearly 90%) of Thalassiosiraceae. The dominant species were *Stephanodiscus invisitatus* Hohn et Hellerman, *S. hantzschii f. tenuis* (Hust.) Håkansson et Stoermer, and *S. minutula* (Kütz.) Round. In summer and autumn the phytoplankton was dominated by species of Thalassiosiraceae with subdominants species of Chlorococcales and Cryptophyceae (Kiss, 1985, 1987).

The seasonal changes in biomass of phytoplankton and in chlorophyll *a* content were nearly parallel with the algal counts (Fig. 1). Calculated values of primary production were on average 36% higher than those of gross primary production measured by the O_2-method (Table 1).

Among 25 cladoceran species *Bosmina longirostris* was dominant, especially in the first part of the growing season. Among 13 copepods the dominant species were *Acanthocyclops robustus*, *Eucyclops serrulatus*, and *Cyclops vicinus*.

Cladocerans appeared in the water at the beginning of April. The first peak period was in May, which was followed by a drastic decrease at the end of June. In the following period three smaller peaks appeared. *B. longirostris* individuals created the greatest part of the first (Fig. 2.). In summer, other cladocerans (e.g. *Alonella rostrata*, *Iliocryptus sordidus*, *Moina micrura*, *Pleuroxus aduncus*, *Daphnia cucullata*) also dominated the zooplankton. The seasonal pattern of the biomass of cladocerans reflects the species diversity of larger individuals (Fig. 3).

Copepods appeared in March. There was a smaller peak in May, and a second, larger one with significant fluctuation between mid-July and mid-September. About 70% of the copepod individuals were in nauplius stage during the growing season (Fig. 4). Two peak periods can also be

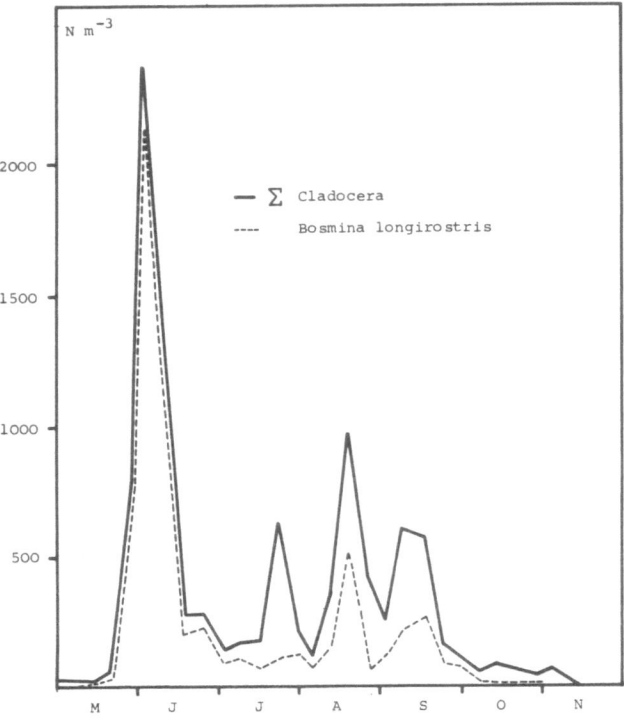

Fig. 2. Abundance of cladocerans and *Bosmina longirostris*.

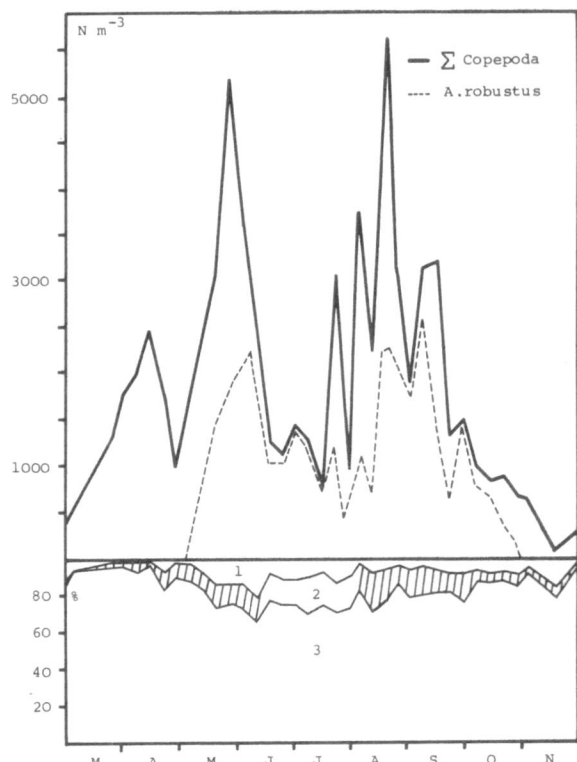

Fig. 4. Abundance of copepods and *Acanthocyclops robustus* (top), % of individuals in adult and larval stages (below), 1 = adult, 2 = copepodite, 3 = nauplius.

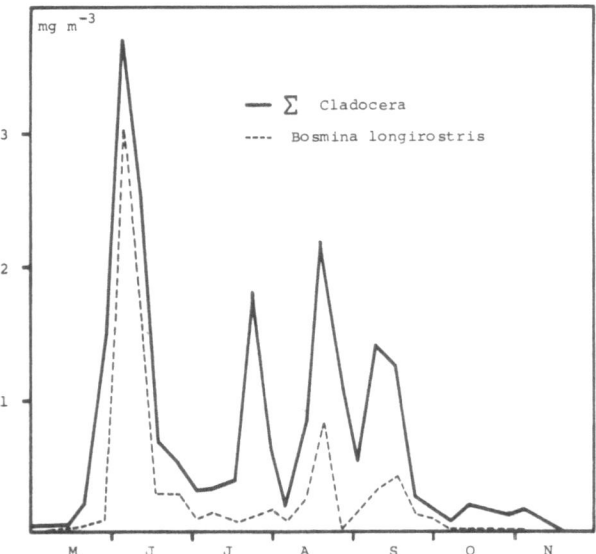

Fig. 3. Biomass of cladocerans and *Bosmina longirostris*.

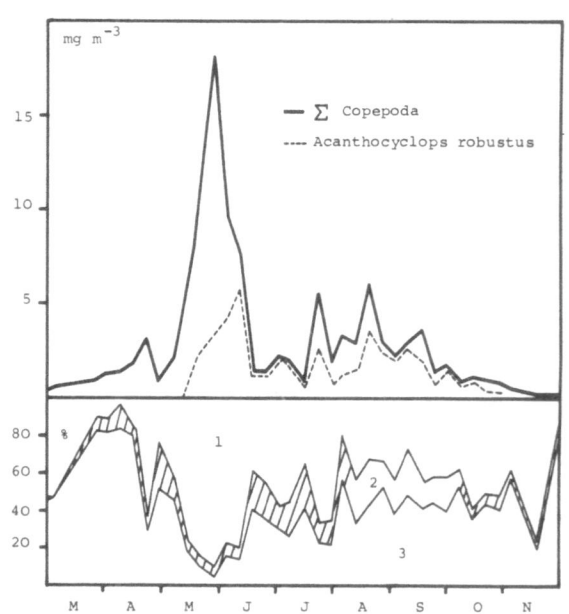

Fig. 5. Biomass of copepods and *Acanthocyclops robustus* (top), % of individuals in adult and larval stages (below), 1 = adult, 2 = copepodite, 3 = nauplius.

observed in the biomass pattern of copepods. The second one is smaller, as compared with the pattern of individual numbers. The main cause of this phenomenon is the occurrence in the first peak of larger populations of *C. vicinus*, whereas in July and September mainly *E. serrulatus* contributed to the copepod biomass, besides *A. robustus*. Adult copepods occurred mainly in May and June (Fig. 5).

Maximum values of abundance and biomass of crustaceans occurred in the periods of low-water discharge, and followed the phytoplankton maxima. This phenomenon, which is also well known in lakes (Patil & Gouder, 1985), demonstrates that crustaceans were supplied with food either in the form of algae or through the detrital chain. One indirect evidence of adequate food supply is the high fecundity values of the dominant cladoceran, *B. longirostris* (Bothár, 1986). For *B. longirostris* and *A. robustus*, maximum values of daily-produced organic carbon referred to unit water volume were found in June (Table 1).

We calculated the efficiency of the secondary producers as the ratio of their production to phytoplankton production, expressed as percentage on a growing season basis. This ratio provides an index of the relative amount of phytoplankton production that is subsequently incorporated into secondary production. This efficiency seems to be analogous to the ecological efficiency of Slobodkin (1960) and Kozlovsky (1967). Calculated on the basis of averaged daily values, 0.09% of the organic carbon content of phytoplankton 'appeared' in the crustacean biomass. Some 0.03% of organic carbon produced by phytoplankton 'appeared' in crustacean production. The total crustacean production was estimated to be twice that of *B. longirostris* and *A. robustus*, which can be considered a realistic estimation according to the biomass data.

Discussion

The trophic state of the River Danube, as judged from individual numbers, biomass, and chlorophyll *a* content of phytoplankton, can be characterized as eu-polytrophic. On the basis of the intensity of primary production, the river is to be categorized as eu-polytrophic to hypertrophic (Rodhe, 1969; Hübel, 1971, Felföldy, 1980; OECD, 1982).

In Table 1, we summarize the daily averages of our productivity data in three forms. The first line of each data-series refers to unit volume, i.e. concentration which is the most commonly used measure. In the second line all parameters are calculated for unit area to make them comparable with data from other water-bodies. In the third line, productivity parameters refer to the volume of water discharged daily through the cross-section of the Danube (V_d). The last calculation provides the most realistic picture about the biological characteristics of a river, in which water-level fluctuations may result in major alterations of discharge. Consequently, the concentration values only would not show the real situation, especially if we consider temporal variations. Thus in a high-water period, primary productivity of small intensity can reach the same productivity referring to the discharge as a more intensive one in a low-water period. Therefore, maximum productivity values that were calculated in different forms appeared in different months, because of changing hydrological characteristics of the river.

Few data are available on the primary productivity of large rivers, referring to their main channel. The majority of investigations were carried out in reservoirs, dead arms, and ox-bow lakes. According to Table 2, where we summarize pri-

Table 2. Average values of gross primary production in various rivers ($g\,C\,m^{-2}\,d^{-1}$).

River	$g\,C\,m^{-2}\,d^{-1}$	References
Ottawa (Canada)	0.30	Rosemarin 1975
Danube (Czechoslovakia)	0.69	Ertl & Juris 1967
Danube (Hungary)	1.22	Dvihally 1975
Vistula (Poland)	1.65	Javornický 1966
Volga (U.S.S.R.)	2.25	Tarasova 1970
Raritan (U.S.A.)	2.55	Flemer 1970
Danube (Hungary)	5.29	present paper

mary productivity data of several rivers, the present primary productivity of the River Danube exceeds all of them. The reason for this phenomenon is intensive eutrophication since the seventies (Kiss, 1985).

Data for secondary planktonic productivity are not available from running waters. Comparing our results with those from standing waters, abundance and production values are smaller in the Danube but turnover times are shorter (Bothár, 1986, 1987).

Our P/B value for phytoplankton (Table 1) is close to the average (113) measured in IBP waterbodies (Le Gren & Lowe-McConnell, 1980). The P/B values of the two zooplankton species investigated are above the IBP average (15.9); further, that of A. robustus is bigger than the upper limit of the IBP range (44). The probable reason is that mainly nauplii and copepodite stages contributed to the biomass of this species in the Danube (Fig. 5). Adult individuals created only 5% of the whole biomass. Therefore egg production was small, not more than 7% of the total production. Because of the long development time of the females, the greater part of egg production appears in a downstream section.

Our ecological efficiency value of 0.03% is about ten times smaller than the lower part of the range (0.1-27.4%) characteristic for IBP lakes. Low efficiency values were also obtained by several authors for eutrophic standing waters (Pederson et al., 1976; Adalsteinsson, 1979).

Our investigations strongly suggest that in running waters the energy transfer efficiency between primary and secondary trophic levels tends to vary more markedly and inversely with trophic state and eutrophication than in standing waters (Hillbricht-Ilkowska et al., 1972; Parsons, 1982). In eutrophicated lakes, the main reason is that the greater part of the phytoplankton production is not passed on to the zooplankton but channelled to different trophic levels (e.g. decomposers), or temporarily stored as a non-desirable algal bloom. These postulates are true in running waters, too. Contributing to this in flowing waters, the hydraulic conditions and the 'lifespan' of the river, sensu Rzóska (1978), have different influences on the phytoplankton and zooplankton.

In the middle section of the Danube, with an average flow velocity of 1-1.2 m s^{-1}, the phytoplankton organisms are capable of creating mass productions because of their rapid development. On the other hand, a planktonic alga moves with the water, so it would seem that it is irrelevant whether the water is moving (Hynes, 1979). Development times of crustaceans are longer than those of algae, so the whole development of an individual can only be completed in a lower river section. Moreover flowing, sometimes turbulent, water creates unfavourable conditions. Consequently they produce populations with smaller individual numbers than in standing waters with similar trophic state and food conditions. In lakes, individual numbers for cladocerans and copepods can reach $10^3 \, l^{-1}$, but generally are rather below than above $10^2 \, l^{-1}$. The highest values are from summer-warm eutrophic lakes and reservoirs in Europe (Le Cren & Lowe-McConnell, 1980). In the Danube, crustacean individual numbers are in the order of magnitude $10^1 \, l^{-1}$. Retention time of the water mass governs the size of standing crops in lakes and reservoirs, too: the more frequent the renewal of water the smaller the standing stock. Brook & Woodward pointed out as early as 1956 that a short retention time reduced plankton variety and the plankton mostly consisted of species with a rapid rate of replacement. This phenomenon is supported by the high P/B values for crustaceans in the Danube.

The elimination effect of flow is also species-selective, which was already shown by such early workers as Reif (1939), who stated that Chlorophyceae, desmids, and actively swimming crustaceans are eliminated more rapidly than are diatoms and Cyanophyta. Consequently we can say with Rzóska (1978) that rivers, like all water bodies, are colonised by organisms intensively but selectively.

References

Adalsteinsson, H., 1979. Zooplankton and its relation to available food in Lake Myvatn. Oikos 32: 162–194.

Bothár, A., 1986. Population dynamics and estimation of production in *Bosmina longirostris* (O. F. Müller) in the River Danube. Hydrobiologia 140: 97–104.

Bothár, A., 1987. The estimation of production and mortality of *Bosmina longirostris* (O. F. Müller) in the River Danube. Hydrobiologia 145: 285–291.

Bothár, A., 1987. Produktionsschätzung von *Acanthocyclops robustus* (G. O. Sars) in der Donau. In D. Müller (ed.), Wissenschaftliche Kurzreferate. 26. Arbeitstagung der IAD, Passau: 339–343.

Brook, A. J. & W. B. Woodward, 1956. Some observations on the effects of water inflow and outflow of the plankton of small lakes. J. anim. Ecol. 23: 101–114.

Dvihally, S. T., 1975. Primary production of the Hungarian Danube. Verh. int. Ver. Limnol. 19: 1717–1722.

Dvihally, S. T., M. Ertl, K. T. Kiss, A. Schmidt & N. Stefková, 1982. Mit dem Sauerstoffhaushalt zusammenhängende Untersuchungen in der mittleren Donau. Wissenschaftliche Kurzreferate. 23. Arbeitstagung der IAD, Wien: 8–15.

Ertl, M. & S. Juris, 1967. Measurements of primary production in the River Danube. Biologia (Bratislava) 22: 654–659.

Felföldy, L., 1980. Biological water qualification. VIZDOK, Budapest, 263 pp. (in Hungarian).

Flemer, D. A., 1970. Primary productivity of the north branch of the Raritan River, New Jersey. Hydrobiologia 35: 273–296.

Gutelmacher, B. L., 1986. Metabolism of plankton. Nauka, Leningrad: 155 pp. (in Russian).

Hillbricht-Ilkowska, A., I. Sponiewska, T. Weglenska & A. Karabin, 1972. The seasonal variation of some ecological efficiencies and production rates in the plankton community of several Polish lakes of different trophy. In Z. Kajak & A. Hillbricht-Ilkowska (eds.), Productivity problems of freshwaters. Polish Scientific Publishers, Warsaw: 111–127.

Hübel, H., 1971. Primärproduktion des Phytoplanktons. [14]C- oder Radiokohlenstoffmethode. In G. Breitig & W. von Tümpling (eds.), Ausgewählte Methoden der Wasseruntersuchung, Bd II. Biologische, mikrobiologische und toxikologische Methoden, C., VEB G. Fisher Verl., Jena: 1–11.

Hynes, H. B. N., 1979. The ecology of running waters. University of Toronto Press, Toronto: 541 pp.

Javornický, P., 1966. Measurements of production and turnover of phytoplankton in four localities of Poland. Ekol. Pol. A 14: 203–214.

Kiss, K. T., 1985. Changes of trophity conditions in the River Danube at Göd. Annls Univ. Scient. bpest. Rolando Eötvös, Sect. Biol. 24-26: 47–59.

Kiss, K. T., 1987. Phytoplankton studies in the Szigetköz

Section of the Danube during 1981-1982. Arch. Hydrobiol. Suppl. 78, Algol. Stud. 47: 247–273.

Kothé, P., 1981. Bestimmung von Sauerstoffproduktion und Sauerstoffverbrauch im Gewässer mit der Hell-Dunkel-flaschenmethode, SPG und SVG. DIN 38412 Teil 13. DEV L 13.

Kozlovsky, D. G., 1967. A critical evaluation of the trophic level concept. I. Ecological efficiencies. Ecology 49: 48–60.

Kristiansen, J., 1971. Phytoplankton of two danish lakes with special reference to seasonal cycles of the nannoplankton. Mitt. int. Ver. Limnol. 19: 253–265.

Le Cren, E. D. & R. H. Lowe-McConnell (eds.), 1980. The functioning of freshwater ecosystems. International Biological Programme 22. Cambridge University Press, 588 pp.

OECD, 1982. Eutrophication of waters, monitoring, assessment and control. OECD Publications Office, Paris, 154 pp.

Parsons, T. R., 1982. Zooplanktonic production. In R. S. Barnes & K. H. Mann (eds.), Fundamentals of aquatic ecosystems. Blackwell Scientific Publications, Oxford: 46–66.

Patil, C. S. & B. Y. M. Gouder, 1985. Ecological study of freshwater zooplankton of a subtropical pond (Karnataka State, India). Hydrobiologia 70: 259–267.

Pederson, G. L., E. B. Welch & A. H. Litt, 1976. Plankton secondary productivity and biomass: their relation to lake trophic state. Hydrobiologia 50: 129–144.

Reif, C. B., 1939. The effect of stream conditions on lake plankton. Trans. Am. microsc. Soc. 58: 398–403.

Rigler, F. H. & J. A. Downing, 1984. The calculation of secondary productivity. In J. A. Downing & F. H. Rigler (eds.), A manual on methods for the assessment of secondary productivity in fresh waters. I.B.P. Handbook 17, Blackwell Scientific Publications, Oxford: 19–58.

Rodhe, W., 1969. Crystallization of eutrophication concepts in Northern Europe. In Eutrophication: causes, consequences, correctives, Nat. Acad. Sci. Washington: 50–64.

Rosemarin, A. S., 1975. Comparison of primary productivity (^{14}C) per unit biomass between phytoplankton and periphyton in the Ottawa River near Ottawa, Canada. Verh. int. Ver. Limnol. 19: 1584–1592.

Rzóska, J., 1978. On the nature of rivers. Dr W. Junk bv Publishers, The Hague: 67 pp.

Slobodkin, L. B., 1960. Ecological energy relationships at the population level. Am. Nat. 94: 213–236.

Tarasova, T. N., 1970. Primary production and organic substance destruction in the place of construction of the Cheboksary power station in 1966. Uchen. zap. Gork. Univ. Ser. Biol. 105: 32–36. (in Russian)

Hydrobiologia **191**: 173–188, 1990.
P. Biró and J. F. Talling (eds), Trophic Relationships in Inland Waters.
© 1990 *Kluwer Academic Publishers.*

Zooplankton structure in the Loosdrecht lakes in relation to trophic status and recent restoration measures

R.D. Gulati
Limnological Institute, Rijksstraatweg 6, 3631 AC Nieuwersluis, The Netherlands

Key words: zooplankton structure, eutrophication, restoration, P-loading, rotifers, crustacea

Abstract

A five-year zooplankton study (1982-86) on three shallow and highly eutrophic lakes in the Loosdrecht area (The Netherlands) did not reveal any significant changes following the considerable reduction in external P-loading (from about 1.0 g to 0.3 g P m^{-2} year^{-1}) since mid-1984.

The recent annual fluctuations in the rotifer and crustacean densities are within the range of those found before the restoration measure became operative. A decrease in the average size of the crustaceans and an absence of large-bodied forms reflects an increased fish predation rather than a change in the quality or quantity of their sestonic food ($< 150 \mu m$) which continues to be dominated by filamentous cyanobacteria and *Prochlorothrix hollandica*, a prochlorophyte discovered in these lakes recently.

Introduction

Until recently there was virtually no published work on the limnology of the lakes in the Loosdrecht area. However, from the data on water balance maintained by the Provincial Water Authority of Utrecht and from a publication by Golterman (1965), it has been known that the Loosdrecht lakes have a negative water budget. This is partly because the evaporative water loss exceeds precipitation especially during long dry spells; also there is a net seepage loss to the surrounding polder country (Kal *et al.*, 1984). To compensate for this deficit and to maintain a constant water level, water from the Vecht, a grossly-polluted river in the vicinity, was let into the lakes in summer. The lakes were thus receiving ca. 1 g P m^{-2} year^{-1} as allochthonous P load up to the early eighties, besides other nutrients and pollu-

tants. Consequently, they became highly eutrophic.

Restoration measures became operative in mid-1984 when inflow of water from the river Vecht was replaced by FeCl$_3$-treated water from the Amsterdam-Rhine Canal. This has led to a significant decrease in the annual external P-loading to 0.3 g m^{-2} year^{-1} (van Liere, 1986), mainly in the summer.

A multidisciplinary study (1982-84), before the restoration measure became effective, provided the first limnological insight into the lakes (see van Liere *et al.*, 1984 and Loogman & van Liere, 1986).

The studies of zooplankton in the lakes (Gulati, 1983, 1984) indicate low specific clearance rates related to high seston concentrations. Comparing with zooplankton data collected about 30 years ago (Geelen, 1955), Gulati (1984) observed an

174

eight- to ten-fold increase in the crustacean and rotifer numbers. This was especially due to species of *Keratella* and *Bosmina* but also to species which appeared between 1955 and 1982, especially the rotifer *Anuraeopsis* sp. Gulati *et al.* (1985) attributed the low zooplankton (> 150 μm) to seston (< 150 μm) biomass ratios to dominance in the lakes of filamentous bluegreens (cyanobacteria). Irvine (1986) explained the co-existence of *Bosmina longirostris* and *B. coregoni* and *Daphnia cucullata* on the basis of differential feeding behaviour and food selectivity.

This paper briefly presents the data of a five-year study (1982-86) on zooplankton structure

(composition, abundance and biomass) in relation to the trophic status of three Loosdrecht lakes and changes in response to the restoration measures.

The lakes

The zooplankton study was started in Loosdrecht Lake (stations 5 and 9) and Lake Breukeleveen (station 6) in August 1981 and in Lake Vuntus (station 7) in May 1983 (Fig. 1).

Important morphometric features of the lakes are summarized in Table 1. For details of the

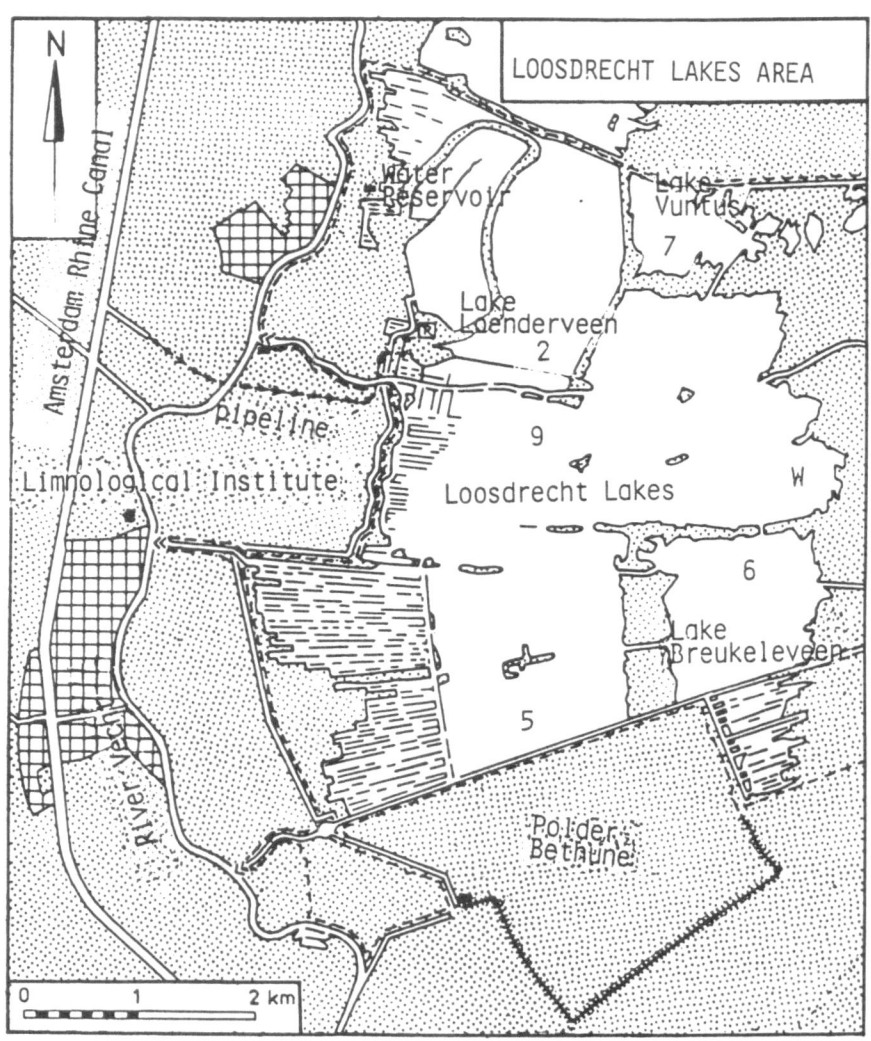

Fig. 1. Location of sampling stations in the Loosdrecht lakes; for other details see text.

Table 1. Morphometry of the three Loosdrecht lakes.

	Lake Breukeleveen	Lake Vuntus	Loosdrecht Lake	Total
Average depth (m)	1.4	1.4	1.8	1.7
Surface area (ha)	179	88	979	1246
Volume 10^6 m^3)	2.58	1.26	17.9	21.74

origin, history of formation, morphology and hydrology see papers in Loogman & Van Liere (1986). The lakes are situated 20 km SE of Amsterdam (lat. 52° 20′, long. 5° 5′ E). They are stripped *Sphagnum* bogs, created since 1633 by peat winning. The presence of connections between the lakes allows exchange of water (Fig. 1).

There is a significant downward seepage loss to the low-lying polder country (Polder Bethune) to the south. The water deficit caused by evaporation losses, leakage and the loss to polders was supplemented in the years up to 1983 by drawing water from the river Vecht, but because of the restoration measures, as mentioned, external P-loading has been considerably reduced.

Methods

Sampling

The water was sampled usually between 09.00 and 11.00 hours, with a 5-litre Friedinger sampler (length 60 cm) in the 10-70 cm stratum. To reduce the effects of horizontal patchiness of zooplankton in the lake six 5-litre samples were taken, from a vessel that was allowed to drift up to about 100 m while sampling, to get a 30-litre composite sample for all the routine measurements.

Zooplankton counting

One- or 2-l water from the well-mixed stock sample was filtered through a 33 μm mesh-size nylon sieve and the zooplankton on the filter preserved in 4% formalin. The mesh size used retains practically all the rotifer species (Duncan & Gulati, 1981). Fresh samples were used for taxonomic work.

The zooplankton samples were subdivided using Kott's splitter (Kott, 1953). Mostly 10% of the original sample in replicate (3 ×) was counted under a Utermöhl inverted microscope and densities expressed as ind. l^{-1}.

Measurement of seston mass (carbon)

The biomass of seston, i.e. zooplankton (> 150 μm) and seston fraction (< 150 μm), was measured using the COD method (Golterman, 1969), with some modifications (Gulati *et al.*, 1982). Oxygen consumed is a measure of carbon oxidized which is obtained by multiplying the oxygen consumed by 0.364 (Winberg & collaborators, 1971).

Results

Background information on the environmental conditions

Physico-chemical data of the lakes (1984-85) are summarized in Table 2. Because of the low mean depth ($\overline{Z}$: < 2.0 m) a thermal heterogeneity is lacking in summer. Wind-induced resuspension, high concentrations of algae and cyanobacteria in the water column and the colour of water have resulted in a poor light climate so that Secchi-disc, transparency values of > 50 cm are exceptional (Gons *et al.*, 1986).

Seston < 150 μm in all the lakes abounds in cyanobacteria, *viz.* three *Oscillatoria* spp. (*O. agardhii*, *O. redekei*, *O. limnetica*) and *Prochlorothrix hollandica*, a dominant prochlorophyte in the lake discovered recently by Burger-Wiersma *et al.* (1986).

Common diatoms are *Diatoma elongatum*, *Asterionella formosa* and *Melosira* spp. Flagellates, especially nanoflagellates, most of which remain

176

Table 2. Summary of physico-chemical conditions (mg l^{-1}), primary production and chlorophyll *a* concentration in the Loosdrecht lakes. The data are annual means, and are from routine measurements at the Limnological Institute except for pH and NH$_4$ which were taken from the Municipal Water Works, Amsterdam.

Lake	Year	Temp. (°C)	pH	Secchi-depth (m)	P-total (mg l^{-1})	NH$_4$-total (mg l^{-1})	SiO$_2$ (mg l^{-1})	Primary production (g C m$^{-2}\cdot$d^{-1})	Chlorophyll *a* (mg l^{-1})
Breukeleveen	'84	11.6	8.7	0.25	0.107	0.33	2.3	1.05	0.11
	'85	12.0	8.7	0.23	0.103	0.41	2.6	1.00	0.11
Vuntus	'84	10.2	8.6	0.25	0.082	0.38	1.7	0.50	0.11
	'85	10.5	8.6	0.30	0.080	0.53	1.8	0.65	0.07
Loosdrecht Lake	'84	11.8	8.6	0.30	0.121	0.28	1.9	1.10	0.12
	'85	11.4	8.7	0.25	0.109	0.32	3.1	1.45	0.11

unidentified, are persistent with densities up to 29000 ind. ml^{-1} in summer.

Primary production rates (^{14}C-technique) are quite high (pers. comm. W.A. de Kloet). The mean daily rates of production in the lakes during May-September range from 0.9 g C m^{-2} to 2.3 mg C m^{-2}.

The seston mean mass of (<150 µm) on an annual basis, is highest in Lake Breukeleveen and lowest in Loosdrecht Lake. The data do not indicate a decrease in seston mass in response to the restoration measures (Fig. 2). On the contrary, seston concentrations increased at stations 5 and 9 in 1986.

The fish biomass is dominated by bream (*Abramis brama*). Although the growth conditions of the fish <30 cm are not optimal (Van Densen *et al.*, 1986), the fish >30 cm feed more efficiently on larvae of *Chironomus plumosus*. The pikeperch *Stizostedion lucioperca* is the main piscivore. The O + pikeperch consumes chiefly cyclopoids and *Leptodora kindtii*.

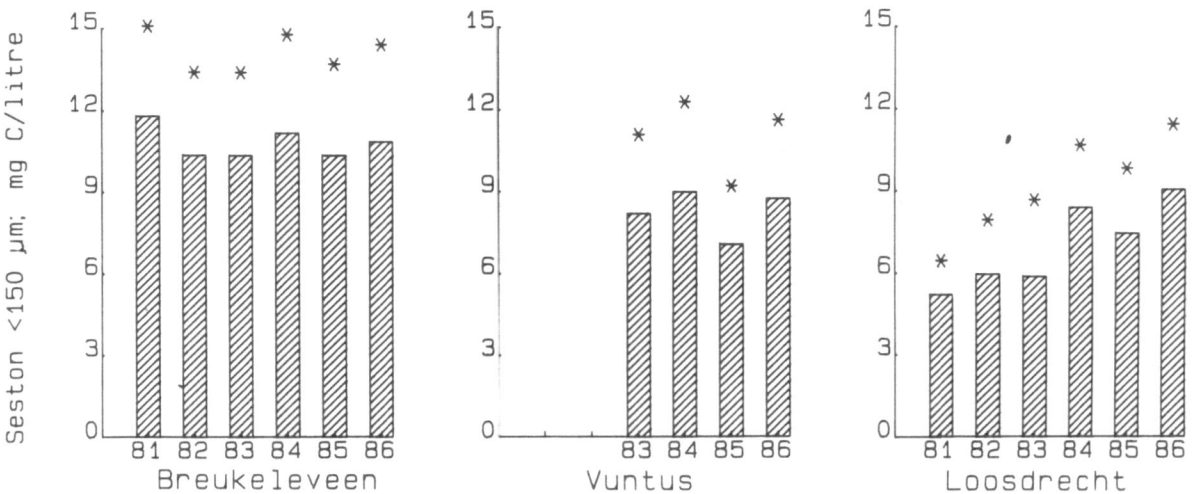

Fig. 2. Annual means of seston mass, with S.D. (*), in the three Loosdrecht lakes derived from monthly means for the different years (1981-86).

Zooplankton

Protozoa

The study of Protozoa is restricted to an inventory of the species composition only. Of the 23 genera of Protozoa (Gulati *et al.*, 1987), 12 belonged to ciliates, 9 to rhizopods and actinopods, and 2 to zooflagellates. The protozoa are dominated by ciliates, especially *Coleps* (*C. hirtus*?). *Tintinnopsis lacustris* is also found regularly although sparse in summer.

Among the rhizopods at least 4 species each of *Difflugia* and *Arcella* occur.

The protozoans may be even more abundant in epipelic collections than encountered by us in planktonic samples. Thus, their position as a link in the decomposer and grazer food chain will be quite important.

Rotifera

Among the 20 genera of rotifers recorded 39 species were identified. Six of these belonged to *Collotheca*, 4 each to *Synchaeta* and *Trichocerca*, 3 each to *Polyarthra*, *Asplanchna* and *Keratella* and 2 species each to *Lepadella* and *Brachionus*. The other 12 genera had one species each per genus.

Most rotifer species are usually planktonic, except *Collotheca*, and solitary. Only *Conochilus* occurred both as solitary and colonial forms. All the species except *Philodina* sp. are representatives of the Monogononta. But for *Filinia* sp., all the common forms belong to the Order Ploima. Belonging to this order, members of the family Brachionidae (*Anuraeopsis* sp., *Keratella* spp.) are the most dominant in the Loosdrecht lakes. Other important species belong to *Polyarthra*, *Filinia* and *Synchaeta*.

The species composition of rotifers and their abundance in general are typical of highly eutrophic water-bodies (Gulati, 1983; Gulati *et al.*, 1985).

Seasonal changes in rotifer densities

In the Loosdrecht lakes there was a steady increase up to 1985 in the annual rotifer maximum in June, caused chiefly by *A. fissa* and *K. cochlearis*. In Breukeleveen, although the maxima also occurred in June, both the lowest (10 500 ind. l^{-1}) and highest values (16 000 ind. l^{-1}) of the annual maxima were recorded in 1986 and 1985, i.e. after the restoration measures (Fig. 3). In Vuntus the rotifers exhibited the highest annual maximum (22 000 ind. l^{-1}) in 1983, but this decreased steadily to one-third this level in 1986, caused mainly by *A. fissa*. These conflicting trends in rotifer dynamics in the 3 adjacent lakes appear to be within the amplitude of annual oscillations within a lake. The sampling interval of a fortnight was too long to record all the population fluctuations.

Summarizing, in all the three lakes *A. fissa* contributed 25-38% to the total rotifer populations in 1986 compared with 29-57% in the years 1982-85.

Annual mean densities of rotifers

Broad generalizations based on annual rotifer means are: 1) in all the lakes *A. fissa* and *K. cochlearis* dominate (Fig. 4); 2) four of the dominant species together constitute two-thirds, or more, of the total rotifer densities; 3) the mean annual densities fluctuate greatly (Fig. 4); and 4) an increase in the annual means from 1984 to 1985, but a decrease in 1986, is attributable chiefly to *A. fissa*.

Crustaceans

Among the 15 crustacean genera found in the lakes, 9 are sporadic (Table 3). The cladoceran diversity is low. *D. hyalina*, a common inhabitant of shallow, eutrophic lakes (Beattie *et al.*, 1978), was virtually absent. It is likely the *D. hyalina* found now and the *D. longispina* reported by Geelen (1955) may belong to the same taxonomic complex (pers. comm. K. Vijverberg). Also, den-

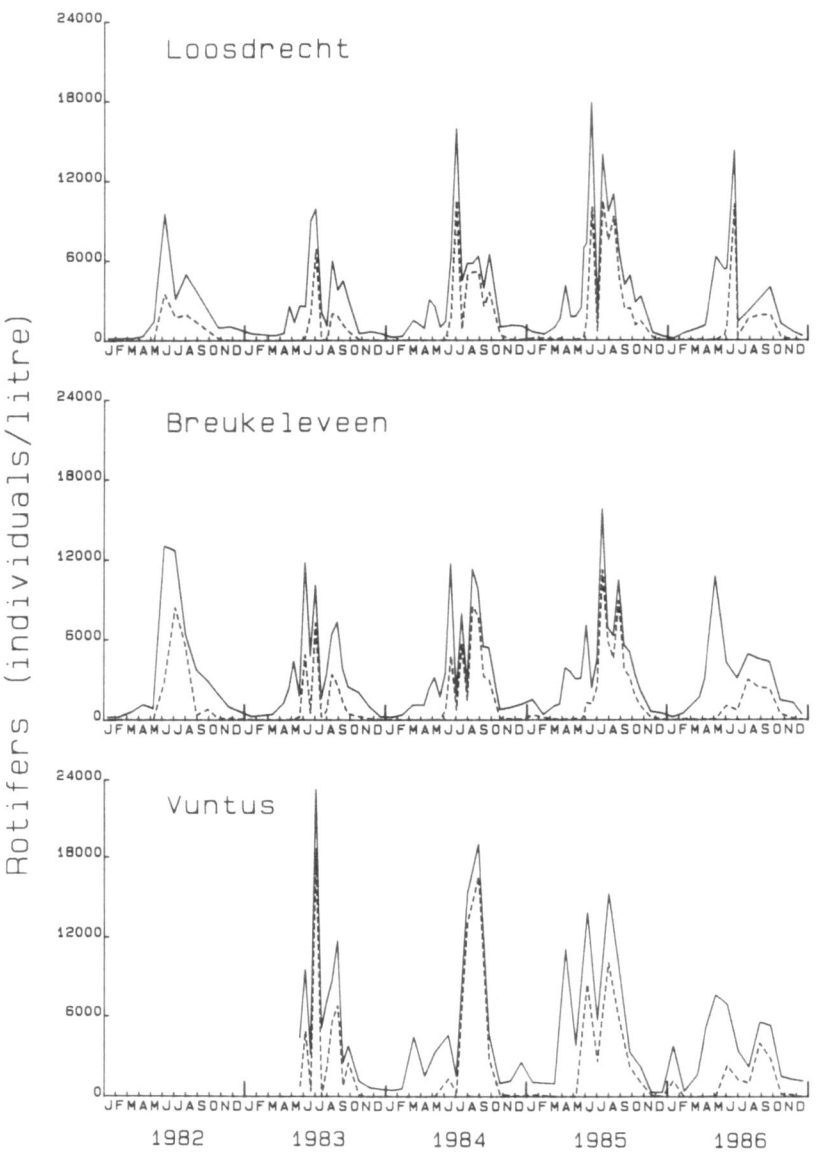

Fig. 3. Seasonal changes in rotifer densities (ind. l^{-1}) in the 3 Loosdrecht lakes; (———) total rotifers; (· · · · ·) *Anuraeopsis fissa*.

sities of *D. cucullata* in the lakes are low and lacking a seasonal continuity. Both *Bosmina longirostris* and *B. coregoni* co-exist in spring and early summer. Thereafter, only the latter species is found. The bosminids contribute the most to cladoceran densities and to the grazers' population (Gulati, 1984).

Ceriodaphnia and *Diaphanosoma* occur only in summer at temperatures above 16 °C.

Leptodora kindtii is the only predatory cladoceran, present from mid-June to late October in densities (1 to 2 ind. l^{-1}) that are somewhat higher than in other shallow water-bodies in The Netherlands.

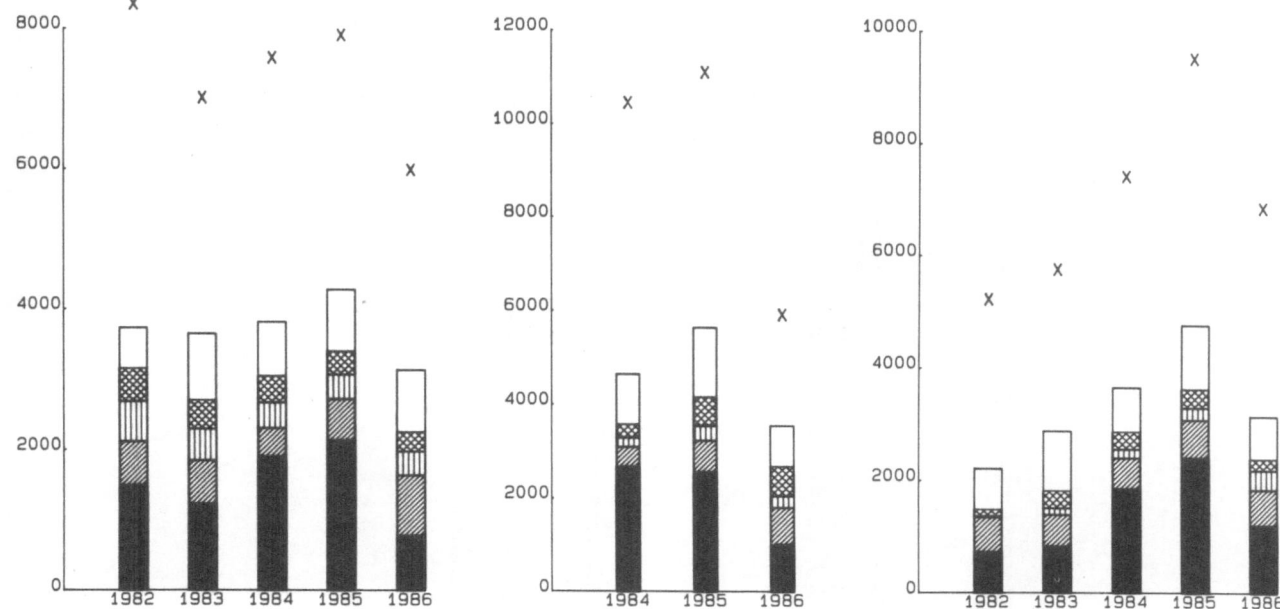

Fig. 4. Histograms of the annual rotifer means (ind. l^{-1}) in the Loosdrecht lakes (panels from left to right: Lake Breukeleveen, Lake Vuntus, and Loosdrecht Lake). The proportions of the important species are indicated as: shaded, *A. fissa*; diagonal stripes, *K. cochlearis*; vertical stripes, *Filinia* sp.; and diagonally cross-hatched, *Polyarthra* spp.; the unshaded area represents all other species combined; X, standard deviation.

The copepods are dominated by cyclopoids. The calanoids (*Eudiaptomus gracilis* and *Eurytemora lacustris*) are very sparse. The cyclopoid juvenile stages, particularly the nauplii, are encountered consistently throughout the year.

Seasonal changes in crustacean densities

In Lake Loosdrecht the annual maxima of crustaceans (800 to 4000 ind. l^{-1}) occurred in

Table 3. Crustacean species found during 1981–'85 in the three Loosdrecht lakes (* = rre, or sporadic forms).

Cladocera	
Daphnia cucullata	Eudiaptomus gracilis*
D. haylina*	Eurytemora lacustris*
Bosmina longirostris	Cyclops vicinus
B. coregoni	Megacyclops viridis*
Ceriodaphnia quadrangula*	Thermocyclops crassus*
Diaphanosoma brachyurum*	Thermocyclops oithonoides*
Chydorus sphaericus	Acanthocyclops robustus
Alona rectangula*	Diacyclops bicuspidatus
Rhynchotalona falcata*	Mesocyclops leuckarti
Leptodora kindtii*	Eucyclops macruroides

May-June (Fig. 5). At station 9 the annual maximum in 1985 was twice as high as in the other years. The cyclopoids, and their nauplii, annually

Table 4. Mean percentage contribution of filter-feeders and total cyclopoid copepods, with % nauplii in parentheses, in the total Crustacea (ind. l^{-1}) of the three Loosdrecht lakes during 1982–86.

Lake		Total Crustacea	% filter-feeders	% total cyclopoids
Breukeleveen	1982	655	42	58 (44)
	1983	777	42	58 (43)
	1984	988	44	56 (38)
	1985	901	43	57 (43)
	1986	482	22	78 (56)
Vuntus	1984	585	23	77 (55)
	1985	657	23	77 (53)
	1986	537	18	82 (56)
Loosdrecht	1982	636	55	45 (30)
	1983	558	39	61 (44)
	1984	714	38	62 (45)
	1985	942	46	54 (38)
	1986	972	38	62 (42)

180

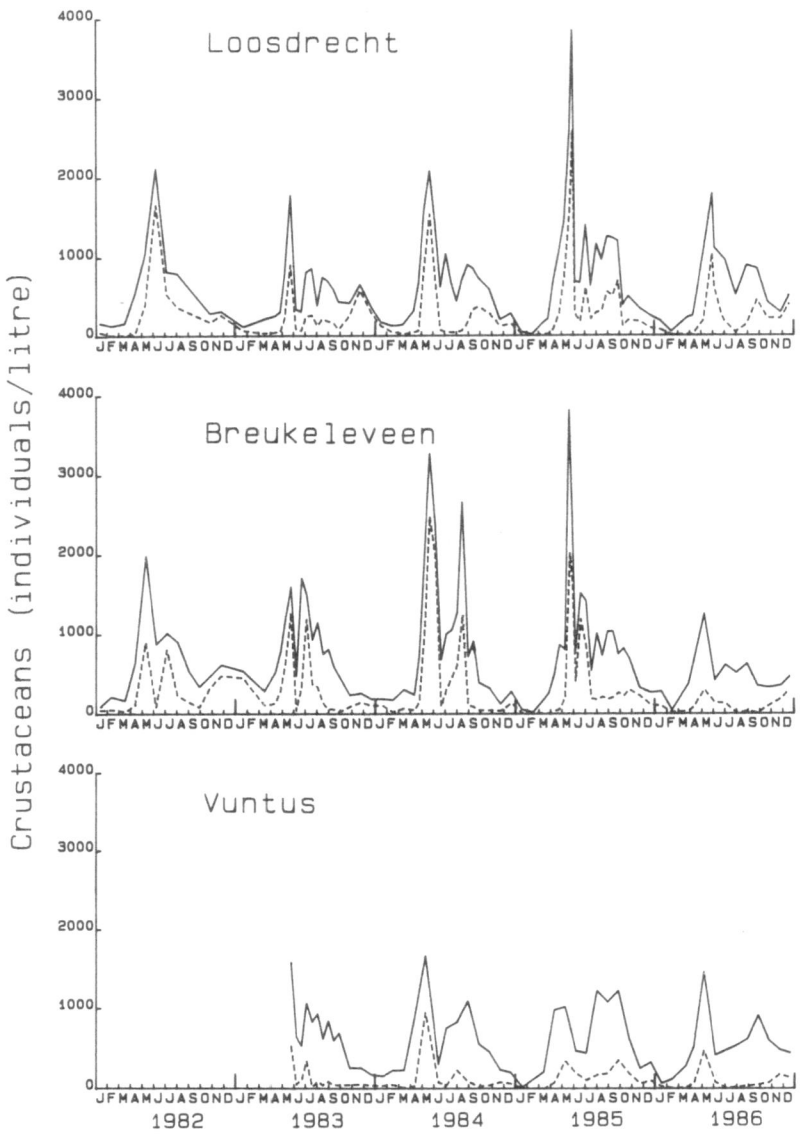

Fig. 5. Seasonal changes in crustacean densities (ind. 1^{-1}) in the three Loosdrecht lakes; (————) total crustaceans, (---------) Cladocera.

contributed from 45 to 62% (Table 4) of the total crustaceans. They chiefly determined the seasonal patterns of the community, their concentrations fluctuating between 50 ind. 1^{-1} and six times as high. Nauplii contributed about two-thirds of the total cyclopoid densities and were thus the most abundant crustaceans in all the lakes.

The most outstanding annual peaks of filter-feeders occurred invariably in May or June,

reaching 2500 ind. 1^{-1}; *Bosmina* spp. accounted for up to 90% of these maxima. The other filter-feeders, mainly *Chydorus* sp. and *Daphnia* sp., contributed the rest. In 1985 the filter-feeders' peak crashed within a week (Fig. 5), mainly because *B. longirostris* densities decreased sharply (Fig. 6). It was replaced by *B. coregoni*, of which the maximum did not exceed 250 ind. 1^{-1}.

D. cucullata, which reached maximally 100-

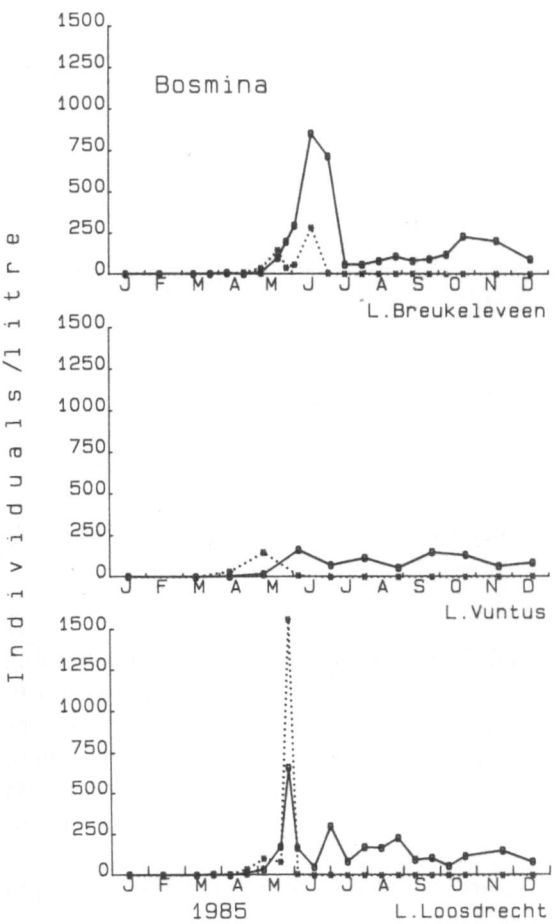

Fig. 6. Densities of *Bosmina longirostris* (············) and *B. coregoni* (———) in the three Loosdrecht lakes in 1985.

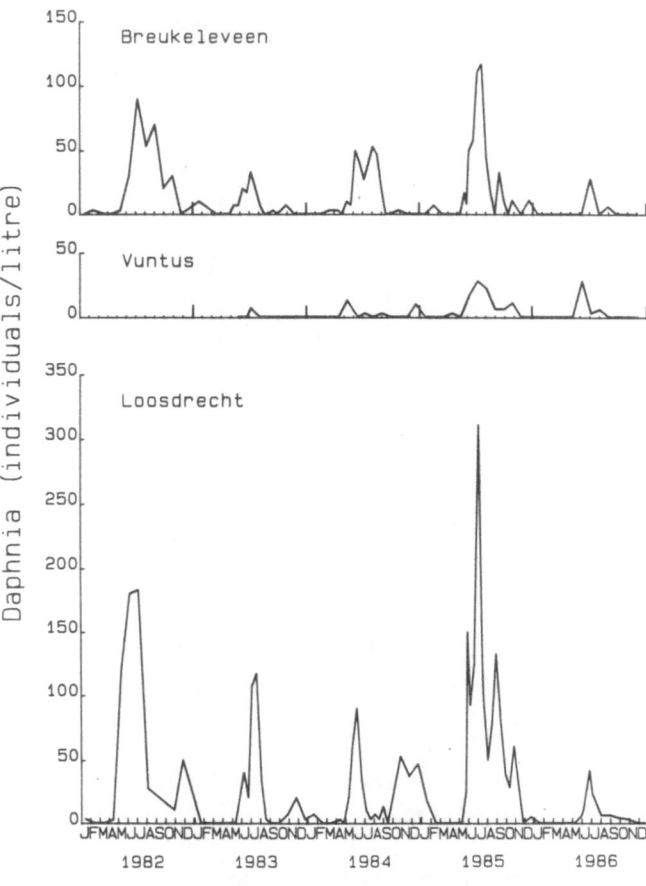

Fig. 7. Densities of *Daphnia* species in the three Loosdrecht lakes, 1982-86.

300 ind. l^{-1} in the different lakes up to 1985, slumped to a maximum of 40 ind. l^{-1} in 1986 (Fig. 7).

The filter-feeders' decline in Lake Breukeleveen in 1986 was even more pronounced than in the Loosdrecht Lake (Fig. 5). All the three filter-feeders had their lowest spring maxima in 1986. Their mean annual densities were then between 22 and 39% of those in the preceding 4 years.

Annual means of crustacean densities

The cyclopoid copepods dominate, especially in Lake Vuntus (Table 4). The mean proportion of the filter-feeders was far more variable in Lake Breukeleveen and in Lake Loosdrecht but was rather constant (18-23%) in Lake Vuntus. The mean densities increased in all the lakes up to 1985, but declined in 1986. *Daphnia cucullata*, which declined drastically in 1986, contributed importantly to the cladoceran decrease in Lake Breukeleveen (Fig. 7).

Zooplankton biomass

The seasonal biomass data of zooplankton (Fig. 8) broadly corroborate the data on population densities. Generalizing, the annual changes are not spectacular, even though the densities and biomass declined in 1986 both in the Loosdrecht Lake and Lake Vuntus.

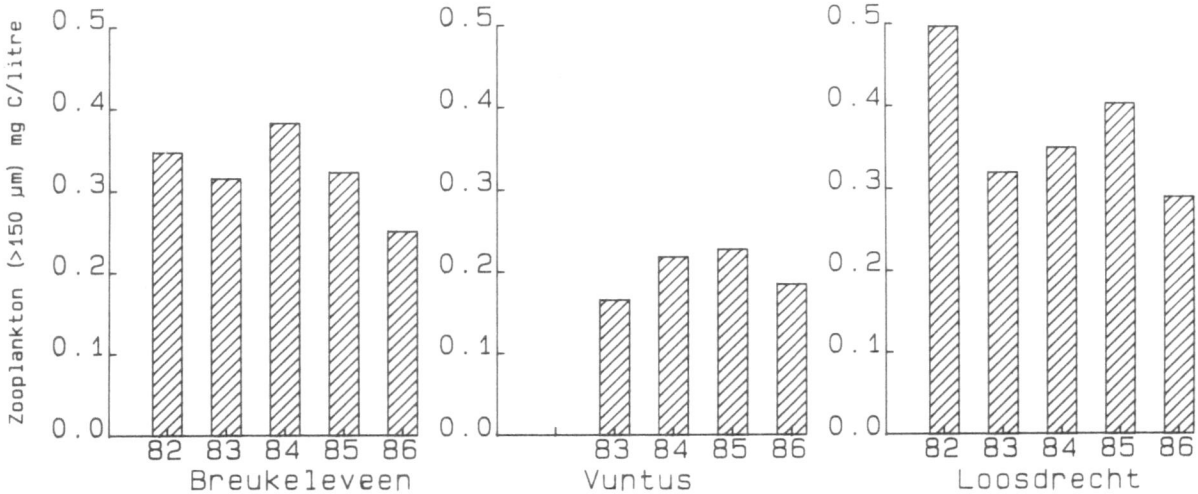

Fig. 8. Annual means of zooplankton mass in the three Loosdrecht lakes, based on monthly means during the five-year study (1982-86).

The share of zooplankton in the total seston mass has decreased from 3-6% up to 1984 to 2-2.5% in 1986. Both an increase in seston <150 μm during 1986 in Lake Loosdrecht (Fig. 2) and decrease in zooplankton mass (Fig. 8) caused this.

Density-biomass relationships: The overestimation of crustacean biomass (>150 μm) by inclusion of some rotifers, or its underestimation because of naupliar exclusion, does not appear to introduce serious errors in the biomass estimates. Therefore, regression relations between zooplankton densities (X) and biomass (Y) provide a useful means to compare the zooplankton structure, *viz.* by deriving the mean individual weights of the organisms in the community and studying their seasonal variations. The linear regression statistics on the un-transformed data (Table 5) show highly significant correlations ($P < 0.001$) between density and biomass. The coefficients of determination (R^2) show that between 55 and 73% of the variation in the pooled biomasses of the different years is explained by the densities.

The linear relationship between the two parameters may be adversely affected by the significant decrease in mean individual biomass of the organisms at relatively high densities, and

vice versa. Other factors which will considerably influence the density and biomass relations are: 1) size selective fish-predation; 2) shift in species composition and in relative species dominance; and 3) changes in length and weight relationships of the species caused by the environmental factors.

The lower values of both slopes and intercepts in the lakes Breukeleveen and Vuntus than in the Loosdrecht Lake indicate a lower biomass per individual irrespective of the prevailing densities in the lakes. This also lends support to the differences observed among the lakes (Table 4) in the proportions of naupliar densities.

Table 5. Constants and other relevant data of linear regression statistics on the untransformed data ($Y = a + bX$) on crustacean densities (ind. l^{-1}) (X) and crustacean biomass (μg C l^{-1})(Y), of the Loosdrecht lakes for 1981–85; $P < 0.001$ in all cases.

Laje period	Breukeleveen '81–'85	Vuntus '83–'85	Loosdrecht '81–'85
Intercept (a)	37.1	46.6	121
Slope (b)	0.399	0.243	0.404
R^2	0.72	0.55	0.73
F ratio	178	42	185
n	72	36	70

Density versus individual biomass: Individual biomass derived by dividing the community biomass by community density appears to be negatively related with the community density (Gulati *et al.*, 1987). Linear regression statistics based on the log-transformed densities (X) and biomass per individual (Y) showed that between 24 and 50% of individual biomass variations in the lakes are caused by changes in density. The computed individual weights show that the animals in the Loosdrecht lakes are the heaviest and in Lake Vuntus the lightest. The mean individual weights derived (excluding nauplii) for the zooplankton community in the lakes were between 0.60 and 0.77 μg C ind.$^{-1}$. Individual weights derived for all the dates, using the respective densities and biomass regressions for the

lakes, vary more than the individual lengths of the zooplankters. Therefore, from the derived individual dry weights (DW = μg C ind.$^{-1}$) lengths (L) were back-calculated using L : W regression equation (W = 9.86 L$^{2.1}$) given in Peters & Downing (1984). The computed lengths (Fig. 9) show a decrease from 1981 to 1985, in Lake Loosdrecht and Lake Breukeleveen, although significance of the decrease has not been statistically tested. The annual length maxima have considerably decreased, especially in Lake Loosdrecht -from *ca.* 0.60 mm in 1982 to 0.45 mm in 1985; even the annual minima decreased relatively to the same extent i.e. from about 0.38 mm to 0.27 mm (Fig. 9). Thus, these changes show an alteration in the zooplankton community structure. The proportion of nauplii has not signi-

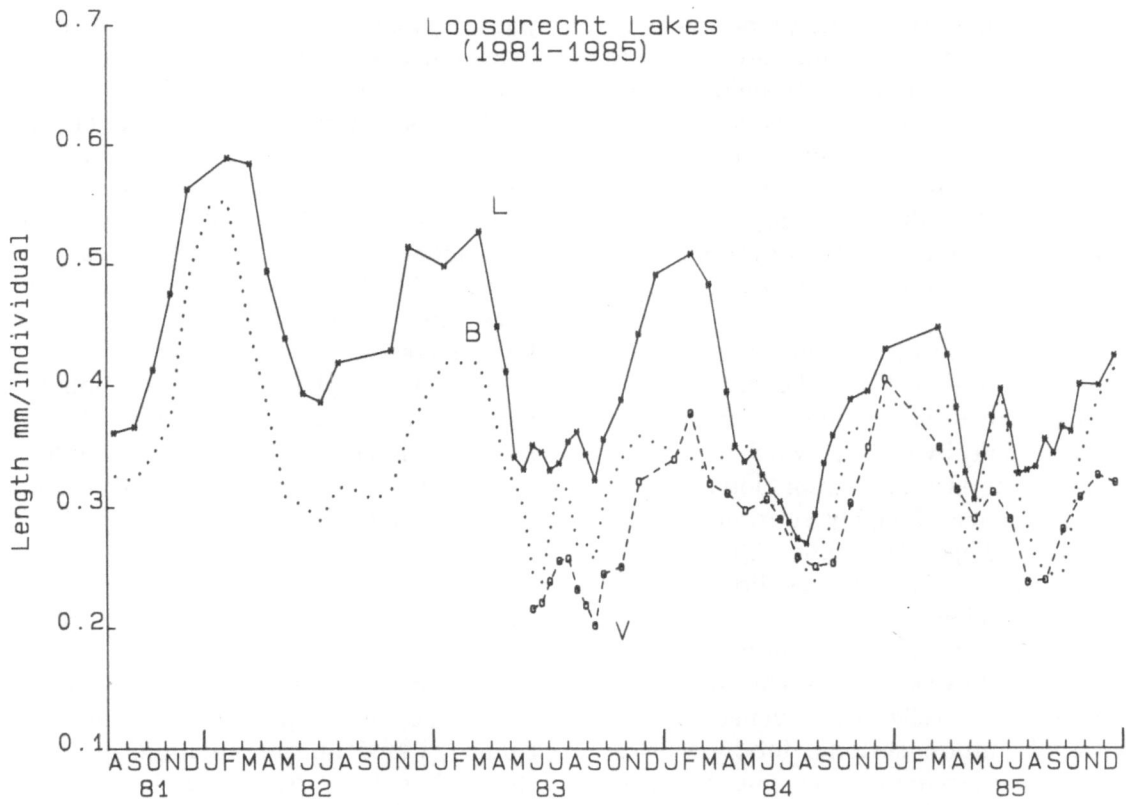

Fig. 9. Mean individual length of crustaceans in the three Loosdrecht lakes during 1981-85; the codes L, B and V are for the three lakes Loosdrecht, Breukeleveen and Vuntus. The lengths were back-calculated from individual weights using L : W regressions in Peters & Downing (1984) and smoothed using a moving average of three successive dates; the data are not corrected for nauplii.

ficantly increased so as to cause such a decrease in the mean individual length in the community. The decrease in individual length appears to be independent of changes in the species composition since there is no evidence of an increase in relatively small-sized species. The role of size-selective fish predation in structuring the community, i.e. a decrease in species average size, is not ruled out.

Discussion

The rotifers

Some generalizations may be drawn on the basis of zooplankton composition and abundance. First, rotifers predominate in the microzooplankton, with density maxima – and even annual means – that are among the highest reported in the literature. The species that contribute chiefly to the high rotifer numbers, K. cochlearis and A. fissa, are both among the smallest of rotifers in the lakes. The former generally dominates in spring and the latter in summer. Their periods of dominance coincide with those of nanoflagellates and bacteria, besides filamentous cyanobacteria which may be too large for these species to consume.

Despite some major declines in the rotifer and crustacean densities in 1986, the zooplankton densities are still high. The dominance in the lakes of rotifers, especially by A. fissa, which comprised ca. one-third the total zooplankton biomass, indicates the prevalence of highly eutrophic conditions in the lakes (Pejler, 1957; Gulati, 1983). This may be because the eutrophic conditions in the lakes promote a high concentration of detritus that enhances bacterial production (Gliwicz, 1969; Hillbricht-Ilkowska, 1977). The abundance of rotifers in general reflects an availability of a wide range of natural sestonic food particles, which rotifers as a group may consume (Pourriot, 1977; Dumont, 1977; Starkweather, 1980). Bacteria may be a more important source of food for rotifers than for the co-occurring cladocerans and copepods (Starkweather, 1980). The dominant

crustacean zooplankters which coincide with the rotifers abundance, B. coregoni and cyclopoid copepods, appear to prefer food in a size range much larger than that of bacteria (pers. data on grazing) so that competition for food by the two groups may be unimportant. Therefore, a niche separation between the rotifers and crustaceans may explain both their co-occurrence and simultaneous dominance.

Besides, according to 'size efficiency hypothesis' (Brooks & Dodson, 1965; Hall et al., 1976), the absence in the Loosdrecht lakes of large-bodied filter-feeding Cladocera may considerably relieve competition for food for the small-bodied rotifers. Also, the ability of some rotifers (e.g. Brachionus calyciflorus) to even consume some species of toxic cyanobacteria (Starkweather & Kellar, 1983) may impart added advantage to the rotifer species over certain sympatric crustaceans which may be inhibited by cyanobacteria (Lampert, 1981).

In addition to food availability and competition for food, predation may be important in regulating the rotifer dynamics and size structure. Several invertebrate taxa which predate on rotifers (Williamson, 1983), namely the cyclopoids and the cladoceran Leptodora kindtii, are present in the Loosdrecht lakes. Nevertheless, they do not seem to effectively reduce the rotifer abundance. Possibly increased predation during summer in Loosdrecht by planktivorous fish on the crustaceans mentioned (Van Densen et al., 1986) may reduce the intensity of predation on rotifers. Thus, the rotifers not only coexist with crustaceans but also achieve the very high densities.

Rotifer and crustacean biomass: some comparisons

The average annual rotifer biomass of 185 μg C l^{-1} (range 130-250 μg C l^{-1}) in the three lakes studied, based on computations from volumes (Bottrell et al., 1976; Ruttner-Kolisko, 1977), is ca. 37% of the total zooplankton biomass. On individual dates this percentage may be as high as 85%, especially in Breukeleveen, which is similar to the extreme proportion reported by Haberman

(1983). A decrease in the contribution of rotifers to zooplankton biomass is not yet discernible, although consistent differences between the lakes are accompanied by differences in trophic interrelationships. For example, Lake Vuntus, in which 50% of the zooplankton biomass is accounted for by rotifers compared with 31-36% in the other two lakes, has only a transient *Daphnia* population, and the lowest *Bosmina* densities (Gulati et al., 1987). Differences in the magnitude of fish predation may explain such differences among the lakes since abiotic conditions and other biotic factors in the lakes are rather similar.

An individual rotifer within the community of the Loosdrecht lakes weighs 0.05 μg C (biomass/density = 185/3768), compared with the crustacean average of 0.44 μg C ind.$^{-1}$ derived similarly. Based on a general body exponent 'b' of 0.75, applied widely in ecology and physiology (Peters & Downing, 1984), the 'activity coefficient A' for the rotifers and crustaceans may be calculated as $A = aW^b$ where a is the intercept, i.e. the activity coefficient for an animal weighing 1 μg C, and W is the average weight of an individual in a rotifer or crustacean community. The computed activity coefficients for individuals in the two communities with mean individual weights of 0.05 and 0.44 μg C ind.$^{-1}$ are 0.11 and 0.54, respectively. These activity coefficients multiplied by the respective, average annual densities of 3770 ind. l^{-1} for rotifers and 700 ind. l^{-1} for crustaceans give 'community coefficients' of 415 and 385, respectively. Thus, despite their significantly lower biomass, the rotifers may be even more important. Pace (1986) drew similar conclusions for Quebec lakes in which microzooplankton (though mostly Protozoa) was also found to be an important component. Therefore, relations based on size structure of the communities may be of great importance in production energetics and in trophic interrelationships (see also Makarewicz & Likens, 1979; Pace & Orcutt, 1981).

The crustaceans

About two-thirds of the 15 crustacean genera recorded in the lakes are sporadic (Gulati et al.,

1987). The cladocerans are dominated by *Bosmina* spp. for only short periods in spring and early summer, rather than also by one or two species of *Daphnia* as is often the case in deep stratifying lakes (Gulati et al., 1982). Second, the cyclopoids, with their 4 species, and nauplii persist almost throughout the year. The virtual absence of calanoids like *Eudiaptomus gracilis* is probably related to the high level of eutrophication and food composition (Gulati, 1983; Gulati et al., 1985) rather than to predation by fish.

The oscillations exhibited by *Daphnia cucullata* during the study period are striking. This species decreased in the successive years up to 1984 (Fig. 7), but recovered in all the lakes in 1985. However, in 1986 the numbers were the lowest for the entire study period. This drastic decline in 1986 in the *Daphnia* dynamics is difficult to explain on the basis of changes in food quality or quantity. Also *D. hyalina*, an important inhabitant of other eutrophic lakes like Tjeukemeer (Vijverberg & Richter, 1982) as well as mesotrophic lakes (Gulati et al., 1982), is sparse in the Loosdrecht lakes. In Tjeukemeer this species has been shown to utilize blue-green algae (Bloem & Vijverberg, 1984), and even though it is an important prey of planktivore fish, it has not disappeared. Reasons for the disappearance or sparse occurrence of *D. hyalina* in the Loosdrecht lakes are still quite obscure.

The success of the two *Bosmina* species may be partly due to their marginal co-existence. Generally *B. longirostris* is more abundant in April-May when cyanobacteria have not yet attained their maximum but both flagellates and diatoms are available. From the co-occurrence of the two species in early summer one is tempted to speculate that a competition for food does not cause disappearance of *B. longirostris*. This is because neither the food concentration is limiting, nor are the concentrations of cyanobacteria in May-June so high that they interfere with or inhibit the food-uptake mechanism and thus cause its extinction. In contrast, one would expect *B. coregoni*, as larger in size (its length range being 290-610 μm versus 245-530 μm for *B. longiros*-

tris), to be more vulnerable to fish-predation, which is high in early summer (van Densen *et al.*, 1986). Therefore, on theoretical grounds, the chance that *B. coregoni* should disappear is greater than that *B. longirostris* disappears. However *B. coregoni* uses both a coarser and wider range of food size fractions than *B. longirostris* and *D. cucullata* (Irvine, 1986). The effect of water temperature on the two *Bosmina* species is not known, although *B. longirostris* was absent in summer when temperature exceeded 16 °C.

An earlier hypothesis (see Hutchinson, 1967) that *B. longirostris* replaces *B. coregoni* in lakes passing from oligotrophy to eutrophy does not seem to be valid in Dutch lakes (Gulati, 1972). Our observations confirm that the two species can coexist, even though for short intervals, in a range of lakes varying from highly eutrophic ones like the Loosdrecht lakes (Gulati, 1983; Gulati *et al.*, 1985) and Friesian lakes (Beattie *et al.*, 1978) to mesotrophic lakes like Lake Vechten (Gulati *et al.*, 1982).

In the case of *Daphnia cucullata* avoidance of competition with the co-occurring *B. coregoni* by differential use of food (Irvine, 1986) as well as the capacity to avoid large, filamentous cyanobacteria by narrowing the carapace gape (Gliwicz & Sieldler, 1980) does not explain its annual oscillation and very low densities as occurred in 1986. Predation on *Daphnia cucullata* may be greater and perhaps more variable, leading to sharp annual differences as noted.

The success of copepods perhaps lies in their omnivorous habits. Our grazing studies (^{14}C-technique; unpublished grazing data) show that despite a lack of linearity or continuity in the uptake rates, particles dominated by filamentous cyanobacteria may supplement the copepod daily food menu.

Changes in size structure

A continuous decrease in the mean (community) length of crustacean zooplankton since 1981, especially in the Loosdrecht lakes (Fig. 9), is accompanied by a parellel increase in the rotifers up to 1985. This suggests: 1) an increased predation on large-sized zooplankters; and 2) an increased consumption of smaller particles, particularly the nonfilamentous forms like flagellates. That these shifts in community length may be caused by size-selective predation is, as already mentioned, well documented.

One major effect of decrease in the animal size may be a change in the pattern and rates of energy or nutrient flow at system level. Bartell (1981) examined this in relation to predation by planktivorous fishes on phosphate release by zooplankton. It appears that a decrease in phosphorus release rates by zooplankton due to decrease in size and abundance of large zooplankters may be considerably offset by an enhanced P-release rate per unit biomass. Thus, if we take into account that with the decrease in crustacean length there is a simultaneous increase in rotifer numbers, the increase in the zooplankton nutrient flux, related to size decrease, may be even more important. Calculations of nutrient release by rotifers in the Loosdrecht lakes in 1983 by Ejsmont-Karabin (see also Ejsmont-Karabin, 1983) show that the rotifers may contribute 45-90% of the total P-regeneration by zooplankton in summer, when they are abundant. These release rates – if included in the P-flow scheme of Van Liere *et al.* (1989) – will largely meet, or even exceed, the primary production requirements of phytoplankton.

Acknowledgements

I am specially indebted to Messrs Guus Postema and Klaas Siewertsen for their technical assistance in both processing raw data and preparing the illustrations; Miss Cecilia Kroon is thanked for the typing work and Mr. E.M. Mariën for photography.

References

Bartell, S. M., 1981. Potential impact of size-selective planktivory on phosphorus release by zooplankton. Hydrobiologia 80: 139–145.

Beattie, D. M., H. L. Golterman & J. Vijverberg, 1978. An introduction to the limnology of the Friesian Lakes. Hydrobiologia 58: 49–64.

Bloem, J. & J. Vijverberg, 1984. Some observations on the diet and food selection of *Daphnia hyalina* (Cladocera) in a eutrophic lake. Hydrobiol. Bull. 18: 39–45.

Bottrell, H. H., A. Duncan, Z. M. Gliwicz, E. Grygierek, A. Herzig, A. Hillbricht-Ilkowska, H. Kurasawa, P. Larsson & T. Weglenska, 1976. A review of some problems in zooplankton production studies. Norw. J. Zool. 24: 419–456.

Brooks, J. L. & S. I. Dodson, 1965. Predation, body size and composition of plankton. Science 150: 28–35.

Burger-Wiersma, M. Veenhuis, H. J. Korthals, C. C. M. van Wiel & L. R. Mur, 1986. A new prokaryote containing chlorophyll *a* and *b*. Nature 320: 262–263.

Densen, W. L. T. van, C. Dijkers & R. Veerman, 1986. The fish community of the Loosdrecht Lakes and the perspectives for biomanipulation. Hydrobiol. Bull. 20: 147–163.

Dumont, H. J., 1977. Biotic factors in the population dynamics of rotifers. Arch. Hydrobiol. Beih. Ergebn. Limnol. 8: 98–122.

Duncan, A. & R. D. Gulati, 1981. Parakrama Samudra (Sri Lanka) Project, a study of a tropical lake ecosystem. III. Composition, density and distribution of the zooplankton in 1979. Verh. int. Ver. Limnol. 21: 1001–1008.

Ejsmont-Karabin, J., 1983. Ammonia nitrogen and phosphorus excretion by planktonic rotifers. Hydrobiologia 104: 231–236.

Geelen, J. F. M., 1955. The plankton of the lakes. Unpublished report no. 38. Amsterdam Municipal Waterworks (in Dutch).

Gliwicz, Z. M., 1969. Studies on the feeding of pelagic zooplankton in lakes with varying trophy. Ekol. Pol. 17: 663–708.

Gliwicz, Z. M. & E. Siedler, 1980. Food size limitation and algae interfering with food collection in *Daphnia*. Arch. Hydrobiol. 88: 155–177.

Golterman, H. L., 1965. Hydrobiologische aspecten van de Vechtplassen. Akademiedagen Kon. Ned. Akad. Wet. 17: 23–36/ (in Dutch).

Golterman, H. L., 1969. Methods for chemical analysis of water; ed. H. L. Golterman with assistance of R. S. Clymo. London, I.B.P., Oxford etc. Blackwell I.B.P. Handbook 8, 172 pp.

Gons, H. J., R. D. Gulati & L. v. Liere, 1986. The eutrophic Loosdrecht lakes: current ecological research and restoration perspectives. Hydrobiol. Bull. 20: 51–60.

Gulati, R. D., 1972. Limnological studies on some lakes in the Netherlands. I. A limnological reconnaissance and primary production of Wijde Blik, an artificially deepened lake. Freshwat. Biol. 2: 37–54.

Gulati, R. D., 1983. Zooplankton and its grazing as indicators of trophic status in Dutch lakes. Envir. Monit. Assessm. 3: 343–353.

Gulati, R. D., 1984. The zooplankton and its grazing as measures of trophy in the Loosdrecht Lakes. Verh. int. Ver. Limnol. 22: 863–867.

Gulati, R. D., K. Siewertsen & G. Postema, 1982. The zooplankton: its community structure, food and feeding and role in the ecosystem of Lake Vechten. Hydrobiologia 95: 127–163.

Gulati, R. D., K. Siewertsen & G. Postema, 1985. Zooplankton structure and grazing activities in relation to food quality and concentration in Dutch lakes. Arch. Hydrobiol. Beih. Ergebn. Limnol. 21: 91–102.

Gulati, R. D., G. Postema & K. Siewertsen, 1987. The hypertrophic Loosdrecht Lakes: zooplankton structure in the periods before (1982-83) and after (1984-86) reduction in external P-loading. Limnological Institute, Nieuwersluis, Internal Report 1987-6. 108 pp.

Haberman, J., 1983. Comparative analysis of plankton rotifer biomass in large Estonian lakes. Hydrobiologia 104: 293–296.

Hall, D. J., S. T. Threlkeld, C. W. Burns & P. H. Crowle, 1976. The size-efficiency hypothesis and size structure of zooplankton communities. Ann. Rev. Ecol. Syst. 7: 177–208.

Hillbricht-Ilkowska, A., 1977. Trophic relations and energy flow in pelagic plankton. Pol. ecol. Stud. 3: 3–98.

Hutchinson, G. E., 1967. A treatise on limnology. 2. Introduction to lake biology and the limnoplankton, Wiley, New York, 1115 pp.

Irvine, K., 1986. Differential feeding behaviour of the dominant Cladocera as an explanation of zooplankton community structure in the Loosdrecht Lakes. Hydrobiol. Bull. 20: 121–134.

Kal, B. F. M., G. B. Engelen & Th. E. Cappenberg, 1984. Loosdrecht lakes Restoration Project. Hydrobiology and physico-chemical characteristics of the lakes. Verh. int. Ver. Limnol. 22: 835–841.

Kott, P., 1953. Modified whirling for the subsampling of plankton. Aust. J. Mar. Freshwat. Res. 4: 387–393.

Lampert, W., 1981. Toxicity of blue-green *Microcystis aeruginosa*: effective defense mechanism against grazing pressure by *Daphnia*. Verh. int. Ver. Limnol. 21: 1436–1440.

Liere, L. van, S. Parma, L. R. Mur, P. Leentvaar & G. B. Engelen, 1984. Loosdrecht Lakes Restoration Project, an introduction. Verh. int. Ver. Limnol. 22: 829–834.

Liere, L. van, 1986. Loosdrecht lakes: origin, eutrophication, restoration and research programme. Hydrobiol. Bull. 20: 9–15.

Liere, L. van, R. D. Gulati, F. G. Wortelboer & E. H. R. R. Lammens, 1990. Phosphorus dynamics following restoration measures in the Loosdrecht lakes. Hydrobiologia 191: 87–95.

Loogman, J. G. & L. van Liere (eds.), 1986. Proceedings of the Waterquality Research Loosdrecht Lakes Symposium on: Restoration of shallow lake ecosystems with special emphasis on Loosdrecht Lakes, 20 (1–2), 259 pp.

Makarewicz, J. C. & G. E. Likens, 1979. Structure and

188

function of the zooplankton community of the Mirror Lake, N.H. Ecol. Monogr. 49: 109–127.

Pace, M. L., 1986. An empirical analysis of zooplankton community size structure across the trophic gradients. Limnol. Oceanogr. 31: 45–55.

Pace, M. L. & J. D. Orcutt Jr., 1981. The relative importance of protozoans, rotifers and crustaceans in a freshwater zooplankton community. Limnol. Oceanogr. 26: 822–830.

Pejler, B., 1957. Taxonomical and ecological studies on plankton Rotatoria from northern Swedish Lapland. K. Svensk. Vet. Akad. Handl. Ser. 4,6: 1–68.

Peters, R. H. & J. A. Downing, 1984. Empirical analysis of zooplankton filtering and feeding rates. Limnol. Oceanogr. 29: 763–784.

Pourriot, R., 1977. Food and feeding habits of Rotifera. Arch. Hydrobiol. Beih. Ergebn. Limnol. 8: 243–260.

Ruttner-Kolisko, A., 1977. Suggestions for biomass calculation of plankton rotifers. Arch. Hydrobiol. Beih. Ergebn. Limnol. 8: 71–76.

Starkweather, P. L., 1980. Aspects of feeding behaviour and trophic ecology of suspension feeding. Hydrobiologia 73: 63–72.

Starkweather, P. L. & P. E. Kellar, 1983. Utilization of cyanobacteria by *Brachionus calyciflorus*: *Anabaena flos-aquae* (NRC-44-1) as a sole or complementary food source. Hydrobiologia 104: 373–377.

Vijverberg, J. & A. F. Richter, 1982. Population dynamics and production of *Daphnia hyalina* Leydig and *Daphnia cucullata* Sars in Tjeukemeer. Hydrobiologia 95: 235–259.

Williamson, C. E., 1983. Invertebrate predation on planktonic rotifers. Hydrobiologia 104: 385–396.

Winberg, G. G. & collaborators, 1971. Symbols, units and conversion factors of freshwater productivity. I.B.P. London, 23 pp.

Hydrobiologia **191**: 189–198, 1990.
P. Bíró and J. F. Talling (eds), Trophic Relationships in Inland Waters.
© 1990 *Kluwer Academic Publishers.*

Ecological background and importance of the change of chironomid fauna (Diptera: Chironomidae) in shallow Lake Balaton

György Dévai
Department of Ecology, L. Kossuth University, H-4010 Debrecen, Hungary

Key words: Lake Balaton, sediment-dwelling chironomids, population-dynamics, water quality, TOC, organic matter circulation

Abstract

The objectives of this research were to record the changes in composition of the open-water, bottom-dwelling chironomid fauna in Lake Balaton between 1978–1984, to examine the causes of these changes, and to discover their significance in the life of the lake.

The spatio-temporal dispersion of larvae is compared with the water and sediment quality of each basin in the lake. It is established that, under present conditions, nutrient status can be regarded as the chief environmental factor.

Studies of population dynamics show that chironomids play a highly important role in preserving sediment quality. Chironomids are an essential element in the organic matter circulation of the lake. They dominate a sub-system that retards water quality degradation, and thus they play a prominent role in the natural prevention of eutrophication.

Introduction

The task of this research, which was carried out between 1978–1984, was to study the qualitative and quantitative composition of the sediment-dwelling chironomid fauna in Lake Balaton. The purpose was to collect basic biological data pertinent to the elaboration of plans for the control of water and sediment quality, and to the preparation of the matter and energy budget of the lake. A presentation of information for the evaluation and prediction of the changes occurring in the quality of water and sediment was also intended. A review of literature on sediment-dwelling organisms, especially chironomids, and a detailed presentation of our investigations, have already been given (Dévai, 1980, 1981, 1984,

1988; Dévai & Moldován, 1983; Dévai *et al.*, 1982, 1983). This paper aims at providing a summarizing evaluation of previous results.

Material and methods

In accordance with the stated objectives and the characteristic gradient of water quality in the lake (Somlyódy *et al.*, 1983), attention was mainly focussed on describing differences along the longitudinal axis of the lake, and on determining the state of the Keszthely-basin, the basin most endangered by water-quality deterioration. Therefore, nine sampling stations (H/1–H/9) were established along a longitudinal axis at the middle part of the bed (Fig. 1). Five more stations

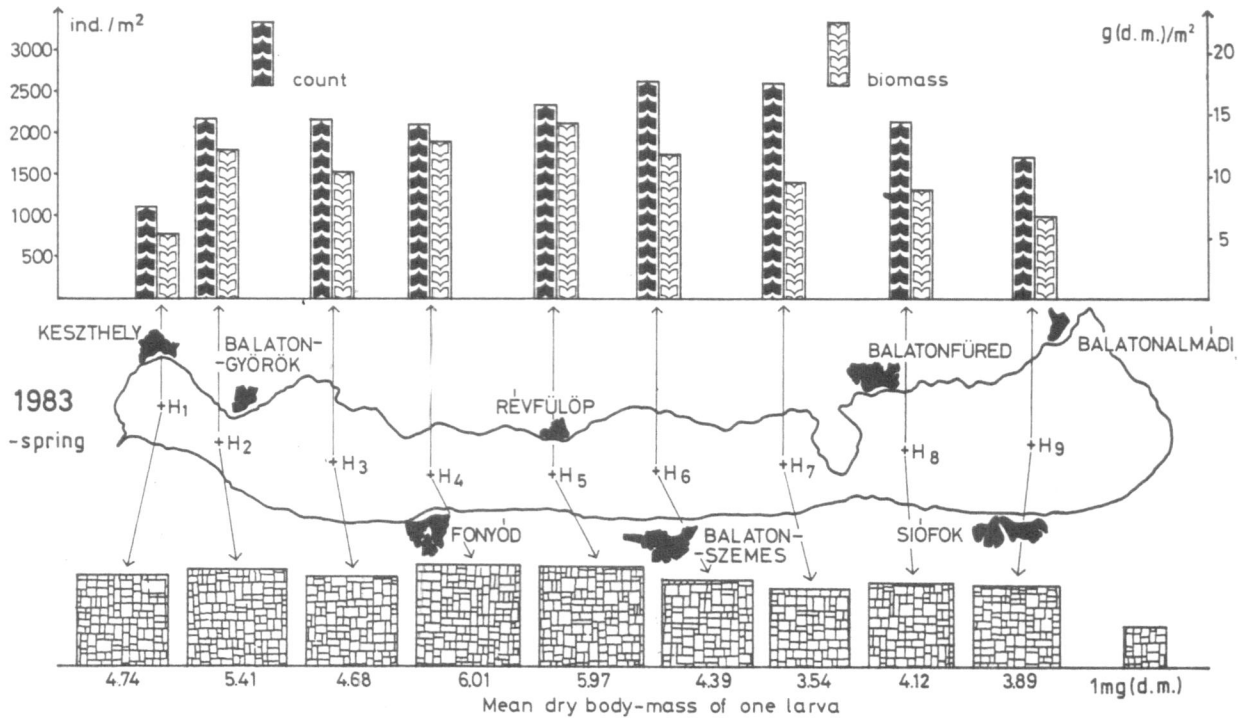

Fig. 1. Average larval count, biomass and mean body mass values of chironomids in Lake Balaton, spring 1983.

(K/1–K/5) were established in the Keszthely-basin; two of these were in its centre and three near-shore (Fig. 5). The characterization of Lake Balaton, and sampling and processing methods, have been described in detail(Dévai, 1984; Heim *et al.*, 1984; Tóth *et al.*, 1986).

Results and discussion

Composition and change of the chironomid fauna

In 1978, numbers of open-water, sediment-dwelling chironomid larvae in Lake Balaton increased from SW to NE, whereas biomass values decreased. This situation was caused by differences in the species composition. In the Keszthely- and Szigliget-basins (Fig. 6) there were relatively high proportions of large-bodied *Chironomus* species, whereas in the Central- and Northeastern-basins *Procladius* and *Tanypus* species, and representatives of small-bodied Chi-ronominae taxa, dominated (Dévai *et al.*, 1982; Dévai, 1984).

Our data were compared with the water quality of Lake Balaton and other similar shallow lakes in Hungary; also with the conclusions of typifying surveys found in the literature (Thienemann, 1954; Brundin, 1958; Stahl, 1966, 1969; Saether, 1975, 1979, 1980; Carter, 1976, 1977, 1978; Wiederholm & Eriksson, 1979; Kajak, 1980; Kansanen & Aho, 1981). Thus, the *Chironomus-Procladius* community of the Keszthely- and the Szigliget-basins indicated moderately eutrophic conditions, at least at the beginning of the study, whereas the dominance of *Procladius* observed between 1978–1982 in the Northeastern-basin reflected oligotrophic or perhaps oligo-mesotro-phic conditions.

Five sediment factors were examined to deter-mine the causes of the characteristic spatial pattern of the chironomid fauna in the lake. These factors were: (1) heavy metals (As, Cd, Co, Cr, Cu, Fe, Mn, Ni, Pb, Zn), (2) chlorinated hydrocarbons (aldrin, dieldrin, alpha-, beta-, gamma-, delta-HCH, heptachlor, pp'-DDE, pp'-DDT, op'-TDE, pp'-TDE), (3) organic carbon content, (4) elemental sulphur content and (5) redox dynamics.

Neither the qualitative nor the quantitative occurrence of heavy metals and chlorinated hydrocarbons, singly or together, were related to the spatial pattern of the chironomid communities (Dévai, 1984).

However, the organic carbon content of the sediment was related to the occurrence of chironomid biomass and numbers along the longitudinal axis of the lake. Thus, the nutrient content of the substratum measured as organic carbon content, i.e. the quantitative and qualitative composition of the detritus, can be regarded as an important environmental factor (Dévai, 1984).

Examination of the taxonomic composition of the recent and subfossil chironomid fauna proved that the spatial differences observed during 1978 in the sediment-dwelling chironomid fauna are secondary, i.e. are consequences of eutrophication, which differs in degree in different basins. The composition of the chironomid fauna also reflects this and is continuously changing in the Keszthely-basin without signs of returning to the former situation. Starting in 1978, the basin was more and more characterized by a chironomid community of stagnant backwaters, well-fertilized fish ponds, and basins of open-air sewage-treatment plants. Unfortunately, this transformation of the fauna in 1978 could be considered rather advanced in the Szigliget-basin too, and signs of the changes were already detectable in other areas of the lake at that time (Dévai & Moldován, 1983; Dévai, 1984). Data from the autumn of 1982 and from 1983 indicate that a similar process has started in the Central-basin and, perhaps, also in the Northeastern-basin. By the autumn of 1982, the faunal composition observed in 1978 had changed (Dévai, 1984). Numbers and biomass were evening-up in such a way that Lake Balaton, over practically its whole length, became similar to the state of the Keszthely-basin. Numbers, biomass and species composition of the chironomid fauna remained similar during the spring of 1983 (Fig. 1), although by the beginning of autumn important changes occurred (Figs. 2, 3). The numbers of larvae, especially in the Central- and the Northeastern-basins, decreased, and the proportion of Procladius species increased greatly. However, conditions did not return completely to those we found in 1978.

On the basis of these facts it can be established that it is the utilizable nutrient content of the sediment that enabled the mass occurrence of Chironomus balatonicus, the dominant species of the open-water sediment (cf. Warwick, 1975; Cole & Weigmann, 1983). Relying upon the latest examinations, the critical limiting value of organic carbon content is about 12 g kg^{-1} (dry matter). For the biomass and permanent occurrence of C. balatonicus larvae, however, the constant presence of about 15 g organic carbon content kg^{-1} (dry matter) seems to be needed (Dévai et al., 1983). The parallel changes that we found between 1978–1983 in the composition of the chironomid fauna and in the organic carbon content, and moreover the striking coincidence between predicted trends (cf. Dévai, 1984) and the subsequent investigation results, supported this idea (Fig. 4).

These results were confirmed by the spatio-temporal changes of the contents of easily mobilizable iron and manganese, organic carbon, elemental sulphur and of redox potential values. During the last few years, the amount of decaying organic matter has increased in the sediment of Lake Balaton. At the same time destructing and decomposing processes have considerably accelerated. Such changes in the sediment of the lake are connected with the acceleration of eutrophication. A change is predicted in which the oxygen conditions become degraded, as a result of progressively stronger decay processes, and will cause a complete change in the fauna. This is proved by the fact that the sediment of the Keszthely-basin already has a sapropelic, slightly mineralized organic content, and in the case of destructing processes, anaerobic pathways have a greater and greater predominance.

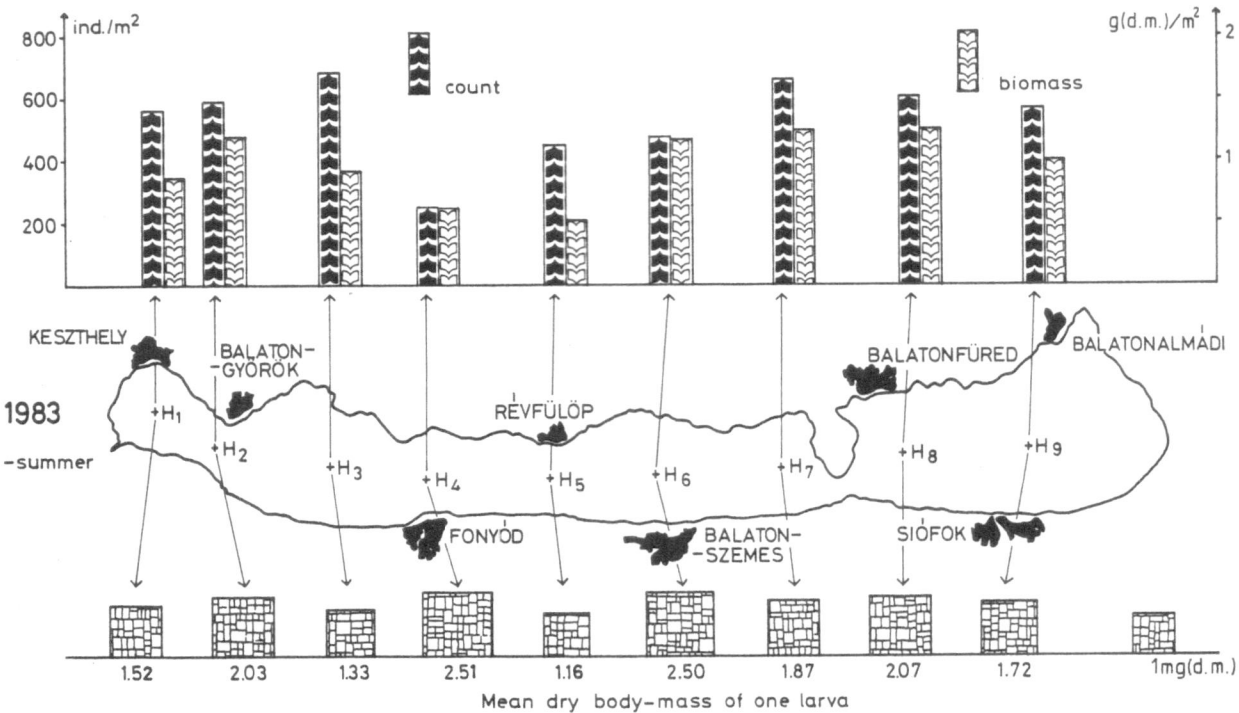

Fig. 2. Average larval count, biomass and mean body mass values of chironomids in Lake Balaton, summer 1983.

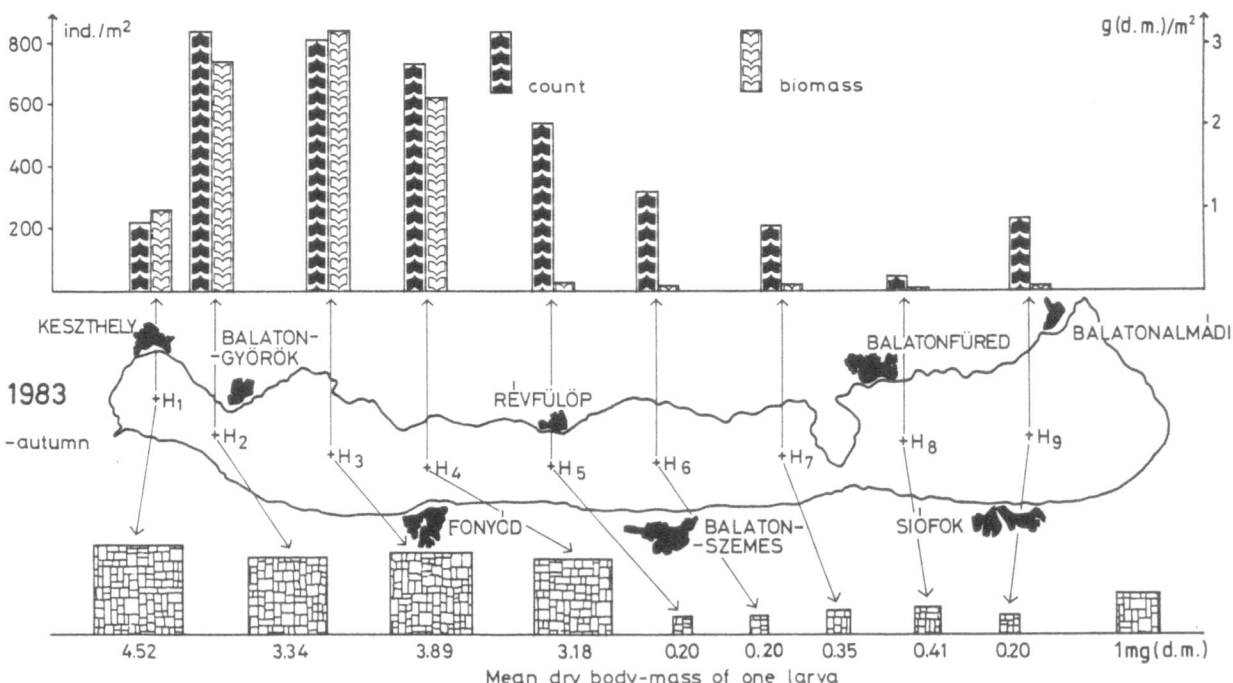

Fig. 3. Average larval count, biomass and mean body mass values of chironomids in Lake Balaton, autumn 1983.

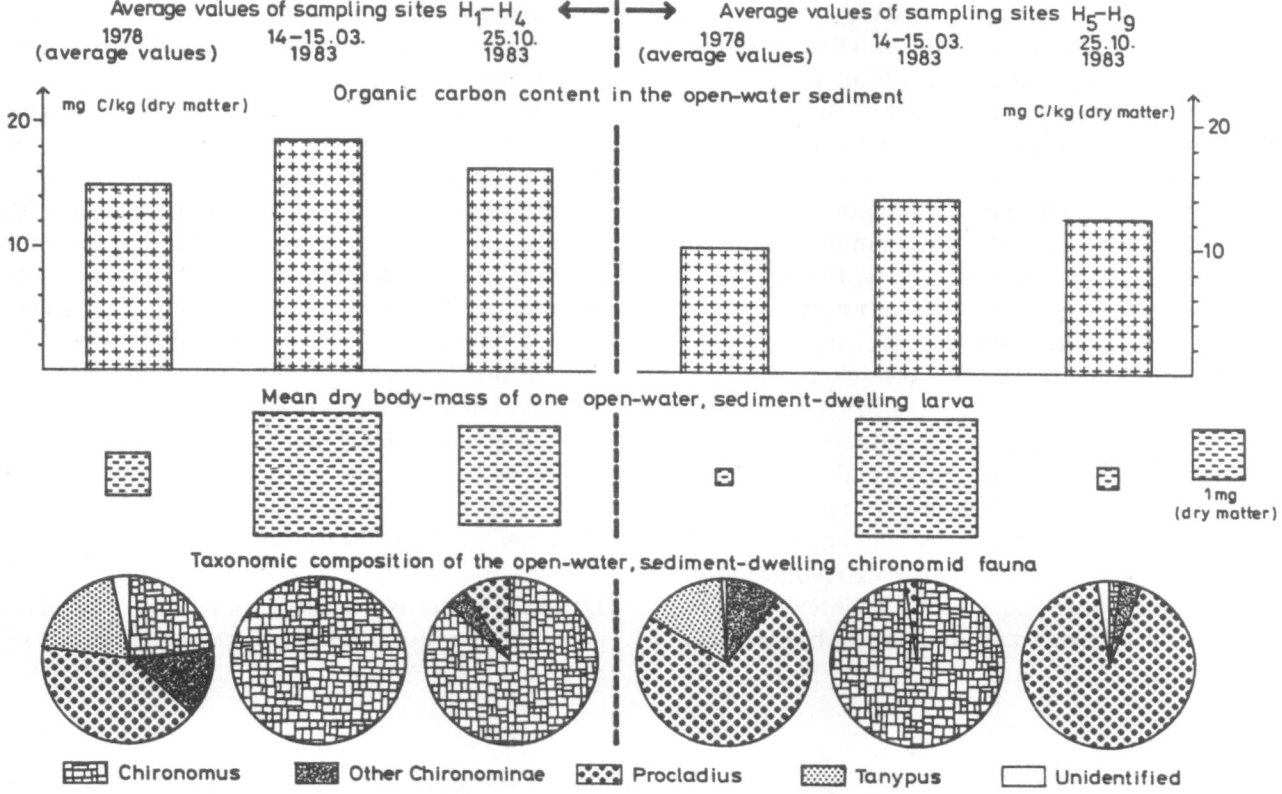

Fig. 4. Relationship between organic carbon content and chironomid faunal composition in Lake Balaton.

This unusually quick and considerable change required that population-dynamics of the sediment-dwelling chironomids be studied in connection with their role in the organic matter circulation of the lake. Using data collected over a period of several years in the Keszthely-basin, a phenological pattern, as shown by the 1983 data, was observed (Fig. 5). Highest numbers of larvae were observed at the beginning of April. This relatively high number continuously decreased until the last third of May. Numbers then increased, but only reached 60% of the maximum counts in spring. This increase stopped in the last third of June and from then, until the middle of July, the numbers of larvae decreased considerably. A new increase was observed again until the end of July, but in the first half of August numbers decreased even more than earlier. Numbers increased again in the second half of August, and remained at the same level between the end of

August and the middle of September. Of the four consecutive peaks, the first was strikingly high, whereas the other three seemed almost the same, although some decrease occurred approaching autumn. After the middle of September, the numbers of larvae drastically decreased in a few days to the lowest level of 1983. Their numbers then gradually began to increase until the middle of November. However, this increase was very slow, and numbers reached only one quarter of their value at the beginning of the year. On the basis of experiences of several years, we may consider this trend as characteristic of Lake Balaton, at least of the areas and periods that have great food abundance.

These results indicated that a detailed analysis of emergence dynamics would be valuable, because the emergence period at Lake Balaton is so extended (from mid-March to early October) and emergences are very frequent within the

194

period; they occur every second day, on average, and are of moderate intensity on every fifth day from May to September (cf. Dévai, 1988).

This peculiar form of emergence dynamics demanded the determination of the time of mass emergences and the degree of synchronization in 1983, simultaneously with examinations of larvae. The emergence in 1983 was similar to earlier years, and was almost continuous. Emergence of high intensity was observed five times: between 11–18 April, 25 June–03 July, 30 July–2 August, 11–16 August, and 12–18 September. The first emergence was the greatest. In May, emergence was of average intensity for almost the whole month, and the sum of individual emergences during this period was similar to high but short emergences during other periods. The two great emergences in the period 30 July – 16 August must be considered only one, which was interrupted by the unusual, quick and unfavourable change in the weather. Moreover, emergences, at least the stronger ones, occurred at the same time over the whole area of the lake.

It is useful to compare the results of larval collections with the emergence observations. Figure 5 shows that the changes in emergence intensity can easily be brought into relation with changes in larval counts. The main emergence periods almost always overlap the same types of sections on the hypothetical curve. Great emergences come after the periods that show the highest numbers of larvae in the given period of time, and then numbers of larvae usually decrease quickly and considerably.

Chironomids in the circulation of organic matter

Detailed analysis of the spatio-temporal dispersion of larvae, and the emergence dynamics of

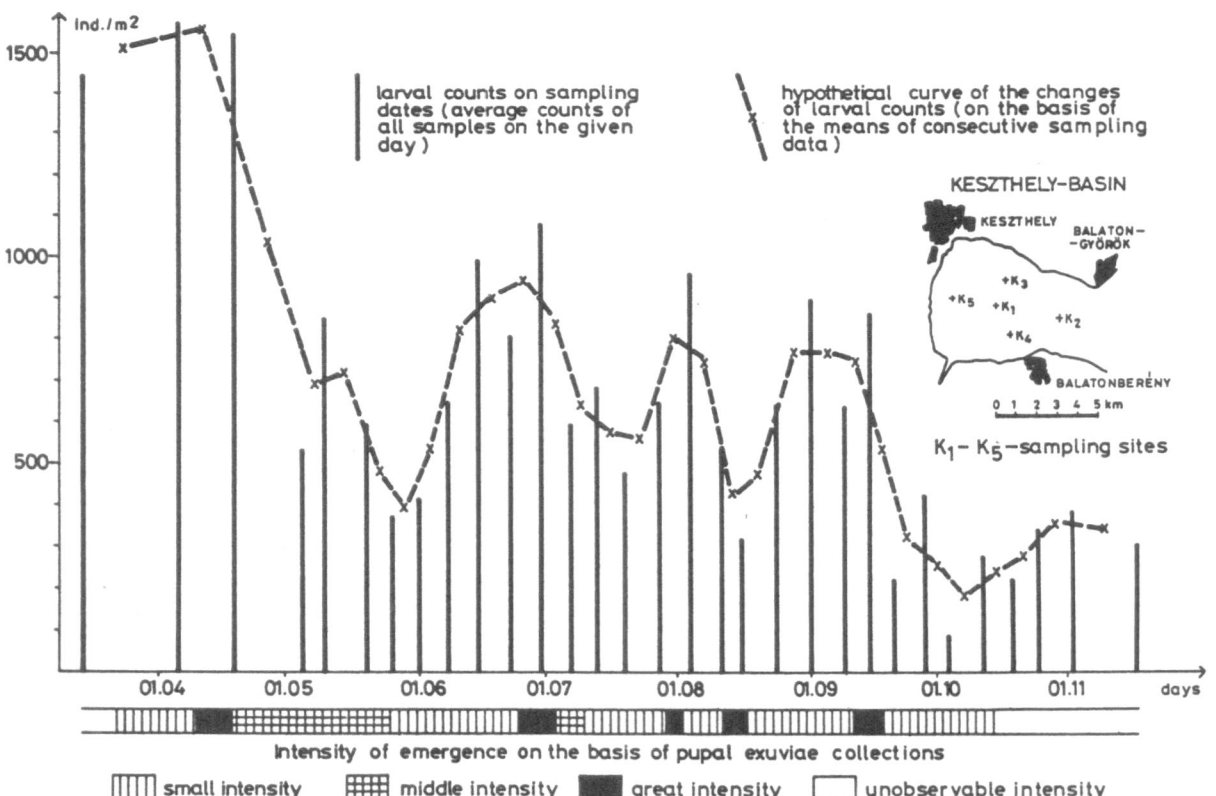

Fig. 5. The changes of larval chironomid counts and intensity of emergence in the Keszthely-basin in 1983.

adults, supported the opinion that chironomid numbers, biomass and production are significant components for the matter and energy budgets and for water quality control of the lake. For example, in 1983 there were an average of 6×10^{11} larvae, calculated for the whole area of the lake; in March, which had the greatest abundance of food, the number was more than double this value (14.5×10^{11}). The biomass values were also staggering. Their yearly average calculated for the whole lake was 2400 tons, but the maximum value in March reached 8100 tons.

This enormous chironomid population deserves attention because natural purification of waters depends largely on the dynamics of biogenic matter circulation and energy flow (bioactivity) of the water-bodies, primarily on the activity of heterotrophic organisms, among which emerging insects are of great importance (Dévai, 1980).

Detritus-feeding larvae play a direct role in the elimination of decayable and putrefiable organic matter, the amount of which is increasing in the sediment because of eutrophication. Although the organic content of the bodies of chironomid larvae is negligible as compared to the amount of organic matter in the sediment (the ratio is on average only 0.1%), the quantity of sediment these larvae circulate is considerable (relying on Entz's results in 1964, it is 0.5–$1.5 \, \mathrm{g\,m^{-2}\,day^{-1}}$, using the average individual density of the lake). Because most of the sediment is composed of detritus, chironomid larvae circulate about 2–3% of the organic compounds of the upper sediment-layer during the year. During digestion the larvae expose one part of the organic matter and make it available for bacterial destruction, whereas another part is utilized in the course of their existence and growth. The importance of predatory larvae, which are abundant, must also be considered. By entering into newer and higher levels of the food-chain, the proportion of metabolism used for existence is steadily increased. Due to these processes, the activity of chironomids contributes considerably to the breakdown of organic matter under natural conditions.

It was probable that the food abundance between the autumn of 1982 and the spring of 1983, at least in the Northeastern- and Central-basins, would not remain permanent. This was because the sudden and excessive increase in chironomid population would probably 'consume' the utilizable nutrient content, and some chironomid species (e.g. *Chironomus balatonicus*) would be temporarily eliminated from these areas. Paleolimnological study of the sediment in the Northeastern-basin verified that there have already been events of this type in the history of the lake (Dévai & Moldován, 1983).

The results obtained in the summer and autumn of 1983 firmly supported this supposition (Figs. 2–4). The question is, when will this – so far – temporary change become constant, i.e. turn into an irreversible process, which will unfortunately restrict the possibilities of treatment. The Keszthely-basin is already in a critical state and the Szigliget-basin is taking this direction. Increasing warning signs indicate that the break-up of natural matter-circulation dynamics of the sediment has already begun in the Central- and also in the Northeastern-basin.

Sediment-dwelling larvae are of great importance in the constant reworking and concomittent aeration of the sediment. With this activity, the larvae contribute to the preservation of aerobic conditions, i.e. to the maintenance of the 'healthy' matter circulation and energy-flow dynamics of the lake.

The amount of organic matter released from the lake by emerging chironomids seems most important in forming water and sediment qualities.

The organic matter released by emergence was not negligible in 1978 either, but in 1983 it was considerable (Fig. 6): about 6300 tons during the first series of emergence, i.e. chironomids removed approximately 3300 tons of organic carbon, 700 tons of nitrogen and 60 tons of phosphorus from the lake. Because there are four emergences in a year on the whole area of the lake, during one vegetation season, chironomids release almost two-thirds of the yearly available phosphorus-loading (i.e. about 100 tons of P).

196

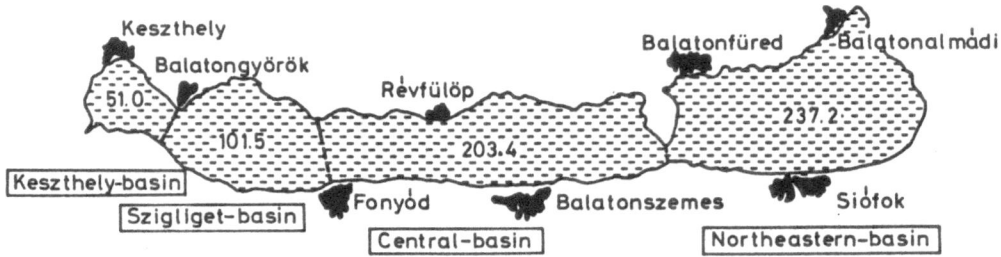

The basins of Lake Balaton and their area (km^2)

The biomass of emerging adults and the estimated C,N and P values released by adults in 1983 (in tons)

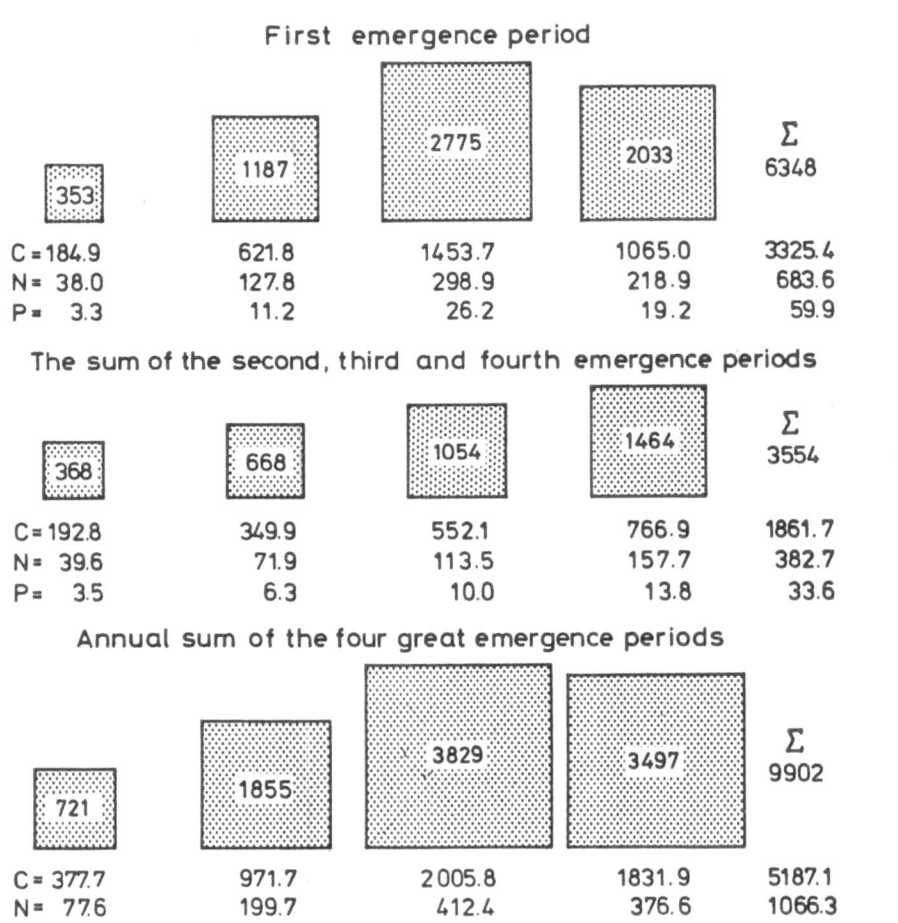

Fig. 6. Contribution of emerging chironomid adults to the organic matter circulation of Lake Balaton and its four basins.

From all these considerations it follows that chironomids are important in the factor-system which counterbalances the increase in trophic status under natural conditions. Investigations of effect-dynamics have shown that natural water systems, by way of self-regulating and self-controlling processes, can maintain their state on a supra-individual level, too, despite external effects (perturbations). However, self-directing processes, which are present in natural ecological systems and manifest in homeostasis, can only partly (i.e. within the ranges of the given state of water and sediment qualities) prevent changes of unfavourable direction. Lake Balaton is very close to this critical state. If the best opportunities are provided for the processes (mainly biological) that oppose water quality deterioration in natural ways, serious impairments can be avoided and even gradual improvement can be expected.

Acknowledgements

I should like to thank the Hungarian Office of Environmental and Nature Protection and the Hungarian Academy of Sciences for moral and financial support. Many thanks are also due to the Institute for Water Quality Control of the Research Centre for Water Resources Development, and personally to Dr László Tóth, for affording sampling possibilities and discussing scientific problems; to Professor Dr Pál Jakucs, head of our Ecological Department, for his constant help and valuable advice; and to my colleagues in the Ecological Department of L. Kossuth University (A. Bagyó, E. Béke, Zs. Dienes, R. Komjáthi, M. Miskolczi, Dr J. Moldován, Dr Zs. Preczner, D. Rónai, E. Szilágyi, Zs. Vajdics and J. Vojnits) and in the Hajdú-Bihar County Water and Canalization Works (Dr I. Czégény, Dr I. Dévai, Dr Cs. Heim and Dr I. Wittner) for their assistance in sampling, preparation and evaluation.

References

Brundin, L., 1958. The bottom faunistical lake type system and its application to the southern hemisphere. Moreover a theory of glacial erosion as a factor of productivity in lakes and oceans. Verh. int. Ver. Limnol. 13: 288–297.

Carter, C. E., 1976. A population study of the Chironomidae (Diptera) of Lough Neagh. Oikos 27: 346–354.

Carter, C. E., 1977. The recent history of the chironomid fauna of Lough Neagh, from the analysis of remains in sediment cores. Freshwat. Biol. 7: 415–423.

Carter, C. E., 1978. The fauna of the muddy sediments of Lough Neagh, with particular reference to eutrophication. Freshwat. Biol. 8: 547–559.

Cole, R. A. & D. L. Weigmann, 1983. Relationships among zoobenthos, sediments, and organic matter in littoral zones of western Lake Erie and Saginaw Bay. J. Great Lakes Res. 9: 568–581.

Dévai, Gy., 1980. Vorstudien zur Bedeutung der sedimentbewohnenden Zuckmücken im Stoffhaushalt des Balatonsees (Ungarn). In D. A. Murray (ed.), Chironomidae. Proc. 7th int. Symp. on Chironomidae, Dublin, August, 1979. Pergamon Press, Oxford: 269–273.

Dévai, Gy., 1981. [A review of sediment studies on Lake Balaton. In I. Kárpáti (ed.), New Results of Balaton research II.] MTA VEAB Monográfiái 16: 201–209 (in Hungarian with English summary).

Dévai, Gy., (ed.), 1984. Studies of the ecological effects of Lake Balaton and River Zala sediments on chironomids (Diptera: Chironomidae). Acta biol. debr. oecol. hung. 1: 1–183, Tab. 1–7, Fig. 1–59.

Dévai, Gy., 1988. Emergence patterns of chironomids in Keszthely-basin of Lake Balaton (Hungary). Spixiana, Suppl. 14: 201–211.

Dévai, Gy., I. Czégény, I. Dévai & F. Máté, (1980) 1982. [Relationship between bottom-dwelling chironomid fauna and the sediment quality of Lake Balaton. Part one. Iron and manganese content of the sediment]. Acta biol. debr. 17: 51–74 (in Hungarian).

Dévai, Gy., & J. Moldován, 1983. An attempt to trace eutrophication in a shallow lake (Balaton, Hungary) using chironomids. Hydrobiologia 103: 169–175.

Dévai, Gy., W. Wülker & A. Scholl, 1983. Revision der Gattung Chironomus Meigen (Diptera). IX. C. balatonicus sp.n. aus dem Flachsee Balaton (Ungarn). Acta zool. hung. 29: 357–374.

Entz, B., 1964. Ernährungs-Untersuchungen an Chironomiden des Balaton. I. Quantitative Ernährungs-Untersuchungen an Larven von Chironomus plumosus L. Annal. biol. Tihany 31: 165–175.

Heim, Cs., I. Dévai & J. Harangi, 1984. Gas chromatographic method for the determination of elemental sulphur in sediments. J. Chromatogr. 295: 259–263.

Kajak, Z., 1980. Role of invertebrate predators (mainly Procladius sp.) in benthos. In D. A. Murray (ed.), Chiro-

198

nomidae. Proc. 7th int. Symp. on Chironomidae, Dublin, August, 1979. Pergamon Press, Oxford: 339–348.

Kansanen, P. H. & J. Aho, 1981. Changes in the macrozoobenthos associations of polluted Lake Vanajavesi, Southern Finland, over a period of 50 years. Ann. zool. fenn. 18: 73–101.

Saether, O. A., 1975. Nearctic chironomids as indicators of lake typology. Verh. int. Ver. Limnol. 19: 3127–3133.

Saether, O. A., 1979. Chironomid communities as water quality indicators. Holarct. Ecol. 2: 65–74.

Saether, O. A., 1980. The influence of eutrophication on deep lake benthic invertebrate communities. Prog. Wat. Tech. 12: 161–180.

Somlyódy, L., S. Herodek & J. Fischer (eds.), 1983. Eutrophication of shallow lakes: modeling and management. The Lake Balaton case study. Collaborative Proc. Ser. CP-83-S3. IIASA, Laxenburg, x + 369 pp.

Stahl, J. B., 1966. Characteristics of a North American *Sergentia* lake. Gewäss. Abwäss. 41–42: 95–122.

Stahl, J. B., 1969. The uses of chironomids and other midges in interpreting lake histories. Mitt. int. Ver. Limnol. 17: 111–125.

Thienemann, A., 1954. *Chironomus*. Leben, Verbreitung und wirtschaftliche Bedeutung der Chironomiden. In A. Thienemann (ed.), Die Binnengewässer 20. E. Schweizerbart'sche Verlagsbuchhandlung, Stuttgart, XVI + 834 pp.

Tóth, J. A., Zs. Preczner & S. Nagy, 1986. [Relationship between redoxpotential, organic carbon content and the number of bacteria in the sediment of shallow waters]. Hidr. Közl. 66: 102–107 (in Hungarian with English summary).

Warwick, W. F., 1975. The impact of man on the Bay of Quinte, Lake Ontario, as shown by the subfossil chironomid succession (Chironomidae, Diptera). Verh. int. Ver. Limnol. 19: 3134–3141.

Wiederholm, T. & L. Eriksson, 1979. Subfossil chironomids as evidence of eutrophication in Ekoln Bay, Central Sweden. Hydrobiologia 62: 195–208.

Hydrobiologia **191**: 199–212, 1990.
P. Biró and J. F. Talling (eds), Trophic Relationships in Inland Waters.
© 1990 *Kluwer Academic Publishers.*

Trophic relationships in the pelagic zone of Mondsee, Austria

Martin Dokulil[1], Alois Herzig[2] & Albert Jagsch[3]
[1] *Institut für Limnologie der Österreichischen Akademie der Wissenschaften, Gaisberg 116, A-5310
Mondsee, Austria*; [2] *Forschungsinstitut für das Burgenland, Biologische Station Illmitz, A-7142, Illmitz,
Austria*; [3] *Bundesanstalt für Fischereiwirtschaft, A-5310 Scharfling, Austria*

Key words: trophic interactions, pelagic zone, seasonal succession, lakes, ecology

Abstract

Data are presented on nutrient concentrations, phytoplankton biovolume development, zooplankton composition and population dynamics, and fish from a deep, stratifying, alpine lake (Mondsee, Austria) during a three-year period between 1982 and 1984. Development of the phytoplankton is closely related to structuring events of the physico-chemical environment. Dissolved silicate and phosphorus concentrations are critical for the summer situation. During summer algal abundance is largely affected by grazing of zooplankton, but no clear-water phase was observed at the end of the spring peak of phytoplankton.

Temperature and food are factors responsible for the timing and growth of the zooplankton populations. Because of close overlap in the epilimnion, exploitative and mechanical interference competition and predation by invertebrate and vertebrate predators are the main structuring forces acting on the zooplankton community, and hence influence phytoplankton indirectly.

Introduction

Freshwater phytoplankton assemblages are influenced greatly by the concentrations of nutrients and the complex relationship between algae and herbivorous zooplankton, which in turn can be influenced by both invertebrate predators and fish.

Most commonly phosphorus concentration is limiting phytoplankton standing crop in nutrient-poor lakes. An increasing input of phosphorus generally results in the dramatic elevation of algal abundance, and may favor certain species over others (Schindler & Fee, 1974; Schindler, 1977; Reynolds, 1986). Moreover, the relative amounts of certain nutrients can influence phytoplankton species-composition. Blue-green algae may domi-

nate at low N : P ratios (Rhee & Gotham, 1980; Smith, 1983), whereas the Si : P ratio may determine the success of different diatom species (Kilham, 1971; Tilman *et al.*, 1982). Under specific conditions other constraints can override these nutrient effects (Reynolds, 1986). Depletion of certain nutrients such as nitrogen or silicate can lead to drastic shifts in phytoplankton species composition (Sommer *et al.*, 1986).

The response of the algal community to herbivores is the net result of effects both negative (grazing by zooplankton) and positive (growth enhancement due to nutrient recycling) (Porter, 1977; Gliwicz, 1977; Carpenter & Kitchell, 1984; Bergquist & Carpenter, 1986; Sommer *et al.*, 1986). The nature and intensity of grazing pressure on the phytoplankton assemblage depend on numerous features of the zooplankton community, such as its composition (Burns, 1968), different feeding modes of taxa (Frost, 1980), and feeding activity (Porter, 1977). Further, the outcome of grazing by zooplankton may be influenced by selective feeding and by interactions between various taxonomic entities within the zooplankton such as food partitioning, various forms of competition and predation (Sommer *et al.*, 1986). Some influences on zooplankton and indirect effects on phytoplankton can often be related to higher trophic levels, mainly fish (Hrbáček, 1982; Lynch, 1979).

The present aim is to identify some of the above-mentioned interactions, and to quantify their importance among the planktonic organisms within the pelagic zone of a mesotrophic lake. Influences on phytoplankton by herbivores will only be demonstrated at the community-level. Species-specific algal responses (Elser *et al.*, 1987) are not considered.

Study site and methods

Mondsee is a moderately large (area 14.2 km^2), deep (maximum depth = 68.3 m) alpine lake. Details of location, morphometry, hydrology and eutrophication history are included in Jagsch (1982), Dokulil & Jäger (1982), Dokulil & Skolaut (1986) and Klee & Schmidt (1987). At present the lake can be classified as mesotrophic due to recovery after restoration measures (Dokulil, 1984).

Samples were collected, at weekly intervals during the years 1982 and 1983 in the western part of the lake, from the surface down to 40 m at the depths indicated in the graphs. During winter and in 1984 the sampling interval was extended to two weeks. Chemical data, except silicate (here expressed as silica), originate from monthly samples at the deepest point of the lake. Standard analytical techniques were used for all chemical quantities (Mackereth *et al.*, 1978).

Secchi transparency was estimated using a white disc of 25 cm diameter. The depth of the 1% level of surface irradiance, usually taken as the lower limit of the euphotic zone, was located from radiation measurements using an underwater quantum sensor (Li-COR, USA) or converted from the Secchi-depth using a factor of 2.5 established from systematically combined measurements.

Phytoplankton biovolume was estimated from cell counts and size estimates on an inverted microscope (Lund *et al.*, 1958) using geometric approximations (Rott, 1981).

Zooplankton was collected in accordance with the above sampling scheme with a modified 10-l Schindler-Patalas trap, sieved through 30 μm net, and counted under a stereo microscope. Rotifers were sedimented in counting chambers and enumerated on the inverted microscope. Occasionally the collection interval was reduced to three days.

Results and discussion

Physico-chemical conditions

Physico-chemical properties of the lake throughout 1982-1984 are summarized in Fig. 1. Soon after vernal overturn, at the onset of thermal stratification, light conditions in the epilimnion improve markedly. At the same time soluble reactive silica (SRS) is strongly depleted to levels below 100 μg l^{-1}. Phosphate-phosphorus is below detection level (< 1 μg l^{-1}) in the epilimnion for the greater part of the year (Fig. 2). These low concentrations even persist through the winter period of 1983/84, a winter with no significant ice cover and much larger phytoplankton biovolumes compared to previous years. Values up to 30 μg PO$_4$-P l^{-1} were observed in autumn in the hypolimnion near the bottom, representing phosphorus release due to de-oxygenation of the sediment surface (Gunatilaka, 1980; Jagsch, 1982; Jagsch & Bruschek, 1982; Danielopol, pers. com.).

Similarly, ammonium-nitrogen concentrations range from 100 to 600 μg l^{-1} near the bottom and from less than 10 to 50 μg l^{-1} in the epilimnion. Nitrate-nitrogen (< 200 to > 600 μg l^{-1}) is

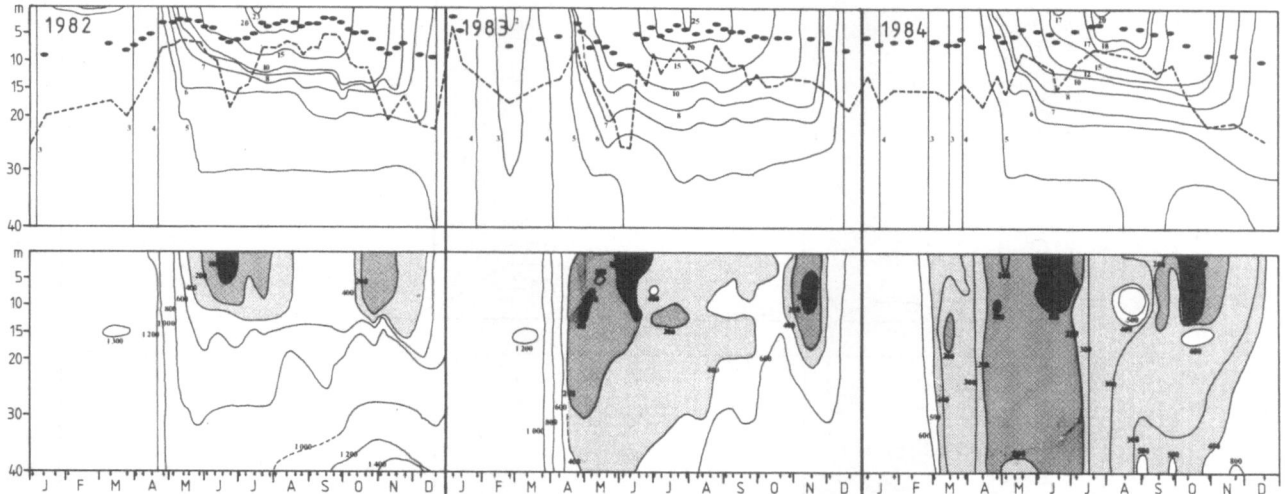

Fig. 1. Some physico-chemical dynamics of Mondsee for the years 1982-1984. Upper panel: depth-time distribution of isotherms, Secchi-depth (–) and the 1% light level (---). Lower panel: isopleths of soluble reactive silica (SRS) as μg l⁻¹. SRS-concentrations below 400 μg l⁻¹ are shaded. Sampling dates are indicated at the bottom of the graph as small arrows.

reduced to concentrations below 200 μg in the epilimnion during late summer and early autumn (Fig. 2). Nitrogen to phosphorus ratios (by weight) always exceeded critical values of 10-17.

Seasonal dynamics of phytoplankton

In response to these environmental variables, phytoplankton biovolume exhibits characteristic bimodal patterns each year (Fig. 3). The spring peak, largely consisting of diatoms (Dokulil & Skolaut, 1986; Dokulil & Kofler, in press), is reached while the lake is still mixed. Thereafter total biovolume in the water column decreases rapidly (Fig. 3, upper panel) and phytoplankton growth is largely restricted to the epilimnion (Fig. 3, lower panel). Considerable quantities of the blue-green alga *Oscillatoria rubescens* D.C. are located in the metalimnion throughout the summer period (cf. Figs. 1 and 3, lower panel and Dokulil, in press). A second maximum of biovolume, largely differing between years, is reached around September and is usually composed of blue-green algae and chrysophyceans. Among other diatoms *Tabellaria flocculosa* (Lyngb.) Kütz. var. *asterionelloides* (Grun.) Knud. dominates at

the end of the year (Dokulil & Skolaut, 1986; Dokulil & Kofler, in press) because of adaptation to moderate temperatures and light intensities (Kofler, 1986; Dokulil, 1987; Kofler & Dokulil, in press). Cryptophyceans appear throughout the year but usually in low quantities since they are heavily preyed upon by zooplankton (Dokulil, 1987). These mobile flagellates dominate the winter season (Dokulil & Skolaut, 1986) largely because of smaller loss rates compared to other algal species (Dokulil, 1987).

Phytoplankton biovolume is persistently correlated with calculated (total minus dissolved phosphorus) particulate phosphorus (P-part) over the years (Fig. 4). The correlation coefficient, $r = 0.56$ for all observations ($n = 128$), is significant at the 95% confidence level. The coefficients for individual years (Fig. 4) show best agreement for the data in 1983. The rather poor correlation must partly be attributed to the independent collection in time and space of the two variables and partly to variable amounts of P-part other than algal. Calculation of algal phosphorus content based on the Redfield ratio yields an average background level of 2.5 μg P-part l⁻¹, a value closely represented by the intercepts of the 1983 and 1984 regression lines (Fig. 4). Preliminary

202

Fig. 2. Isopleths of nutrient concentrations in Mondsee, 1982-1984, as $\mu g\,l^{-1}$. (A) total-phosphorus, (B) phosphate-phosphorus, (C) ammonium-nitrogen, (D) nitrate-nitrogen.

203

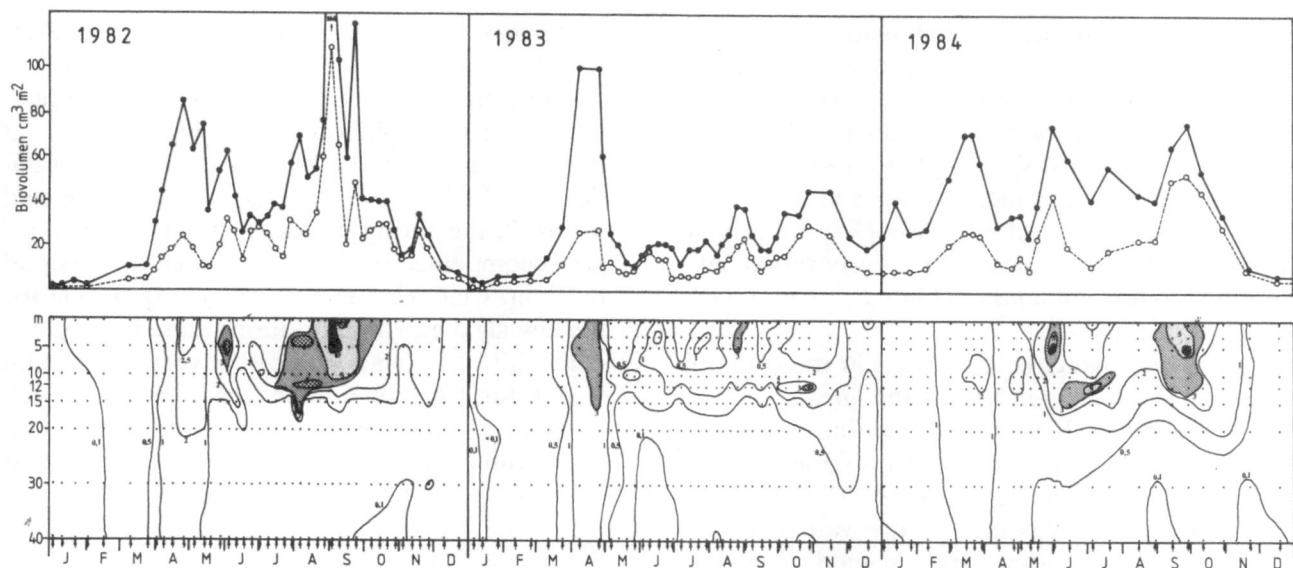

Fig. 3. Development of the phytoplankton assemblage in Mondsee, 1982-1984. Upper panel: Biovolume within the total water column (●) and the euphotic zone (o–o) as cm³ m⁻². Lower panel: Depth-time distribution of total phytoplankton biovolume as 10⁶ μm³ l⁻¹. Higher biovolumes are shaded.

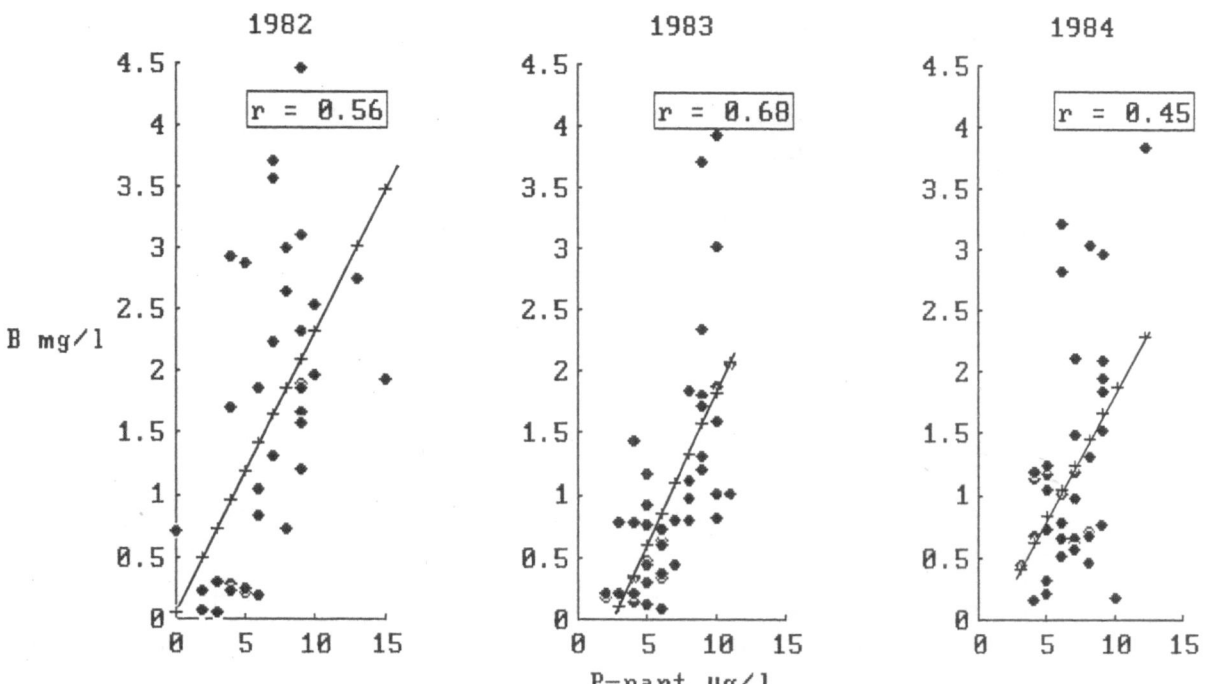

Fig. 4. Correlation of phytoplankton biomass (B) as mg l⁻¹ to particulate phosphorus (P-part) as μg l⁻¹ for the individual years investigated. Correlation coefficients (r) are inserted.

experiments of nutrient enrichment with natural phytoplankton reveal phosphorus as the limiting nutrient during the summer months (Calderon & El Higzi, 1983). In some experiments nitrate further stimulated growth and photosynthetic rates.

The observed depletion of nitrate in the epilimnion in the course of the year (Fig. 2) can be attributed to some extent to incorporation into phytoplankton biomass (Fig. 5; $r = -0.44$, $n = 128$, $p \leq 0.005$).

Average biovolume of diatoms in the euphotic zone is inversely related to the mean concentration of SRS (Fig. 6). Exponential increase of diatoms occurs in spring when soluble reactive silica concentrations are high. As diatom crops increase the silicate content of the water declines. Thermal stratification and silicate depletion (Fig. 1) thereafter result in a drastic decrease of diatom populations (Fig. 6) associated with shifts in species composition (cf. Fig. 10 in Dokulil & Skolaut, 1986). A second growth phase of diatoms is initiated in autumn when silicate con-

centrations are replenished to 0.4-0.5 mg SiO_2 l^{-1} (Fig. 6).

The annual development, magnitude and fluctuation of total phytoplankton biovolume is at times largely influenced by the presence and abundance of herbivorous zooplankton. As indicated in Fig. 7, the spring peaks of the plankton algae are accompanied or shortly followed by copepod maxima. Cladocerans are partly responsible for the low algal biovolumes during summer (Fig. 7). Although the relationship between individual values for total phytoplankton and crustacean biomass shows considerable scatter, seasonal and annual averages are closely correlated (Fig. 8).

Zooplankton dynamics

Qualitative aspects

About 40 rotifer and 13 crustacean species comprise the zooplankton of Mondsee. According to

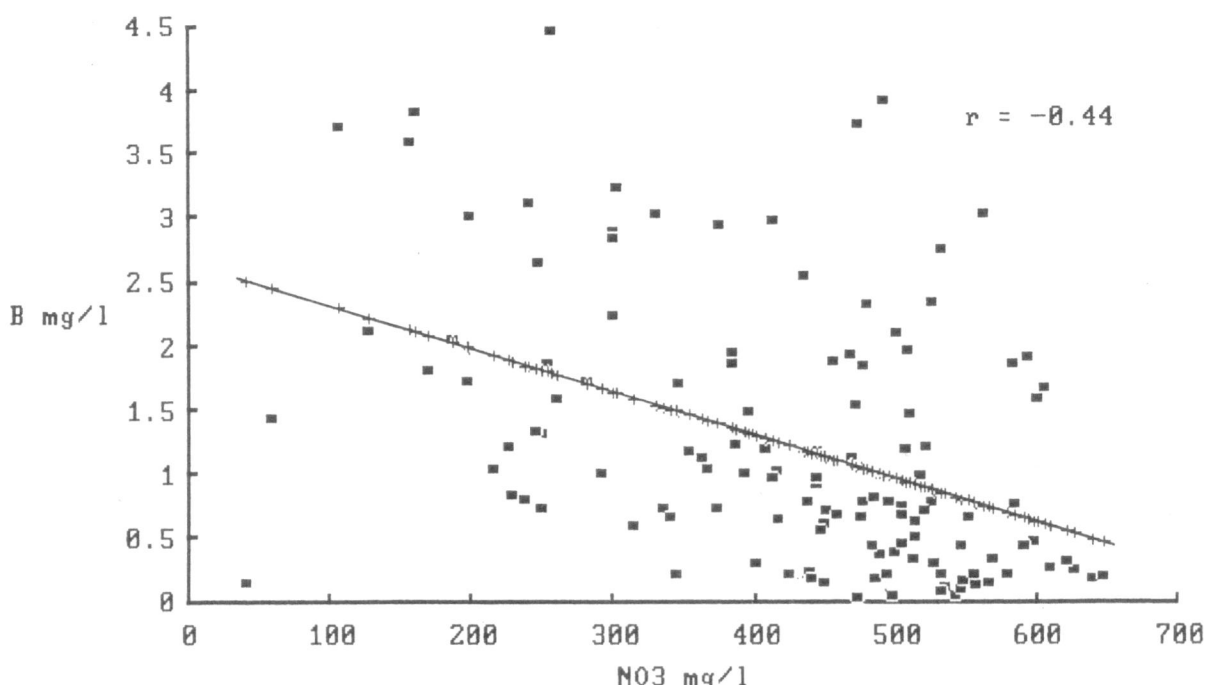

Fig. 5. Correlation of phytoplankton biomass (B; mg l^{-1}) to nitrate concentration (NO_3-N; mg l^{-1}). The correlation coefficient (r) is inserted.

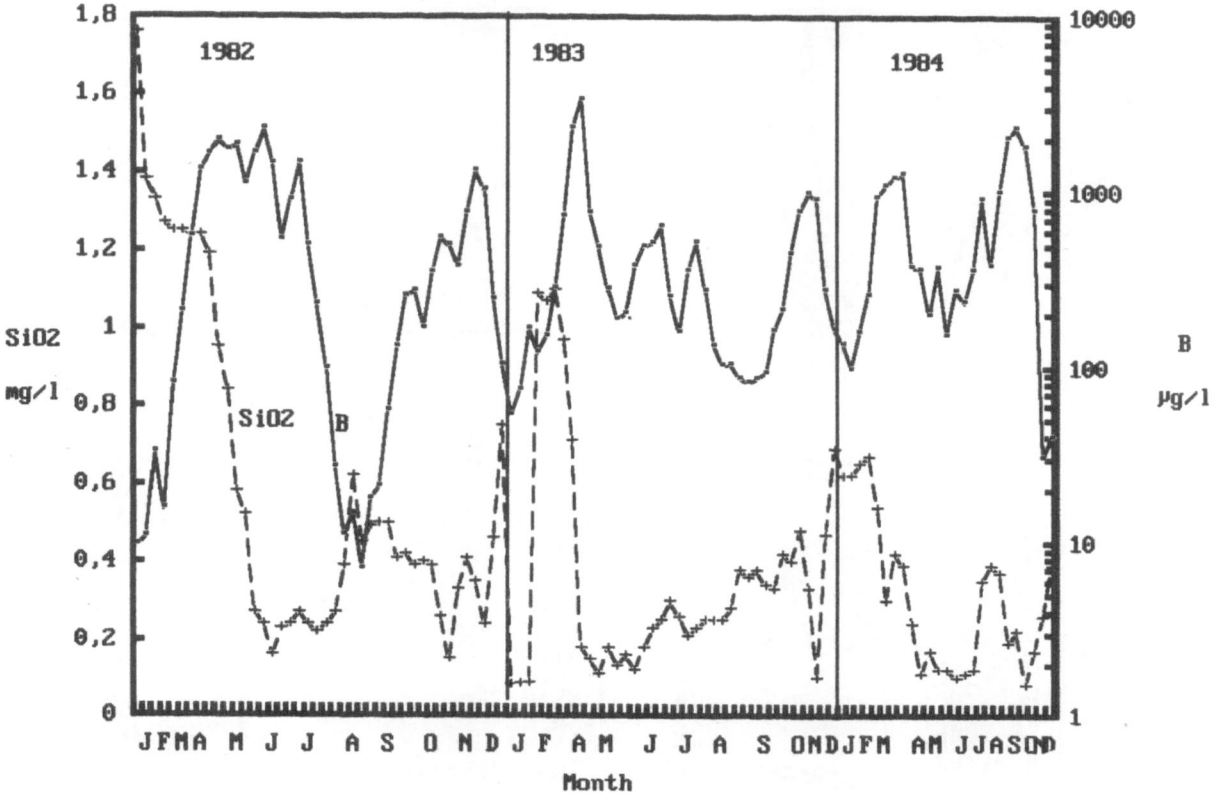

Fig. 6. Average soluble reactive silica concentration (SiO₂; mg l^{-1}) and average total diatom biomass (B; $\mu\text{g l}^{-1}$, logarithmic scale) in the euphotic zone of Mondsee for the years 1982-1984.

their seasonal occurrence some 10-16 rotifer species are prevailing. As in other Austrian and European lakes the quantitatively important species are *Keratella cochlearis* (15-95% of total numbers), *Kellicottia longispina, Polyarthra vulgaris, P. dolichoptera* and *Synchaeta* spp. (*oblonga-pectinata* group). In addition the summer plankton is characterized by *Gastropus stylifer, Ascomorpha ovalis, A. saltans, Pompholyx sulcata* and *Trichocerca* spp. (*similis, rousseleti* and others) (Sharma, 1983; Herzig, 1987). The crustacean component consists of 7 Cladocera and 6 Copepoda. Numerically important are *Bosmina coregoni, Daphnia hyalina, D. cucullata, Eudiaptomus gracilis, Cyclops abyssorum* and *Mesocyclops leuckarti.* In addition the two raptorial-feeding Cladocera, *Leptodora kindti* and *Bythotrephes longimanus,* form important components of the summer/early autumn plankton.

Quantitative aspects

In Fig. 7 numbers of the zooplankton individuals are plotted against time from October 1982 until August 1984. Copepods are most abundant in autumn, winter and spring, whereas the Cladocera reach highest numbers in late spring and summer. The rotifers develop a maximum in summer at a time where the daphnids occur in low numbers. At the time of the copepod maximum nauplii are prevalent in the populations. Older developmental stages (copepodites, adults) dominate the populations late spring and summer.

Controlling factors for the development of zooplankton

The main physical limitation is temperature. In Mondsee, as in other alpine lakes, perennial, con-

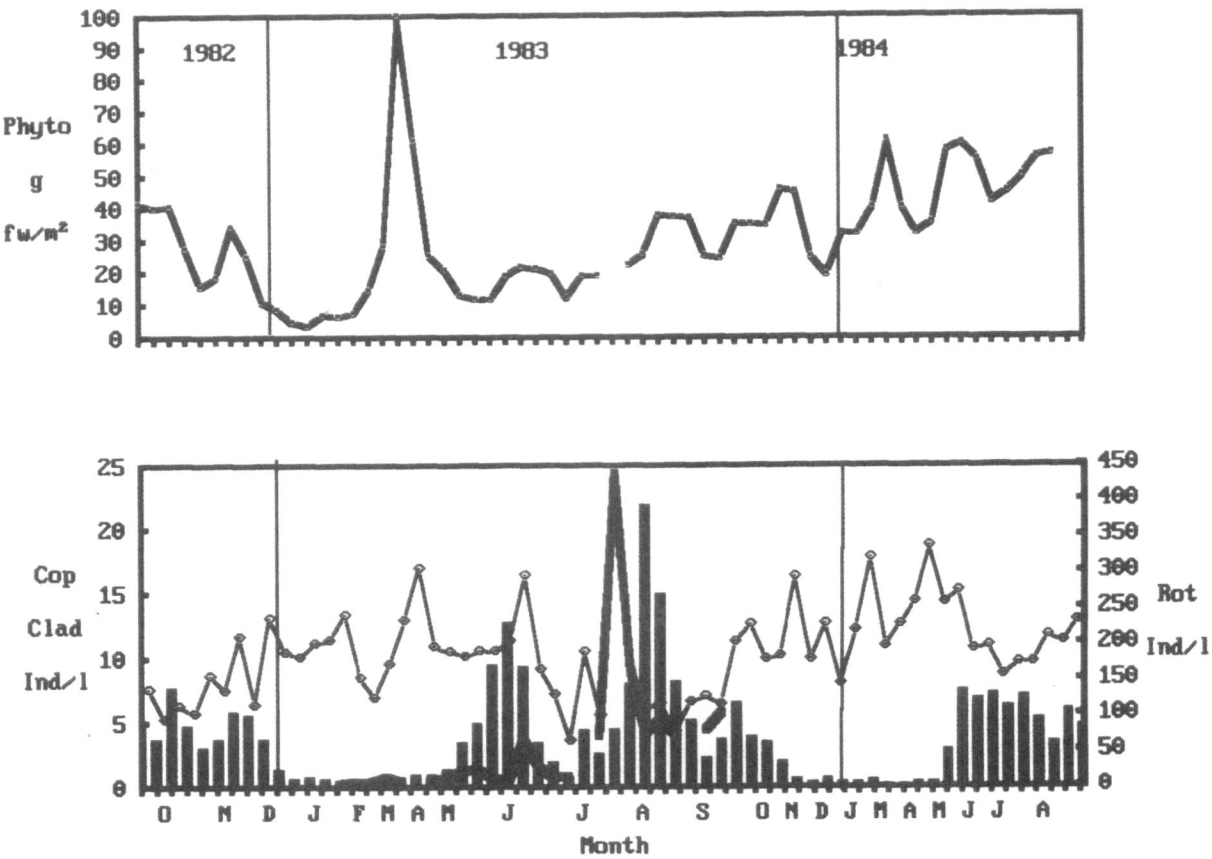

Fig. 7. Development of the zooplankton community, October 1982 – August 1984 (lower panel). Copepods (Cop, o), cladocerans (Clad, bars) and rotifers (Rot, thick line) as individuals per liter (ind. l^{-1}). Upper panel: Total phytoplankton in the water column as g fresh-weight m^{-2}.

tinuously reproducing, eurythermal species (i.e. *Eudiaptomus gracilis*, *Cyclops abyssorum*, *Keratella cochlearis*) are numerically important throughout the year. During the warmer months of the year the Cladocera develop very well; they produce resting eggs in autumn/winter and hatch out of these from February onwards.

For the development of all the organisms and the build-up of successful populations, food plays an obvious role. In simple terms, the role of food supply is indicated by comparing phytoplankton and zooplankton biomass. In deep stratified lakes the two trophic levels are fairly directly connected via the 'grazing food chain' (Herzig, 1979). Phytoplankton biomass is plotted against zooplankton biomass in Fig. 8 (actual values per sampling day,

seasonal and annual averages) and despite some scatter in the diagram this rather direct coupling of the two trophic levels becomes obvious.

The effects of herbivory by zooplankton may result in clear-water periods, as described by Lampert & Schober (1978) for Lake Constance and experimentally tested by Lampert *et al.* (1986). Such an effect could not be observed in Mondsee.

The various species of herbivorous zooplankton and different developmental stages may display distinct food preferences (Bogdan & Gilbert, 1987), but all of them feed on the smaller phytoplankton which in Mondsee rarely comprises more than 20% of the total biovolume (Dokulil & Skolaut, 1986). There is evidence from field

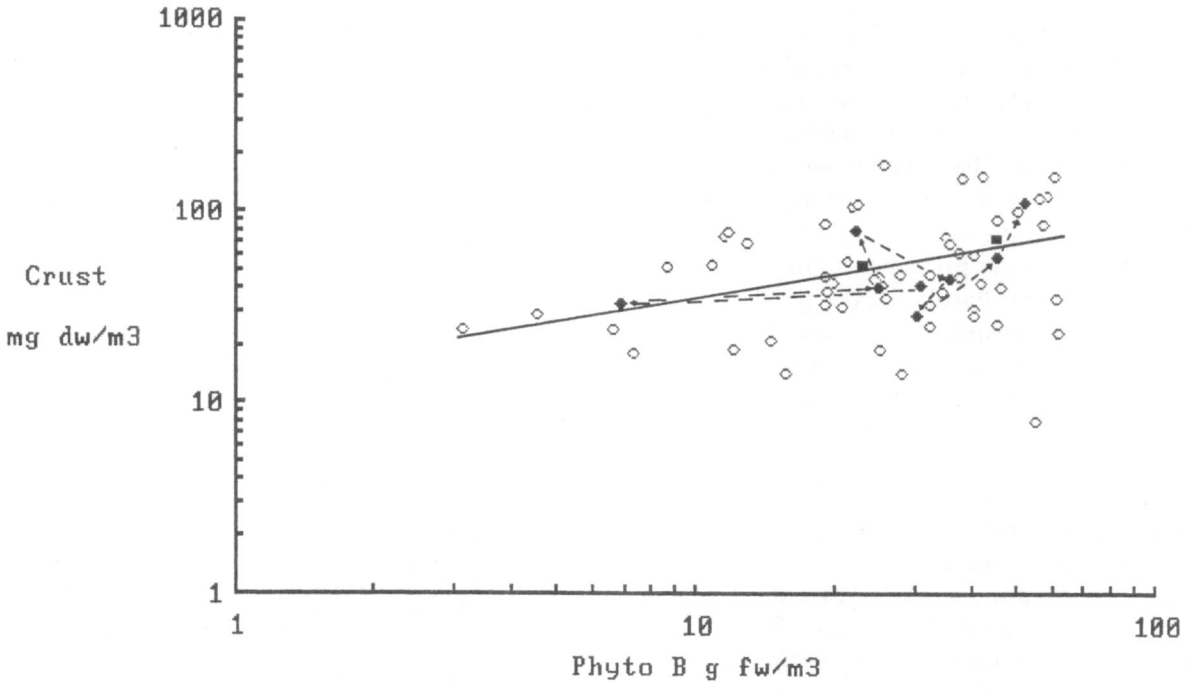

Fig. 8. Correlation of total phytoplankton biomass (*g* fresh-weight m^{-3}) to total crustacean biomass (mg dry weight m^{-3}), based on individual pairs of data (○), seasonal averages from autumn 1982 to summer 1984 (●), and annual mean values (■).

studies and laboratory experiments that Cladocera will outcompete Rotifera (Neill, 1984; Gilbert & Stemberger, 1985; Herzig, 1987). Most frequently this effect is attributed to exploitative competition. Nevertheless support is increasing for the idea that mechanical interference competition is one agent for the evolution of inverse relationships between the abundance of Rotifera and Cladocera (e.g. *Daphnia* spp. – *Keratella cochlearis*: Gilbert & Stemberger, 1985; Burns & Gilbert, 1986). In this case *Keratella* is swept into the branchial chamber of the daphnids and is either killed instantly, mortally wounded, or is loosing the attached egg. Occasionally *Keratella* can even be found in the gut of *Daphnia*.

In Mondsee we find a well-expressed inverse relationship between Rotifera (mainly *Keratella cochlearis*) and the Cladocera (mainly *Daphnia hyalina* and *D. cucullata*) in late spring and summer (see Figs. 7, 10). During this period 'good food' (e.g. Cryptophyceeae, Chrysophyceae) is scarce and competitive competition may occur.

On the other hand *K. cochlearis* was found within the thicket of the filtering limbs of a number of *D. hyalina*. According to the experiments of Burns & Gilbert (1986) and the actual numerical abundance of rotifers and daphnids in Mondsee (400-800 rotifers l^{-1} and 10-50 daphnids l^{-1}), instantaneous mortality rates of *Keratella cochlearis* of 0.1-0.25 day^{-1} can be attributed to the interference with daphnids, and hence this interaction may act as a structuring agent on the plankton community. Another interesting aspect here is the role of predation. It has often been shown that species of Cladocera are the major food supply for invertebrate and vertebrate planktivorous predators. The effect of predation varies: vertebrates select larger prey, invertebrates the smaller one (for review see de Bernardi *et al.*, 1987). In Mondsee *Leptodora kindti* and *Bythotrephes longimanus* prey very efficiently on daphnids. Preliminary enclosure experiments in the laboratory (at 18.0 °C) revealed mean elimination rates of 3.8 daphnids day^{-1} *Leptodora*

208

indiv. $^{-1}$ and 5.6 daphnids. day $^{-1}$. *Bythotrephes* indiv. $^{-1}$. These values would mean that during summer the daily elimination rate by the *Leptodora*-population can vary between 3.4% and 38% and that by the *Bythotrephes*-population between 2.2% and 8.0% (Herzig, unpubl. results). As long as more than 100 raptorial-feeding Cladocera are present per cubic metre the *Daphnia* spp. can hardly develop, and peak-development of *Daphnia* spp. is closely followed by peak-development of *Leptodora* and *Bytho-trephes* (Fig. 9). These results compare well with findings by de Bernardi (1974) and Duncan (1975).

This predator-prey relationship is strongly modified by planktivorous fish. The biology of one characteristic species of Mondsee – *Chal-calburnus chalcoides mento* (Cyprinidae) – was studied by Orellana (1985). Its main food source is copepods and *Bosmina* in early spring; as soon as daphnids appear the larger specimens are selected by *Chalcalburnus*, and the occurrence of *Leptodora* and *Bythotrephes* leads to a preference of this diet (some results of fish gut analysis are given in Fig. 9). *Leptodora*, *Bythotrephes* and *Daphnia* spp. are most electively selected by *Chal-calburnus*; the results also indicate that size-specific selectivity exist within *Daphnia* and *Bosmina* (Orellana, 1985). In Lago Maggiore whitefish (*Coregonus* sp.) show a comparable feeding behaviour and impact on the zooplankton (Guissani, 1974; de Bernardi & Guissani, 1975).

Invertebrate and vertebrate predation seem to control the *Daphnia* populations. A more detailed analysis showed that the population dynamics of *D. hyalina* and *D. cucullata* clearly reflect this for July 1984 (3-5 days sampling interval) (Santos-Borja, 1984) (Fig. 10). Both daphnids show rather high instantaneous birth rates and hence food limitation is not very likely. The high instantane-ous death rates, however, point towards preda-tion being the cause for these losses.

For all the above-mentioned interactions the animals need the opportunity for interference, and that means the populations of the various inter-acting species need to overlap in space for at least some time each day. In the summer months the

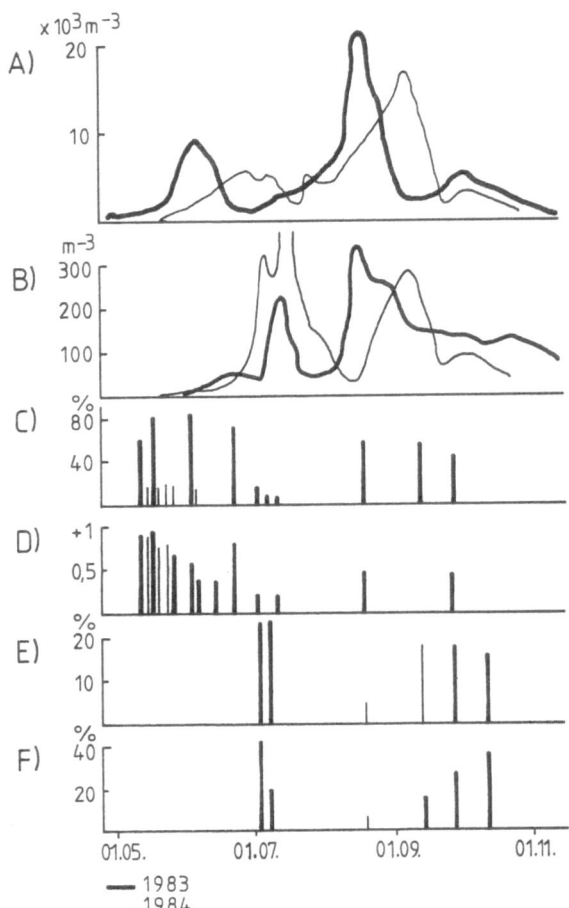

Fig. 9. Dynamics of the *Daphnia* species (A), the inverte-brate predators, *Leptodora kindti* and *Bythotrephes longimanus* (B) and the % found in fish guts of *Daphnia* (C), *Leptodora* (E) and *Bythotrephes* (F). For the daphnids in the fish gut the electivity index is given (D).

vertical distribution (the horizontal dimension is not considered) of rotifers, crustaceans and pelagic fish reveal that even during daytime most of the individuals are concentrated within 5-10 m (except *D. hyalina*), and during night, zooplank-ton and pelagic fish are concentrated in the upper-most 5 to 8 m of water (Fig. 11); the maximum densities are found within a 2 m zone of water. Therefore the animals have ample opportunity to meet and to interact with each other.

To summarize, development, growth and species succession of the phytoplankton depend on the physico-chemical structuring forces of the

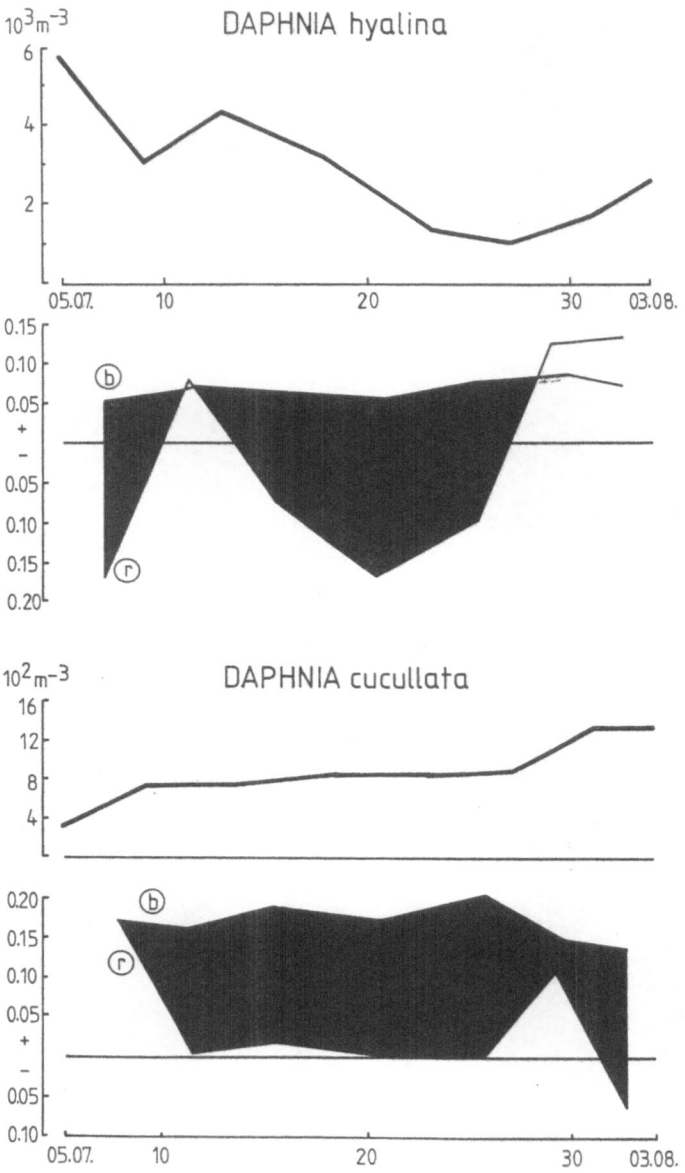

Fig. 10. Dynamics of *Daphnia hyalina* and *Daphnia cucullata* within a period of one month (July-August) as 10^3 individuals m^{-3}; b = birth rate, r = rate of population change (day^{-1}).

environment. Duration of mixing, onset of thermal stratification and temperature have great influence on the timing and sequence of algal species. Phytoplankton populations during summer can be divided into an epilimnetic assemblage and a metalimnetic community largely consisting of *Oscillatoria rubescens*. At this time of year phosphorus in general and silicate specifically are

limiting for diatoms; both are exhausted because of uptake and incorporation into algal cells. Moreover the abundance of small-sized phytoplankton in the epilimnion is controlled by herbivorous zooplankton. However no clear-water phase was observed after the spring peak of phytoplankton.

Temperature and food are responsible for the

210

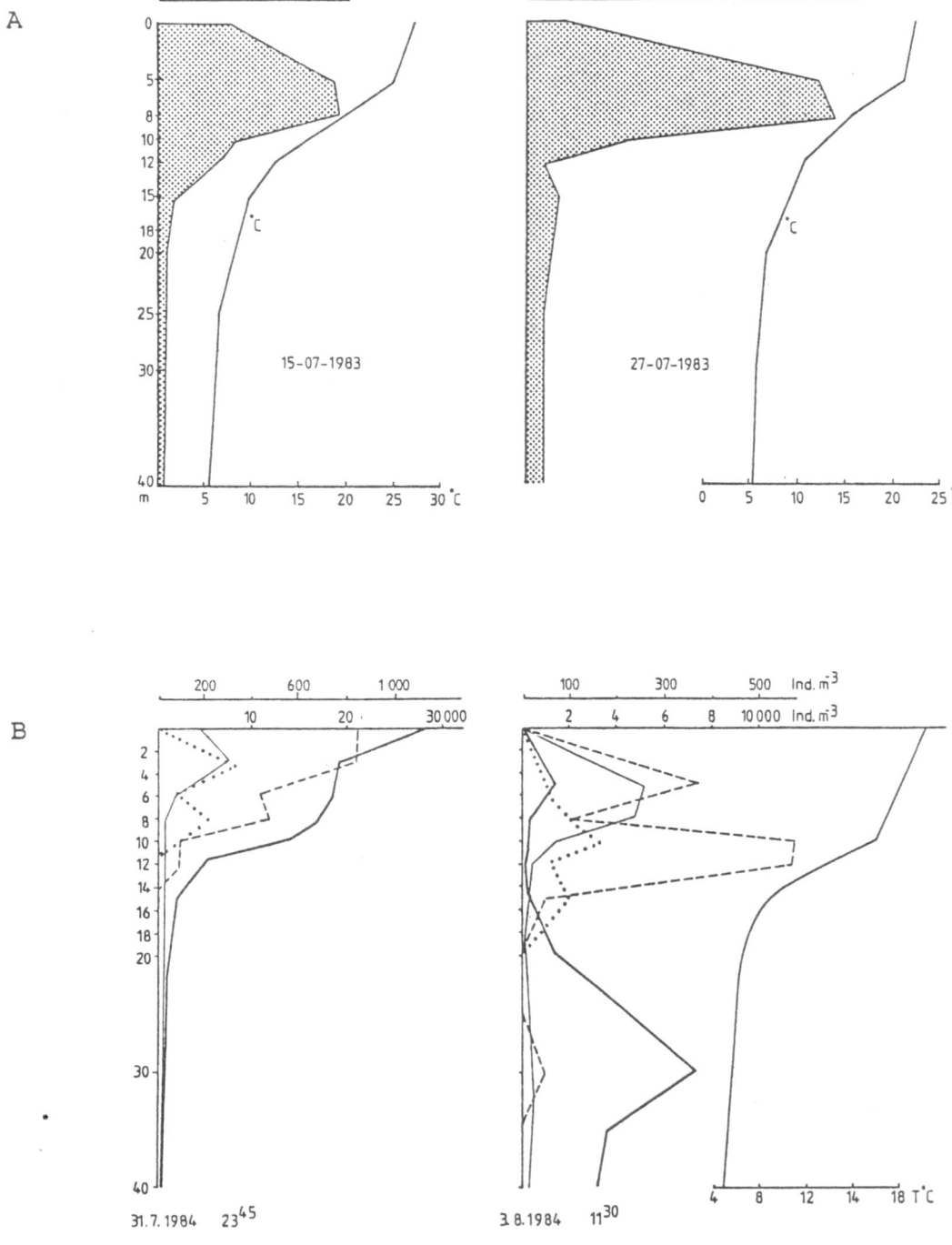

Fig. 11. Vertical distribution of zooplankton and fish. (A) Distribution of rotifers (ind. l^{-1}) and temperature on two sampling occasions in 1983. (B) Distribution of *Daphnia cucullata* (thick line), *Daphnia hyalina* (thin line), *Leptodora kindti* (–––), and *Bythotrephes longimanus* (...) during night (left) and during day (right) as ind. m^{-3}. (C) Distribution of *Calcalburnus calcoides mento* during night (left) and day (right) as percentage of total fish for two sampling occasions. (Data pers. comm. Bobek, echo-sounder recordings, Bio-Sonics 105).

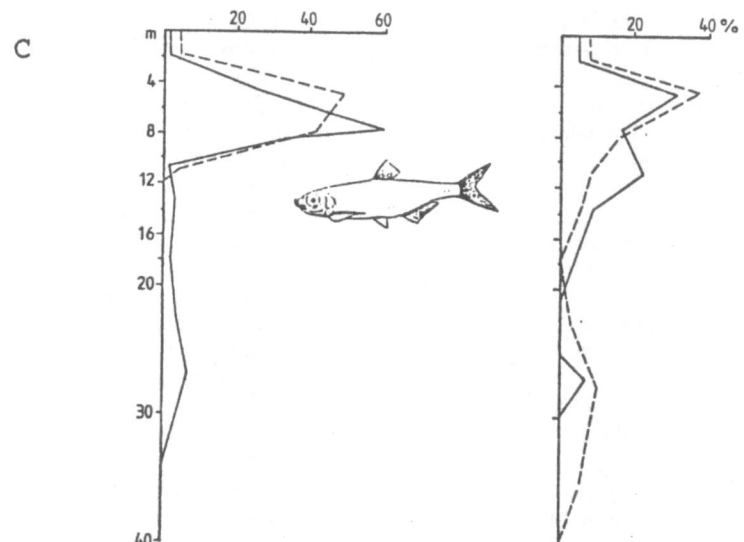

timing and the growth of the zooplankton populations. Competition (exploitative and interference) and predation (invertebrate and vertebrate) are the main structuring forces.

All these interactions are possible because of
close interactions due to overlap in space and
time.

Acknowledgements

We would like to thank C. Skolaut, L. Eisl, G.
Bruschek, and R. Niedereiter for technical help
both in field and laboratory, K. Mayrhofer for
data reduction, K. Maier for drawing figures, and
I. Gradl for typing the manuscript.

References

Bergquist, A. M. & S. R. Carpenter, 1986. Limnetic herbivory: effects on phytoplankton populations and primary
production. Ecology 67: 1351–1360.
Bogdan, K. G. & J. J. Gilbert, 1987. Quantitative comparison
of food niches in some freshwater zooplankton. Oecologia
72: 331–340.
Burns, C. W., 1968. The relationship between body size of
filter-feeding Cladocera and the maximum size particle
ingested. Limnol. Oceanogr. 13: 675–678.
Burns, C. W. & J. J. Gilbert, 1986. Effects of daphnid size

and density on interference between Daphnia and Keratella
cochlearis. Limnol. Oceanogr. 31: 848–858.
Calderon, L. S. & F. A. R. El Higzi, 1983. Response of phyto-
plankton to artificial enrichment with nitrogen and phos-
phorus in enclosures in Mondsee, Austria. Unpubl. Report
UNESCO Training Course in Limnology, 18 pp.
Carpenter, S. R. & J. F. Kitchel, 1984. Plankton community
structure and limnetic primary production. Am. Nat; 124:
159–172.
De Bernardi, R., 1974. The dynamics of a population of
Daphnia hyalina Leydig in Lago Maggiore, Northern Italy.
Mem. Ist. Ital. Idrobiol. 31: 221–243.
De Bernardi, R., G. Guissani & M. Manca, 1987. Cladocera:
predators and prey. Hydrobiologia 145: 225–243.
Dokulil, M., 1984. Die Reoligotrophierung des Mondsees.
ANL 2/84: 46–53.
Dokulil, M., 1986. Faktoren der Steuerung und der Regu-
lation in Phytoplanktonpopulationen. ANL 2/86: 60–65.
Dokulil, M., 1987. Seasonal and spatial distribution of
cryptophycean species in the deep stratifying, alpine lake
Mondsee and their role in the food web. Hydrobiologia.
Dokulil, M. (in press). Population dynamics of a metalimnic
blue-green alga, Oscillatoria rubescens, in an alpine lake.
Schweiz. Z. Hydrol. (in press).
Dokulil, M. & P. Jäger, 1985. General limnological charac-
terization of the Trumer Lakes, Mondsee, Attersee and
Traunsee. In D. Danielopol, R. Schmidt & E. Schultze
(eds.), Contributions to the paleolimnology of the Trumer
Lakes (Salzburg) and the lakes Mondsee, Attersee and
Traunsee (Upper Austria). Limnol. Inst. ÖAW, Mondsee:
16–24.
Dokulil, M. & S. Kofler (in press). Population dynamics and
autecology of Tabellaria flocculosa var. asterionelloides I.
Seasonal and spatial distribution. Diatom Res. (in press).

212

Dokulil, M. & C. Skolaut, 1986. Succession of phytoplankton in a deep stratifying lake: Mondsee, Austria. Hydrobiologia 138: 9–24.

Duncan, A., 1975. The importance of zooplankton in the ecology of reservoirs. Proc. Symp. 'The effects of storage in water quality', Univ. Reading: 247–272.

Elser, J. J., N. C. Goff, N. A. MacKay, A. L. St. Amond, M. M. Elser & S. R. Carpenter, 1987. Species-specific algal responses to zooplankton: experimental and field observations in three nutrient-limited lakes. J. Plankton Res. 9: 699–717.

Frost, B. W., 1980. Grazing. In I. Morris (ed.), The physiological ecology of phytoplankton. Univ. Calif. Press., Berkeley, CA: 465–491.

Gilbert, J. J. & R. S. Stemberger, 1985. Control of *Keratella* populations by interference competition from *Daphnia*. Limnol. Oceanogr. 30: 180–188.

Gliwicz, Z. M., 1977. Food size selection and seasonal succession of filter feeding zooplankton in a eutrophic lake. Ecol. Polska, 25: 179–225.

Guissani, G., 1974. Predazione selettiva del coregone bondella (*Coregonus* sp.) de Lago Maggiore. Mem. Ist. Ital. Idrobiol. 31: 181–203.

Gunatilaka, G., 1981. Note on bioavailable phosphorus from surficial sediments of Mondsee. Arb. Lab. Weyregg 5: 119–125.

Herzig, A., 1979. The zooplankton of the open lake. In H. Löffler (ed.), Neusiedlersee. Limnology of a shallow lake in Central Europe. Dr. W. Junk bv Publ., The Hague, Boston, London, Monogr. Biol. 37: 281–335.

Herzig, A., 1987. The analysis of planktonic rotifer populations: a plea for long-term investigations. Hydrobiologia 147: 163–180.

Hrbáček, J., 1982. Species composition and the amount of zooplankton in relation to fish stock. Roz. Česk. Akad., Ved Rada, Mat.-Prir. Ved 72: 1–116.

Jagsch, A., 1982. Mondsee. In H. Sampl, R.-E. Gusinde & H. Tomek (eds.), Seenreinhaltung in Österreich. Limnologie – Hygiene – Maßnahmen – Erfolge. BM Land- u. Forstwirtschaft, Wien: 155–163.

Jagsch, A. & G. Bruschek, 1982. Zustand vom Mondsee, Irrsee und Mondseezuflüssen: Ergebnisse der Wasserchemie 1981. Arb. Lab. Weyregg 6: 113–122.

Kilham, P., 1971. A hypothesis concerning silica and the freshwater planktonic diatoms. Limnol. Oceanogr. 16: 10–18.

Klee, R. & R. Schmidt, 1987. Eutrophication of Mondsee (Upper Austria) as indicated by the diatom stratigraphy of a sediment core. Diatom Res. 2: 55–76.

Kofler, S., 1986. Temperatur und Strahlung als bestimmende Faktoren für Wachstum und Morphologie von *Tabellaria flocculosa* var. *asterionelloides* (Bacillariophyceae) in Kultur. Ph.D. Thesis, Univ. Wien: 146 pp.

Kofler, S. & M. Dokulil (in press). Population dynamics and autecology of *Tabellaria flocculosa* var. *asterionelloides* II. Interaction of temperature and light on the growth in culture. Diatom Res. (in press).

Lampert, W. & U. Schober, 1978. Das regelmäßige Auftreten von Frühjahrs-Algenmaximum und 'Klarwasserstadium' im Bodensee als Folge von klimatischen Bedingungen und Wechselwirkungen zwischen Phyto- und Zooplankton. Arch. Hydrobiol. 82: 364–386.

Lampert, W., W. Fleckner, H. Ray & B. E. Taylor, 1986. Phytoplankton control by grazing zooplankton: a study on the spring clear-water phase. Limnol. Oceanogr. 31: 478–490.

Lund, J. W. G., C. Kipling & E. D. Le Cren, 1958. The inverted microscope method of estimating algal numbers and the statistical basis of estimation by counting. Hydrobiologia 11: 143–170.

Lynch, M., 1979. Predation competition and zooplankton community structure: an experimental study. Limnol. Oceanogr. 24: 253–272.

Mackereth, F. J. H., J. Heron & J. F. Talling, 1978. Water analysis: some revised methods for limnologists. Freshwat. Biol. Ass., Sci. Publ. 36: 1–120.

Neill, W. E., 1984. Regulation of rotifer densities by crustacean zooplankton in an oligotrophic montane lake in British Columbia. Oecologia 61: 175–181.

Orellana, C. P., 1985. Nahrungserwerb und Biologie der Seelaube *Chalcalburnus chalcoides mento* (Agassiz) in Mondsee. Diplomarbeit Univ. Salzburg, 69 pp.

Porter, K. G., 1977. The plant-animal interface in freshwater ecosystems. Science 65: 159–170.

Reynolds, C. S., 1986. Experimental manipulations of the phytoplankton periodicity in large limnetic enclosures in Blelham Tarn, English Lake District. Hydrobiologia 138: 43–64.

Rhee, G.-Y. & I. J. Gotham, 1980. Optimum N : P. ratios and coexistence of planktonic algae. J. Phycol. 16: 486–489.

Rott, E., 1981. Some results from phytoplankton counting intercalibrations. Schweiz. Z. Hydrol. 43: 34–62.

Santos-Borja, A. C., 1984. Populations dynamics and production of *Daphnia hyalina* and *Daphnia cucullata*. Unpubl. Report UNESCO Training Course Limnology, Austria, 25 pp.

Schindler, D. W., 1977. Evolution of phosphorus limitation in lakes. Science 195: 260–262.

Schindler, D. W. & E. J. Fee, 1974. Experimental lakes area: whole lake experiments in eutrophication. J. Fish. Res. Bd. Canada 31: 937–953.

Sharma, B. K., 1983. The rotifer community of Mondsee, an alpine lake in Austria. Unpubl. Report UNESCO Training Course Limnology Austria, 103 pp.

Smith, V. H., 1983. Low nitrogen to phosphorus ratios favor dominance by blue-green algae in lake phytoplankton. Science 221: 669–671.

Sommer, U., Z. M. Gliwicz, W. Lampert & A. Duncan, 1986. The PEG-Model of seasonal succession of planktonic events in fresh waters. Arch. Hydrobiol. 106: 433–471.

Tilman, D., S. S. Kilham & P. Kilham, 1982. Phytoplankton community ecology: the role of limiting nutrients. Ann. Rev. Ecol. Syst. 13: 349–372.

Hydrobiologia **191**: 213–221, 1990.
P. Biró and J. F. Talling (eds), Trophic Relationships in Inland Waters.
© 1990 *Kluwer Academic Publishers.*

Trophic relationships between primary producers and fish yields in Lake Balaton

Péter Biró & Lajos Vörös
*Balaton Limnological Research Institute of the Hungarian Academy of Sciences, H-8237 Tihany,
Hungary*

Key words: chlorophyll *a*, morphometry, fish yields, energy transfer, Lake Balaton

Abstract

Relationships between chlorophyll *a* content of the water, the shoreline-length : water area ratio and the annual total fish yield as catch per unit effort (CUE: kg ha^{-1} 100 h^{-1} as annual mean values) have been calculated by multivariable regression. The determination coefficient ($r^2 = 0.913$) showed a significant dependence of fish yield on morphometry of different lake areas. Accordingly, fish carrying capacity of the open water areas, calculated from chlorophyll *a* content and S/A, ranged from 12 to 34%, but that of the littoral zone between 66 and 88%. These findings have also been supported by echo-sounding records of the horizontal distribution of fish.

Bream (*Abramis brama* L.) contributes the majority (70-80%) of fish stock and yield. Its food mainly consists of zooplankton and benthic invertebrates in ratios that are widely variable with season and depend on the age of fish. Average daily food consumption of individuals (age group 3 + and over) varies between 2 and 5 g. Bream consumes two- to three-times more food in the SW basin than in the NE one. This means that the present stocks inhabiting areas from NE to SW consume annually 13 249-20 085 t yr^{-1} of food. According to estimated calorific values, the annual energy consumption of local populations along the longitudinal axis of the lake varies between 93 and 141 kJ m^{-2} yr^{-1}. The efficiency of energy transfer from primary producers to fish is low and varies from 0.04 to 0.1%.

Introduction

Lake Balaton has undergone changes induced by man-made eutrophication during the last two to three decades. The shallow Lake Balaton (area 595 km², average depth 3.3 m) is characterized by intensive metabolism and mineralization (Biró, 1984). Due to possible feed-back mechanisms, especially at lower energy levels, the energy flux is 'suppressed' and 'narrow' (Oláh *et al.*, 1977). It means that only a fragment of the primarily pro-

duced material gets to the energy levels of invertebrates and fishes, because the overwhelming majority of this material is inaccessible for these levels due to intensive metabolism, fast bacterial decomposition and significant mineralization. In consequence, the efficiency of energy transfer from primary producers to fish remains low (Biró, 1984).

Relationships between production rates of the primary producers and fish, the two endpoints of the food web, have been described by second degree polynomials (Biró & Vörös, 1982). Bream (*Abramis brama* L.) comprises about 70-80% of the fish stock and because of great variation of the polynomial relationship, a power function has been calculated between the mean chlorophyll *a* content of the water, the shoreline-length/water area ratio and the annual bream yields at distinct fishing areas (Biró & Vörös, 1988). This relationship closely fitted the statistical data. Accordingly, calculations based on the above indices and the total annual commercial fish harvest were made.

Until now there were no reliable data concerning the food turnover of bream. Consequently, the aim of this review paper is to characterize the food consumption and energy transformation of the local bream stocks. Special attention is paid to the efficiency of energy transfer from primary producers to fish, as a general approach to the energy flow along a simplified food-web.

Material and methods

Measurements of chlorophyll *a* content were made bi-weekly from samples collected along the longitudinal axis of the lake during 1971-80 (Vörös, 1982; and unpublished). Primary produc-

tion was measured with the ^{14}C technique (Herodek, 1977; Vörös, MS; Vörös *et al.*, 1983). Fisheries statistics were obtained from the Balaton Fishing Company (Siófok). These included the annual fishing efforts (hours), and the amounts of fish species caught at five fishing areas of the lake (Fig. 1). The fish yields in weight were converted into catch per unit effort (CUE: kg ha^{-1} 100 h^{-1} as annual mean values) (Table 1). Using the mean annual chlorophyll *a* contents of the water (X_1), the total fish yields (Y) and the ratio shoreline-length/water area (X_2 = S/A = km km^{-2}), a multivariable regression has been calculated.

The density and horizontal distribution of fish in Lake Balaton were observed with a SIMRAD echo-sounder during 1985-87 (Paulovits & Biró, unpubl. MS). Intestinal contents of 3 + to 8 + bream were analyzed in 377 specimens collected from the NE and SW basins of the lake during 1982-83. The quantity of food was measured in cylinders by the water displacement technique (Hunt, 1960). Mean weights of food organisms were also obtained from the literature (Biró *et al.*, unpubl. MS). The total weight of food consumed was calculated according to the mean weight of food organisms and their frequency of occurrence.

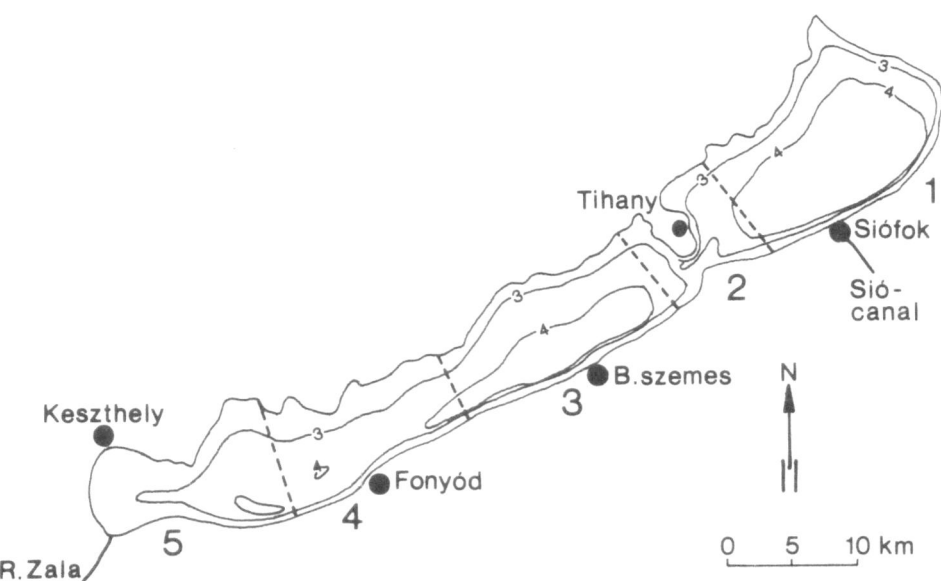

Fig. 1. Fishing areas (1-5) in Lake Balaton.

Table 1. Data used for the calculation of the relationship between the total fish yields (Y = CUE), the chlorophyll a content of the water (X_1 = mg m^{-3}) and the shoreline-length/water surface ratio (X_2 = S/A = km km^{-2}).

Years	No. of fishing area									
	1		2		3		4		5	
	$X_2 = 0.324$		$X_2 = 0.425$		$X_2 = 0.366$		$X_2 = 0.362$		$X_2 = 0.514$	
	Y	X_1	Y	X_1	Y	X_1	Y	X_1	Y	X_1
1971	1.80	3.60	7.38	–	4.38	2.95	5.40	–	7.59	5.70
1972	2.45	2.50	4.78	–	3.00	2.70	4.45	–	8.68	8.35
1973	2.38	3.95	6.82	–	3.58	6.20	3.49	9.30	9.30	13.30
1974	2.25	3.10	5.20	4.00	2.96	9.05	4.50	10.75	9.72	33.75
1975	2.05	3.51	5.35	5.10	3.67	8.33	3.62	10.33	11.15	12.68
1978	3.06	6.05	5.14	8.56	4.70	11.46	5.04	14.92	12.81	13.39
1979	3.52	5.10	4.93	8.78	5.20	15.29	5.19	25.47	11.91	39.64
1980	2.80	7.32	4.57	11.46	4.74	16.69	5.17	21.02	10.59	25.61

Calorific values (1 kcal = 4.184 kJ) of food invertebrates were obtained from the literature (Winberg, 1971; Le Cren & Lowe-McConnell, 1980). The energy contents of the total food consumed by different individuals were derived from their qualitative composition, assessed weight and calorific values. Based on estimated parameters of population dynamics and stock density, the annual food turnover and energy budget of the bream populations have been calculated per unit area. The efficiency of energy transfer was assessed in %.

Results

During the years 1971-80 the annual mean chlorophyll a concentration (X_1) increased significantly along the longitudinal trophic gradient of the lake from NE to SW. Chl a varied between 2.5 and 39.6 mg m^{-3} (Table 1). Apart from its horizontal fluctuations, it generally increased all over the lake with time, although it has tended to decrease during the last two years (1986-87). The period between 1980 and 1986 has not been analysed, because great annual and seasonal oscillations appeared in the chlorophyll a content of the water making the calculations obscure.

Mean values of annual gross primary produc-

tivity during the 1970s were 182 g C m^{-2} yr^{-1} (7 615 kJ m^{-2} yr^{-1})* at Tihany and 274 g C m^{-2} yr^{-1} (11 464 kJ m^{-2} yr^{-1}) at the central part of the lake and 810 g C m^{-2} yr^{-1} (33 890 kJ m^{-2} yr^{-1}) at the SW basin (Herodek, 1977). The same gross measurements in 1986 were 180 g C m^{-2} yr^{-1} (7 531 kJ m^{-2} yr^{-1}) at Tihany (NE basin), and 870 g C m^{-2} yr^{-1} (36 401 kJ m^{-2} yr^{-1}) at Keszthely (SW basin).

The annual commercial fish harvest (Y) showed a two- to four-fold difference between the extreme basins of the lake: 3.18-4.52 kJ m^{-2} yr^{-1} (7.6-10.8 kg ha^{-1}) were obtained for the NE-basin, 5.44-11.5 kJ m^{-2} yr^{-1} (13-27.5 kg ha^{-1}) for the central part and 16.3-19.37 kJ m^{-2} yr^{-1} (39-46.3 kg ha^{-1}) for the SW-basin. Expressed in CUE (kg ha^{-1} 100 h^{-1}) the yield values varied between 1.8-3.5 kg at area no. 1, 4.57-6.82 kg at area no. 2, 2.96-5.2 kg at area no. 3, 3.49-5.4 kg at area no. 4, and 7.59-12.81 kg at area no. 5 (Table 1, Fig. 1).

The calculated regressions of total fish yields, average chlorophyll a content and the shoreline-length/water area ratios of the various parts of Lake Balaton are presented in Figs. 2-3. The regression coefficient ($r = 0.955$) and the determination coefficient ($r^2 = 0.913$) were very high,

---

* 1 g C = 41.84 kJ (Le Cren & Lowe-McConnell, 1980).

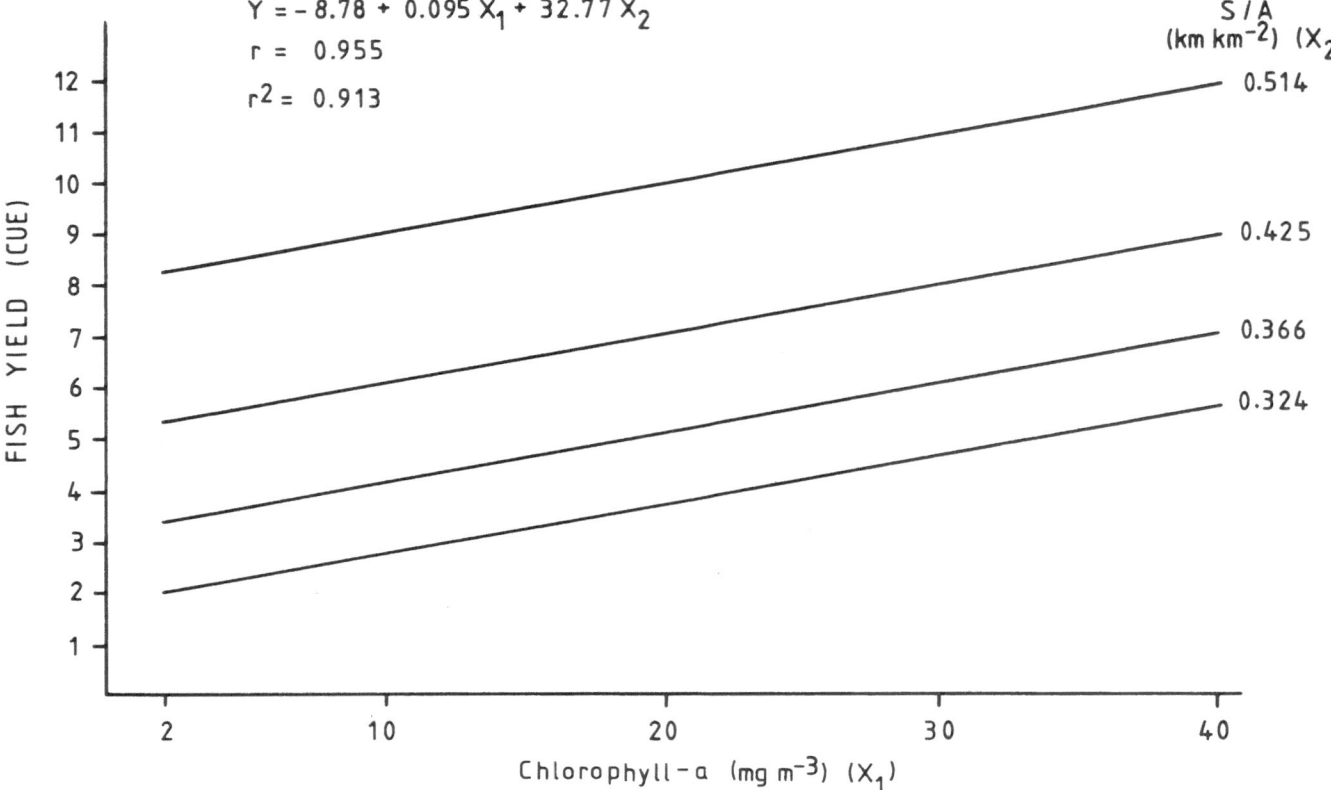

Fig. 2. Calculated multivariable regressions for the relationship between total annual fish yield (Y = CUE), chlorophyll *a* concentration of the water (X_1) and shoreline-length/water area ratio at different fishing areas of Lake Balaton.

Table 2. Averge values of parameters at various fishing areas calculated from the multivariate regression.

Years 1974–80	No. of fishing area				
	1	2	3	4	5
Average chlorophyll *a* content	5.18	7.58	12.16	16.49	26.21
Average CUE	2.73	5.03	4.24	4.70	11.23
Calculated CUE	2.32	5.85	4.35	4.63	10.54
CUE if $X_1 = 0$	1.82	5.13	3.20	3.07	8.05
CUE in % of S/A	78.40	87.70	73.60	66.30	76.70

indicating that the morphometry (X_2) strongly influences the total fish yield of the lake. Considering the most critical period of 1974-80 when the eutrophication became more advanced, the average chlorophyll *a* content of the water derived from the regression varied between 5.18 and 26.21 mg m^{-3} in different areas. The average CUE ranged between 2.73 and 11.23 kg ha^{-1} 100 h^{-1}, to which variation the morphometry of different lake areas contributed 66.3-87.7% (Table 2).

From echograms obtained along the N-S segment of the deepest part of the lake (11.2 m at 'Tihany-well'), the mosaic-like pattern of fish distribution is evident. In offshore areas the average density of fish varied between 200 and 300 ind. ha^{-1}, but inshore (littoral zone) 1200-3000 ind. ha^{-1} was characteristic (Figs. 4-5).

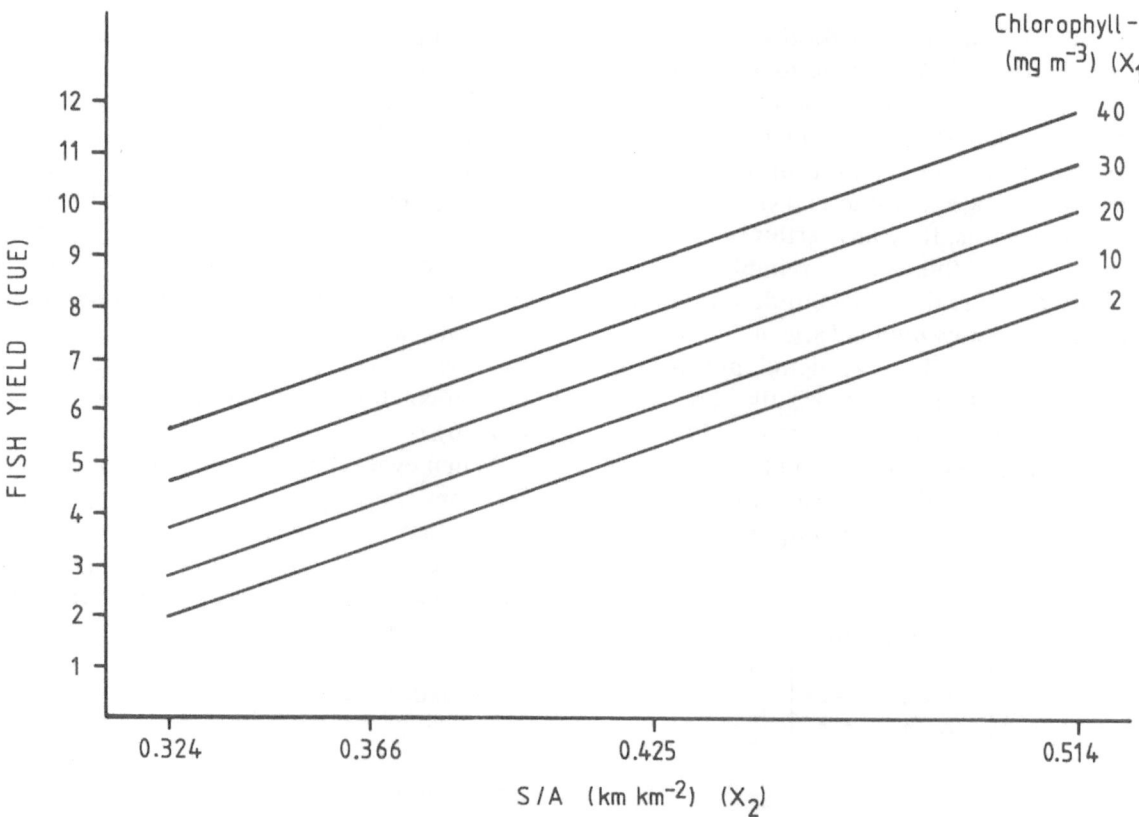

Fig. 3. Regressions between fish yields (CUE) and shoreline-length/water area ratio at different levels of chlorophyll *a* concentration of the water.

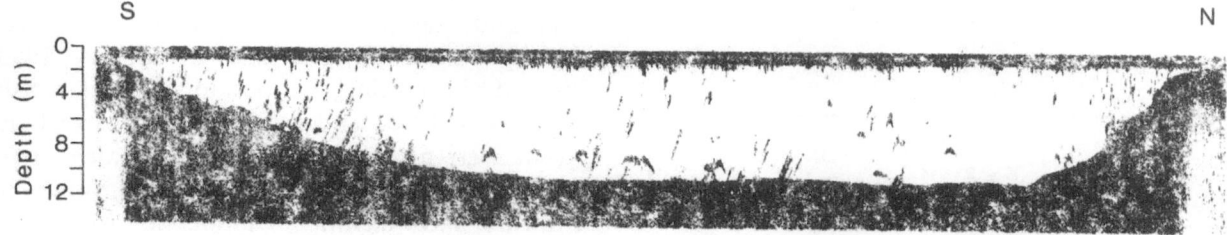

Fig. 4. Sonar recording along the N-S section of the deepest part of the lake ('Tihany-well') showing the distribution of fish in offshore (200-300 ind. ha^{-1}) and inshore (1000-3000 ind. ha^{-1}) areas.

Fig. 5. Detail of the same recording (southern part) showing the littoral zone densely inhabited by fish.

In the energy flow along the food-web, probably the dense bream (*Abramis brama* L.) stocks ($\bar{B}$ = 160-180 kg ha^{-1}) and populations of the top-predator pikeperch (*Stizostedion lucioperca* L.) ($\bar{B}$ = 1.6-9.7 kg ha^{-1}) play important roles and they are the major commercial yield. Together with other cyprinids, they are certainly the major biomass. However, there are no reliable estimates for the forage fish and the accumulated biomass of silver carp (*Hypophthalmichthys molitrix* Val.), which is supposed to be very significant in Lake Balaton. The food of bream mainly consists of zooplankton (42-95%), chironomids (17-83%), detritus (20-35%) and molluscs (11-72%). Their ratios are widely variable in space and time and with the age of fish (Biró *et al.*, unpubl. MS). The

daily amount of food consumed by individuals varied between 2 and 5 g. This suggests that the whole bream stock of Lake Balaton consumes c. 13000-20000 t yr^{-1} of food. Based on estimates of calorific values, the energy turnover of local bream populations varies between 93 and 141 kJ m^{-2} yr^{-1} along the longitudinal axis of the lake. The food of pikeperch consists of 0 + bream from 2 to 10% (Biró, 1973) and its annual food (energy) turnover was assessed between 2.22 and 6.7 kJ m^{-2} yr^{-1}. The values of energy flow along this simplified food web need further verification (Fig. 6).

When evaluating the efficiency of energy transfer from primary producers to fish, small differences between areas have been observed. The efficiency of energy transfer from gross primary production to fish (yield) ranged from 0.04 to 0.06% in the NE-basin, from 0.047 to 0.1 in the central part and from 0.048 to 0.057% in the SW-basin of the lake (Fig. 7).

Discussion

During the last 20-30 years, an enhanced primary production, sequential blue-green blooms (Herodek, 1977; Vörös, 1982), and increased secondary and fish production have been observed (Ponyi *et al.*, 1983; Dévai & Moldován, 1983; Biró, 1983). During this period, two mass fish-kills (1965, 1975) and regular collapses of blue-greens were also observed. The fish-kill in 1975 mainly concentrated to the area No. 4, however; it did not have significant influence on the total annual commercial fish yield (see Table 1).

Some 47 of the 74 fish species inhabiting the Carpathian Basin are known in our lake and its drainage area. Thirteen non-native species were introduced since the last century, or some of them appeared by chance (Biró, 1984). Some 20-24 species are quite frequent, but the fish capture generally consists only of 15-17 species. Pikeperch (*Stizostedion lucioperca*), carp (*Cyprinus carpio*) (restocked for anglers), bream (*Abramis brama*), razor fish (*Pelecus cultratus*), asp (*Aspius aspius*), pike (*Esox lucius*), sheatfish (*Silurus gla-*

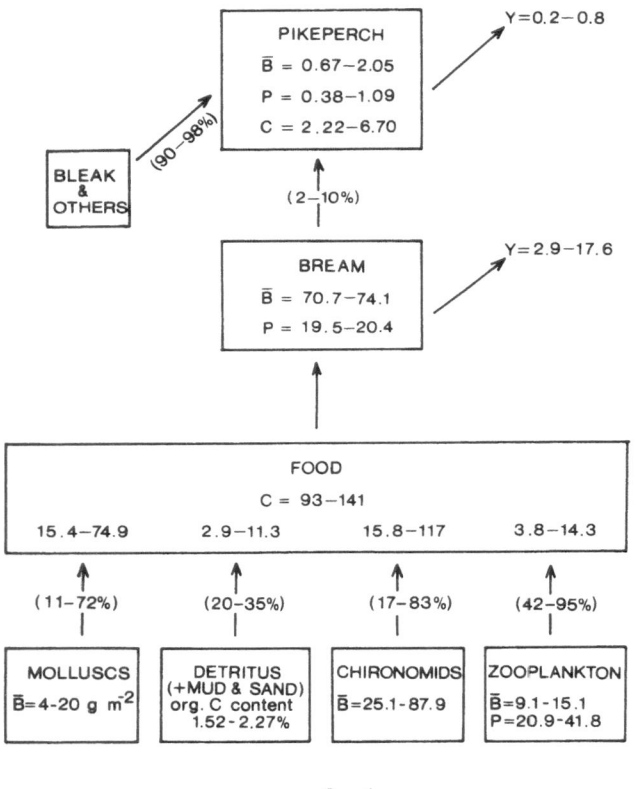

Fig. 6. Assessed annual energy flow along a simplified food-web of bream and pikeperch in kJ m^{-2} yr^{-1} in Lake Balaton. $\bar{B}$ = average biomass, P = production, C = annual food consumption, Y = yield. Figures in brackets indicate the range of variation of different food items consumed.

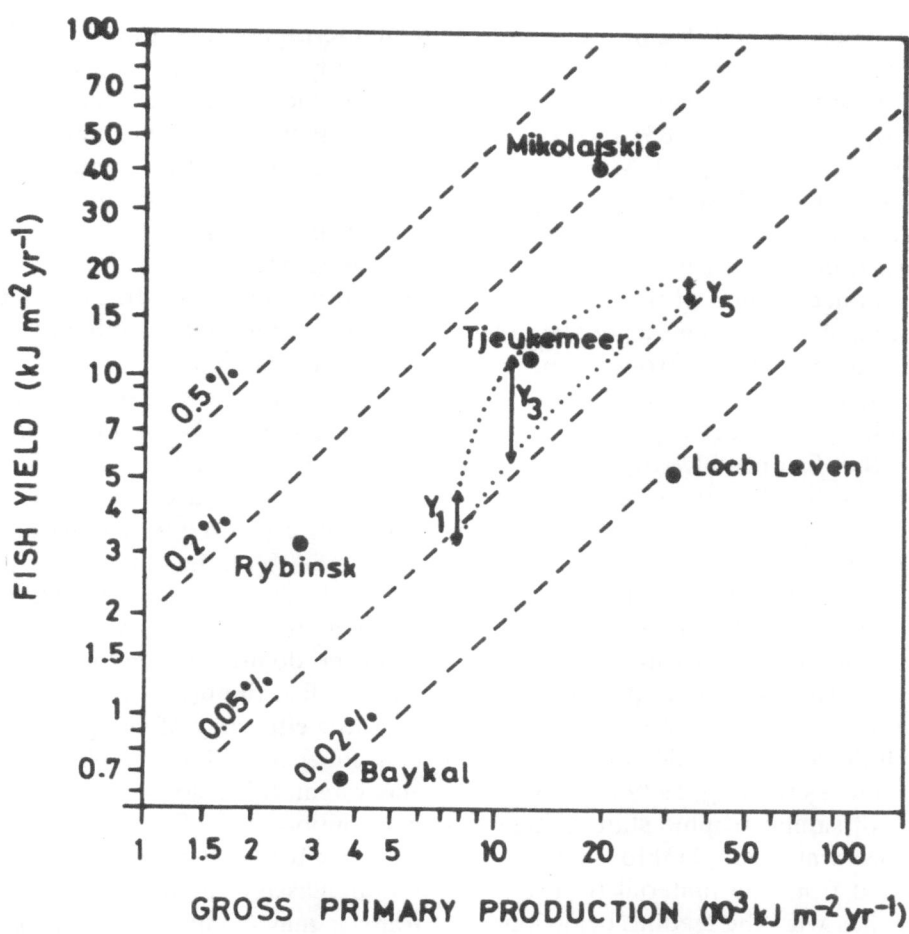

Fig. 7. Efficiency of transfer of energy from primary production to fish at different fishing areas of Lake Balaton, in relation to that in other lakes. Y_1 = NE-basin, Y_3 = central area of the lake, Y_5 = SW-basin (Figure is modified from Le Cren & Lowe-McConnell, 1980, p. 331, Fig. 6.10).

nis) etc. are of crucial importance. Eel (*Anguilla anguilla*) was introduced in 1961 and silver carp (*Hypophthalmichthys molitrix*) in 1972 in order to increase the fish production of the lake, which (as commercial yield) varied between 7.6 and 46.3 kg ha^{-1} along the longitudinal trophic gradient from NE to SW.

Fish populations in Lake Balaton have undergone changes from the individual or combined effects of eutrophication, overfishing (commercial + sport), interspecific competition from native and non-native species, and loading of nutrients and toxic compounds, as well as different environmental perturbations and human impacts (Biró, 1977). Some endangered and former-

ly believed – extinct fish species (e.g. *Perca fluviatilis*, *Micropterus salmoides*) disappeared from the lake itself and at present have self-sustaining stocks in different rivulets, as refugium areas (Zalewski *et al.*, MS). At present about 20-24 species of fishes inhabit the lake, 7-8 occurring in great numbers and having economic significance. For the distribution of fish, a mosaic-like pattern is characteristic (Paulovits & Biró, unpubl. MS). Due to eutrophication and introduction of exotic species (Tóth & Biró, 1985), with the formation of new trophic connections, increased intra- and inter-specific competition has been observed especially in the littoral zone (Biró, 1983). Population dynamics, food and feeding of pikeperch

(*S. lucioperca* L.), bream (*A. brama* L.), eel (*Anguilla anguilla* L.), and silver carp (*Hypophthalmichthys molitrix* Cuv.) were investigated by several authors (Biró, 1974, 1983, 1985; Biró *et al.*, unpubl. MS; Herodek *et al.*, unpubl. MS; Paulovits & Biró, unpubl. MS; Tátrai & Ponyi, 1976).

According to our recent findings, the total fish yields in Lake Balaton are statistically related to the chlorophyll *a* concentration of the water and (mainly) to the morphometric characteristics of the different lake areas. These strong relationships suggest the dependence of fish production on the current primary productivity of the lake. This also implies that fish productivity (or carrying capacity) follows the alterations occurring in primary and secondary productivity, with an indirect relevance of physico-chemical cycles and biological processes at the mud/water interface.

The results confirm that phytoplanktonic production has been utilized by fish with a very low degree of efficiency. The increase of phytoplankton biomass on the eutrophic areas of Lake Balaton has not been followed by a significant increase of zooplankton biomass (Herzig, 1979). Compared to other lakes of similar trophic state, a huge amount of the open water zooplankton biomass becomes eliminated from the material turnover. The main reason lies with the feeding behaviour of the zooplankton. During frequent blue-green blooms, the phytoplankton is utilized by the zooplankton community in a very low efficiency (by about 70% less efficiency), while the community grazing rate on average varies between 2.9 and 13.4% day^{-1} (Zánkai & Ponyi, 1986). Of the total fish carrying capacity, the open water areas contribute much less (12-34%) than the littoral zone (66-88%). Open water algal and zooplankton communities may cover the energy requirement of pelagic fish, such as the razor fish (*Pelecus cultratus* L.) and silver carp (*H. molitrix* Cuv.), but the main fish food stocks are those of the littoral zone.

The energy flux in Lake Balaton appears more stable and less sensitive to internal effects than to the external impacts heavily influencing the whole trophic system. These influences, however, are varied in effectiveness. Some, such as P-loading, can interrupt or alter the energy flow between the two endpoints of the food-web, or speed up the matter and energy circulation at lower trophic levels. Figures on efficiency of energy transfer show a great energy dissipation within the ecosystem. Considering the simplified food-web (Fig. 6) it appears that the energy flux through benthos and periphyton is more important than that of the plankton community. In this context, the littoral zone plays the most significant role for fish production of the lake.

Recent average values of the annual total fish catch (commercial + sport fishing) of Lake Balaton vary between 20 and 24 kg ha^{-1} (8.4-10.0 kJ m^{-2} yr^{-1}). Because the overall average rate of exploitation (E) of different fish species has reached about 40% of the potential values (Biró, 1983), the total fish production of the lake can be assessed as about 20-25 kJ m^{-2} yr^{-1}. These figures, however, do not contain the production of forage and 0^+ fish (natural recruitment).

A low efficiency of energy transfer may be the consequence of high respiratory activity in the ecosystem. It is also partly due to the mosaic-like distribution of the fish stocks, as well as varying fishing effort and population dynamics. In addition, biased estimates for the energy budget and transfer may occur as a result of highly variable parameters in space and time. The efficiencies may be higher for the total biomass and production of fish, although the fractions of the younger generations are not yet satisfactorily known.

Acknowledgements

This work was financially supported by AKA Programme ('Akadémiai Kutatási Alap', theme no. 30054).

References

Biró, P., 1973. The food of pike-perch (*Lucioperca lucioperca* L.) in Lake Balaton. Annal. Biol. Tihany 40: 159–183.
Biró, P., 1974. Observations on the food of eel (*Anguilla anguilla* L.) in Lake Balaton. Annal. Biol. Tihany 41: 133–151.

Biró, P., 1977. Effects of exploitation, introductions and eutrophication on percids in Lake Balaton. J. Fish. Res. Bd Can. 34: 1678–1683.

Biró, P., 1983. On the dynamics of fish populations in Lake Balaton. Rocz. nauk. Roln. Ser. H. 100z. 3: 55–64.

Biró, P., 1984. Lake Balaton: a shallow Pannonian water in the Carpathian Basin. In: Taub, F. B. (ed.), Ecosystems of the world. 29, Lakes and reservoirs, 231–245. Elsevier Sci. Publ. Amsterdam.

Biró, P., 1985. Dynamics of pikeperch, *Stizostedion lucioperca* (L.) in Lake Balaton. Int. Revue ges. Hydrobiol. 70: 471–490.

Biró, P., S. E. Sadek & G. Paulovits (MS). The food of bream, *Abramis brama* L., in two outside basins of different trophic state of Lake Balaton. 21 pp.

Biró, P. & L. Vörös, 1982. Relationships between phytoplankton and fish yields in Lake Balaton. Hydrobiologia 97: 3–7.

Biró, P. & L. Vörös, 1988. Relationship between the yield of bream, *Abramis brama* L., chlorophyll-*a* concentration and shore-length: water area ratio in Lake Balaton, Hungary. Aquacult. Fish. Managem. 19: 53–61.

Dévai, Gy. & J. Moldován, 1983. An attempt to trace eutrophication in a shallow lake (Balaton, Hungary) using chironomids. Hydrobiologia 103: 169–175.

Herzig, A., 1979. The zooplankton of the open lake. In: Löffler, H. (ed.), Neusiedlersee: the limnology of a shallow lake in central Europe. Dr. W. Junk b.v. Publ., The Hague – Boston – London: 281–335.

Herodek, S., 1977. Recent results of phytoplankton research in Lake Balaton. Annal. Biol. Tihany 44: 181–198.

Herodek, S., K. Györe & A. Zsigri (MS). A fehér busa telepítés hatása a Balaton vizminőségére és élővilágára. Haltenyésztési Kutató Intézet (Szarvas). Mimeo, 10 pp.

Hunt, B. P., 1960. Digestion rate and food consumption of Florida gar, warmouth and large mouth bass. Trans. am. Fish. Soc. 89: 206–210.

Le Cren, E. D. & R. H. Lowe-McConnell (eds), 1980. The functioning of freshwater ecosystems. IBP Handbook No. 22, Cambridge University Press, 588 pp.

Oláh, J., E. O.-Tóth & L. Tóth, 1977. A Balaton foszfor anyagcseréje. Magy. Tud. Akad. Biol. Tud. Oszt. Közl. 20: 111–139.

Paulovits, G. & P. Biró (MS). A Balatonba telepített fehér busa tavon belüli megoszlásának és vándorlásának vizsgálata. 31 pp.

Ponyi, J. E., I. Tátrai & A. Frankó, 1983. Quantitative studies on Chironomidae and Oligochaeta in the benthos of Lake Balaton. Arch. Hydrobiol. 97: 196–207.

Tátrai, I. & J. Ponyi, 1976. On the food of pike-perch fry (*Stizostedion lucioperca* L.) in Lake Balaton in 1970. Annal. Biol. Tihany 43: 93–104.

Tóth, J. & P. Biró, 1984. Exotic fish species acclimatized in Hungarian natural waters. FAO/EIFAC Techn. Papers 42 (Suppl.) 2: 550–554.

Vörös, L., 1982. Quantitative and structural change of phytoplankton in Lake Balaton between 1965-1978. Aquacultura Hungarica (Szarvas) 3: 137–144.

Vörös, L., É. Vizkelety, F. Tóth & J. Németh, 1983. Trofitás vizsgálatok a Balaton Keszthelyi medencéjében. Hidrológiai Közlöny 62: 390–395.

Vörös, L., (MS). The importance of picoplankton in Lake Balaton. 12 pp.

Zalewski, M., P. Biró, M. Przybilski, I. Tátrai & P. Frankiewicz (MS). The structure of fish communities in streams of north part of the catchment area of Lake Balaton. 16 pp.

Zánkai, P. N. & J. E. Ponyi, 1986. Consumption, density and feeding of crustacean zooplankton community in a shallow, temperate lake (Lake Balaton, Hungary). Hydrobiologia 135: 131–147.

Winberg, G. G. (ed.), 1971. Symbols, units and conversion factors in studies of freshwater productivity. London: IBP Central Office, 23 pp.

Hydrobiologia **191**: 223–231, 1990.
P. Biró and J. F. Talling (eds), Trophic Relationships in Inland Waters.
© 1990 *Kluwer Academic Publishers.*

223

Predation pressure from above: observations on the activities of piscivorous birds at a shallow eutrophic lake

Ian J. Winfield
University of Ulster Freshwater Laboratory, Traad Point, Ballyronan, Co. Londonderry BT45 6LR, Northern Ireland, U.K.

Key words: piscivorous birds, freshwater fish, predation pressure, environmental structure, distribution, Sovdeborgssjon

Abstract

The foraging activities of piscivorous birds at a small eutrophic lake in southern Sweden were observed during the ice-free months of 1984. Species-specific patterns of abundance and distribution were apparent. Great crested grebes and red-breasted mergansers were present for periods of months and weeks, respectively, while grey herons, black-headed gulls and red kites made numerous shorter visits. Both cyprinids and percids were seen to be captured. Speculative calculations suggest that the amount of fish removed by the birds is significant when compared with that taken by the lake's piscivorous fish, constituting 34% of total consumption in summer and at 99% becoming by far the more important component in late autumn. The peculiar characteristics of this source of predation pressure and their implications for fish populations are discussed.

Introduction

Historically, one of the areas within the field of trophic relationships in inland waters which has received a major research effort has been that of predation. Studies at the levels of individuals, populations, and communities are now being synthesised into a general understanding of the role of predation in freshwater ecosystems. Zaret (1980) gives a comprehensive review of much of the work done so far.

Such studies of predation contribute not only to our knowledge of freshwater systems, but also to our understanding of ecological communities in general. They are particularly relevant to current controversies concerning the roles of various forms of abiotic and biotic factors in the determi-nation of community structure (e.g. see Strong *et al.*, 1984). Rigorous experimental work has shown that the situation in nature is not as straightforward as some recent arguments suggest it to be. For example, a series of investigations involving the bluegill sunfish (*Lepomis macrochirus* Rafinesque) shows that the influences of competition and predation, including the threat of predation, are intimately linked through their interdependent effects on habitat use and hence foraging conditions (Werner *et al.*, 1981; Werner *et al.*, 1983a; Werner *et al.*, 1983b).

Numerous studies of predation in fish communities published over the last few decades have, like the sunfish work cited above, been concerned solely with the activities of piscivorous fish. However, several works in more recent years suggest that the effects of piscivorous birds can, at least on occasion, also be of importance. Such findings have been reported from inland waters as diverse in nature as a stream (Power, 1984), a river (Kennedy & Greer, 1988), a marsh (Britton & Moser, 1982) and a lake (Leah *et al.*, 1980), although the actual frequency of the importance of this component of predation pressure in freshwaters remains to be determined. If predation pressure from birds is a commonly important aspect of the ecology of freshwater fishes it cannot be ignored. While descriptive and experimental studies of fish communities are proceeding with speed and developing sophisticated predictive mathematical models (see Dill, 1983), such work must incorporate the true complexity of reality. Fryer (1987) counsels of the need to found mathematical models on a comprehensive familarity with the basic natural history of the organism in question.

Given our relative ignorance of the qualitative and quantitative importance of piscivorous birds in the ecology of freshwater fishes, this study was designed to produce a purely descriptive account of the activities of piscivorous birds at a small eutrophic lake in southern Sweden. A series of observations at the lake gathered data on the nature of the predation pressure resulting from these avian populations which allowed their likely effects on the previously well-studied fish populations to be evaluated.

Methods

Highly mobile bird populations have a greater potential than most animals to show both short-term (over minutes and hours) and long-term (over weeks and seasons) changes in abundance at a specific site. Consequently, a direct observation sampling programme was designed to cover both of these elements of potential variability.

Between dawn and dusk of one day each month between April and November 1984 (specifically the 18th, 17th, 11th, 16th, 17th, 10th, 18th, and 15th of the month respectively), one or two observers recorded the activities of piscivorous birds at the shallow (maximum depth 3.5 m), eutrophic (summer chlorophyll *a* 25-36 mg m^{-3}) lake Sovdeborgssjon in southern Sweden (55° 35′ N, 13° 42′ E). Observations were not made during the months of January, February, March and December as at these times the lake was completely and consistently covered by ice, and hence devoid of piscivorous birds. The shape and relatively small size (0.11 km^2) of the lake allowed its entire surface, including both open water and littoral water lily (*Nymphaea alba* L. and *Nuphar luteum* L.) zones, to be reliably monitored from a point on the north-west shore by the use of 10 × 50 binoculars and a × 20-45 zoom lens telescope (see Fig. 2). Further details of the lake and its fish populations may be found in Persson (1983a).

The duration of each day's dawn-dusk observations varied with seasonal changes in daylength between a maximum of 17.75 h in June (04.00 to 21.45 hours) and a minimum of 8.50 h in November (07.30 to 16.00 hours). During each observation, details were recorded on the identity, abundance, location and activity of all piscivorous birds. Particular attention was paid to their location within the lake, the timing of their activities, and the identity and approximate size (by comparison with the known bill length of the bird) of fish taken.

In addition to the above recordings, the lake's small population of great crested grebes (*Podiceps cristatus* L.), when present, was subjected to further observation. At 0.5 h intervals throughout the day one bird was selected at random and watched continuously by one observer for a period of 5 minutes. During this period the exact location and number of dives, and their outcomes, were noted.

For the purpose of this investigation the lake's piscivorous birds were assigned to one of two groups: residents, defined as birds which spent more than one day at the lake (and hence can be assumed to obtain their entire daily ration at Sovdeborgssjon), and visitors, defined as birds which spent only part of a day at the lake (and hence may have also been amassing their daily ration from other locations).

Results

Species present and their abundance through the year

The two resident piscivorous birds of Sovde-borgssjon, the great crested grebe and the red-breasted merganser (*Mergus serrator* L.), both showed marked seasonal changes in abundance (Fig. 1a). The initial grebe population, which arrived soon after ice-melt in late March, consisted of two pairs of breeding adults. One of the grebes was subsequently found dead on its nest in the first half of June and its partner subsequently left

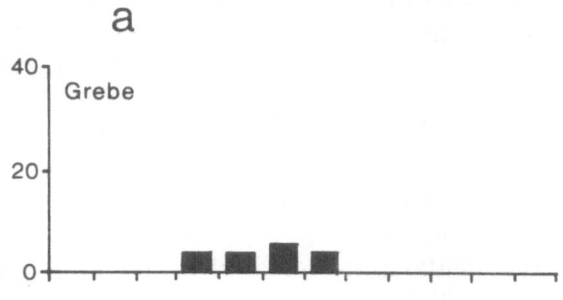
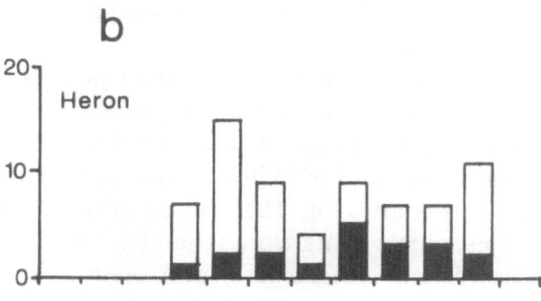
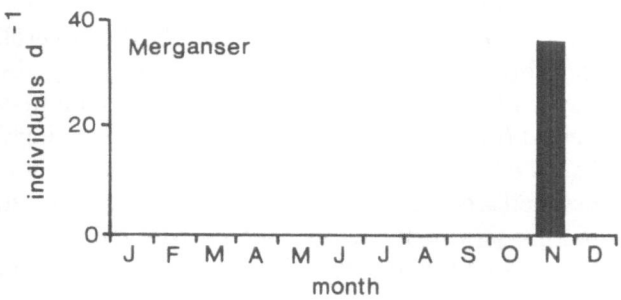
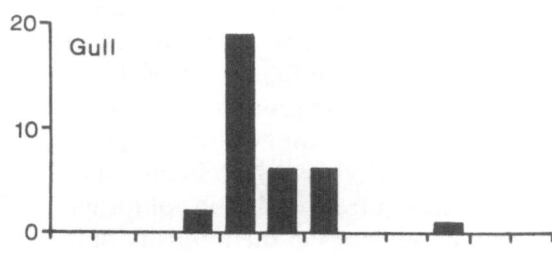
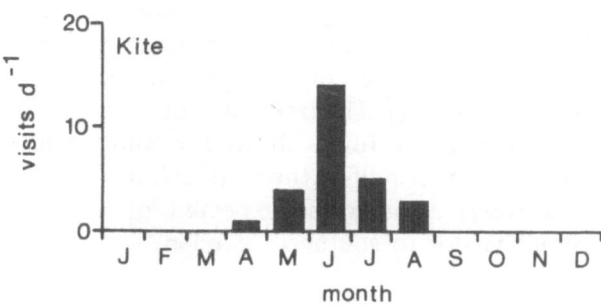

Fig. 1. Abundance of (a) resident and (b) principal visiting piscivorous birds in terms of individuals present and visits day^{-1} respectively. For the heron, total visits are shown divided into foraging (closed part of bar) and roosting (open part of bar) visits.

the lake, while the other pair hatched out four young of which two were successfully reared to independence. The grebes left the lake in the first half of August having been present continuously for approximately four months. Later in the year a flock of 36 mergansers used Sovdeborgssjon as a resting stage during their southward autumn migration and were present for a number of weeks during November and early December prior to ice-formation.

Seasonal changes in the frequency of visits made by the lake's principal visitors, the grey heron (*Ardea cinerea* L.), black-headed gull (*Larus ridibindus* L.) and red kite (*Milvus milvus* L.) were also apparent (Fig. 1b). In addition to these species, single visits were made by two common terns (*Sterna hirundo* L.), one osprey (*Pandion haliaetus* L.) and one kingfisher (*Alcedo atthis* L.) during the July observation, although only one of these birds (a common tern) was seen to catch and eat one fish. The durations of these visits ranged between a few seconds for some of the black-headed gulls to over three hours by a grey heron. Herons were present at the lake throughout the ice-free portion of the year, using it as both a foraging ground and a daytime roost (foraging visits at the water's edge comprised 13-56% of total visits which included individuals perched high in trees and thus not foraging). They showed no changes in the intensity of their visits between first (April-July) and second (August-November) halves of the ice-free year (35 and 34 total visits respectively, G test (Sokal & Rohlf, 1981), G = 0.014, $p > 0.05$; 6 and 13 foraging visits respectively, G = 2.710, $p > 0.05$). Black-headed gulls visited significantly more frequently in the spring and early summer, and were rare visitors towards the end of the year (33 and 1 visits respectively). The frequency of visits by red kites, peaking in June, showed a similar bias towards the first half of the year (24 and 3 visits respectively) as is to be expected of an early summer visitor to southern Sweden.

Activity and distribution patterns

The great crested grebes displayed an extensive foraging period and were observed to catch fish at all times of the day. Feeding periodicity during the daylight hours was analysed by allocating the 5-minute observations to one of three equal periods 'morning', 'midday' and 'evening'. Such analysis showed no significant variation in average diving rate in April, May or July (Single-classification ANOVA: April F = 1.545, $p > 0.05$; May F = 2.503, $p > 0.05$; July F = 0.909, $p > 0.05$), with an overall mean of 0.64 dives min^{-1}. However, in June there was a significant change during the day (F = 8.419, $p < 0.001$) as the diving rate was significantly higher during the midday period (1.40 dives min^{-1} compared with 0.38 dives min^{-1}, analysis by T-method of Sokal & Rohlf, 1981). The relatively small number (31) of fish seen to be consumed by the grebes during the entire study precludes any similar monthly analysis of capture rates. However, pooling of data from all four months into morning, midday and evening periods shows that overall the capture rate varied significantly during the day, with most fish being taken during the midday period (morning 7 fish, midday 19 fish, evening 5 fish; G = 10.435, $p < 0.01$). While no comparable detailed records were made, the red-breasted mergansers of November were observed to dive throughout the day with the exception of a period of loafing between 11.00 and 12.00 hours.

Potential within-day changes in the foraging activities of Sovdeborgssjon's visiting piscivorous birds were investigated by a consideration of the precise times of their visits. Analysis with visits again pooled from all months into morning, midday and evening periods showed no significant changes in the frequency of visits during the daylight hours (grey heron G = 1.980, $p > 0.05$; black-headed gull G = 0.823, $p > 0.05$; red kite G = 2.140, $p > 0.05$).

Several species-specific distribution patterns were observed during the study. During June (Fig. 2a), when the consumption rate of piscivorous fish (see below) was at its maximum, the great crested grebes foraged exclusively in the

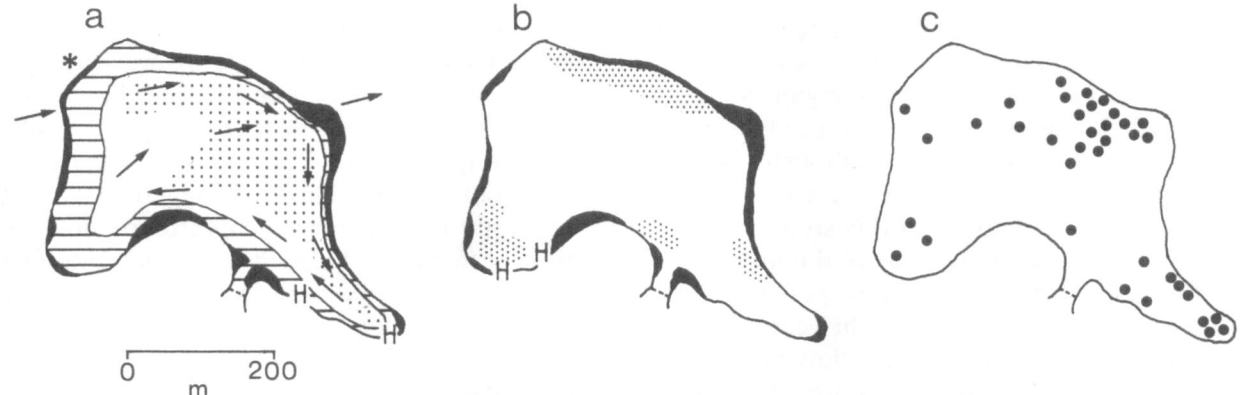

Fig. 2. Distributions of foraging piscivorous birds (a) in June and (b) November, and (c) fish captures of all months. Grebes are represented by a slightly-stippled area, mergansers by more heavily-stippled areas, herons by the letter H, and the typical foraging flight of a gull by a series of arrows. Water lilies are shown as a hatched area, while closed areas define extensive reedbeds. The broken line represents a fish-proof barrier and the observation point is indicated by an asterisk.

open water area of the lake and never dived in the water lily zone nor in any of the fringing reedbeds. In contrast, foraging herons were restricted by water depths to littoral sites well within the water lily band and near to the reedbeds, while black-headed gulls circled the perimeter of the lake 2-4 m above the outer margin of the water lily zone. Similar distributions were observed during the other months that these species were present. In November (Fig. 2b), grey herons again frequented littoral foraging sites although the decline of the water lilies meant that they were now feeding in more open water. The red-breasted mergansers foraged exclusively in the now largely structureless littoral zone, and in contrast to the grebes did not utilise the centre of the lake. The overall distribution of observed fish captures (Fig. 2c) largely reflected the distribution of foraging locations described above.

Diet composition and consumption rates

Identification of captured fish to species was difficult throughout the study, although both cyprinids (i.e. roach *Rutilus rutilus* L. and rudd *Scardinius erythrophthalmus* L.) and perch (*Perca fluviatilis* L.) were positively identified, and the majority of all captures were 10-14 cm in length (Fig. 3). The great crested grebes took 31 fish in

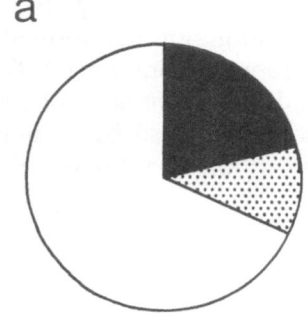

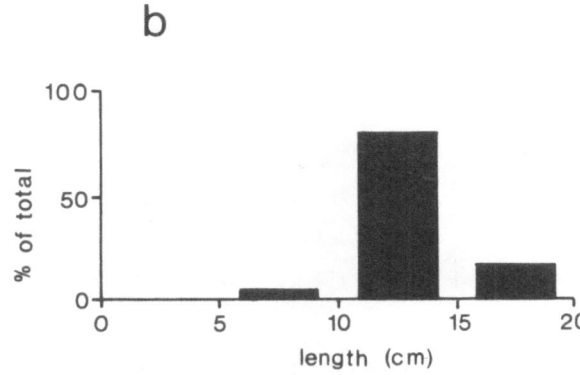

Fig. 3. Diet composition by prey type with cyprinid, percid and unidentified fish represented by closed, stippled and open segments respectively.

the size range 6-17 cm, while the few fish seen to be taken by the grey heron, black-headed gull, common tern and red-breasted merganser (2, 1, 1 and 1 individuals respectively) were all approximately 10 cm in length. The single fish taken by the red kite was approximately 15 cm long.

Using the observations of this study and published accounts of piscivorous bird consumption rates it is possible to calculate the potential levels of predation occurring through the year in Sovdeborgssjon (Fig. 4a), and hence allow an exploratory analysis of the relative importance of predation pressure arising from the activities of birds and fish in this lake. By their definition, resident birds take all of their daily ration from the lake and thus the almost exclusively piscivorous great crested grebes and red-breasted mergansers can be assumed to consume 200 g and 235 g wet weight of Sovdeborgssjon fish individual^{-1} d^{-1} respectively (Cramp & Simmons, 1977).

In contrast, during the course of each day the visiting piscivores may also be foraging at other water-bodies in addition to Sovdeborgssjon, thus

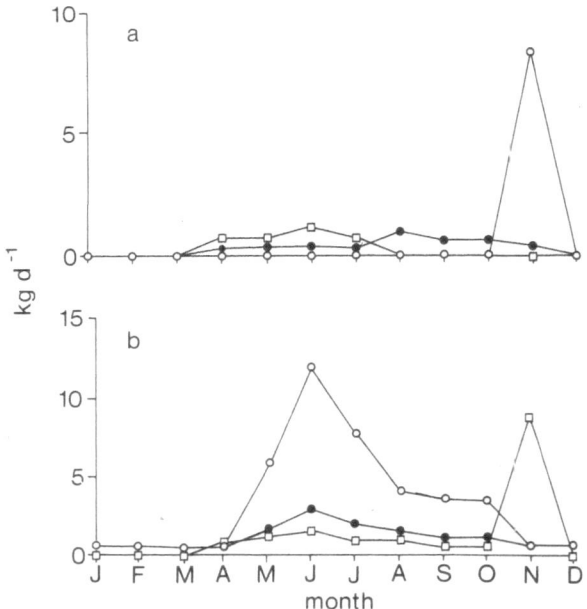

Fig. 4. Calculated daily consumption rates of Sovdeborgssjon fish by (a) grebes (squares), mergansers (open circles) and herons (closed circles), and (b) all piscivorous birds (squares) and fish (closed circles and open circles assuming pike biomass of 130 kg and 500 kg respectively).

complicating any calculations of amounts of fish removed from the study site. However, the present observations indicate that the actual amounts of fish removed from Sovdeborgssjon by black-headed gulls, red kites and common terns are so low (only 3 individuals throughout the entire study, during which such captures are unlikely to have been missed due to their obvious nature) that they are considered negligible. Note that the same assumption is not necessarily true for the threat of predation that the activities of these piscivores generates (see discussion). The amounts of fish likely to be removed from Sovdeborgssjon by the more cryptically feeding visiting grey herons, which also consume significant amounts of non-fish prey (Cramp & Simmons, 1977), were calculated assuming a conservative 50% of their daily intake of 415 g (Cramp & Simmons, 1977) to be Sovdeborgssjon fish.

The above calculations suggest that predation by the great crested grebes amounted to a maximum of 1.20 kg d^{-1} in June, whereas consumption by herons peaked in August at 1.04 kg d^{-1}. The red-breasted mergansers were consuming 8.46 kg d^{-1} in November, which is equivalent to 705 individual 10-cm fish weighing 12 g each.

Similar calculations can be made for Sovdeborgssjon's only significant piscine piscivore, the pike (*Esox lucius* L.) using published consumption rates, which vary markedly with seasonal changes in temperature, and prey calorific values (Diana, 1979 and Jorgensen, 1979 respectively). In 1983 the biomass of piscivorous pike in Sovdeborgssjon was approximately 130 kg, although this represented a significant decline from a previously stable level of 500 kg (S.F. Hamrin, University of Lund, *pers. comm.*). Consequently, calculations were made using both of these biomass values. Consumption rates by pike peaked in June at 3.19 or 12.25 kg d^{-1} and, due to lower temperatures, declined to only 0.08 or 0.33 kg d^{-1} in winter (calculations based on 130 and 500 kg of pike respectively). These seasonal changes in consumption rate by the poikilothermic fish contrast with those of the migrating homiothermic bird populations (Fig. 4b). In June, conservative calculations suggest that 34 or 12% of the total weight of fish consumed was taken by birds, while in November this value rose to 99 or 96% (calculations assuming heron diet is only 50% Sovdeborgssjon fish, and pike biomasses of 130 and 500 kg respectively).

Discussion

The observations of this study revealed that Sovdeborgssjon is frequented by several species of piscivorous birds, each with a unique pattern of abundance, seasonality and foraging behaviour, and suggest that the amounts of fish consequently removed are significant when compared with the more commonly acknowledged effects of piscivorous fish. If this situation is not peculiar to Sovdeborgssjon the role of piscivorous birds in the functioning of the lake ecosystem may be greater than has so far been appreciated.

Numerous characteristics of the biology and predatory activities of birds differ greatly from those of piscivorous fish, and are likely to have significant consequences for the ecology of their prey. One of the most important differences between these two groups of piscivores is their contrasting metabolic demands, and hence fish consumption rates. While the peak consumption rate of poikilothermic pike does not exceed 3% of its own body weight per day (Diana, 1979), for homiothermic piscivorous birds such as grebes and mergansers this value is an order of magnitude higher at around 30% (Cramp & Simmons, 1977). Thus even very small populations of piscivorous birds, such as the grebes in the present study, may consume significant amounts of fish.

Another fundamental difference between piscivorous birds and fish is the former's powers of flight and hence great mobility. While both groups of predators can potentially move amongst the various habitats of a single water body (see below), piscivorous birds can also move over periods of minutes or weeks to forage at other, potentially distant, locations. Thus predation pressure arising from piscivorous birds can change tremendously during both the day and season, posing considerable methodological problems to investigators. While, with the exception of the grebes, there was little change in the activities of Sovdeborgssjon's piscivorous birds during the day, seasonal patterns were evident for both resident (*sensu* this study) and visiting birds. Kennedy & Greer (1988) found an even more marked seasonal exploitation of migrating riverine salmonids by cormorants (*Phalacrocorax carbo* L.) which was further largely restricted to the first few hours after dawn. Rowe (1984) reports a similarly opportunistic exploitation by this species of a lake salmonid following the loss of previously-protecting macrophytes. Clearly, the large-scale distribution ecology of piscivorous birds is a complex subject and is beyond the scope of the present study.

However, even within a single lake the activities of piscivorous birds can be distributed amongst a variety of habitat types such as open water, submerged and emergent macrophytes. Resource partitioning on the basis of habitat use is of great importance in freshwater fish communities (Werner, 1984), and thus the disturbing activities of avian predators could have far-reaching consequences for foraging fish balancing the conflicting demands of finding food and evading predators. In the summertime Sovdeborgssjon, perch and roach are found to varying extents in both the open water and water lily zones (Persson, 1983a; Persson, 1983b), distributions which are due at least in part to feeding conditions (Persson, 1986). In contrast, during November feeding rates are much lower due to reduced temperatures and the fish concentrate in a few dense shoals along the shore which at this time of year is devoid of significant water lily growth (Persson, 1983b). Predation risk from both fish and birds may be a second factor important in producing such distributions, as suggested by Bohl (1980) in a study of the fish populations of another lake. Clearly predation risk and feeding requirements are intimately linked in fish, as Werner *et al.* (1983b) and Power (1984) give examples of predator-induced restrictions to sub-optimal foraging habitats leading to reduced growth rates of a lake-dwelling carnivore and a stream-dwelling herbivore respectively.

During summer, significant predation pressure

around Sovdeborgssjon from birds is diffuse; grebes forage in the open water, where they can dive to the bottom of the relatively shallow water column with ease, and herons stalk the water lilies and reedbeds. Low foraging flights by gulls and terns may also contribute to the threat of predation at the outer edge of the lake's littoral water lily zone, although the water lilies themselves are likely to offer considerable protection against aerial predators. Certainly roach, perch and rudd apparently perceive a lower threat of predation risk when in a laboratory model of this habitat than when in simulated open water (Winfield, 1986). A significant degree of predation pressure must also arise at this time of year from Sovdeborgssjon's pike population, although its precise distribution within the lake is unknown. The distribution of predation pressure is simpler during late autumn when the major component of the total predation pressure probably arises from the mergansers and herons, and is focused exclusively around the edge of the lake. It is possible that this extremely clumped distribution of the fish near to the refuge-affording structure of the reedbeds is a protective mechanism against the mergansers, which in contrast to grebes are known to forage cooperatively (Cramp & Simmons, 1977).

A final distinction which will be drawn between the predation pressure arising from piscivorous birds and that from predatory fish lies in the nature of the attack itself. A fundamental difference arises because while attacks by piscivorous fish are launched from within the same medium as their target, birds must temporarily enter the water from the air above it. Giles & Huntingford (1984) differentiate between two categories of avian piscivore on the basis of the tactics that they employ when making such attacks: 'underwater chasers' which includes grebes and mergansers, and 'stalking/aerial ambushers' such as herons, gulls and terns. While this change of medium imposes an absolute restriction on the areas and depths that a bird can exploit, in contrast to the potentially wider ranging predatory fish, for the stalking/aerial ambushers it increases the chances of a surprise attack being successfully made. Thus the foraging tactics of herons, gulls, terns and larger stalk-ing/aerial ambushers effectively bypass the complex defensive behaviours relating to predator detection, inspection and evasion which are so ubiquitous and important in the interactions of fish with their piscine piscivores (see Pitcher, 1986).

While the underwater chasers can be defended against by shoaling, as perhaps Sovdeborgssjon's fish do when attacked by mergansers, the only defense against the surprise attacks of the ambushers is to avoid the areas that they frequent. Post-capture escape is unlikely, as Recher & Recher (1986) observed in over 3000 captures of 24 genera of fish by 8 species of herons. They noted that in the absence of no more elaborate post-capture defense mechanisms than violent flexions of the body, which is the only feature possessed by the majority of freshwater fishes, only 0.5% of captured fish successfully escaped. In addition to a consequently very wide spectrum of potential fish species which can be exploited, by virtue of their relatively large body and hence gape size, birds can also exploit a very wide range of prey sizes. Thus the sizes of fish consumed in Sovdeborgssjon simply largely reflect the size distributions of its stunted roach (Persson, 1983b) and perch (Persson, 1983a) populations. Escape by size is unlikely to be an important factor in the interaction between freshwater fish and their avian predators.

All of the above features of predation pressure arising from piscivorous birds will only be of importance for the ecology of their prey if evolutionary significant amounts of fish are removed. The calculated consumption rates, while being admittedly speculative, suggest that this may be the case in Sovdeborgssjon as predation by birds removes amounts of fish comparable with or greater than those taken by piscivorous fish, despite the fact that only a relatively small number of birds forage at the lake most of the year. Clearly, the effects of piscivorous birds are much greater than might be expected from a casual inspection of their numbers and cannot be routinely dismissed as negligible, as were the effects of planktivorous fish before the classic studies of Hrbáček *et al.* (1961) and Brooks & Dodson (1965). The relative importance and extent of the role of piscivorous birds in the freshwater ecosystem can only be determined by many more studies.

Acknowledgements

This work was carried out at the Institute of Limnology, University of Lund (Sweden) during a research fellowship in the European Science Exchange Programme. I thank the Board of the Institute and its Fish Ecology group for their generous provision of facilities during my stay, and I am also indebted to Stellan Hamrin for the use of his pike biomass data. I also thank Eva Bergstrand for her constructive criticism of an earlier draft of this paper. I am particularly grateful to my wife Denise for the many long, and often wet, hours that she spent in the field.

References

Bohl, E., 1980. Diel pattern of pelagic distribution and feeding in planktivorous fish. Oecologia 44: 368–375.

Britton, R. H. & M. E. Moser, 1982. Size-specific predation by herons and its effect on the sex ratio of natural populations of the mosquito fish *Gambusia affinis* Baird and Girard. Oecologia 53: 146–151.

Brooks, J. L. & S. I. Dodson, 1965. Predation, body size and composition of plankton. Science 150: 28–35.

Cramp, S. & K. E. L. Simmons, 1977. The birds of the western Palearctic, 1. Oxford University Press, Oxford, 720pp.

Diana, J. S., 1979. The feeding pattern and daily ration of a top carnivore, the northern pike (*Esox lucius*). Can. J. Zool. 57: 2121–2127.

Dill, L. M., 1983. Adaptive flexibility in the foraging behaviour of fishes. Can. J. Fish. aquat. Sci. 40: 398–408.

Fryer, G., 1987. Quantitative and qualitative: numbers and reality in the study of living organisms. Freshwat. Biol. 17: 177–189.

Giles, N. & F. A. Huntingford, 1984. Predation risk and interpopulation variation in anti-predator behaviour in the three-spined stickleback, *Gasterosteus aculeatus* L. Animal Behaviour 32: 264–275.

Hrbáček, J., Dvořáková, V. Kořinek & L. Procházková, 1969. Demonstration of the effect of the fish stock on the species composition of zooplankton and the intensity of metabolism of the whole plankton association. Verh. int. Ver. Limnol. 14: 192–195.

Jørgensen, S. E., 1979. Handbook of environmental data and ecological parameters. Pergamon Press Ltd, Oxford, 1162 pp.

Kennedy, G. J. A. & J. E. Greer, 1988. Predation by cormorants (*Phalacrocorax carbo* L.) on the salmonid populations of the R. Bush. Aquacult. Fish. Managem. 19: 159–170.

Leah, R. T., B. Moss & D. E. Forest, 1980. The role of predation in causing major changes in the limnology of a hyper-eutrophic lake. Int. Revue ges. Hydrobiol. 65: 223–247.

Persson, L., 1983a. Food consumption and competition between age classes in a perch *Perca fluviatilis* population in a shallow eutrophic lake. Oikos 40: 197–207.

Persson, L., 1983b. Food consumption and the significance of detritus and algae to intraspecific competition in roach *Rutilus rutilus* in a shallow eutrophic lake. Oikos 41: 118–125.

Persson, L., 1986. Effects of reduced interspecific competition on resource utilization in perch (*Perca fluviatilis*). Ecology 67: 355–364.

Pitcher, T. J., 1986. The functions of shoaling behaviour. In T. J. Pitcher (ed.), The behaviour of teleost fishes. Croom Helm, London: 294–337.

Power, M. E., 1984. Depth distributions of armored catfish: predator-induced resource avoidance? Ecology 65: 523–528.

Recher, H. F. & J. A. Recher, 1968. Comments on the escape of prey from avian predators. Ecology 49: 560–562.

Rowe, D. K., 1984. Some effects of eutrophication and the removal of aquatic plants by grass carp (*Ctenopharyngodon idella*) on rainbow trout (*Salmo gairdnerii*) in Lake Parkinson, New Zealand. N. Z. J. Ma. Freshwat. Res. 18: 115–127.

Sokal, R. R. & F. J. Rohlf, 1981. Biometry (2nd ed.). W. H. Freeman & Company, San Francisco, 859 pp.

Strong, D. R., D. Simberloff, L. G. Abele & A. B. Thistle (eds.), 1984. Ecological communities: conceptual issues and the evidence. Princeton University Press, Princeton (N.J.), 613 pp.

Werner, E. E., 1984. The mechanisms of species interactions and community organization in fish. In D. R. Strong, D. Simberloff, L. G. Abele & A. B. Thistle (eds.), Ecological communities: conceptual issues and the evidence. Princeton University Press; princeton (N.J.): 360–382.

Werner, E. E., G. G. Mittelbach, D. J. Hall & J. F. Gilliam, 1983a. Experimental tests of optimal habitat use in fish: the role of relative habitat profitability. Ecology 64: 1525–1539.

Werner, E. E., J. F. Gilliam, D. J. Hall & G. G. Mittelbach, 1983b. An experimental test of the effects of predation risk on habitat use in fish. Ecology 64: 1540–1548.

Winfield, I. J., 1986. The influence of simulated aquatic macrophytes on the zooplankton consumption rate of juvenile roach, *Rutilus rutilus*, rudd, *Scardinius erythrophthalmus*, and perch, *Perca fluviatilis*. J. Fish. Biol. 29 (Suppl. A): 37–48.

Zaret, T. M., 1980. Predation and freshwater communities. Yale University Press, New Haven, 187 pp.

Hydrobiologia **191**: 233–240, 1990.
P. Biró and J. F. Talling (eds), Trophic Relationships in Inland Waters.
© 1990 *Kluwer Academic Publishers.*

The distribution of waterfowl in relation to mollusc populations in the man-made Lake Zegrzyńskie

Anna Stańczykowska, Przemysław Zyska, Andrzej Dombrowski, Henryk Kot & Ewa Zyska
Agricultural-Pedagogical University, 08-110 Siedlce, ul.B.Prusa 12, Poland

Key words: dam reservoir, molluscs, waterfowl, trophic relations, distribution

Abstract

Preliminary investigations on the occurrence of molluscs and waterfowl at the man-made Lake Zegrzyńskie were begun in 1986.

The numbers, biomass and dominance structure of molluscs were analysed at different stations in the Lake. Some mollusc species were observed in huge numbers. Waterfowl, especially benthivorous species, were found in big flocks all the year round, but reached highest numbers in autumn. The possible effects of predation pressure from waterfowl on mollusc communities were analysed.

Introduction

High densities of molluscs observed on Lake Zegrzyńskie, and correspondingly high numbers of molluscivorous birds, provide suitable conditions for ecological studies of dependencies between two groups. Determination of chosen population features in these two trophic levels, and their mutual relationships, are the subject of studies started in 1986. The data presented here have been gathered during the first two years of the extensive CPBP Programm (No. 04.10.08) entitled 'Functioning of freshwater ecosystems, their protection and restoration', planned to be continued for the next three years. That is why the results are not conclusive and at this stage cannot be fully interpreted. A long-term objective of this research is to assess the magnitude of the impact of waterfowl on the molluscan prey populations. We believe that such work is required because in hydrobiology the majority of investigations on predator–prey relationships refer to planktonic organisms (Karfoot & Ship, 1987). Interactions between predators and their prey as determined by the impact of fish or birds on benthos are problems relatively rarely examined.

Freshwater molluscs are known to be consumed by waterfowl such as coot (*Fulica atra*), tufted duck (*Aythya fuligula*), pochard (*Aythya ferina*), goldeneye (*Bucephala clangula*) and other benthos feeders (Leuzinger & Schuster, 1970; Willi, 1971; Jacoby & Leuzinger, 1972; Stempniewicz, 1974; Suter, 1982; Crame & Simons, 1977; and others).

There are significant differences between estimations of the degree to which waterfowl utilize molluscs. Wiktor (1969) has stated that in Szczecin Firth the bivalve *Dreissena polymorpha* was only slightly utilized although its density was extremely high. It does not seem probable that waterfowl would affect the numbers of *D. polymorpha* in Masurian lakes. In Lake Gopło (central Poland) consumption of molluscs (especially of

D. polymorpha) by birds was much higher: it amounted to about 32 per cent in summer (Stempniewicz, 1974) and in winter coot itself consumed 93 per cent of the total biomass of the molluscs (Mikulski *et al.*, 1975).

In Swiss lakes, only recently invaded by *D. polymorpha*, significant changes were noted in populations of birds, most probably due to the recent appearance and mass development of this mollusc. Invasion of *D. polymorpha* affected not only the number of waterfowl but also the distribution of birds – they began to gather in places where molluscs were most numerous (Leuzinger & Schuster, 1970).

Most investigations on molluscs as food for birds were connected with one, mass-occurring species, mostly *D. polymorpha*. A specific pattern of distribution of molluscan fauna in Lake Zegrzyńskie, with a more diverse community dominated by different species in different parts of the lake, gave us the opportunity to examine food selectivity in benthivorous birds and its relation to their overall feeding strategies.

Description of study-site

The man-made Lake Zegrzyńskie arose as a result of the impoundment of the Narew and Bug rivers by a dam constructed on the Narew river at Debe in 1962-64. The total capacity of the lake is 100×10^6 m³ and the working draw of water is 11×10^6 m³, resulting in a variation in depth of 0.5 m.

Lake Zegrzyńskie is not deep: generally 3-4 m, being deeper (6-8 m) only in the old river beds near the dam. The lake is supplied with water by two rivers (the Bug and the Narew) carrying different pollution loads has a small water exchange rate in the main basin during low and medium water-flow, and is characterized by a great diversity of communities in different zones. The communities were found to vary considerably both in quality and numbers (Dusoge *et al.*, 1985). The phytoplankton was rich, with a maximum biomass of 32 mg wet wt l⁻¹. Zooplankton showed a high species diversity with low numbers and

biomass of 0.15-1.69 mg wet wt l⁻¹. The bottom fauna (without Mollusca) consisted mainly of Oligochaeta and larvae of Chironomidae, and was very abundant (over 200 g wet wt m⁻²). The communities of molluscs in the littoral zone in 1980-81 were rich in quality and numbers (Jurkiewicz-Karnkowska, in press).

The fish populations were also very rich in numbers. Some 23 fish species, dominated by roach (*Rutilus rutilus*), bream (*Abramis brama*), bleak (*Alburnus alburnus*) and stickleback (*Gasterosterus aculeatus*), have been recorded (Grudniewski, in prep.).

Material and methods

The distribution of molluscs in Lake Zegrzyńskie has been examined in May, July and October 1986 and in May and July 1987. Each time the samples were taken by means of a bottom dredge and a bottom Ekman-Birge sampler at 8 to 9 stations in the littoral and in the middle of the lake. In July 1987 the samples were taken from 3-5 stations set up on 17 profiles, at various depths and at various distances from the shore. The material was fixed with 4% formalin. Molluscs were identified to species; all specimens were measured, weighed, dried at about 60 °C for 24 hrs and then weighed again, to obtain dry weight with and without shell.

Observations on the distribution and numbers of waterfowl were carried out from October 1986 to August 1987. In total 18 censuses were made, 2-6 in each phenological period (breeding period, period of dispersal, autumn wintering, spring passage). The censuses were made by four observers in one day. The position and activity of the birds were mapped during the counting. When analyzing results, the reservoir was divided into 8 basins according to limnological characteristics (Fig. 5).

The average number of each species living on particular basins in the phenological periods mentioned above served as the basis, in terms of bird-day units, for calculating the exploitation of a given area by diving ducks (Nilsson, 1972). The energy requirement of benthivorous species in the non-breeding period was assessed using available published data on daily basal metabolic rate (Suter, 1982). The differences in energy requirement between the two sexes were examined. Changes of the benthivorous guild energy requirement and the mollusc energy values (Stańczy-kowska, 1977; Winberg, 1971), together with data on the dynamics of the mollusc number, have been used for estimating the pressure of predator on prey.

Results

Molluscs – a total of 51 species were encountered in Lake Zegrzyńskie, including 28 snail and 23 bivalve species. All have been recorded in 1980-81 in the littoral zone of the lake by Jurkiewicz-Karn-kowska (in press). From the present study it is clear that the occurrence of most species is not restricted to the littoral zone, as they are also found offshore. The number of species in Lake Zegrzyńskie is high, both as compared with other dam reservoirs (see Jurkiewicz-Karnkowska, in press) and with several tens of natural lakes situated in northern Poland (Stańczykowska *et al.*, 1983).

Studies have proved that, in spite of large species richness, only several species play a substential role in terms of numbers and biomass. Especially abundant were *Viviparus viviparus* L., *Dreissena polymorpha* (Pall.) and representatives of the Sphaeridae. The dominance of various species is connected with different lake parts and to a much lesser degree with depth (Fig. 1). *Viviparus viviparus* was especially numerous in the discharge section of the Narew river, in the western part of the lake and in some parts near the dam. In the latter environment there were places where *D. polymorpha* was especially abundant. The representatives of Sphaeridae were especially predominant in the mouth of the Bug river.

Maximum numbers and biomass of molluscs in the Lake can be considered as very high; average values are also high. Relatively few habitats have a relative paucity of molluscs. The numbers of molluscs are similar in the littoral zone and off-shore (Fig. 2).

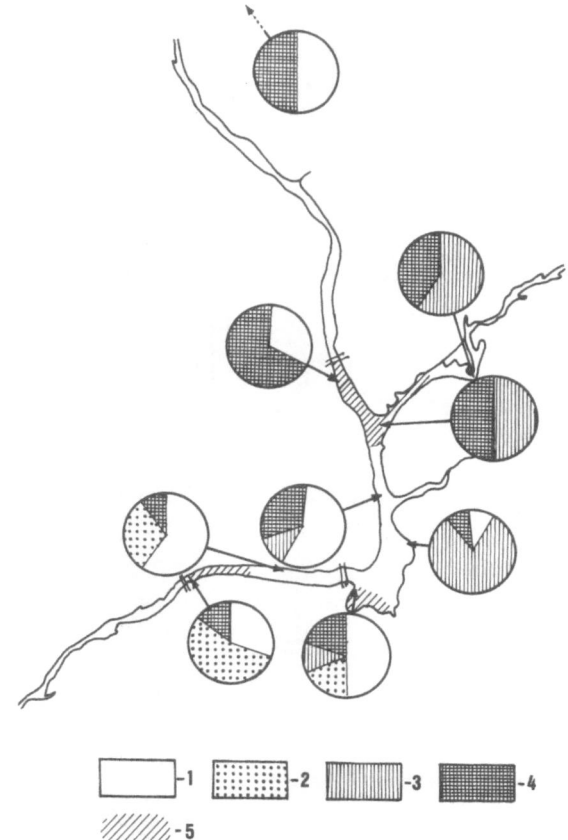

Fig. 1. Percentage composition (in terms of biomass) of the mollusc community at different sites in Lake Zegrzyńskie. 1. *Viviparus viviparus*, 2. *Dreissena polymorpha*, 3. Sphaeridae, 4. others/. Places of foraging by birds are shown by hatched areas.

Birds – during census of Lake Zegrzyńskie 31 species of birds were recorded. The number of species varied depending on the phenological periods: lowest in the breeding season – 12 species, highest in autumn passage – 22 species. Average numbers in the bird community varied from 2900 individuals in spring to 13600 birds in autumn (Fig. 3).

The whole bird community was divided into trophic groups or foraging guilds: herbivores, benthivores, piscivores, and omnivores. Benthivores were represented by 7 species: tufted duck (*Aythya fuligula*), pochard (*Aythya ferina*), goldeneye (*Bucephala clangula*), scaup (*Aythya marila*), coot (*Fulica atra*), common scoter (*Mellanitta*

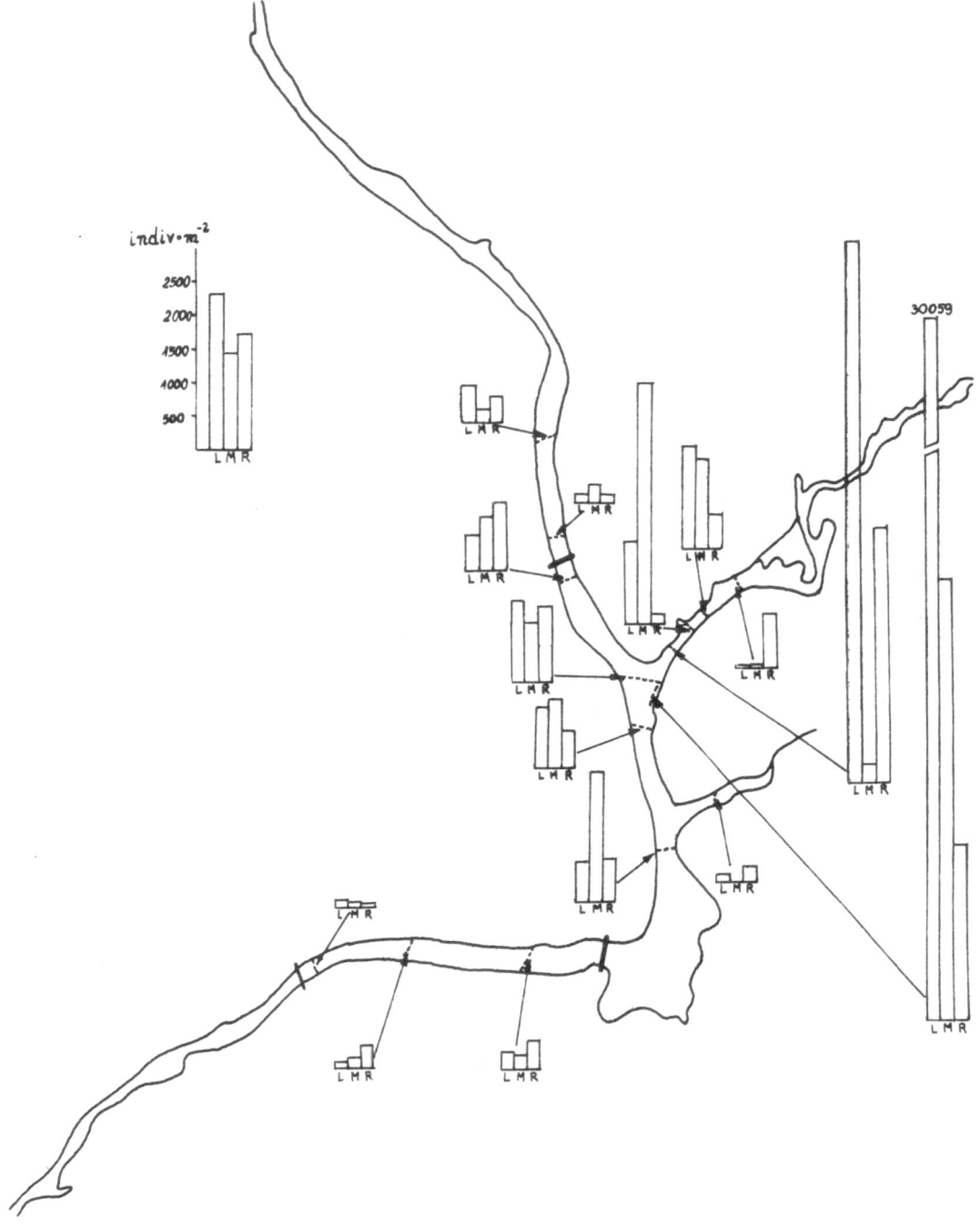

Fig. 2. Distribution of molluscs (indiv. · m⁻²) along profiles sampled in July 1987. L – left side of the Lake, M – middle of the Lake, R – right side of the Lake.

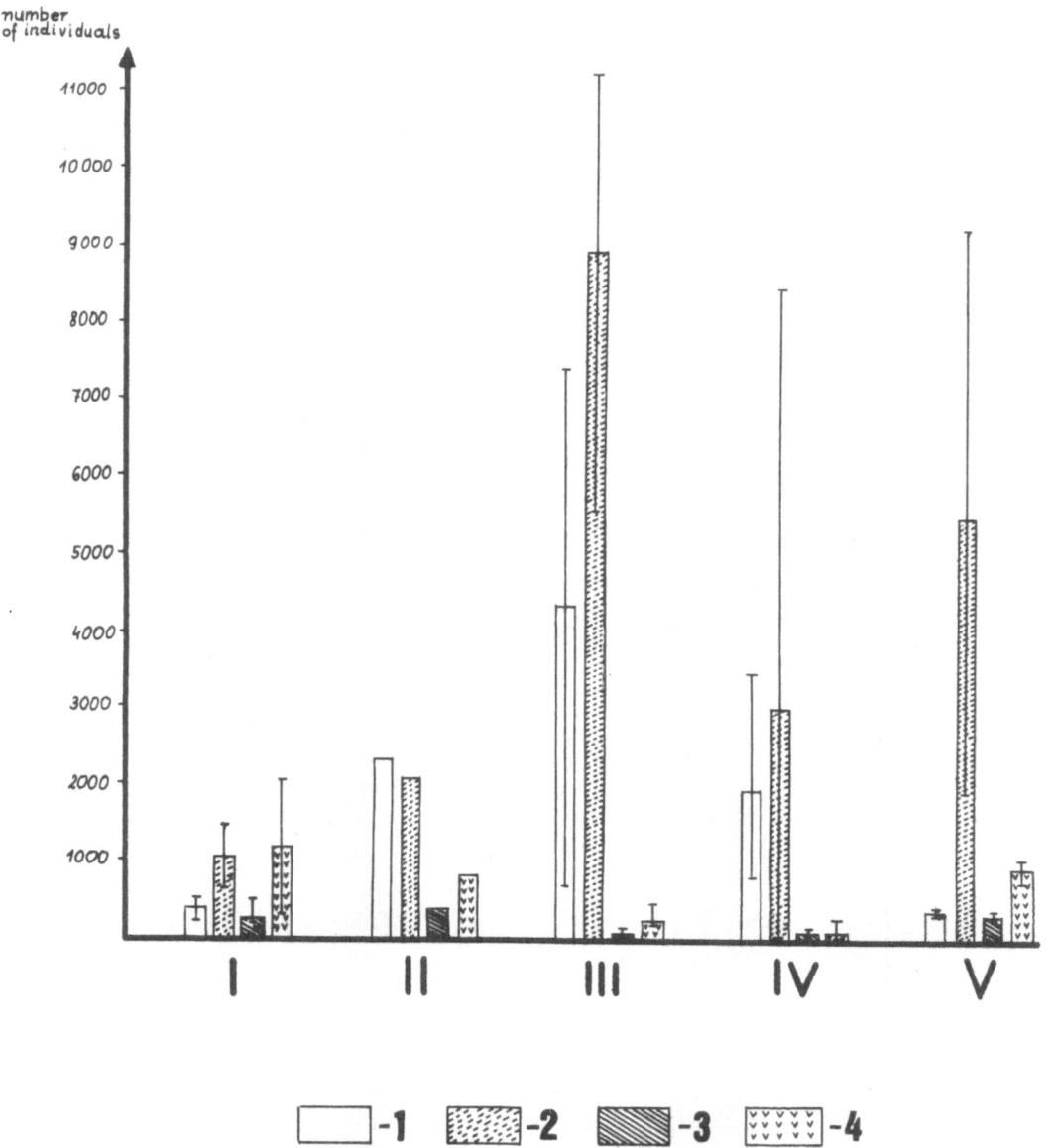

Fig. 3. The seasonal fluctuations of abundance of avian foraging guilds on Lake Zegrzyńskie. I – breeding period, II – period of dispersal, III – autumn passage, IV – wintering period, V – spring passage, 1. herbivores, 2. benthivores, 3. piscivores, 4. omnivores. The histograms indicate the average number of birds, error bars the minimum and maximum numbers.

nigra) and long tailed duck (*Clangula hyemalis*). The highest number of benthivorous species (6) was observed in autumn, with 3-4 species in the remaining seasons. The species composition of the benthivorous guild changes with season. Tufted duck was most frequent (up to 59% of the foraging guild) in the breeding season and autumn passage, whereas in other seasons it occurred only

as en accessory species. The share of pochard in the benthivorous guild was clearly low during the whole season (1-22%). Goldeneye was especially abundant in autumn and in winter when its share was 26% and 31%, respectively. Coot was the most frequent species nearly all year round, varying between 9% and 90% of the guild (Fig. 4).

During the whole study period a variability of

238

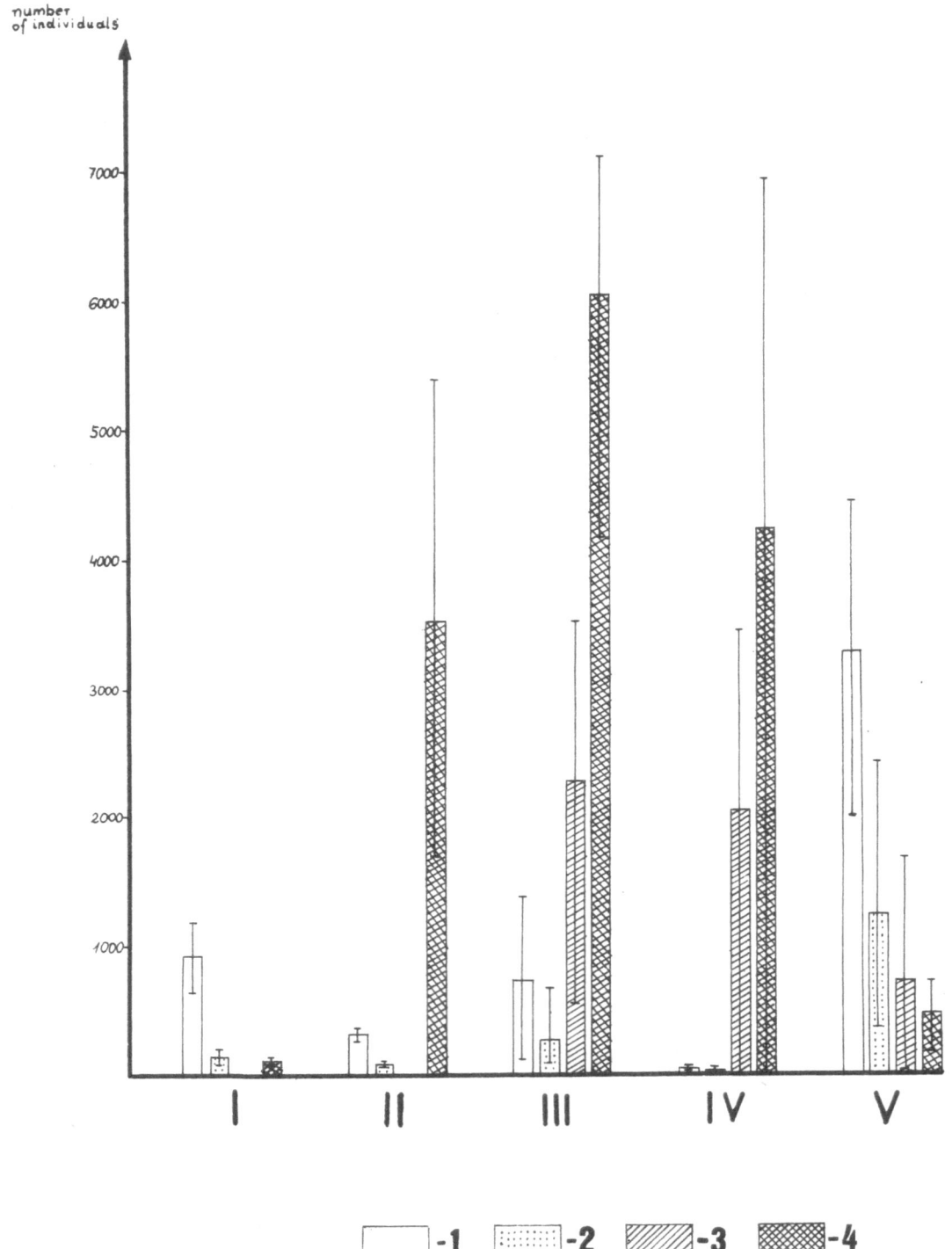

Fig. 4. The seasonal fluctuations of abundance of benthivorous species on Lake Zegrzyńskie. I-V as in Fig. 3. 1. *Aythya fuligula*, 2. *Aythya ferina*, 3. *Bucephala clangula*, 4. *Fulica atra*.

distribution of various bird species in the lake was observed. Tufted duck occurred most abundantly during autumn passage in basins 4 and 6, during spring in basins 4 and 7, and during the breeding period in basin 7. Pochard was observed in basins 4 and 7 both in autumn and in spring. Goldeneye preferred basins 2, 3, 4 and 7 in autumn passage, basins 2 and 4 in winter, and basin 7 in spring. Frequent occurrence of coot was observed during autumn in basins 2, 3, 4 and 5, during wintering in basins 2 and 4, and at spring passage in basins 2 and 5.

Taking into account the above findings it was concluded that basins 1, 2, 3, 4 and 7 were most frequently occupied by the whole foraging group during the non-breeding period. Locations of especially intensive foraging are marked in Figs. 1 and 2. The ratio of the energy requirements of the whole benthivorous bird community to the energy equivalent of mollusc biomass abundance was calculated (Fig. 5). It was found that the birds can remove 0.2% to 0.4% of the annual production of the prey population (basins 3, 4, 5, 6 and 8) or even 10-20% (basins 2 and 7) (Table 1). The weakest pressure was observed in the region of the mouth of the Bug river and in the region close to the dam.

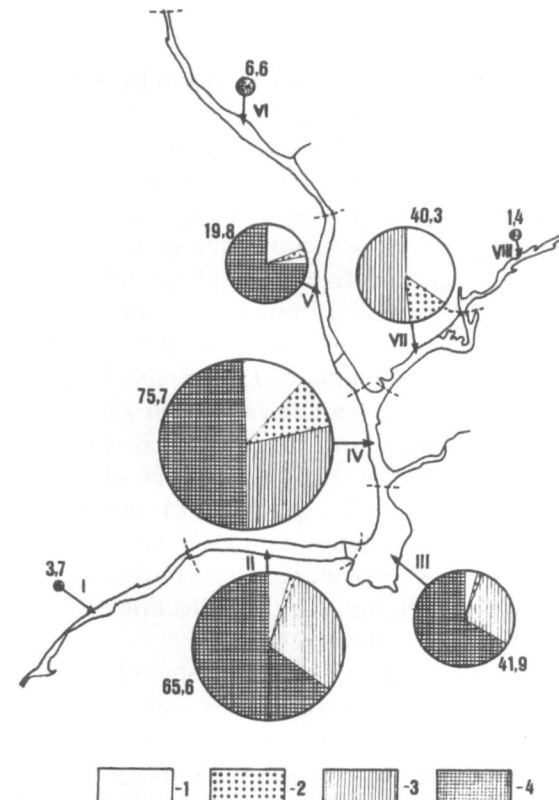

Fig. 5. The energy requirement (in 10^6 kcal year^{-1}) of benthivorous species in particular basins of Lake Zegrzyńskie during the non-breeding season. 1-4, groups as in Fig. 4.

Table 1. The estimated pressure of benthivores on molluscs as an energy requirement (in terms of mollusc biomass) in relation to their food supply.

	Basins							
	1	2	3	4	5	6	7	8
benthivorous guild energy requirement (10^6 kcal year^{-1})	3,7	65,6	41,9	75,7	19,8	6,6	40,3	1,4
A. benthivorous guild energy requirement in terms of mollusc biomass (10^6 g) during the non-breeding period	14,9	262,5	167,4	302,8	79,4	26,6	161,3	5,7
area of basins (km^2)	2,5	6,0	9,0	3,5	3,0	5,0	2,5	2,5
B. estimated mollusc biomass (10^6 g) in the basins	?	2520	10710	6650	2760	1250	800	2000
estimated benthivores pressure on molluscs as the proportion A/B	?	10,4%	1,6%	4,6%	2,9%	2,2%	20,2%	0,2%

Conclusion

The investigations revealed the selectivity of foraging places by benthivorous birds. These species foraged only in some places in spite of abundance of food (molluscs) in other parts of the lake. High activity of goldeneye was observed in the places of mass occurrence of *D. polymorpha* and Sphaeridae. In places of abundant occurrence of Sphaeridae tufted duck and goldeneye also foraged there. Coot, on the other hand, occurred in the region of *D. polymorpha* colonies. In most places where birds foraged *Viviparus viviparus* was abundant: however there were places where, in spite of mass occurrence of this snail, the birds did not forage. These observations corroborate the hypothesis of active selection of specific foraging places by these birds. A next step in this research must be the collection of diet data through gut analysis.

High variability of distribution and numbers of benthivores in Lake Zegrzyńskie during the annual cycle is important in influencing the changing pressure of birds on molluscan prey populations. This pressure is additionally complicated by the diverse abilities of collecting food by different species of birds (different depth for diving, anatomical differences in beak and stomach).

Moreover, fish living in the Lake also consume molluscs. Other investigations were carried out at the same time that aimed at evaluating quantitatively the food consumed by the dominant fish, namely roach and bream, which chiefly feed on molluscs. Molluscs were also often found in the guts of other fish living there (Terlecki, in prep.).

We hope that further investigations included in the Programme, in particular those aimed at examining the bioenergetic value of molluscs as food for birds, and those concerned with determining the influence of various other factors (e.g. fish) on the mollusc populations, will permit us to determine how they function in the ecosystem.

References

Cramp, S. & E. L. Simmons, eds, 1977. Handbook of the birds of Europe, the Middle-East and North Africa. Vol. 1. Oxford.

Dojlido, J., L. Jakubowska, J. Moraczewski & A. Praszkiewicz, 1967. Charakterystyka limnologiczna wód Narwi i Bugu przed i po utworzeniu Jeziora Zegrzyńskiego (Limnological characteristics of waters of Narew and Bug before and after formation of Lake Zegrzyńskie). Pr. Inst. Gosp. Wod. 4: 57–85.

Dusoge, K., L. Bownik-Dylińska, J. Ejsmont-Karabin, I. Spodniewska, T. Węgleńska, 1985. Plankton and benthos of man-made Lake Zegrzyńskie. Ekol. pol. 33, 3: 455–479.

Jacoby, H. & H. Leuzinger, 1972. Die Wandermuschel *Dreissena polymorpha* als Nahrung der Wasservögel am Bodensee. Anz. orn. Ges. Bayern. II, 26–35.

Kerfoot, W. C. & A. Sih, 1987. Predation – direct and indirect impacts of aquatic communities. Univ. Press of New England, Hanover and London, pp. 386.

Leuzinger, H. & S. Schuster, 1970. Auswirkungen der Massenvermehrung der Wandermuschel *Dreissena polymorpha* auf die Wasservögel des Bodensees. Ornit. Beob. 67: 269–274.

Mikulski, J., B. Adamczak, L. Bittel, R. Bohr, D. Bronisz, W. Donderski, A. Giziński, M. Luścińska, M. Rejewski, E. Strzelczyk, N. Wolnomiejski, W. Zawiślak & R. Żytkowicz, 1975. Basic regularities of productive processes in the Iława Lakes and in Gopło Lake from the point of view of utility values of the water. Pol. Arch. Hydrobiol. 22: 102–122.

Nilsson, L., 1972. Local distribution, food choice and food consumption of diving ducks on a South Swedish lake. Oikos 23: 82–91.

Stańczykowska, A., 1977. Ecology of *Dreissena polymorpha* (Pall.) (Bivalvia) in lakes. Pol. Arch. Hydrobiol. 24: 461–530.

Stańczykowska, A., E. Jurkiewicz-Karnkowska & K. Lewandowski, 1983. Ecological characteristics of lakes in north-eastern Poland versus their trophic gradient. X. Occurrence of molluscs in 42 lakes. Ekol. pol. 31, 2: 459–475.

Stempniewicz, L., 1974. The effect of feeding of coot (*Fulica atra* L.) on the character of the shoals of *Dreissena polymorpha* Pall. in the Gopło Lake. Acta Univ. N. Copernici. se. mat.-przyr. 34: 84–103.

Suter, W., 1982. Vergleichende Nahrungsökologie von überwinternden Tauchenten (*Bacephala*, *Aythya*) und Blasshuhn (*Fulica atra*) am Untersee-Ende (Hochrhein) (Bodensee). Orn. Beob. 79: 225–254.

Wiktor, J., 1969. Biologia *Dreissena polymorpha* (Pall.) i jej ekologiczne znaczenie w Zalewie Szczecińskim. Stud. Mater. Morsk. Inst. Ryb. Gdynia A, 5: 1–88.

Willi, P., 1971. Wasservögel rienigen den Bodensee. Ringiers Unterhaltungs – Blätter 'Das gelbe Heft' nr 34.

Winberg, G. G. (ed.), 1971. Methods for the estimation of production of aquatic animals. Academic Press. London. New York, pp. 175.

Wojciechowska, J., A. Praszkiewicz & J. Dojlido, 1979. Analiza jakości wód Jeziora Zegrzyńskiego – w perspektywie do 1990 i 2000 r. (An analysis of man-made Lake Zegrzyńskie water quality in future – till the year 2000). Instytut Meteorologii i Gospodarki Wodnej. Warszawa, 100 pp.

Hydrobiologia **191**: 241–248, 1990.
P. Biró and J. F. Talling (eds), Trophic Relationships in Inland Waters.
© 1990 *Kluwer Academic Publishers.*

Light-limited algal growth in Lake Loosdrecht: steady state studies in laboratory scale enclosures

Machteld Rijkeboer & Herman J. Gons
Limnological Institute, Vijverhof Laboratory, Rijksstraatweg 6, 3631 AC Niewersluis, The Netherlands

Key words: Loosdrecht lakes, eutrophication, *Prochlorothrix hollandica*, growth kinetics of algae, continuous cultures, light attenuation

Abstract

Phytoplankton growth in the shallow, turbid Lake Loosdrecht (The Netherlands) is importantly influenced by light availability, and thus the concentrations of the various light-attenuating materials. The system is highly eutrophic and supports an algal biomass of ca. 160 mg Chl m^{-3}. A model is proposed here which predicts algal growth in the lake as a function of the light received and subsequent attenuation in the water column by phytoplankton, tripton and background colour. The model is based on an energy balance which relates growth rate to the 'true' growth yield on light energy and the energy demand for cell maintenance. The coefficients for energy conversion ($Y = 0.002$ gDW kJ^{-1}) and cell maintenance ($\mu_e = 0.031$ day^{-1}) were determined from steady state growth kinetics of *Prochlorothrix hollandica* in light-limited laboratory flow systems with the same depth as the lake and receiving summer average conditions of irradiance. Light attenuation by phytoplankton and tripton were quantified using specific attenuation coefficients: 0.011 m^2 mg^{-1} Chl for the phytoplankton and 0.23 m^2 g^{-1} DW for tripton.

The growth studies demonstrated that Lake Loosdrecht can support a much higher algal biomass in the absence of non-algal particulate matter. The proposed model is used to predict chlorophyll *a* concentrations in dependence on growth rate and levels of tripton. Since approximately 75% of the sestonic dry weight in Lake Loosdrecht may be attributed to tripton, it is concluded that the algal biomass is markedly lowered by the abundance of tripton in the water column. A knowledge of the sources and fate of tripton in the lake is thus of fundamental importance in modelling phytoplankton dynamics.

Introduction

Due to excessive external nutrient loading, the shallow Lake Loosdrecht (The Netherlands) has become highly eutrophic. The lake is part of a system situated 20 km south-east of Amsterdam; its area is 9.8 km^2, and average depth ca. 2 m. For details on the lake's origin and eutrophication see Van Liere (1986a). Although drastic measures for reducing external phosphorus loading have been effective since 1984, the water quality did not improve sofar. Phosphate-uptake studies in 1983 (Riegman & Mur, 1986) indicated that the growth of the phytoplankton was phosphorus-limited, but in 1985 chlorophyll *a* and sestonic dry weight concentrations changed independently of phosphorus supply in laboratory scale enclosures (Sweerts *et al.*, 1986; Van Liere, 1986b).

In summer the phytoplankton is dominated by filamentous prokaryotic species, among which the recently isolated *Prochlorothrix hollandica* (Burger-Wiersma et al., 1986) is likely to be the most important. From May until October in recent years the population density was rather stable. In the years 1984 to 1986 the average seston dry weight and chlorophyll *a* concentration during June until August were 39 ± 11.1 g m^{-3} and 159 ± 40.7 mg m^{-3}, respectively.

Considering the physiological range, the growth rates must be low, and the phytoplankton seemed to be growing close to steady state (Gons et al., 1986a). For light-limited ('nutrient-saturated') growth conditions it has been assumed earlier that natural phytoplankton systems are capable of approaching a stable steady state (e.g. Bannister, 1974).

The upper limit to population density at steady state will be determined by a balance between production and loss. From this point of view growth rate of the phytoplankton is controlled by the loss factors zooplankton grazing and sedimentation, and low rates of these processes allow for high algal biomass.

Several more or less complicated models have been proposed for the prediction of algal population density from photosynthesis and underwater light characteristics (e.g. Bannister, 1974, 1979; Wofsy, 1983). Gons & Mur (1975) proposed a simple energy balance for light-limited algal growth using chemostat data (see below). In principle, this relationship is suitable for predicting biomass concentration in mixed layers from growth kinetics studied in the laboratory. However, in a lake the light irradiance will not be absorbed by the phytoplankton component only, since significant parts of the light will be attenuated by tripton (i.e. inanimate particulate matter).

For analyzing the phytoplankton dynamics and predicting population density in L. Loosdrecht, a simple steady state model is derived using growth kinetics of *P. hollandica* in laboratory enclosures, and measurements of light attenuation due to this alga, sestonic matter, and sediment particles from the lake. Model predictions of phytoplankton biomass are compared to chlorophyll *a* concentrations in L. Loosdrecht. The role of the high, variable concentrations of tripton in the seston is emphasized.

A comparison of the growth rate predictions from the model with the rates of changes in measured concentrations of chlorophyll *a* will be discussed in Gons & Rijkeboer (1990).

Theory

Energy balance

For growth of algae in general, an inverse relationship exists between population density and specific growth rate. In case of light-limited growth this can be explained by the energy balance as proposed by Gons & Mur (1975);for a water column wherein all incoming radiation (neglecting escaping upwelling light) is absorbed by algae only:

$$E_d(0) \cdot c = (\mu + \mu_e) \cdot X_{ph} \cdot z \qquad (1)$$

where $E_d(0)$ is de downward light irradiance just below the water surface (kJ m^{-2} time^{-1}),

> c the 'true' efficiency of light energy conversion into biomass (dimensionless),
> μ the specific growth rate (time^{-1}),
> μ_e the specific maintenance rate constant (time^{-1}),
> X_{ph} the algal biomass concentration in energy units (kJ m^{-3}),

and z depth of the water column (m).

$\mu \cdot X_{ph}$ represents that part of energy fixed in biomass synthesis and $\mu_e \cdot X_{ph}$ the energy which as a result of maintenance processes is diverted from growth.

In steady state continuous cultures μ is equal to the dilution rate and $E_d(0)$ and X_{ph} are easily measured. Values for c and μ_e can be obtained as the slope and y-intercept, respectively of a regression of μ on $E_d(0)/X_{ph}$ plot.

Since incident light in the lake will be absorbed not only by the phytoplankton, but also by tripton and gilvin (i.e. dissolved 'yellow' substances) an alternative model, based on the former principle, has been derived.

Underwater light climate

In order to quantify the role of the different light scattering and absorbing components in natural water, K_d, the vertical attenuation coefficient for downward irradiance, may be split into partial attenuation coefficients for water (W), gilvin (G), phytoplankton (PH) and tripton (TR), respectively (for discussion see Kirk, 1983):

$$K_d = K_W + K_G + K_{PH} + K_{TR} \qquad (2)$$

The sum of K_W and K_G may further be defined as the background attenuation coefficient (K_B).

The partial attenuation coefficients for phytoplankton and tripton will be determined by their concentration X (per unit volume) and a specific attenuation coefficient k, e.g.

$$K_{PH} = X_{ph} \cdot k_{ph} \qquad (3a)$$
$$K_{TR} = X_{tr} \cdot k_{tr} \qquad (3b)$$

where X_{ph} is the phytoplankton dry weight concentration (g m^{-3}),
   X_{tr} the tripton dry weight concentration (g m^{-3}),
   k_{ph} the specific light attenuation coefficient for phytoplankton (m^2 g^{-1}),
and k_{tr} the specific light attenuation coefficient for tripton (m^2 g^{-1}).
From eq. (2) and (3) it follows that:

$$K_d = X_{ph} \cdot k_{ph} + X_{tr} \cdot k_{tr} + K_B \qquad (4)$$

Since the fraction of the underwater light absorbed by the algae is given by $X_{ph} \cdot k_{ph}/K_d$ (eq. 4), eq. (1) can be extended to:

$$E_d(0) \cdot Y \cdot X_{ph} \cdot k_{ph}/K_d = (\mu + \mu_e) \cdot X_{ph} \cdot z,$$
hence
$$E_d(0) \cdot Y \cdot k_{ph}/(K_d \cdot z) = \mu + \mu_e \qquad (5)$$

where $K_d \cdot z$ is the optical depth (dimensionless), and Y the 'true' growth yield for algal dry weight (g kJ^{-1})
   Eq. (5) predicts a linear relation between growth rate and subsurface irradiance divided by the optical depth, in which the slope depends on two algal physiological parameters, i.e. those for harvesting of light and conversion of the absorbed light into biomass, respectively. The intercept on the ordinate represents the negative value of the specific maintenance rate constant, μ_e.

For adjustment to the field situation, the biomass of algae will be expressed in units of chlorophyll a and the subsurface irradiance derived from meteorological data.

Thus the concentration of chlorophyll a is given by:

$$CHL = Q \cdot p \cdot Y/\{(\mu + \mu_e) \cdot z\} - TR \cdot k_{tr}/k_c - K_B/k_c \qquad (6)$$

where CHL is chlorophyll a concentration (mg m^{-3}),
   TR the tripton concentration (g m^{-3}),
   Q the isolation (kJ m^{-2} time^{-1}),
   p the constant for conversion of Q to net downward irradiance (PAR) just below the watersurface,
   Y the 'true' growth yield for chlorophyll a (mg kJ^{-1}),
and k_c the specific light attenuation coefficient for chlorophyll a (m^2 mg^{-1}).
After re-arrangement, for computing growth rate from measured chlorophyll a in the lake we obtain:

$$\mu = \frac{Q \cdot p \cdot Y}{(CHL + TR \cdot k_{tr}/k_c + K_B/k_c) \cdot z} - \mu_e \qquad (7)$$

Materials and methods

P. hollandica was grown in laboratory scale enclosures (LSE), under light-limiting conditions. The LSE consisted of a horizontal perspex cylinder with a length of 2 m, i.e. approximately the average depth in the lake, and a volume of 128 l. The cylinder was immersed in a thermostated waterbath and temperature was kept constant at 20 °C. The suspension was completely mixed, using powerful stirrers. For more details on the LSE, see Rijkeboer & Gons (1988) and Van Liere (1986b).

Illumination was provided by high intensity daylight lamps (HMI 1200 W, Osram, Germany), having a spectral composition comparable with sunlight (Flik & Keyzer, 1981). Irradiance (PAR) was set according to the average insolation during June until August for the previous 10 years (Royal Netherlands Meteorological Institute, De Bilt) and a water surface reflection of 20% (L. Van Liere, pers. comm.). Incident light for the three steady state experiments was 100 to 111 W m^{-2} during the 16 h light period. The light measurements were made in W m^{-2} PAR, using a R 3001 photometer with SD 101 head unit (Macam, Livingston, Scotland).

At one side synthetic medium was fed using a peristaltic pump (Minipuls 2, Gilson, France). The medium has been described by Van Liere (1979), but except for $NaHCO_3$, FeEDTA and trace-elements, the concentrations were half the original ones. The dilution rates were 0.042, 0.081 and 0.169 day^{-1}. Steady state was defined when the biomass was constant during at least 4 subsequent days.

Chlorophyll a and phaeopigment were determined according to Moed and Hallegraeff (1978). Dry weights were determined after filtration over Whatman (GF/F) filters, and drying at 80 °C.

In order to determine the specific light attenuation coefficient of tripton, epipelon (sediment particles) from a sandy stretch of bottom in L. Loosdrecht was collected. Field work has established that the complete layer of this particulate matter is subject to resuspension (Gons et al., 1986b). In the laboratory the epipelon was filtered over 150 μm mesh. For the light measurements the fraction with low settling rates was used.

Results and discussion

Energy balance for growth of P. hollandica

The curve for light-limited growth (Fig. 1) gives the relationship between steady state population

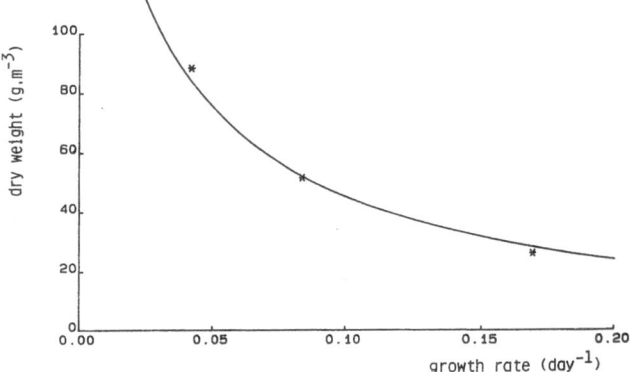

Fig. 1. The predicted relationship (eq. 1) between steady state dry weight concentration of *P. hollandica* and growth rate in the LSE. The asterisks indicate the measured data.

density as dry weight and growth rate in the LSE. At the different growth rates, the densities of the algal cultures were such that no transmission of photosynthetic active radiation could occur, although some lateral loss of light occurred. Thus, an approximately constant amount of energy was consumed by the cultures, regardless of the growth rate. From this, it follows that the biomass must decrease at increasing growth rate, since the latter must be proportional to the amount of energy consumed per unit of biomass over time, which is the specific light uptake rate $E_d(0) \cdot 1/X_{ph}$.

Linearization of the energy balance, and expressing both X_{ph} and $E_d(0)$ in units of energy, provided the values $c = 0.04$ and $\mu_e = 0.031$ day^{-1}. A conversion factor of 22 kJ per g dry weight was used (Gons & Mur, 1975).

These values are similar to those for the blue-green alga *Oscillatoria agardhii* (Van Liere & Mur, 1979).

Biomass and light attenuation

For the conversion of light into algal biomass the alternative eq. (5) was derived. Of course, this relation holds only in conditions where k_{ph} is constant and K_d does not change with depth, but at least adequately describes light attenuation in that part of the water column where irradiance allows positive net oxygen production.

Applying eq. (5) to the steady state data (Table 1), and the measured vertical attenuation coefficients a non-linear relationship was found. This may be explained by a variable k_{ph}, since its value is known to be influenced by change in light quality due to selective light absorption and increase in scattering (Atlas & Bannister, 1980; Kirk, 1983).

For dry weight concentrations up to 25 g m^{-3}, the relationship between dry weight and light attenuation coefficient was linear; the specific attenuation coefficient k_{ph} was 0.23 m^2 g^{-1} (Table 2). When K_d values at higher concentrations were included the data were fit better using a parabolic function. Substitution of k_{ph} = 0.23 m^2 g^{-1} in eq. (3) gave 'true' vertical attenuation coefficients and using these values a linear relation between growth rate and the quotient of subsurface light and optical depth was found (Fig. 2).

Based on chlorophyll a, the specific attenuation coefficient k_c for *P. hollandica* was 0.011 m^2 mg^{-1}. This value is somewhat lower than those reported by most other authors, ranging between 0.012 and 0.020 (e.g. Bannister, 1974; Verduin, 1982; Kirk, 1983). However, k_c for green algae may vary with depth from 0.012 to 0.005 m^2 mg^{-1} due to selective absorption (Atlas & Bannister, 1980). In its pigmentation *P. hollandica* closely resembles the green algae (Burger-Wiersma et al., 1986).

In suspensions from L. Loosdrecht we measured a very rapid decrease of blue light (Rijkeboer et al., 1986), which may be attributed to selective absorption by both gilvin and phytoplankton, and possibly by tripton. It is most likely that these conditions account for both population density and depth dependency of the specific attenuation coefficient for the complete water column.

Table 1. Steady state concentrations of dry weight and chlorophyll a of *Prochlorothrix hollandica* in the LSE.

growth rate	(day^{-1})	0.042	0.081	0.169
dry weight	(g m^{-3})	88.5	51.7	26.2
chlorophyll a	(mg m^{-3})	1531	1068	545

Table 2. Partial light attenuation coefficients (m^{-1}) for *P. hollandica*, tripton and sestonic matter. For symbols, see Table 3.

		(n):	(r^2):
K_{PH}	$= 0.23 \cdot X_{ph} + 0.90$	4	0.994
K_{TR}	$= 0.23 \cdot TR + 0.84$	9	0.996
K_{SES}	$= 0.23 \cdot X_{ses} + 0.85$	14	0.986

Biomass, light attenuation and growth rate in L. Loosdrecht

From eq. (5) (Fig. 2) it followed that Y, the conversion factor of light into biomass, was 0.002 g kJ^{-1} and that the specific maintenance constant was 0.031 day^{-1}. Expressed per unit chlorophyll a, the value of Y was 0.044 mg kJ^{-1}. As the chlorophyll a content per unit dry weight was nearly constant, the steady state chlorophyll a concentrations were also fitted very well using the 'true' vertical attenuation coefficients.

In order to predict the chlorophyll a levels during summer in L. Loosdrecht, we used the values given in Table 3.

The light uptake rate by the water column – $E_d(0)$ – was based on the insolation – Q – and corrected with a constant – p – for conversion to PAR and a correction for surface reflection. The average value during June, July and August in 1984 until 1986 was 6300 kJ m^{-2} day^{-1}.

Since the measured sestonic dry weight con-

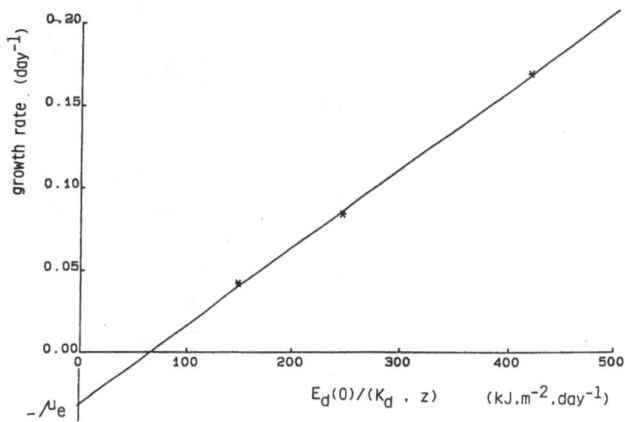

Fig. 2. The relationship between growth rate and the quotient of subsurface light and optical depth (see text for explanation) in the LSE.

Table 3. List of symbols.

Symbol	Explanation	Units	Value
$E_d(0)$	downward light irradiance just below the water surface	$kJ\ m^{-2}\ day^{-1}$	–
Q	insolation	$kJ\ m^{-2}\ day^{-1}$	–
p	constant for conversion to PAR and correction for surface reflection	no dimension	0.38
Y	'true' growth yield for chlorophyll a	$mg\ kJ^{-1}$	0.044
X_{ph}	concentration of phytoplankton	$g\ m^{-3}$	–
CHL	concentration of chlorophyll a	$mg\ m^{-3}$	–
TR	concentration of tripton	$g\ m^{-3}$	–
K_d	vertical attenuation coefficient for downward irradiance	m^{-1}	–
K_B	partial light attenuation coefficient due to water and dissolved compounds	m^{-1}	1.2
k_{ph}	specific light attenuation coefficient of algal biomass	$m^2\ g^{-1}$	0.23
k_c	specific light attenuation coefficient for chlorophyll a	$m^2\ mg^{-1}$	0.11
k_{tr}	specific light attenuation coefficient of tripton	$m^2\ g^{-1}$	0.23
μ	specific growth rate	day^{-1}	–
μ_e	specific maintenance rate constant	day^{-1}	0.031
z	depth of the mixed layer	m	1.85

centration in the lake represents the sum of algae and tripton, which cannot be completely separated physically, the tripton concentration in the lake was estimated from the chlorophyll a content per unit dry weight, as applied to the growth of *P. hollandica* (Table 1). The value of 0.021 g chlorophyll a/g dry weight was intermediate compared to contents under light-limiting conditions at low growth rates of the blue-green alga *Oscillatoria agardhii* (Van Liere, 1979) and of the green alga *Scenedesmus protuberans* (Gons, 1977).

The share of tripton in the sestonic dry weight was obtained by subtraction of the derived algal dry weight concentration. During the summers 1984 until 1986 the average tripton concentration was $31.5 \pm 10\ g\ m^{-3}$; on the average more than three-quarters of the total dry weight concentration may have been detritus and inorganic particles. For summer 1984, the same conclusion has been drawn from estimates of the algal biovolumes (Gons, 1987). The specific attenuation coefficients for both epipelic and sestonic matter from L. Loosdrecht (see Methods) were $0.23\ m^2\ g^{-1}$ for dry weight concentrations up to $25\ g\ m^{-3}$ (Table 2). Here the same phenomenon occurred as found with *P. hollandica*; at higher concentrations the slope between dry weight and light atte-

nuation coefficient decreased. The dry weight based specific attenuation coefficients of the three classes of particles were not significantly different; however, we need more data on the low population densities of *P. hollandica*.

The background attenuation coefficient in L. Loosdrecht during summer has been estimated to be $1.2\ m^{-1}$ (Gons, 1987).

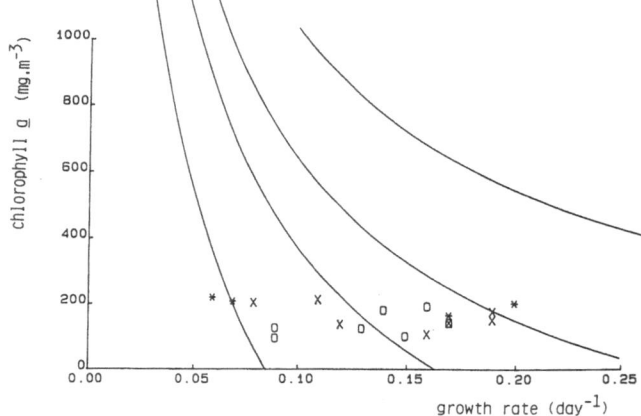

Fig. 3. The upper limits of chlorophyll a concentration as dependent on growth rate and tripton concentration (see text), predicted (eq. 6) for the average insolation during June, July and August in 1984 until 1986. The symbols refer to the measured chlorophyll a concentrations in L. Loosdrecht at the predicted growth rates (eq. 7) in 1984 (circles), 1985 (squares) and 1986 (asterisks).

The upper right curve in Fig. 3 shows the limit of chlorophyll *a* concentration as dependent on growth rate in the absence of tripton. The remaining curves have been computed for minimum (18.7), average (31.5) and maximum (56 g m^{-3}) concentration of tripton estimated for the summers 1984 until 1986, and thus predict the ranges in which chlorophyll *a* may vary for the given set of environmental parameters.

For the sampling dates in June, July and August growth rates of the phytoplankton were determined (eq. 7) from the average insolation value during the two weeks before sampling, the measured chlorophyll *a* and the calculated tripton concentrations. The predicted growth rates were 0.13 ± 0.03, 0.15 ± 0.04, and 0.13 ± 0.07 (day^{-1}) in 1984, 1985 and 1986, respectively.

Growth rate varied between close limits, having much lower values than the maximum specific growth rate of 0.5 day^{-1} (T. Burger-Wiersma, pers. comm.). The ^{14}C-fixation and overall oxygen balance of the water column also indicated low growth rates (Gons *et al.*, 1986a; Rijkeboer & Gons, 1988).

The measured chlorophyll *a* concentrations have been plotted on the scale of predicted growth rates in Fig. 3. Two values only were found outside the given ranges; these may be ascribed to deviation of the light conditions during the sampling interval from the used average insolation. With respect to the insolation and the growth of *P. hollandica* in the absence of other light absorbing and scattering components, a much higher algal biomass can be achieved than was found in L. Loosdrecht. Obviously, the comparatively low levels of chlorophyll *a* have been due mainly to the competition for the available light between the algae and the high concentration of tripton in the lake.

A knowledge of the sources and fate of tripton in the lake is thus of fundamental importance in modelling phytoplankton dynamics.

Acknowledgements

Part of this research was funded by the Ministry of Housing, Physical Planning and Environment. Thanks are due to Frans de Bles, Loes Breebaart and Ronny van Keulen for assistance. Tineke Burger-Wiersma, Andien van den Heuvel and Luuc Mur (Laboratory for Microbiology, University of Amsterdam) are thanked for their kind help in growing *P. hollandica* in the LSE. Dr Martin T. Auer (Michigan Technological University, Houghton, Mich., USA) and Dr L. Somlyody (Research Center for Water Resources Development, Budapest, Hungary) are thanked for valuable comments.

References

Atlas, D. & T. T. Bannister, 1980. Dependence of mean spectral extinction coefficient of phytoplankton on depth, water color, and species. Limnol. Oceanogr. 25: 157–159.

Bannister, T. T., 1974. A general theory of steady state phytoplankton growth in a nutrient saturated mixed layer. Limnol. Oceanogr. 19: 13–30.

Bannister, T. T., 1979. Quantitative description of steady state, nutrient-saturated algal growth, including adaptation. Limnol. Oceanogr. 24: 76–96.

Burger-Wiersma, T., M. Veenhuis, H. Korthals, C. C. M. Van Der Wiel & L. R. Mur, 1986. A new prokaryote containing chlorophylls *a* and *b*. Nature, 320: 262–264.

Flik, B. J. G. & A. Keyzer, 1981. Estimation of the primary production in the Lake Maarsseveen I with an incubator technique. Hydrobiol. Bull., 15: 41–50.

Gons, H. J., 1977. On the light-limited of *Scenedesmus protuberans* Fritsch. Thesis, University of Amsterdam, 120 pp.

Gons, H. J., 1987. De relatie tussen doorzicht en slib in de Loosdrechtse Plassen in verband met de zwemwaternorm. Limnological Institute, Niewersluis. WQL-report 1987-3.

Gons, H. J., R. D. Gulati & L. Van Liere, 1986a. The eutrophic Loosdrecht Lakes: current ecological research and restoration perspectives. Hydrobiol. Bull. 20: 61–75.

Gons, H. J. & L. R. Mur, 1975. An energy balance for algal populations in light-limiting conditions. Verh. int. Ver. Limnol. 19: 2729–2733.

Gons, H. J. & M. Rijkeboer, 1990. Algal growth and loss rates in Lake Loosdrecht: first evaluation of the roles of light and wind on a basis of steady state kinetics. Hydrobiology 191: 129–138.

Gons, H. J., R. Veeningen & R. Van Keulen, 1986b. Effects of wind on a shallow lake ecosystem: redistribution of particles in the Loosdrecht Lakes. Hydrobiol. Bull. 20: 109–120.

Kirk, J. T. O., 1983. Light and photosynthesis in aquatic ecosystems. Cambridge University Press, Cambridge, 401 pp.

Moed, J. R. & G. M. Hallegraeff, 1978. Some problems in the estimation of chlorophyll-a and phaeopigment from pre-

and post acidification spectrophotometric measurements. Int. Revue ges. Hydrobiol. 63: 787–800.

Riegman, R. & L. R. Mur, 1986. Phosphate uptake kinetics of the natural phytoplankton population from the Loosdrecht lakes. Limnol. Oceanogr. 31: 983–988.

Rijkeboer, M., W. A. De Kloet & H. J. Gons, 1986. A comparison of primary production measurements using two laboratory systems with differences in light quality. Hydrobiol. Bull. 20: 93–99.

Rijkeboer, M. & H. J. Gons, 1988. The relationship between oxygen exchange and changes in seston in laboratory scale enclosures. Verh. int. Ver. Limnol. 23: 756–761.

Sweerts, J-P. R. A., H. J. Gons & M. Rijkeboer, 1986. Phosphate uptake capacity of summer phytoplankton of the Loosdrecht Lakes in relation to phosphorus loading and irradiance. Hydrobiol. Bull. 20: 101–107.

Van Liere, L., 1979. On *Oscillatoria agardhii* Gomont. Thesis, University of Amsterdam. 97 pp.

Van Liere, L., 1986a. Loosdrecht Lakes, origin, eutrophication, restoration and research programme. Hydrobiol. Bull. 20: 9–15.

Van Liere, L., 1986b. Water quality research Loosdrecht Lakes; studying and modelling the impact of water management measures on the internal nutrient cycle. Limnological Institute, Nieuwersluis. WQL-report. 155 pp.

Van Liere, L. & L. R. Mur, 1979. Growth kinetics of *Oscillatoria agardhii* Gomont in continuous culture, limited in its growth by the light-energy supply. J. gen. Microbiol. 115: 153–160.

Verduin, J., 1982. Components contributing to light extinction in natural waters: method of isolation. Arch. Hydrobiol. 93: 303–312.

Wofsy, S. C., 1983. A simple model to predict extinction coefficients and phytoplankton biomass in eutrophic waters. Limnol. Oceanogr. 28: 1144–1155.

Hydrobiologia **191**: 249–254, 1990.
P. Bíró and J. F. Talling (eds), Trophic Relationships in Inland Waters.
© 1990 *Kluwer Academic Publishers.*

Stir-up effect of wind on a more-or-less stratified shallow lake phytoplankton community, Lake Balaton, Hungary

Judit Padisák [1], László G.-Tóth [2] & Miklós Rajczy [1]
[1] *Botanical Department of the Hungarian Natural History Museum, H-1476 Budapest, Pf. 222, Hungary;*
[2] *Balaton Limnological Research Institute of the Hungarian Academy of Sciences, H-8237 Tihany, Hungary*

Key words: shallow lakes, phytoplankton, seasonal succession, short-term changes, mixing, filamentous algae, loss processes

Abstract

Microstratification of phytoplankton in the large shallow Lake Balaton (Hungary) was studied during a 24 h period. Dissolved O_2 showed biological stratification; flagellates exhibited a definite circadian rhythm. In the middle of the investigation a heavy storm broke out which was followed by the disappearance of differences between different layers of water. Storm-induced destratification is described by cluster-analysis. Abundances of dominant species changed differently in connection with the storm. Numbers of *Nitzschia* sp. increased due to stirring up from the sediment surface. Numbers of single-celled or colony-forming species (*Cyclotella comta*, *Crucigenia quadrata*, *Coelosphaerium kuetzingianum*) practically did not change. Numbers of all the three dominant filamentous species (*Aphanizomenon flos-aquae* f. *klebahnii*, *Lyngbya limnetica*, *Planctonema lauterbornii*) significantly decreased, which might be attributed to an unknown loss process and was followed by a competitive displacement by algae of small cell size.

Introduction

Numerous studies have been carried out on the pattern and environmental background of phytoplankton seasonal succession (e.g. Sommer, 1981; Crumpton & Wetzel, 1982; Dokulil & Skolaut, 1986). It is becoming increasingly recognised that besides nutrient conditions physical stability of the water column, mixing periods and the related variables as light penetration, turbidity, etc. are important in understanding the major changes in phytoplankton seasonal sequences (e.g. Reynolds *et al.*, 1983).

Most of the studies were carried out in stratified deep lakes, where, due to the stable thermal stratification and long retention time of water, there is no considerable mixing and nutrient supply during the summer (cf. Sommer, 1981). The basic situation in shallow, polymictic lakes is different. During summer, storms from time to time stir-up the sediment of the lake making the water more turbid, and phytoplankton can utilize the higher nutrient content of pore water (Istvánovics, 1988) throughout the ice-free period.

Phytoplankton of shallow lakes also exhibit seasonal behaviour as observed from monthly

(Tamás, 1975) and weekly (Vörös & Kiss, 1985) samples. Competitive displacements between species have been described (Squires *et al.*, 1979), and in another case the community sequence was tested by means of Margalef's (1960) succession theory (Vörös & Kiss, 1985). In spite of these and other efforts it is still not clear whether or not there is a definite separation between two phytoplankton assemblages e.g. the late summer and the autumn ones, and what are the main environmental variables the above-mentioned phenomenon depends on.

On 9–10 August 1977 we carried out a 24-hour study with 3–4 hour sampling intervals in Lake Balaton to obtain detailed data on the phytoplankton diurnal changes. In the middle of the investigation a heavy storm broke out. This paper describes the storm-induced changes in phytoplankton, and discusses the effect of these changes in the seasonal succession of phytoplankton.

Material and methods

Lake Balaton (latitude: 47° 3′ 50″ – 46° 42′ 6″ N, longitude: 17° 14′ 58″ – 18° 10′ 28″ E, altitude: 104.5 m a.s.l.) is the largest shallow lake in Central Europe. The lake has a surface area of 596 km², a length of 77.9 km, a width of $15_{max} - 9_{mean}$ km and a maximal depth (in a very small deep area) of 11 m. Its average depth is 3.14 m, and the theoretical retention time is 3–8 years.

Samples were taken at Tihany, in front of the Limnological Institute in open water, using 1 l Friedinger sampler on 8 occasions during 9–10 August 1977 from four (surface, 1, 2 and 3 m) layers. Water depth at the sampling point was 3.3 m (Fig. 1).

Quantitative algal samples were fixed in Lugol's iodine solution. Algal species in their maximal dimension larger than 10 μm were counted in an inverted microscope, and smaller ones were counted by a membrane-filter method. A minimum of 800 individuals was counted with the two methods, giving a counting accuracy with 95%

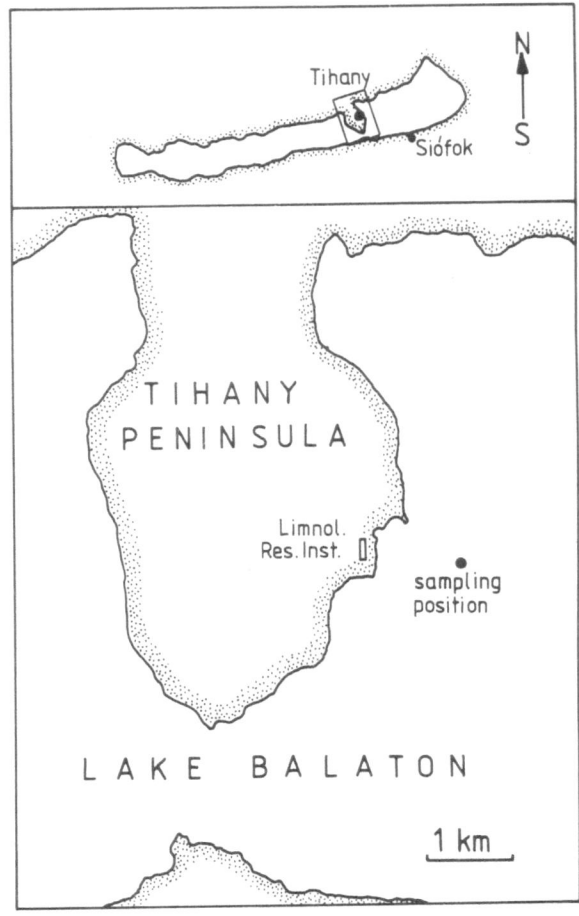

Fig. 1. Location of sampling point.

confidence limits < 5% for the whole phytoplankton community.

The dissolved O_2 content of the water was determined by the classical Winkler method (Szebellédy, 1970).

Wind velocity data were measured at lake level and derive from the records of Siófok Observatory (see Fig. 1) of the Hungarian Meteorological Service.

Similarities between phytoplankton samples were estimated by the function of Hummon (1974), and the similarity matrix was clustered by the WPGMA fusion algorithm (Sneath & Sokal, 1973).

Results

For better understanding of the storm-induced changes in the phytoplankton of this shallow lake, we first have to outline some previous results.

We carried out short-term studies during one month in each of the summers of 1976, 1977, 1978 and 1982, with daily samplings of phytoplankton (G.-Tóth & Padisák, 1978, 1986; Padisák, 1980; G.-Tóth, 1982; Padisák et al., 1988), to investigate the problem mentioned in the introduction. The main conclusion was that the continuous or jump-like nature of phytoplankton succession is strongly dependent on meterological factors. Storms occurring after 5-15 days of calm weather can cause sudden and jump-like changes in phytoplankton community structure, whereas in other years, when storms occur frequently, continuous changes are characteristic. Only in the former case is there a possibility for a shift from physical to biological control of community structure. This result derives from a very detailed diversity- and cluster-analysis study (Padisák et al., 1988) in which two basic types of cluster formation were observed (Fig. 2). Small algae form dense cluster-groups just after storms, and these groups disintegrate on the 4th-5th day after the storm. Large algae begin to cluster on the 3rd-6th day after the storm, and their grouping continues until the next storm. These previous studies pointed to the basic role of storms in the summer succession of shallow lake phytoplankton.

Between 6-8 August 1977, before the time of the

24 h study, the weather was calm. Average wind velocity never reached 4 m s^{-1}. On 9 August, from morning to evening, the weather was also calm (four-hour average wind velocity: 2.1-3.5 m s^{-1}). At about 19 h the weather changed and became stormy. The storm was unexpected and it was not forecast. The four-hour average wind velocity rose to 8 m s^{-1}, the maximum wind velocity reached 25 m s^{-1}.

The dissolved O_2 content of the water regularly showed biological stratification (Fig. 3).

During the calm hours *Ceratium hirundinella* (O.F. Müller) Bergh showed a regular circadial rhythm (more details in Padisák, 1985), but vertical migration of *Rhodomonas minuta* Skuja was also characteristic (see G.-Tóth, 1982).

Similarities between phytoplankton samples, as can be expected in such a closed spatial-temporal study, were strong. The strongest similarity was 0.77, the weakest one was 0.53 (similarity in the function used runs from 0.00 to 1.00). Results of cluster-analysis are given in Fig. 4. Cross-connections in the three upper water strata between 7.00 and 19.00 h are due to the cumulative effect of the above-mentioned vertical migrations. The bottom stratum seemed to be independent. After 19.00 h, when the storm began, algae existing near the bottom were stirred up and mixed with others living in the upper layers. On 10 August stratification did not appear.

Four depth-averages for numbers of the most

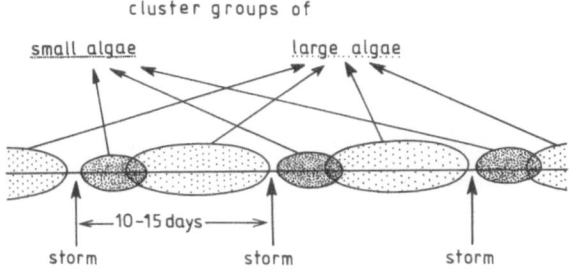

Fig. 2. Summarized and simplified scheme of cluster analyses of phytoplankton sampled day-by-day in Lake Balaton (see text for details).

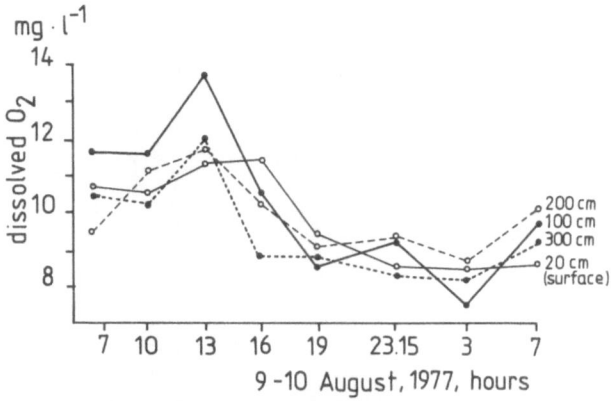

Fig. 3. Changes in dissolved O_2 content in Lake Balaton during 9-10 August 1977.

252

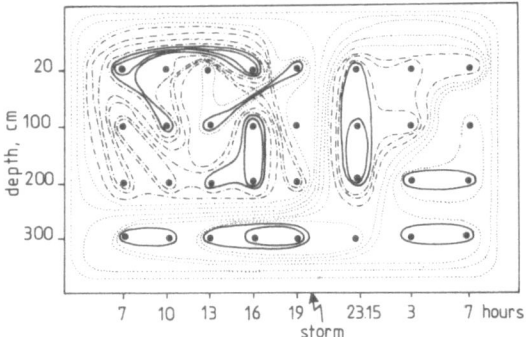

Fig. 4. Result of cluster analysis of phytoplankton. Explanation: Total number of samples was 32, so the number of all fusions is 31. The 12 most tight (similarity range = *sr* = 0.77-0.68) were drawn with continuous lines, the next 9 connections (*sr* = 0.69-0.62) with broken lines and the 10 most weak connections (*sr* = 0.61-0.53) with dotted lines.

abundant phytoplankton species are given in Fig. 5. Numbers of *Coelosphaerium kuetzingianum* Näg., *Cyclotella comta* (Ehr.) Kütz. and *Crucigenia quadrata* Morren did not change significantly (see Lund *et al.*, 1958 for statistical basis) from 7.00 h on 9 August to 7.00 h on 10 August. The numbers of *Nitzschia sp.* increased significantly. This increase is clearly correlated with the storm, which means, considering that the storm began at about sunset, that this population was stirred up from the unsampled layer (3-3.3 m) or from the bottom surface. Numbers of *Aphanizomenon flos-aquae* f. *klebahnii* Elenk., *Lyngbya limnetica* Lemm. and *Planctonema lauterbornii* Schmidle decreased significantly.

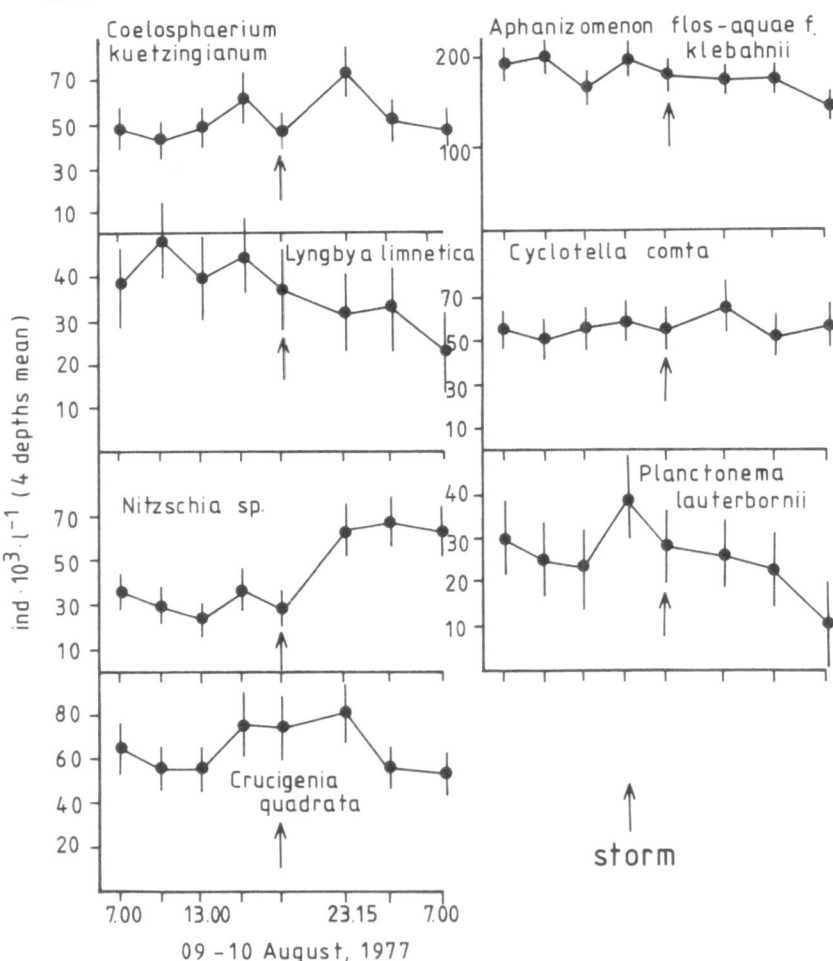

Fig. 5. Numbers (four-depth averages) of most abundant phytoplankton species in Lake Balaton on 9-10 August 1977.

Discussion

Seasonal phytoplankton communities seldom appear as distinct units. In most cases closely allied communities intergrade one with another both spatially and temporally. This is especially true for shallow lake algal communities, where trends in seasonal changes often cannot be described without special efforts and/or by the extensive use of numerical, statistical and mathematical techniques.

Physical and chemical micro-stratification developing in relatively calm periods in Lake Balaton was studied by Entz (1980). Stratification also appears in phytoplankton communities, as demonstrated in this study. The above-bottom stratum of water had sufficient light to permit photosynthesis (see Padisák, 1985 for light data), and the nutrient supply of algae might be better here due to sediment release (Istvánovics, 1988) than in the upper layers, where phosphorus limitation is characteristic in summer (Istvánovics et al., 1986). However, primary production in well-illuminated strata (0.5-2.5 m) was intensive as appears from O_2 profiles. Storms from time-to-time destroy and randomize these two compartments in open water.

Besides that, storms have another effect on the phytoplankton. As can be seen from Fig. 5, numbers of all the three dominant filamentous algae significantly decreased to the morning after the storm, while numbers of single-celled or colony-forming species did not change.

Decreases of filament number were observed in connection with other storms in Lake Balaton. On 9 August 1978 a storm prevented A. flos-aquae f. klebahnii to become overwhelmingly dominant, and after the storm another species of blue-green, Coelosphaerium kuetzingianum, began to develop (Rajczy & Padisák, 1983). In another case, on 23 July 1982, filament number of Anabaenopsis raciborskii Wolosz. decreased from $4.0 \times 10^6 \, l^{-1}$ to $1.9 \times 10^6 \, l^{-1}$ after a storm, but after a 7-day lag the number of filaments began to increase, first fluctuating, then accelerating until the population peaked at 83.7×10^6 filaments l^{-1} (G.-Tóth & Padisák, 1986).

Increase of Anabaena flos-aquae Bréb. ex Born. et Flah. was relatively faster under 'quiescent', stratified conditions than within a mixed water column (Reynolds et al., 1983), although, also in this study mixing at times favoured the net increase of the also filamentous Oscillatoria agardhii var. isothrix.

According to the above-mentioned observations, filamentous algae seem to be poorer competitors under physically unstable conditions. This might be related, at least partly, to the fact that their daily rate of increase is less than might be expected from their individual volumes (Sommer, 1981).

Recent studies have dealt with the effect of mixing (mainly artificial mixing in large enclosures) on the population dynamics of different phytoplankton species. Melosira italica ssp. subarctica O. Müll. (Lund, 1971), Fragilaria crotonensis Kitton (Reynolds, 1983a) and in another study diatoms, desmids and Oscillatoria (Reynolds et al., 1984) grew better within more turbid environments, whereas Eudorina, Sphaerocystis (Reynolds et al., 1984) and Volvox (Reynolds, 1983b) reached their population maxima during quiescent periods. The growth rates of 'other summer species' (like Anabaena, Ceratium and Volvox) were depressed during mixing although they generally maintained their existing biomass (Reynolds et al., 1984).

The importance of loss processes in phytoplankton succession has been the subject of many recent studies (Crumpton & Wetzel, 1982; Reynolds et al., 1982; Reynolds, 1984; Dokulil & Skolaut, 1986). However, it is not clear which loss processes are responsible for the decrease of filament numbers during storms. Attachment to surface, sinking and differences in floating abilities can be ruled out. Grazing by zooplankton or parasitism are not likely to cause such a sudden effect. The most probable factor is some kind of physiological death and then, taking the other short-term results from Lake Balaton into consideration, replacement through nutrient competition.

As the decrease of filament number took place during the night, we may suppose that the mixing and/or the 'blowing effect' of inorganic particles provide physiologically unfavourable conditions.

Considering the role of filamentous algae in zooplankton grazing or in water quality problems (including public health), the question is important not only from the theoretical but also from the practical standpoint, and needs further research.

References

Crumpton, W. G. & R. G. Wetzel, 1982. Effects of differential growth and mortality in the seasonal succession of phytoplankton populations in Lawrence Lake, Michigan. Ecology 63: 1729–1739.

Dokulil, M. & C. Skolaut, 1986. Succession of phytoplankton in a deep stratifying lake: Mondsee, Austria. Hydrobiologia 138: 9–24.

Entz, B., 1980. Physical and chemical microstratification in the shallow Lake Balaton and their possible biotic and abiotic aspects. In Dokulil, M., H. Metz & D. Jewson (eds). Shallow lakes. Contribution to their limnology. Developments in Hydrobiology 3. 63–72. W. Junk Publ. The Hague, Boston, London.

G.-Tóth, L., 1982. Numbers, biomass and production of algae smaller than 10 μm in Lake Balaton. Aquacultura Hungarica (Szarvas) 3: 145–158.

G.-Tóth, L., J. Padisák, 1978. Short-term investigations on the phytoplankton of Lake Balaton at Tihany. Acta Botanica Acad. Sci. Hung. 24: 187–204.

G.-Tóth, L. & J. Padisák, 1986. Meteorological factors affecting the bloom of *Anabaenopsis raciborskii* Wolosz. (Cyanophyta: Hormogonales) in the shallow Lake Balaton, Hungary. J. Plankton Res. 8: 353–363.

Hummon, W. D., 1974. A similarity index based on shared species diversity used to assess temporal and spatial relations among marine Gastrotricha. Oecologia (Berl.) 17: 203–220.

Istvánovics, V., 1988. Seasonal variation of phosphorus release from sediments of shallow Lake Balaton, Hungary. Wat. Res. 22: 1473–1481.

Istvánovics, V., L. Vörös, S. Herodek, L. G.-Tóth & I. Tátrai, 1986. Changes of phosphorus and nitrogen concentrations and of phytoplankton in enriched lake enclosures. Limnol. Oceanogr. 31: 798–811.

Lund, J. W. G., 1971. An artificial alteration of the seasonal cycle of the plankton diatom *Melosira italica* subsp. *subarctica* in an English Lake. J. Ecol. 59: 521–533.

Lund, J. W. G., C. Kipling & E. D. Le Cren, 1958. The inverted microscope method of estimating algal numbers and the statistical basis of estimations by counting. Hydrobiologia 11: 143–170.

Margalef, R., 1960. Temporal succession and spatial heterogeneity in phytoplankton. In Buzzati-Traverso, A. A. (ed). Perspectives in marine biology: 329–349. Univ. Calif. Press, Berkeley, Los Angeles.

Padisák, J., 1980. Short-term studies on the phytoplankton of Lake Balaton in the summers of 1976, 1977 and 1978. Acta Botanica Acad. Sci. Hung. 26: 397–416.

Padisák, J., 1985. Population dynamics of the dinoflagellate *Ceratium hirundinella* in the largest shallow lake of Central Europe, Lake Balaton, Hungary. Freshwat. Biol. 15: 43–52.

Padisák, J., L. G.-Tóth & M. Rajczy, 1988. The role of storms in the summer succession of the phytoplankton community in a shallow lake (Lake Balaton, Hungary). J. Plankton Res. 10: 249–265.

Rajczy, M. & J. Padisák, 1983. Divdrop analysis – a new method for interpretation of species importance in diversity changes. Annal. hist-nat. Mus. natn. Hung. 75: 97–105.

Reynolds, C. S., 1983a. A physiological interpretation of the dynamic responses of populations of a planktonic diatom to physical variability of the environment. New Phytol. 95: 41–53.

Reynolds, C. S., 1983b. Growth rate responses of *Volvox aureus* Ehrenb. (Chlorophyta, Volvocales) to variability in the physical environment. Br. Phycol. J. 18: 433–442.

Reynolds, C. S., 1984. The ecology of freshwater phytoplankton. Cambridge Univ. Press, Cambridge.

Reynolds, C. S., H. R. Morison & C. Butterwick, 1982. The sedimentary flux of phytoplankton in the south basin of Windermere. Limnol. Oceanogr. 27: 1162–1175.

Reynolds, C. S., S. W. Wiseman & M. J. Clarke, 1984. Growth- and loss rate responses of phytoplankton to intermittent artificial mixing and their potential application to the control of planktonic algal biomass. J. Appl. Ecol. 21: 11–39.

Reynolds, C. S., S. W. Wiseman, B. M. Godfrey & C. Butterwick, 1983. Some effects of artificial mixing on the dynamics or phytoplankton populations in large limnetic enclosures. J. Plankton Res. 5: 203–234.

Sneath, P. H. A. & R. R. Sokal, 1973. Numerical taxonomy. Freeman, San Francisco.

Sommer, U., 1981. The role of r- and K-selection in the succession of phytoplankton in Lake Constance. Acta Oecol., Oecol. gen. 2: 327–342.

Squires, L. E., M. C. Whiting, J. D. Brotherson & S. Rushforth, 1979. Competitive displacement as a factor influencing phytoplankton distribution in Utah Lake, Utah. Great Basin Naturalist 39: 245–252.

Szebellédy, L. (ed.), 1970. KGST Egységes vizvizsgálati módszerek (COMECON methods of water chemical analyses) I. Vituki, Budapest.

Tamás, G., 1975. Horizontally occurring phytoplankton investigations in Lake Balaton, 1974. Annal. Inst. Biol. Acad. Sci. Hung. 42: 219–279.

Vörös, L. & N. Kiss, 1985: A fitoplankton szezonális periodicitása és annak összefüggése az eutrofizálódással. Irodalmi áttekintés és esettanulmány (Seasonal periodicity of phytoplankton and its correlation with the eutrophication. Review and case study). In Fekete, G. (ed.) A cönológiai szukcesszió kérdései (Problems of the coenological succession). Biológiai Tanulmányok 12: 121–134, Akadémiai Kiadó, Budapest.

Hydrobiologia **191**: 255, 1990.
P. Bíró and J. F. Talling (eds), Trophic Relationships in Inland Waters.
© 1990 *Kluwer Academic Publishers.*

Heterotrophic flagellates and ciliates in the hypertrophic Lake Søbygaard, Denmark – seasonal trends and methodological difficulties

Annette Petersen*
*National Environmental Research Institute, Freshwater Ecology, Lysbrogade 52, DK-8600, Silkeborg, Denmark (*present address: Danish Development Service, Dept. Videbæk, 2, Hjejlevej, DK-6920 Videbæk, Denmark)*

Key words: ciliates, flagellates, hypertrophic lake, methods

Summary

The seasonal variation in densities of heterotrophic flagellates and ciliates was examined for one year in the shallow hypertrophic Lake Søbygård, Denmark.

Several peaks in numbers of bacteria, heterotrophic flagellates and ciliates were seen in spring and late autumn. The peak of heterotrophic flagellates occurred 3-9 days after the peak in bacteria. Ciliates peaked 3-6 days later than the heterotrophic flagellates. This pattern was not found in summer and autumn, probably due to predation and grazing influences by the macro-zooplankton.

For enumeration of the heterotrophic flagellates a modification of the proflavine staining technique of Haas (1982) was used.

Ciliates were counted on the $< 20 \ \mu m$ and $< 10 \ \mu m$ Lugol-fixed samples after live filtration on monofile nylon nets. During the investigation, however, it was shown that some ciliates were damaged by filtration.

To investigate whether the filtration had any significant effect on total numbers, a series of filtration experiments was performed (Fig. 1).

In this experiment, the live filtration showed a 30-fold underestimation. Several tests during the season confirmed this error tendency.

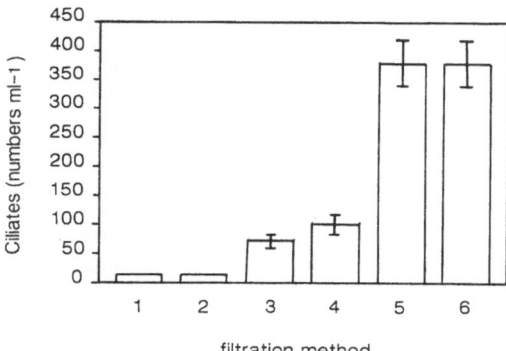

Fig. 1. Effect of filtration on the density of ciliates (mean $\pm$ SD). 1: 1-litre live filtered samples ($< 10 \ \mu m$) fixed in formalin. 2: 1-litre live filtered samples ($< 10 \ \mu m$) fixed in Lugol. 3: Prefixation in formalin followed by filtration (1 litre, 10 μm). 4: Prefixation in Lugol followed by filtration (1 litre, 10 μm). 5: Unfiltered 100-ml sample. 6: Unfiltered 1-liter sample.

In my opinion it is, therefore, only possible to enumerate ciliates in unfiltered samples.

References

Haas, L. W., 1982. Improved epifluorescence microscopy for observing planktonic micro-organisms. Ann. Inst. Oceanogr. 58: 261–266.

Hydrobiologia **191**: 257–263, 1990.
P. Biró and J. F. Talling (eds), Trophic Relationships in Inland Waters.
© 1990 *Kluwer Academic Publishers.*

An assessment of the importance of emergent and floating-leaved macrophytes to trophic status in the Loosdrecht lakes (The Netherlands)

T.J. Malthus [1, a], E.P.H. Best [1, b] & A.G. Dekker [2]

[1] *Limnological Institute, Vijverhof Laboratory, Rijksstraatweg 6, 3631 AC Nieuwersluis, The Netherlands;* [2] *Department of Hydrogeology and Geographical Hydrology, Institute of Earth Sciences, Free University, Postbox 7161, 1007 MC Amsterdam, The Netherlands;* [a] *Present address: Department of Physiology and Environmental Science, School of Agriculture, University of Nottingham, Sutton Bonington, Loughborough, United Kingdom;* [b] *Present address: Centre for Agrobiological Research, Bornsesteeg 65, 6700 AA Wageningen, The Netherlands*

Key words: aquatic macrophytes, eutrophication, lake restoration, phosphorus, aerial photography, shallow lakes

Abstract

The potential importance of the six major emergent and floating-leaved macrophyte species in recycling of sediment phosphorus in the Loosdrecht lakes was studied. Representative plant samples were collected at the time of maximum biomass, and analysed for biomass and carbon, nitrogen and phosphorus contents. Species cover was determined by aerial photography.

Total cover in the seven lakes studied ranged between 2 and 26 percent. For the four main species, biomass per unit area increased with lake trophic status. Consistent differences in C, N and P contents per unit biomass were not observed. Although cover values were small, significant amounts of C, N and P were contained in the macrophytes when compared with maximum sestonic content.

Potential P loads from macrophyte decay were calculated. In Lake Loosdrecht, the P load represented 15 percent of current external P inputs. The potential importance of macrophyte decay to P recycling in the other lakes is greater.

Decay of macrophyte species at the end of the growing season appears to affect autumnal nutrient and chlorophyll *a* levels in the water column of some lakes. The re-establishment of submerged species following lake restoration may increase the importance of this pathway in the lakes.

Introduction

There is now considerable evidence that aquatic macrophytes can take up large quantities of their phosphorus and nitrogen requirements from the sediments via their root systems (Carignan & Kalff, 1980; Barko & Smart, 1980; Smith & Adams, 1986). Since only small fractions of the

nutrients absorbed by aquatic plants are lost via excretion from intact healthy shoots, most nutrient release occurs during senescence and decay (Barko & Smart, 1980). Decomposition studies have shown that large amounts of tissue N and P may be rapidly released during decay which are in a form readily available for direct uptake by phytoplankton (Nichols & Keeney,

1973; Mason & Bryant, 1975; Best *et al.*, 1982; Polunin, 1984; Brock *et al.*, 1985). Thus, aquatic plants represent a potentially important pathway for sediment nutrient recycling in inland waters. Decay-related losses of P from macrophytes have been often overlooked in P budget studies but shown to contribute substantially to internal phosphorus loading in some lakes (Carpenter, 1980; Landers, 1982; Smith & Adams 1986).

In the shallow and eutrophic Loosdrecht lakes, submerged aquatic macrophytes are now rare – a result of low light penetration caused by phytoplankton shading (Best *et al.*, 1984). The macrophyte system is now dominated by emergent and floating-leaved species. Current restoration measures have been targeted at a reduction of the external phosphorus loading (Kal *et al.*, 1984). However, internal recycling of phosphorus within the system appears to be maintaining phytoplankton production at pre-restoration levels (Van Liere *et al.*, 1986). Possible pathways for return of phosphorus lost to the sediments are direct release or via macrophyte growth and decay. Although the direct release of phosphorus from sediments of the Loosdrecht lakes has been studied (Boers, 1986), little is known of the potential contribution from macrophytes.

The aim of this work was to assess the importance of emergent and floating-leaved macrophyte species to the Loosdrecht ecosystem in terms of their cover, biomass and nutrient contents. Because of the complex morphology of the Loosdrecht system, aerial photography was used to determine plant cover. A permanent photographic record can also serve as the basis for long-term comparisons of plant distributions.

Methods

The Loosdrecht lakes consist of seven lakes ranging in trophic status from an oligotrophic water reservoir to the hypertrophic Lakes Breukeleveen and Vuntus (Fig. 1). For a detailed description of the lakes and current restoration measures, see Van Liere & Gulati (1990).

Determination of Plant Cover

On 29 August 1985, at the time of maximum macrophyte biomass, aerial true-colour photographic images were taken over the Loosdrecht Lakes (Rollei SLX camera fitted with a 2C pale yellow filter, Kodak 400 ASA film and 60% stereo overlap). A set of colour prints, measuring 30 × 30 cm, were developed to give a nominal scale of 1:3000. True scale (1:2730) was determined by comparison of selected control points on twenty of the photographs with those on 1:10000 scale topographic maps of the Loosdrecht area.

Lake shoreline and plant distributions were traced onto transparent overlays which served as the basis for detailed plant verification in the field, carried out in each lake during September 1986. The six major plant species, *Phragmites australis*, *Typha angustifolia*, *Schoenoplectus lacustris*, *Nuphar lutea*, *Nymphaea alba*, and *Nymphoides peltata*, were found to be easily distinguishable on the photographs through differences in colour and textural appearance. Stereographic analysis further aided species differentiation by highlighting height differences between the emergent and floating-leaved species. Cover by each species was determined directly from the photographs by planimetry using a digitizing tablet.

Plant samples, biomass and tissue analysis

Plant samples of the six major emergent and floating-leaved species were obtained from three sites, one each in Lakes Eastern Loenderveen, Breukeleveen and Loosdrecht (Fig. 1). The numbers of plants per m² were counted at maximum biomass. Samples consisted of three whole plants excavated by SCUBA divers. In the laboratory the plant mass was divided into above-ground (stems and leaves) and below-ground (roots, rhizomes and root stocks) portions. All samples were rinsed thoroughly and dried to constant weight at 70 °C. Biomass, expressed as ash-free dry weight, was calculated after duplicate subsamples of ground tissue were combusted at

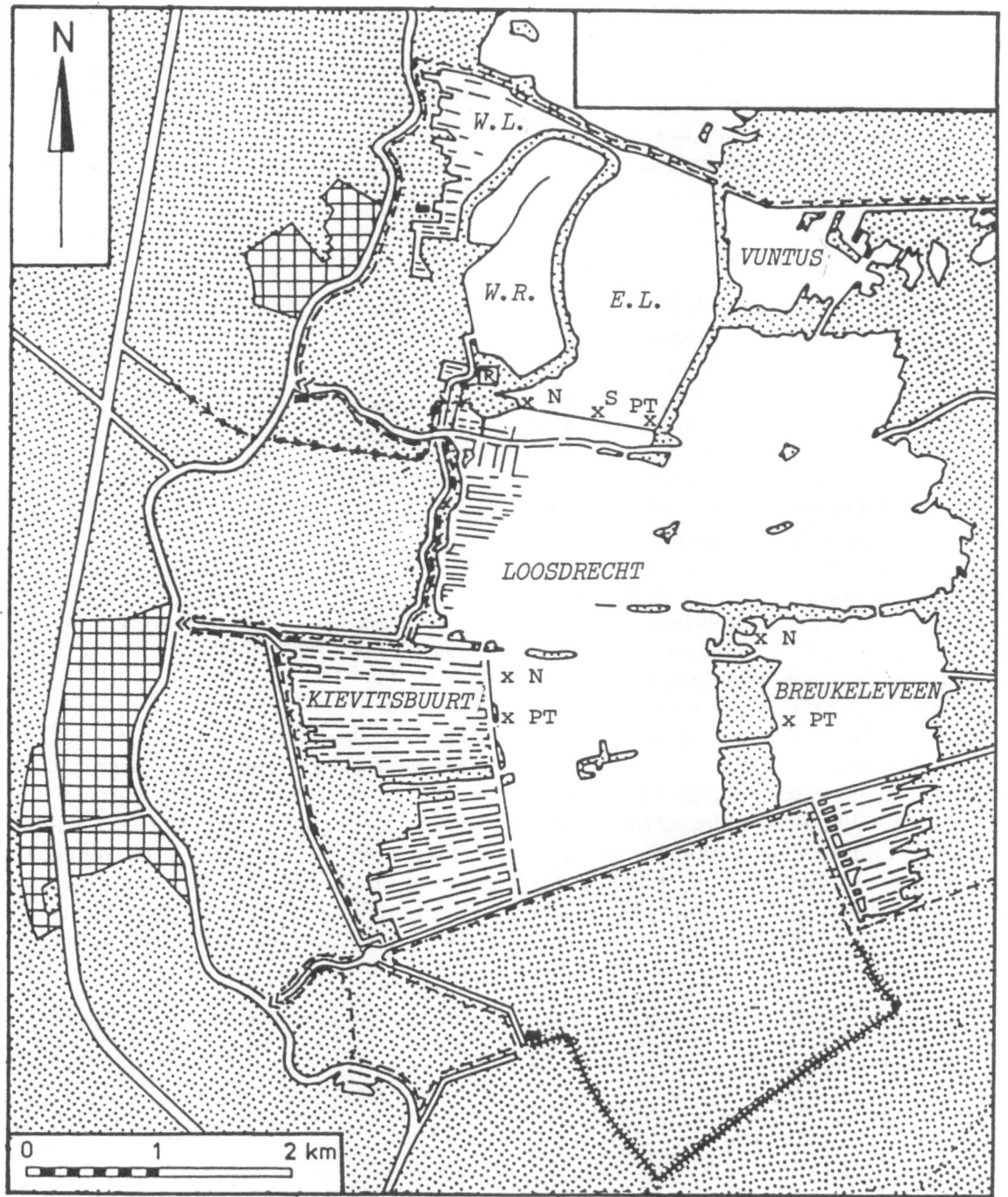

Fig. 1. The Loosdrecht lakes, with plant sampling sites indicated (×). E.L., Lake Eastern Loenderveen; W.R., Water Reservoir; W.L., Lake Western Loenderveen. N, Nymphaeid species; P, *Phragmites*; T, *Typha* and S, *Scirpus*.

450 °C. Biomass was converted to a square metre basis by multiplying the plant weight (average of three) by the number of plants per m². Carbon, phosphorus and nitrogen contents were measured following the methods outlined in Best (1982).

Results

Total cover by aquatic macrophytes in the Loosdrecht Lakes ranged from 2 to 26% of lake surface area (Table 1). Cover by four species, *Phragmites*, *Typha*, *Nuphar* and *Nymphaea*, accounted for more than 95% of total cover in all lakes. There was a high variation in the relative importance of floating-leaved and emergent species in the total cover. For example, in Lakes Eastern Loenderveen and Breukeleveen, cover was predominantly by *Phragmites* and the other emergent macrophytes. In contrast, in Lakes Western Loenderveen and Vuntus floating-leaved species were most abundant. Cover in Lake Loosdrecht was low, perhaps because less of the shoreline was available for colonisation as large areas of shoreline have been developed for harbours and jetties.

For the four major species sampled at all three sites, values of total biomass reflected differences in lake trophic status (Table 2). Biomass was consistently highest in hypertrophic Lake Breukeleveen and lowest in mesotrophic Lake Eastern Loenderveen. Of the individual species,

Table 2. Total plant biomass (ash-free dry weight, AFDW) and average (± SE) above-ground content of carbon, nitrogen and phosphorus of the six major emergent and floating-leaved species. Figures in parentheses are percentages of plant biomass above-ground. EL, Lake Eastern Loenderveen; LL, Lake Loosdrecht; LB., Lake Breukeleveen.

Species		BIOMASS AFDW ($g\,m^{-2}$)	C	N	P
			($mg\,g^{-1}$ AFDW)		
Phragmites australis	EL	4780 (65)			
	LL	5760 (62)	452	21	1.3
	LB	7570 (68)			
Thypha angustifolia	EL	2910 (63)			
	LL	3390 (63)	444	15	1.3
	LB	5020 (78)			
Nuphar lutea	EL	309 (76)			
	LL	332 (72)	421	21	3.2
	LB	742 (75)			
Nymphaea alba	EL	395 (77)			
	LL	500 (62)	418	22	2.3
	LB	533 (45)			
Nymphoides peltata	LB	375 (73)	–	–	1.8
Schoenoplectus lacustris	EL	–	–	–	2.1

biomass values of the two emergent species, *Phragmites* and *Typha*, were among the highest maxima recorded for these species (Scholten, 1985; Meulemans & Roos, 1987). Carbon content per unit biomass was similar for each species at the three sites. Similarly, there were no consistent differences in N and P contents and thus, for clarity, these data have been averaged.

Table 1. Lake morphometric data, and macrophyte coverage expressed as percentages of lake area occupied in the Loosdrecht lakes. Figures in parentheses are total cover in hectares.

Lake	Lake area (ha)	Mean depth (m)	Macrophyte coverage		
			Total	Emergent	Floating
Western Loenderveen	60	1.5*	25.5 (15.3)	5.9	19.5
Water Reservoir	123	10	8.3 (10.2)	4.0	4.3
Eastern Loenderveen	218	2.35	7.8 (16.9)	7.7	0.1
Loosdrecht	979	1.85	1.8 (17.2)	1.1	0.6
Breukeleveen	179	1.45	8.0 (14.3)	7.3	0.7
Vuntus	88	1.45	6.7 (5.9)	3.1	3.6
Kievitsbuurt	128	2.00	1.9 (2.5)	0.2	1.7

* Estimated mean depth

Table 3. Calculated maximum plant biomass and amounts of carbon, nitrogen and phosphorus contained in macrophytes in the Loosdrecht lakes. Figures in parentheses represent total content in the water column calculated from maximum mid-summer sestonic concentrations (Breebaart & Van Liere, 1986). All figures $\times 10^5$ g.

Lake	AFDW	C	N	P
Western Loenderveen	184	54	2.4 (0.6)	0.21 (0.3)
Water Reservoir	238	70	3.2 (11.1)	0.23 (1.4)
Eastern Loenderveen	701	204	9.1 (1.8)	0.59 (0.8)
Loosdrecht	515 (650)	143 (273)	6.3 (33)	0.45 (2.1)
Breukeleveen	807 (126)	260 (110)	10.7 (5.3)	0.77 (0.3)
Vuntus	164 (31)	53 (13)	2.0 (1.7)	0.17 (0.1)
Kievitsbuurt	31	10	0.4	0.05

Macrophyte biomass and nutrient contents were converted to a whole lake basis (Table 3). The biomass data for Lake Eastern Loenderveen (Table 2) were used to calculate whole-lake values in Lakes Western Loenderveen and the Water Reservoir and the Lake Breukeleveen data were used as the basis for Lakes Vuntus and Kievitsbuurt. Total nutrients contained in the above-ground macrophyte system at the time of maximum biomass were calculated from the averaged data on nutrient content (Table 2) and from above-ground plant biomass. The calculated sestonic contents in Table 3 do not represent nutrients only held in phytoplankton. Nevertheless, the data do illustrate that although cover by macrophytes in lakes may be low, a large proportion of C, N and P may be bound within macrophyte tissues. For example, cover in Lakes Breukeleveen and Vuntus is less than 10% of lake surface area, but macrophytes contained a higher content of nutrients than existed in dissolved, algal and detrital fractions in the water column.

Discussion

The results of this study into the importance of present macrophyte cover to the Loosdrecht lakes demonstrates that although cover is low, the emergent and floating-leaved aquatic macrophytes represent an important part of the ecosystem in terms of their biomass and the amounts of N and P that they contain. As such, these macrophytes can act as a mechanism for internal cycling of N and, in particular, P, when they senesce at the end of the growing season.

As an indication of the potential annual load of P from macrophyte decomposition, the above-ground P contents in Table 3 were converted to a loading per unit lake area (Table 4). Although the potential P load from macrophyte decay to Lake Loosdrecht is only 15% compared with external P inputs (external load 300 mg P m^{-2} y^{-1}: Kal *et al.*, 1984), internal recycling of P via macrophyte decomposition is likely to be more important in the other, smaller lakes of the Loosdrecht system.

The calculated P loading rates assume the release of all above-ground P contained in macrophyte tissues and may, therefore, be an overestimation of P release to the water column. However, decomposition studies suggest that between 50 to 90% of bound P is leached as

Table 4. Calculated potential phosphorus loadings from macrophyte decay in the Loosdrecht Lakes.

Lake	P-loading (mg P m^{-2} yr^{-1})
Western Loenderveen	350
Water Reservoir	190
Eastern Loenderveen	270
Loosdrecht	45
Breukeleveen	430
Vuntus	190
Kievitsbuurt	40

inorganic P within the first month of senescence of emergent and floating-leaved species (Best *et al.*, 1982; Polunin, 1984; Brock *et al.*, 1985). Rates of decay, and hence release of nutrients, may also depend on such factors as temperature, oxygen concentration, and trophic status (Anderson, 1978; Polunin, 1984). Without further experimentation it is not known how much of this P enters the water column and what fraction is taken up by decomposers. From studies of submerged macrophytes much of this P appears to be rapidly taken up by phytoplankton in P-limited lakes (Landers, 1982; Smith & Adams, 1986).

The calculated P loads from macrophyte decay may also be an overestimate if translocation of P to roots occurs prior to senescence. Translocation has been shown to occur in emergent species and may be a more important process for P than for N (Spangler *et al.*, 1977; Davis & van der Valk, 1983). Thus decomposition experiments using plant fragments contained in litter bags will be likely to overestimate nutrient release if translocation is a significant process in whole plants. In many studies, however, in which translocation may have occurred, it is difficult to differentiate between P re-allocation from above-ground parts and active root uptake from the sediments after above-ground growth has ceased.

Although the calculated loads from macrophyte decomposition are extreme and assume a 100% release of bound P, there is evidence that macrophyte decay influences processes in the water column in the Loosdrecht lakes system. In Lake Western Loenderveen average chlorophyll *a* concentrations during Sept-Nov 1986 increased seven-fold to 103 mg m^{-3} over previous mid-summer concentrations (14 mg m^{-3}: Amsterdam Municipal Water Authority, unpublished data). Similarly, chlorophyll *a* concentrations in Lake Eastern Loenderveen during the same period increased from 25 to 96 mg m^{-3}. In addition, total phosphorus concentrations were high in both lakes during this period.

No similar major increases in chlorophyll *a* concentrations are evident in routine monitoring data from the eutrophic Lake Loosdrecht and hypertrophic Lakes Breukeleveen and Vuntus. In these lakes, however, chlorophyll *a* concentra-

tions remain high until December and it is possible that macrophyte decomposition helps to maintain high autumn levels. Further, the effect of phosphorus on phytoplankton growth in these lakes may be reduced or even lacking as light limitation appears to limit further increases in phytoplankton growth (Gons & Rijkeboer, 1990). In a mass-balance study of N and P in these lakes, Kal (1986) found a tendency toward higher concentrations of both nutrients during the late growing season (Aug-Oct 1984). he was unable to attribute these increases to any of the measured inputs of P and N to the lakes. An internal release of these nutrients therefore seems likely. Although direct P release from the sediments cannot be discounted (Boers, 1986), macrophyte decomposition may have contributed to increased P and N levels at this time.

From this preliminary study, the present cover of emergent and floating-leaved macrophytes in the Loosdrecht lakes appears to be important for the movement of P from sediments to water column. Further experimental research is required to more accurately quantify the amounts of plant-P that may become available to pelagic phytoplankton from the six species studied. Given that leaf and shoot turnover may be essentially a continuous process (Brock, 1985; Smith & Adams, 1986), it may also be necessary to establish the importance of P supply from decay during the whole growing season.

The implications of the re-establishment of submerged species following lake restoration must also be considered. Submerged macrophytes may be of benefit to lake restoration objectives, particularly through their direct competition with phytoplankton for P in the water column and through the provision of refugia for large-bodied herbivorous zooplankton species (Timms & Moss, 1984). However, macrophyte re-establishment will also further enhance P recycling from the sediments via their decay and via the ability of the plants to lower dissolved oxygen concentrations above the now P-rich sediments (Moss *et al.*, 1986). Thus restoration objectives and the methods of their achievement must be carefully planned.

Acknowledgements

The authors would like to thank Ten Dekkers for carrying out the C, N and P analyses and A. Van Gool for technical assistance. The comments made by Herman Gons, Ramesh Gulati, and Louis Van Liere on earlier drafts are gratefully acknowledged. Partial funding was provided in a grant to T.J.M. from the Beyerinck-Popping Fund.

References

Anderson, F. O., 1978. Effects of nutrient level on decomposition of *Phragmites communis* Trin. Arch. Hydrobiol. 84: 42–54.

Barko, J. W. & R. M. Smart, 1980. Mobilization of sediment phosphorus by submersed freshwater macrophytes. Freshwat. Biol. 10: 229–238.

Best, E. P. H., 1982. The aquatic macrophytes of Lake Vechten. Species composition, spatial distribution and production. Hydrobiologia 95: 65–77.

Best, E. P. H., D. De Vries & A. Reins, 1984. The macrophytes in the Loosdrecht Lakes: a story of their decline in the course of eutrophication. Verh. int. Ver. Limnol. 22: 868–875.

Best, E. P. H., M. Zippin & J. H. A. Dassen, 1982. Studies on decomposition of *Phragmites australis* leaves under laboratory conditions. Hydrobiol. Bull. 16: 21–33.

Boers, P. C. M., 1986. Studying the phosphorus release from the Loosdrecht lakes sediments, using a continuous flow system. Hydrobiol. Bull. 20: 51–60.

Breebaart, L. & L. van Liere, 1986. Waterkwaliteitsonderzoek Loosdrechtse Plassen basisgegevens 1985. Internal Report, Limnological Institute, LI 1986-13, 113 pp.

Brock, Th. C. M., 1985. Ecological studies on Nymphaeid water plants with emphasis on production and decomposition. Ph.D thesis, Catholic University, Nijmegen. 204 pp.

Brock, Th. C. M., M. J. H. De Lyon, E. M. J. M. Van Laar & E. D. M. M. Van Loon, 1985. Field studies on the breakdown of *Nuphar lutea* (L.) Sm. (Nymphaeaceae) and a comparison of three mathematical models for organic weight loss. Aquat. Bot. 21: 1–22.

Carignan, R. & J. Kalff, 1980. Phosphorus sources for aquatic weeds: water or sediments? Science 207: 987–989.

Carpenter, S. R., 1980. Enrichment of Lake Wingra, Wisconsin, by submersed macrophyte decay. Ecology 61: 1145–1155.

Davis, C. D. & A. G. van der Valk, 1983. Uptake and release of nutrients by living and decomposing *Typha glauca* Godr. tissues at Eagle Lake, Iowa. Aquat. Bot. 16: 75–89.

Gons, H. J. & M. Rijkeboer, 1990. Algal growth and loss rates in Lake Loosdrecht: first evaluation of the roles of light and wind on a basis of steady state kinetics. Hydrobiologia 191: 129–138.

Kal, B. F. M., 1986. Monthly mass balances for compartments of the Loosdrecht lakes system: approach and preliminary results. Hydrobiol. Bull. 20: 27–39.

Kal, G. F. M., G. B. Engelen & Th. E. Cappenberg, 1984. Loosdrecht Lakes Restoration project: hydrology and physico-chemical characteristics of the lakes. Verh. int. Ver. Limnol. 22: 835–841.

Landers, D. H., 1982. Effects of naturally senescing aquatic macrophytes on nutrient chemistry and chlorophyll *a* of surrounding waters. Limnol. Oceanogr. 27: 428–439.

Mason, C. F. & R. J. Bryant, 1975. Production, nutrient content and decomposition of *Phragmites communis* Trin. and *Typha angustifolia* L. J. Ecol. 63: 71–95.

Meulemans, J. T. & P. J. Roos, 1987. Production and decomposition in the emergent littoral zone of Lake Maarsseveen. 1, an overview. Hydrobiol. Bull. 21: 61–69.

Moss, B., H. Balls, K. Irvine & J. Stansfield, 1986. Restoration of two lowland lakes by isolation from nutrient-rich water sources with and without removal of sediment. J. appl. Ecol. 23: 391–414.

Nichols, D. S. & D. R. Keeney, 1973. Nitrogen and phosphorus release from decaying water milfoil. Hydrobiologia 42: 509–525.

Polunin, N. V. C., 1984. The decomposition of emergent macrophytes in fresh water. Adv. ecol. Res. 14: 115–116.

Scholten, H. M., 1985. Water purification by aquatic macrophytes: a literature survey. Student report LI-1985-3, Limnological Institute, Nieuwersluis, The Netherlands.

Smith, C. S. & M. S. Adams, 1986. Phosphorus transfer from sediments by *Myriophyllum spicatum*. Limnol. Oceanogr. 31: 1312–1321.

Spangler, F. L., C. W. Fetter & W. E. Sloey, 1977. Phosphorus accumulation – discharge cycles in marshes. Wat. Res. Bull. 13: 1191–1201.

Timms, R. M. & B. Moss, 1984. Prevention of growth of potentially dense phytoplankton populations by zooplankton grazing, in the presence of zooplanktivorous fish, in a shallow wetland ecosystem. Limnol. Oceanogr. 29: 472–486.

Van Liere, L. & R. D. Gulati, 1990. Phosphorus dynamics following restoration measures in the Loosdrecht Lakes. Hydrobiologia 191: 87–95.

Van Liere, L., L. Van Ballegooijen, W. A. De Kloet, K. Siewertsen, P. Kouwenhoven & T. Aldenberg, 1986. Primary production in the various parts of the Loosdrecht lakes. Hydrobiol. Bull. 20: 77–85.

Hydrobiologia **191**: 265–268, 1990.
P. Bíró and J. F. Talling (eds), Trophic Relationships in Inland Waters.
© 1990 *Kluwer Academic Publishers.*

The effect of *Daphnia* on filament length of blue-green algae

Piotr Dawidowicz
Department of Hydrobiology, University of Warsaw, Nowy Swiat 67, 00-046 Warsaw, Poland

Key words: *Daphnia magna*, filamentous blue-green algae, food availability

Abstract

Water from a hypertrophic lake rich in filamentous blue-green algae was passed through a continuous-flow system of aquaria containing *Daphnia magna*, and a control system without *Daphnia*. *Daphnia* caused a significant decrease in the blue-green algal density, and a two-fold reduction in filament length. It is suggested that feeding activity of *Daphnia* may result in an increase in the availability of blue-green filaments to filter-feeding cladocerans.

Introduction

Blue-green algae are poorly utilized by cladocerans due to their low nutritional adequacy (Lampert, 1977; Briand & Mc Cauley, 1978), toxicity (Porter & Orcutt, 1980; Lampert, 1981a, b) and inconvenient size and shape (Gliwicz, 1977; Porter, 1977). Blue-green algal filaments interfere with the filtering processes in cladocerans, thus decreasing food collection efficiency (Gliwicz, 1977; Webster & Peters, 1978; Gliwicz & Siedlar, 1980). Despite these characteristics the density of certain blue-greens (e.g. *Aphanizomenon flos-aquae*) were observed to be effectively controlled by large *Daphnia* (Lynch, 1980). The purpose of the present experiment was to test the hypothesis that large *Daphnia* can break up long algal filaments and make them consequently more available for filter-feeding cladocerans.

Material and methods

Water from the hypertrophic Loosdrecht Lake (The Netherlands) containing up to 1.5×10^5 filaments ml^{-1} of *Oscillatoria limnetica*, *O. agardhii*, *O. redekei*, *Aphanizomenon flos-aquae* and *Lyngbya* sp. (Gulati *et al.*, 1985) was prefiltered through 200 μm mesh net to remove native herbivores, and then pumped at a constant speed to two sets of five 5-l interconnected aquaria. *Daphnia magna* from a laboratory culture was introduced to one set, while the second set served as a control. Every second day the algal filament density and filament length were measured in the inflowing water and in the last aquarium of each set. Two 12-day experiments were performed which differed in flow rate (5.0 and 2.5 l per day) and in the density of the *Daphnia* inoculum (15 and 30 individuals l^{-1}, respectively). Both experiments were done at constant temperature (20 °C) and under artificial light conditions (13 W m^{-2}, 12 h: 12 h light and dark period). At the end of the second experiment, gut contents of *Daphnia* taken from the first and the last aquarium were examined. Ivlev's index of electivity was calculated for different size fractions of blue-green algal filaments.

Results

The concentration of blue-greens in water passing through the system decreased markedly in both experimental and control sets (Table 1). However, the effect of *Daphnia* grazing was also clearly evident and the density of filaments leaving the experimental set was significantly lower than in the control ($p = 0.05$ and $p = 0.001$ for exp. 1 and

Table 1. Density and length of blue-green filaments in water inflowing and outflowing the system (means and standard deviations).

			density · 10^4 ml^{-1}	length (μm)
	inflow		10.4 ± 1.4	124 ± 96
Exp. 1		control	1.8 ± 1.4	129 ± 87
	outflow			103 ± 84
		Daphnia	1.0 ± 0.6	
	inflow		11.0 ± 2.1	101 ± 73
Exp. 2		control	1.5 ± 1.2	108 ± 84
	outflow			42 ± 44
		Daphnia	0.1 ± 0.1	

2 respectively, Student t test). Losses of algae in the control sets were certainly caused by settling, but did not influence the size distribution of filaments (Fig. 1).

The average length of trichomes leaving the control sets was slightly greater than their intial size, probably due to growth of algal colonies in the well illuminated system (Table 1). In the experimental sets filaments longer than 300 μm (1st experiment) or 240 μm (2nd experiment) completely disappeared, the mean filament length decreased and the share of shortest trichomes ($< 60 \mu$m) increased (Table 1, Fig. 1). The effect of *Daphnia* was greater in the second experiment, in which the water residence time and *Daphnia* density were doubled.

Gut content analysis showed that *Daphnia* avoided trichomes longer than 60 μm. Only the shortest filaments ($< 30 \mu$m) were positively selected (Table 2).

Discussion

Filaments of common planktonic blue-green algae are thought to be well protected against grazing by filter-feeding zooplankton, due to their length. Indeed, examination of gut contents indi-

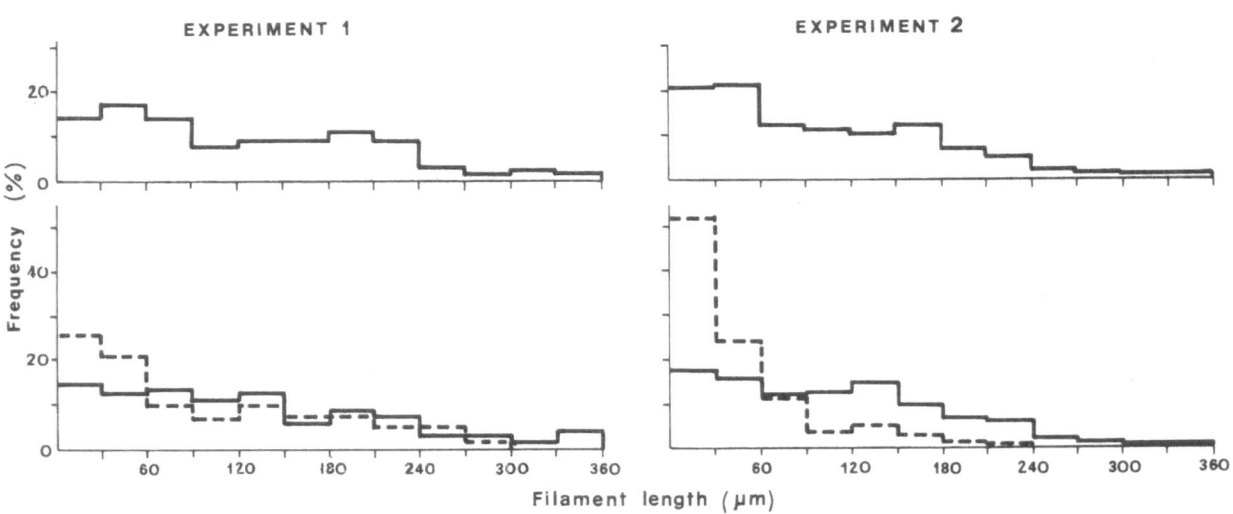

Fig. 1. Frequency distribution of blue-green algal filament length in lake water (top), and in water leaving the control (bottom, solid line) and the *Daphnia* set (bottom, broken line).

Table 2. *Daphnia* feeding on blue-greens.

	Filament length (μm)	% of filaments in gut content of *Daphnia*	Ivlev electivity index
first aquarium	<30	56	+ 0.44
	30–60	18	+ 0.05
	>60	26	– 0.40
last aquarium	<30	83	+ 0.23
	30–60	13	– 0.29
	>60	4	– 0.70

cates that even large *Daphnia* feed mainly on seston particles smaller than 50 μm (Gliwicz, 1969; Nadin-Hurley & Duncan, 1976). Algal filaments longer than 60 μm are strongly avoided both in natural (Bloem & Vijverberg, 1984) and laboratory conditions (the present study). These observations are supported by detailed analysis of the *Daphnia* filtering apparatus, which showed that it is adapted for gathering food particles smaller than 30 μm (Geller & Müller, 1981).

On the other hand, there are data demonstrating the possibility of efficient control of filamentous *Aphanizomenon* blooms by large *Daphnia* populations (Lynch, 1980; Lynch & Shapiro, 1981). According to Holm *et al.* (1983) *Daphnia pulex* may feed on filaments and even flake colonies up to 1.5 mm long. These results were obtained from laboratory experiments with radioactively labelled food. This method proves that long trichomes are removed from the water and ingested, but not necessarily as whole, intact pieces. Data presented here demonstrate the possibility of a shortening of blue-green filaments by *Daphnia*. As a result of being broken, previously inedible algal filaments may be efficiently filtered as small, convenient particles. Short fragments of trichomes can also be utilized by other cladocerans, having a more restricted food size spectrum than *Daphnia*.

Algal colonies larger than the size of the carapace gape in *Daphnia* should not enter the filtering chamber and could not be broken inside. If this assumption is true, carapace gape finally limits the size of colonial, non-toxic blue-greens which could be controlled by *Daphnia*.

Acknowledgments

This work was performed in the Limnological Institute, 'Vijverhof' Laboratory, Rijksstraatweg 6, 3631 AC Nieuwersluis, The Netherlands, and was supported by a grant from the Institute. I thank R.D. Gulati and Z.M. Gliwicz for help and comments. I thank also M.C. Swift for his valuable suggestions and linguistic improvements.

References

Bloem, J. & J. Vijverberg, 1984. Some observations on the diet and food selection of *Daphnia hyalina* (*Cladocera*) in an eutrophic lake. Hydrobiol. Bull. 18: 39–45.

Briand, F. & E. Mc Cauley, 1978. Cybernetic mechanism in lake plankton system: how to control undesirable algae. Nature 273: 228–230.

Geller, W. & H. Müller, 1981. The filtration apparatus of *Cladocera*: filter mesh size and their implications on food selectivity. Oecologia (Berl.) 49: 316–321.

Gliwicz, Z. M., 1969. Studies on the feeding of pelagic zooplankton in lakes varying trophy. Ekol. pol. A. 17: 663–707.

Gliwicz, Z. M., 1977. Food size selection and seasonal succession of filter feeding zooplankton in an eutrophic lake. Ekol. pol. 25: 179–225.

Gliwicz, Z. M. & E. Siedlar, 1980. Food size limitation and algae interfering with food collection in *Daphnia*. Arch. Hydrobiol. 88: 155–177.

Gulati, R. D., K. Siewertsen & G. Postema, 1985. Zooplankton structure and grazing activities in relation to food quality and concentration in Dutch lakes. Arch. Hydrobiol. Beih. Ergebn. Limnol. 21: 91–102.

Holm, N. P., G. G. Ganf & J. Shapiro, 1983. Feeding and assimilation rates of *Daphnia pulex* fed *Aphanizomenon flos-aquae*. Limnol. Oceanogr. 28: 677–687.

Lampert, W., 1977. Studies on the carbon balance of *Daphnia pulex* de Geer as related to environmental conditions. II. The dependence of carbon assimilation on animal size, temperature, food concentration and diet species. Arch. Hydrobiol. Suppl. 48: 310–335.

Lampert, W., 1981a. Inhibitory and toxic effect of blue-green algae on *Daphnia*. Int. Revue ges. Hydrobiol. 66: 285–298.

Lampert, W., 1981b. Toxicity of the blue-green *Microcystis aeruginosa* effective defence mechanism against grazing pressure by *Daphnia*. Verh. int. Ver. Limnol. 21: 1436–1440.

268

Lynch, M., 1980. *Aphanizomenon* blooms: alternate control and cultivation by *Daphnia pulex*. In W. C. Kerfoot (ed.), Evolution and ecology of zooplankton communities. The University Press of New England, Hanover (N.H.); Lond.: 299–304.

Nadin-Hurley, C. M. & A. Duncan, 1976. A comparison of daphnid gut particles with sestonic particles in two Thames Valley reservoirs throughout 1970 and 1971. Freshwat. Biol. 6: 109–123.

Porter, K. G., 1977. The plant-animal interface in freshwater ecosystems. Am. Sci. 65: 159–170.

Porter, K. G. & D. Orcutt, 1980. Nutritional adequacy, manageability, and toxicity as factors that determine the food quality of green and blue-green algae for *Daphnia*. In W. C. Kerfoot (ed.), Evolution and ecology of zooplankton communities. The University Press of New England, Hanover (N.H.); Lond.: 268–281.

Webster, K. E. & R. H. Peters, 1978. Some size-dependent inhibition of larger cladoceran filterers in filamentous suspensions. Limnol. Oceanogr. 23: 1238–1245.

Hydrobiologia **191**: 269–274, 1990.
P. Biró and J. F. Talling (eds), Trophic Relationships in Inland Waters.
© 1990 *Kluwer Academic Publishers.*

Qualitative and quantitative relationships of Amphipoda (Crustacea) living on macrophytes in Lake Balaton (Hungary)

Ilona B. Muskó
Balaton Limnological Research Institute of the Hungarian Academy of Sciences, Tihany, Hungary

Key words: Amphipoda, Lake Balaton, submerged macrophyte, qualitative and quantitative studies

Abstract

Analysis was made of the species composition, egg number per female and the size-frequency of Amphipoda living on the dominating submerged macrophytes (*Potamogeton perfoliatus* and *Myriophyllum spicatum*) at 10 sampling stations of the northern and southern shoreline of Lake Balaton. The dominating Amphipoda at each sampling station was *Corophium curvispinum* (85.9-99.8%, mean: 96.6%). Besides this, two other species – *Dikerogammarus haemobaphes* and *D. villosus* – were also found in the samples. The two *Dikerogammarus* species were found in highest percentage near Keszthely (14.1%); at the other places studied they were under 5%. The number of Amphipoda individuals per g macrophyte dry weight ranged from 5 to 574; the lowest value was found near Keszthely, the highest near B. Mária. The developmental stages of the *C. curvispinum* population differ in different parts of Lake Balaton: the adult (male and female) specimens occurred in highest percentage near Keszthely. The mean number of eggs per *C. curvispinum* female ranged from 2.4 to 6.3, showing differences at the different sampling stations. Regarding the two *Dikerogammarus* species, there were many more *D. haemobaphes* than *D. villosus* individuals at almost every station (mean: 75%). The mean number of eggs per egg-carrying female of *D. haemobaphes* was 11.7, being 19.1 for *D. villosus*.

The total biomass of amphipods (in mg animal dry weight/g macrophyte dry weight) ranged from 1.2 to 59.8. The lowest value was observed near Keszthely (the most hypertrophic basin of Lake Balaton), the highest value near B. Mária.

Introduction

Amphipods play an important role in aquatic ecological systems, participating in self-purification of the water (Dedyu, 1980) and being important food organisms for several fishes (Tölg, 1960; Biró, 1977). As amphipods use detritus as food (Entz, 1943), and the easily stirred up water of Lake Balaton continuously offers detritus for the amphipods living in the littoral zone (Oláh, 1973), the knowledge of the present status of the amphi-

pod population in Lake Balaton may contribute to the estimation of food relationships in the lake.

During our previous investigations it has been established that the dominating species of Amphipoda living on submerged macrophytes was *Corophium curvispinum* sampled around the Tihany-peninsula in August (Muskó, 1986) when the number of individuals to be expected is the greatest as compared with that in other seasons (Biró & Gulyás, 1974).

The aim of the present work was to analyse the

270

species composition, the egg number per female and the size-frequency of Amphipoda living on the dominating submerged macrophyte at 10 sampling stations of the northern and southern shoreline of Lake Balaton in August 1986.

Material and methods

Collecting stations are shown in Figs. 1-5. The sampling points were 10-80 m from the shore, except for station 1 which was 150 m. The dominating submerged macrophyte at every station was *Potamogeton perfoliatus*, except for station 5 where the dominant was *Myriophyllum spicatum*. The stations were situated near the following settlements: 1 – near Alsóörs, 2 – near Balatonudvari, 3 – near Révfülöp, 4 – near Szigliget, 5 – near Keszthely, 6 – near Balatonmária, 7 – near Fonyód, 8 – near Boglárlelle, 9 – near Szántód, 10 – near Zamárdi.

Collections were made in the following way. As the amphipods are quite strongly attached to the substrata (Entz, 1947), the macrophytes were cut carefully by hand-operated lawn mower from a rowing boat, and were put on a metal sieve (mesh size 300 μm) into a bucket with lake water for transport to the laboratory. There the animals were cleaned from plants by tapwater until no specimens could be seen in the tapwater used. The fresh and dry weight of macrophytes was then measured.

The animals were fixed in formalin and after rough selection stored in 70% alcohol. Selection

was according to length (measured from the tip of the rostrum to the base of the telson to the nearest 1 mm). *C. curvispinum* of 3 mm length and longer, as well as the two *Dikerogammarus* species of 6 mm length and longer, were further selected as follows: male, female, egg-carrying female, female with empty brood pouch. The number of eggs present in the brood pouch of the females was counted. The total material was processed in this way, except for station 6 where only ¼ of the total sample was processed. Altogether the data of 32 930 specimens (amphipods) were recorded. To determine the biomass, the dry weight of 3 times 10 or 20 individuals was determined from each size group after drying them at 105 °C for 24 hours. The biomass data were calculated as mg dry weight of animal per dry weight of macrophyte.

Results and discussion

The dominating amphipod at each sampling station was *Corophium curvispinum* (85.9-99.8%, mean 96.6%) (see Fig. 1). Besides this, two other species: *Dikerogammarus haemobaphes* and *D. villosus* were also found in the samples. The latter two species were found in highest percentage near Keszthely (14.1%), whereas at the other places studied they were under 5%. There were many more individuals of *D. haemobaphes* than of *D. villosus* at almost every station (mean: 75% of total *Dikerogammarus* individuals). The dominance of *C. curvispinum* among the amphi-

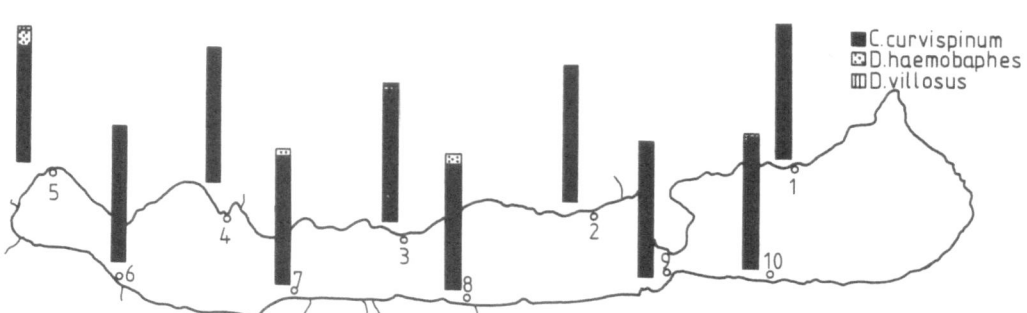

Fig. 1. Distribution of the three amphipod species, as fractions of their total, at the different stations.

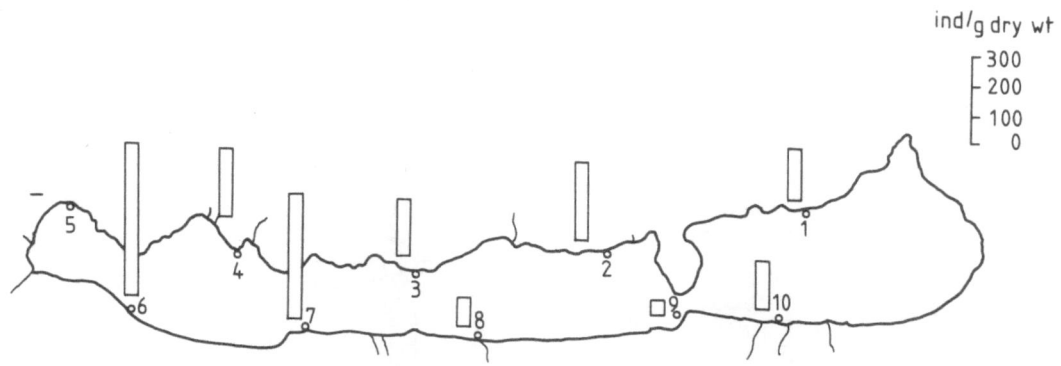

Fig. 2. The total number of amphipoda individuals per g macrophyte dry weight at the different stations.

pods occurring in Lake Balaton has been established since 1934 (Sebestyén, 1934; Entz, 1947; Ponyi, 1957; Biró & Gulyás, 1974).

Our present quantitative analyses showed (see Fig. 2) that the number of amphipod ind. per g macrophyte dry weight ranged from 5 to 574. The lowest value was observed near Keszthely (station 5) and the highest value was found near station 7 (B. Mária). These values are in the same order of magnitude as the respective data obtained at the Tihany-peninsula in August 1983 and 1985 (Muskó, 1986). The present data are also comparable with the respective data of Biró & Gulyás (1974) obtained on *Potamogeton perfoliatus* stands in the open water of Lake Balaton near some of our present sampling stations. There is, however, a tendency for increase of amphipod individuals per macrophyte unit since 1974 (Biró & Gulyás,

1974) until the present investigations in the same seasons, except for station 5 (Keszthely – the hypertrophic part of Lake Balaton), where the number of amphipod individuals reduced sharply (by one order of magnitude: from 7.14 to 0.57 ind. g^{-1} plant wet weight).

From the size-frequency analyses of *C. curvispinum*, at almost every station the juveniles were dominating as compared to the adults, except for station 5 where there were more adults than juveniles (Fig. 3). This means that the stages of development of these amphipods are different at the different sampling stations. According to Entz (1943) the highest number of egg-carrying females of *C. curvispinum* occurred in June. Later the number of young animals was increasing in comparison with adults. The highest number of young animals was found in October when the

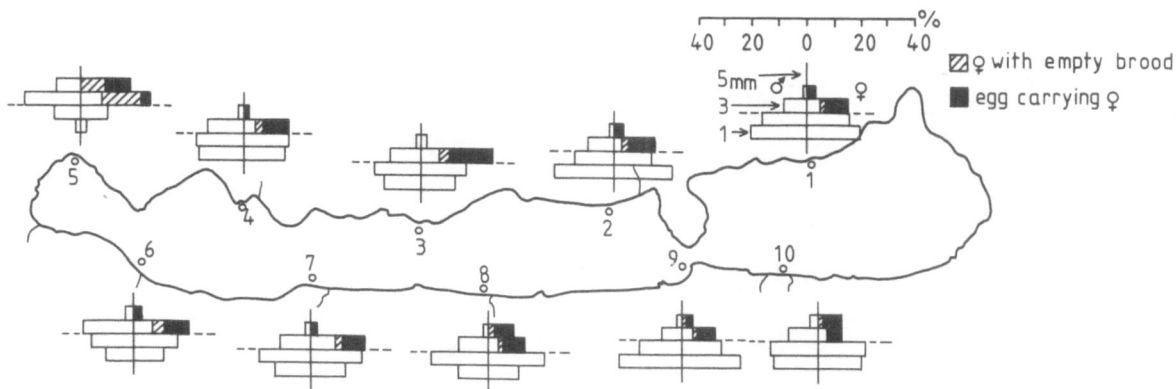

Fig. 3. Size distribution of *C. curvispinum* at the different stations (based on numbers of individuals). Juveniles are plotted below the dashed line.

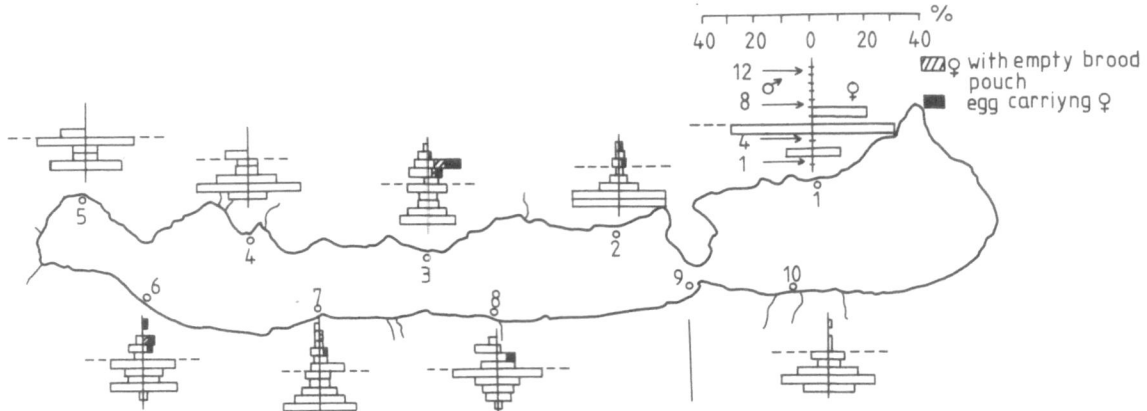

Fig. 4. Size distribution of *D. haemobaphes* at the different stations (based on numbers of individuals). Juveniles are plotted below the dashed line.

number of egg-carrying females was very low. The present life history of the amphipods in Lake Balaton is now under our investigation.

The dominating length measurements of the two *Dikerogammarus* species were 1-4 mm, the percentage of adult specimens being very low at almost every station (Figs. 4, 5). These results are similar to those obtained at the Tihany-peninsula in August 1983 and 1985 (Muskó, 1989).

The mean number of eggs per *C. curvispinum* female ranged from 2.4 to 6.3 (Table 1), with differences at the different sampling stations. The larger females had more eggs (3.3 to 7.6) than the smaller ones (2.0 to 4.6). The relationship between the egg number per egg carrying female (N) and the length of the female (L) can be described by the following equation :

$$N = -5.82 + 2.84L \quad (r = 0.5190, \quad p < 0.1\%)$$

The present values of eggs per egg carrying female are lower than the corresponding data of Entz (1943) and Muskó (1986) obtained around Tihany-peninsula. The former author observed 6 to 7 (mean) eggs per female and during our previous investigations we obtained a mean of 5.8 eggs per female. It is possible that this low value of the egg number per female resulted from the larger proportion of smaller females to larger ones at every station, because the smaller females carry fewer eggs in their brood pouch (as in other species of Amphipoda – see Hynes & Harper,

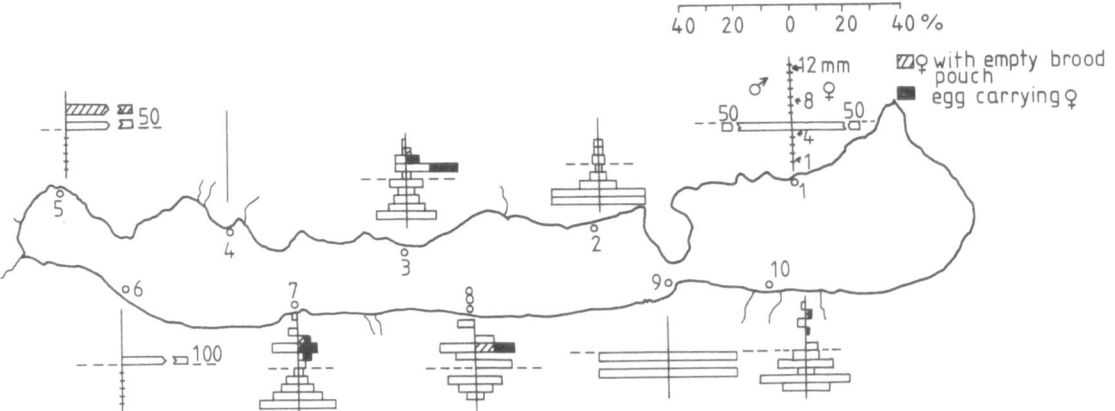

Fig. 5. Size distribution of *D. villosus* at the different stations (based on numbers of individuals). Juveniles are plotted below the dashed line.

Table 1. The mean number of eggs per female of the different amphipod species at the different stations. C.c. = *Corophium curvispinum*, D.h. = *Dikerogammarus haemobaphes*, D.v. = *Dikerogammarus villosus*.

Stations of collections	C.c. mean ± S.D.	D.h. mean*	D.v. mean*
1	3.5 ± 1.98	–	–
2	2.8 ± 1.63	13.5	–
3	2.5 ± 1.29	12.0	13.4
4	2.5 ± 1.45	–	–
5	3.6 ± 1.74	–	–
6	3.5 ± 1.78	9.0	–
7	2.7 ± 1.45	15.7	13.8
8	5.2 ± 2.62	10.0	17.0
9	2.4 ± 1.39	–	–
10	6.3 ± 3.42	–	32.0

* Because of the low number of egg-carrying females found in the samples, statistical data were not available.

Table 2. The biomass data of different amphipod species in mg animal dry weight/g macrophyte dry weight at the different stations. Symbols as in Table 1.

Stations of collections	C.c.	D.h.	D.v.	Total
1	13.8	0.1	0.03	13.9
2	20.7	0.6	–	21.3
3	18.8	3.1	2.5	24.4
4	21.1	0.2	–	21.3
5	0.7	0.4	0.1	1.2
6	57.9	1.8	0.05	59.8
7	38.9	5.8	3.2	47.9
8	10.1	2.8	1.2	14.1
9	3.9	–	0.1	4.0
10	15.4	0.7	1.4	17.5

1972; Hackstein *et al.*, 1986). The reason of such a low number of eggs per female might also be the unsatisfactory feeding conditions.

The mean number of eggs per egg-carrying female of *D. haemobaphes* was 11.7, and 19.1 for *D. villosus*. Egg-carrying females were not found at every station. The present data for *D. haemobaphes* are lower than the respective data obtained around the Tihany-peninsula in 1983 and 1985, whereas the corresponding data for *D. villosus* are the opposite (Muskó, 1986). The relationships between the egg number per egg-carrying female (N) and the length of the female (L) can be described by the following equations:

D. haemobaphes: $N = -25.0 + 4.83L$
$(r = 0.5890, p < 1\%)$
D. villosus: $N = -32.7 + 6.34L$
$(r = 0.7075, p < 0.1\%)$

The total biomass of amphipods (in mg animal dry weight g^{-1} macrophyte dry weight) ranged from 1.2 to 59.8 (Table 2). The lowest value was observed near Keszthely (station 5) and the highest value near B. Mária (station 6). As a consequence of the dominance of *C. curvispinum* at every station, the predominant amphipod

biomass was composed of *C. curvispinum*. The present biomass data are very similar to those obtained by us in 1983 and 1985 around the Tihany-peninsula, where the corresponding values were 2.05 and 2.65 on the *P. perfoliatus* stands, whereas the present value at the nearest station 2 is 2.39 (in the same units).

The qualitative and quantitative relationships, the developmental stages and the egg number of amphipods occurring on submerged macrophytes showed differences in different parts of the littoral zone at the time of our sampling. The most exceptional station was found near Keszthely, where the water is very hypertrophic (Herodek, 1984). In order to estimate the precise trophic relationships of amphipods in Lake Balaton, further investigations are necessary.

To summarize,
(i) The dominant species of amphipod on the submerged macrophyte at each sampling station was *Corophium curvispinum*,
(ii) The hypertrophic Keszthely basin is a very exceptional station because the amphipod biomass was very low there,
(iii) The egg number per female of *C. curvispinum* was very low at each station,
(iv) The value of biomass (in mg animal dry weight/g macrophyte dry weight) ranged from 1.2 to 59.8 at the different stations.

274

References

Biró, K. & P. Gulyás, 1974. Zoological investigations in the open water *Potamogeton perfoliatus* stands of Lake Balaton. Annal. Biol. Tihany 41: 181–203.

Biró, P., 1977. Recent results of ichthyological research of Lake Balaton and its perspectives. Annal. Biol. Tihany 44: 166–171.

Dedyu, I. I., 1980. Amphipods of fresh and salt waters of the south-west part of the USSR. Ya. I. Starobogatov (ed.), Shtiintsa Publishers, Kishinev, 220 pp (In Russian).

Entz, B., 1943. Adatok a magyarországi *Corophium curvispinum* G. O. Sars forma devium alaktanához és biológiájához. MBKM 15: 3–41.

Entz, B., 1947. Qualitative and quantitative studies in the coatings of *Potamogeton perfoliatus* and *Myriophyllum spicatum* in Lake Balaton. MBKM 17: 17–37.

Hackstein, E., M. Schirmer & H. Liebsch, 1986. Untersuchungen zur Populationsdynamik von *Gammarus tigrinus* Sexton (Crustacea: Amphipoda) in der Weser bei Bremen. Arch. Hydrobiol. 105: 443–458.

Herodek, S., 1984. The eutrophication of Lake Balaton: measurements, modeling and management. Verh. int. Ver. Limnol. 22: 1087–1091.

Hynes, H. B. N. & F. Harper, 1972. The life histories of *Gammarus lacustris* and *G. pseudolimnaeus* in Southern Ontario. Crustaceana Suppl. 3: 329–341.

Muskó, I. B., 1986. Qualitative and quantitative studies on the Amphipoda in Lake Balaton at Tihany. BFB-Bericht 58: 61–65.

Muskó, I. B., 1989. Amphipoda (Crustacea) in the littoral zone of Lake Balaton (Hungary). Qualitative and quantitative studies. Int. Rev. ges. Hydrobiol. 74: 1–11.

Oláh, J., 1973. A szerves törmelék kilugozódása, benépesülése és stabilizációja a detrituszformálódás folyamatában: a detritusz mennyisége és tápértéke a Balatonban. Hidrológiai Közlöny 1973 (11): 514–520.

Ponyi, J. E., 1957. Untersuchungen über die Crustaceen der Wasserpflanzbestände im Plattensee. Arch. Hydrobiol. 53: 537–551.

Sebestyén, O., 1934. A vándorkagyló (*Dreissena polymorpha* Pall.) és a szövóbolharák (*Corophium curvispinum* G. O. S. f. *devium* Wundsch) megjelenése és rohamos térfoglalása a Balatonban. MBKM 7: 190–204.

Tölg, I., 1960. Untersuchung der Nahrung von Kaulbarsch-Jungfischen (*Acerina cernua* L.) im Balaton. Annal. Biol. Tihany 27: 147–164.

Hydrobiologia **191**: 275–284, 1990.
P. Biró and J. F. Talling (eds), Trophic Relationships in Inland Waters.
© 1990 *Kluwer Academic Publishers.*

Impact of cyprinids on zooplankton and algae in ten drainable ponds

M-L. Meijer[1], E.H.R.R. Lammens[2], A.J.P. Raat[3], M.P. Grimm[3, 4] & S.H. Hosper[1]
[1] *Institute for Inland Water Management and Waste Water Treatment, P.O. Box 17, 8200 AA Lelystad, The Netherlands;* [2] *Limnological Institute, de Akkers 47, 8536 VD Oosterzee, The Netherlands;* [3] *Organization for Improvement of the Inland Fisheries, P.O. Box 433, 3430 AK Nieuwegein, The Netherlands;* [4] *Present address: Witteveen & Bos, Consulting Engineers, P.O. Box 233, 7400 AE Deventer, The Netherlands*

Key words: biomanipulation, cyprinids, zooplankton, algae, drainable ponds

Abstract

To study the impact of cyprinids on algae, zooplankton and physical and chemical water quality, ten drainable ponds of 0.1 ha (depth 1.3 m) were each divided into two equal parts. One half of each pond was stocked with 0 + cyprinids (bream, carp and roach of 10-15 mm), the other was free of fish. The average biomass of the 0 + fish at draining of the ponds was 466 kg ha^{-1}, to which carp contributed about 80%.

The fish and non-fish compartments showed significant differences. In the non-fish compartments the density of *Daphnia hyalina* was 10-30 ind. l^{-1} and that of *Daphnia magna* 2-4 ind. l^{-1}, whereas in the fish compartments densities were c. 1 ind. l^{-1}. Cyclopoid copepods and *Bosmina longirostris*, however, showed higher densities in the fish compartments. The composition of algae in the two compartments differed only slightly, but the densities were lower in the non-fish compartments. The significant difference in turbidity was probably caused by resuspension of sediment by carp. No significant difference in nutrient concentration between the compartments was found.

Introduction

During the last 30 years the turbidity of the water of most lakes in The Netherlands has increased due to eutrophication. Increased nutrient loading has resulted in large phytoplankton biomasses and the decline of submerged vegetation. This situation led to the disappearance of the predatory fish pike (*Esox lucius*) in many waters (Grimm, 1981). Consequently highly enriched waters contain large biomasses of planktivorous and benthivorous fish, mainly bream (*Abramis brama*) (Lammens, 1986).

Reduction of nutrient-loading is the main policy for eutrophication control. However, the present fish stock seems to hamper the recovery of the lakes. Biomanipulation, involving reduction of the bream biomass, may speed up the recovery process (Shapiro, 1980; Hosper, 1989).

In recent years much attention has been paid to the role of planktivorous fish in the freshwater system. Planktivorous fish feed on large zooplankton species and within one species the larger individuals are preferred (Hbráček *et al.*, 1961; Brooks & Dodson, 1965). Large zooplankters are the most efficient predators on algae. In lakes containing a large biomass of planktivorous fish, grazing on algae by zooplankton may be too low to control algal biomass. Beside zooplankton some fish also feed on the benthic fauna (such as chironomids), causing resuspension of the sediment which leads to a higher turbidity and an increased release of nutrients from the bottom. Furthermore fish may prevent young macrophytes from settling (Ten Winkel & Meulemans, 1984).

Enclosure experiments have shown the effects of fish on the mean individual length of the zooplankton and on the turbidity of the water (Andersson et al., 1978).

In small lakes also, removal of fish has resulted in a shift to large zooplankton species, lower turbidity and lower algal concentrations (de Bernardi & Guissani, 1978; Reinertsen & Olsen, 1984; Shapiro & Wright, 1984; Lazarro, 1987).

Most of the biomanipulation experiments up to now have been done in deep waters. In the shallow Dutch lakes there will be also an impact of resuspension of the bottom.

In order to obtain a better insight into the impact of the removal of planktivorous fish from shallow waters, a number of experiments has been carried out in drainable fishponds. The preliminary results are reported in this paper.

Methods

The experiments took place at the experimental station of the Organization for the Improvement of the Inland Fisheries in Beesd during the period May – November 1986. Ten drainable ponds of 0.1 ha (40 × 25 m, 1.3 m depth) were divided into two compartments by screens (plastic foil/gauze-material). One compartment of each pond was stocked with fish (Table 1): roach (*Rutilus*), bream (*Abramis brama*) and common carp (*Cyprinus carpio*). The other compartment was not stocked with fish.

Table 1. Fish stocking with 0 + individuals in the ponds.

Species	Date	Size	Number
Roach	3 June	10 mm	3150
Bream	3 June	8 mm	6300
Carp	3 July	19 mm	3000

The ponds were inundated by pumping in water from an adjacent polder through a net (2 mm) that excluded fish but allowed algae and zooplankton to enter the ponds. Nitrogen (urea) and phosphate (superphosphate) were added four times to eliminate problems due to nutrient depletion. In total 3.35 g N m^{-2} and 0.5 g P m^{-2} were added to the water during the experimental period.

Oxygen concentration, pH and Secchi transparency were measured *in situ*. Water samples were taken with a transparent tube of 1.5 m length and 5 cm diameter, which sampled the entire water-column. For each compartment samples taken in 25 places were mixed. In the samples nutrient concentrations (silicom, nitrate, nitrite, ammonium, total nitrogen, orthophosphate and total phosphate), chlorophyll *a* concentrations, and species and numbers of algae and zooplankton were determined. Methods employed were according to Dutch Standard Methods with occasional minor modifications. Generally, Dutch Standard Methods (NEN) are in compliance with International Standards (ISO). The nutrients were determined by automated colorimetric methods. For chlorophyll *a* an ethanol extraction was used. To determine zooplankton 25 l of water was filtered over a 120 μm filter. The samples were immediately fixed in 4% formalin. In general subsamples of $^1/_{10}$ were taken with a Kott-sub-sampler.

To determine phytoplankton 1 liter of water was fixed with Lugol solution. The samples were concentrated from 1 liter to 10 ml by sedimentation.

At biweekly intervals were taken in all ponds with a lift net (1.5 × 1.5 m). The fork- and total length were measured. On average 5-20 individuals were caught in each compartment. In four ponds the gut content of the fish was examined. The gut content was fixed in 4% formalin. Individuals of zooplankton, insect larvae and snails were counted; the biomass of the vegetation was estimated.

Monthly, the composition and degree of cover of the submerged vegetation was monitored in each compartment.

In the first week of November the ponds were drained and the fish removed.

Results

The results are illustrated by the time-series of the average of all fish and non-fish compartments. A sign test (Sokal & Rohlf, 1981) was used to quantify the significance of the differences between the compartments for the average of all ponds and for the individual ponds.

Fish

Growth of stocked 0 + cyprinids took place in the first three months of the experiment. At draining of the ponds their total weight ($\pm$ standard deviation) was $466 \pm 74\,\mathrm{kg}\ \mathrm{ha}^{-1}$. Two length classes of common carp were distinguished by length frequency analyses. Carp constituted

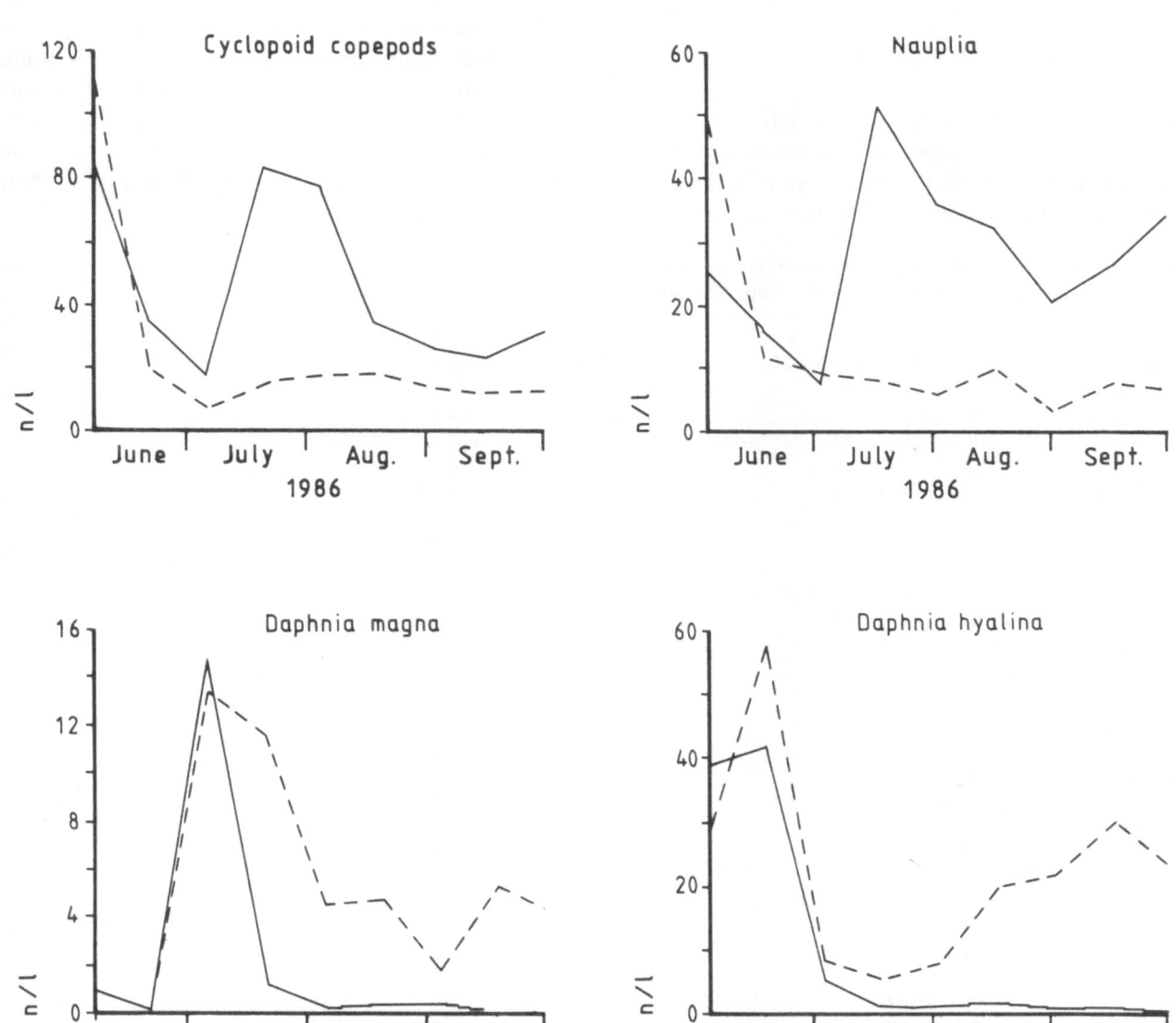

Fig. 1. Density (ind. l^{-1}) of the large cladocera *Daphnia magna* and *hyalina*, cyclopoid copepods and nauplii in fish and non-fish compartments throughout the experiments. Average values of ten ponds.

84.3% of the total fish biomass (large carp, 8-14 cm; 92 ± 32 kg ha^{-1}; small carp, 5.5-8 cm: 304 ± 74 kg ha^{-1}). The biomass of roach (7.4 ± 0.4 cm) at draining was 22 ± 6 kg ha^{-1}, the biomass of bream (6.2 ± 0.4 cm) was 52 ± 26 kg ha^{-1}. Mortality among the stocked bream and roach was high (87% and 95% respectively) but 83% of the stocked common carp survived till the end of the experiment.

Zooplankton

Zooplankton densities were much alike in all ponds. Figure 1 shows the development of the average densities of the main zooplankton species over the 10 ponds. In the fish compartments the densities of large zooplankton species like *Daphnia hyalina* and *Daphnia magna* are significantly lower (Table 2) than in those without fish. In June high densities of the large Cladocera (65 ind. l^{-1}) were found in both compartments. From July on the density of *Daphnia hyalina* rose to 10-30 ind. l^{-1} in the non-fish compartments and that of *Daphnia magna* to 2-4 ind. l^{-1}, whereas in the fish compartments densities were about 1 ind. l^{-1}. However, smaller species like *Bosmina longirostris* and the cyclopoid copepods reached significantly higher numbers (20-30 and 30-80 ind. l^{-1} respectively) in the fish compartments than in those without fish (Table 2). In the non-fish compartments the density of the copepods was 5-20 ind l^{-1}, that of *Bosmina* about 15 ind l^{-1}.

Table 2. Significance of difference between fish and non-fish compartments from 10 July to 30 October (sign test, + = higher in fish compartment, − = lower in fish compartment).

Ponds	1	2	3	4	5	6	7	8	9	10	AVG
Zooplankton											
D. magna	0	−	− − −	− −	0	− −	− − −	0	−	0	− − −
D. hyalina	− −	− − −	− − −	− − −	− − −	− − −	− − −	− − −	− −	− − −	− −
Bosmina	0	0	+ + +	+ +	+ +	0	+ +	+ +	+ +	0	+
Cycl. copep.	+ +	+ +	+ + +	+ +	+ +	0	+ +	+ + +	+ + +	+ +	+ + +
Nauplii	+ + +	+	+ + +	+	+	+	+ + +	+ + +	+ + +	+ + +	+ +
Phytoplankton											
Chlorophyll	+ + +	+ + +	+ + +	+ +	0	0	+	+ + +	+	+ +	+ + +
Tot. number	+ +	+	+ +	+	+	0	+ +	+ +	+ + +	+ + +	+ +
Tot. diatoms	+ +	0	+ +	+ +	−	0	+ +	+	+ + +	+ +	+ + +
Tot. chryso	+ +	+ +	+ + +	0	+ +	+ + +	0	+ + +	+ + +	+	+ +
Tot. crypto	− −	0	0	0	+	0	0	0	0	+	0
Tot. chloroc	+ +	+ +	+ +	+ + +	+ +	0	+ + +	+ +	+ + +	+ +	+ + +
Tot. cyano	−	0	0	0	0	0	− −	0	+ +	0	0
% diatoms	0	0	0	0	0	− −	0	0	0	0	0
% chryso	+ + +	0	+	+	0	+ +	0	+	+	0	+ + +
% crypto	− − −	0	− −	0	−	0	0	−	0	0	−
% chloroc	+ +	+ +	+	0	− −	+	+ +	+	+ +	+ + +	+ +
% cyano	− − −	0	− −	− − −	0	0	− − −	− − −	0	− −	− −
Turbidity	+ + +	+ + +	+ + +	+ + +	+ + +	+ + +	+ + +	+ + +	+ + +	+ + +	+ + +
Nutrients											
orthop	0	0	− −	0	− −	0	0	0	0	0	0
Tot. P	0	0	0	0	0	0	0	0	0	0	0
Sol. N	+ +	+	+ +	0	0	0	−	0	0	0	0
Tot. N	0	0	0	0	0	0	0	0	0	0	0

+ + + $p = 0.005$; + + $p = 0.01$; + $p = 0.05$; 0 = no significant difference.

Phytoplankton

In most ponds chlorophyll *a* concentrations were significantly higher in the fish compartments than in those without fish (Fig. 2, Table 2). However, the difference is small. The chlorophyll *a* concentrations were generally low in both compartments ($5\text{-}15$ mg m^{-3}).

Figure 4 shows the development of the different groups of algae in the ponds. In June the phytoplankton consisted mainly of Cryptophyceae (50%), in July and August diatoms became dominant ($40\text{-}70\%$ abundance), whereas in the final months mainly Cryptophyceae, Chlorococcales and diatoms were present. The diversity of the phytoplankton was high in both compartments; as many as 70 different species were found. The relative abundance of the groups differed slightly from pond to pond. In general the same groups were dominant in all ponds but not at the same time. The difference in the total numbers of algal cells is mainly attributed to diatoms, Chrysophyceae and Chlorococcales (Table 2). The differences in Chrysophyceae are caused by the growth of *Dinobryon divergens* and *D. sertularia*. Several *Navicula*, *Gomphonema* and *Nitzschia* species contributed to the difference in total number of diatoms. *Monoraphidium* was the main species of the Chlorococcales during the whole period. No difference between the two compartments was found in the total number of Cryptophyceae (Table 2). This is probably due to the equal presence in the compartments of the abundant species *Rhodomonas minuta* and *Cryptomonas erosa*. Cyanophyceae show no consistent response to the presence of fish in the ponds (Table 2).

It appears that in these experiments the proportion of different algal groups in the total cell numbers was hardly influenced by the presence of fish. The abundance of the most important algal groups was similar in both compartments (Table 2). Only the relative abundance of the Cyanophyceae, Chrysophyceae and Chlorococcales differed slightly between the compartments.

Turbidity

After the first month the turbidity of the water was lower in the non-fish compartments than in those with fish. In the non-fish compartments Secchi transparency reached the bottom (Fig. 3). In the fish-compartments the Secchi depth varied from

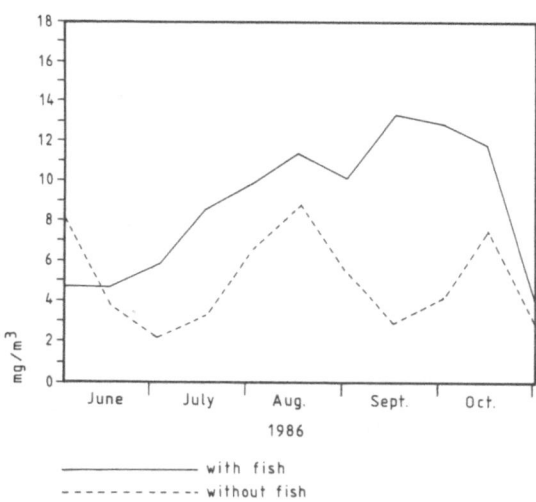

Fig. 2. Chlorophyll *a* concentration (mg m^{-3}) throughout the experiments. Average values of ten ponds.

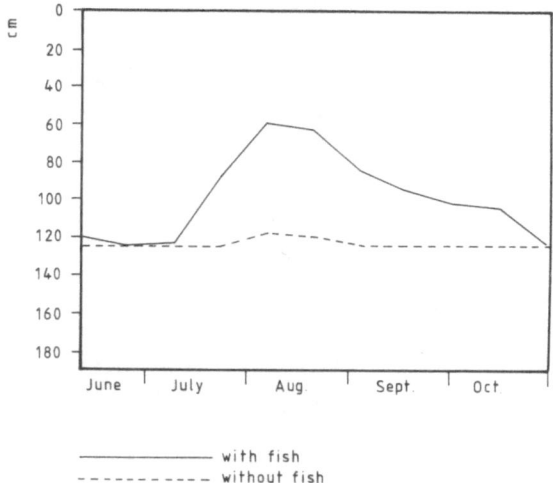

Fig. 3. Secchi disc transparency (cm). Average values of ten ponds.

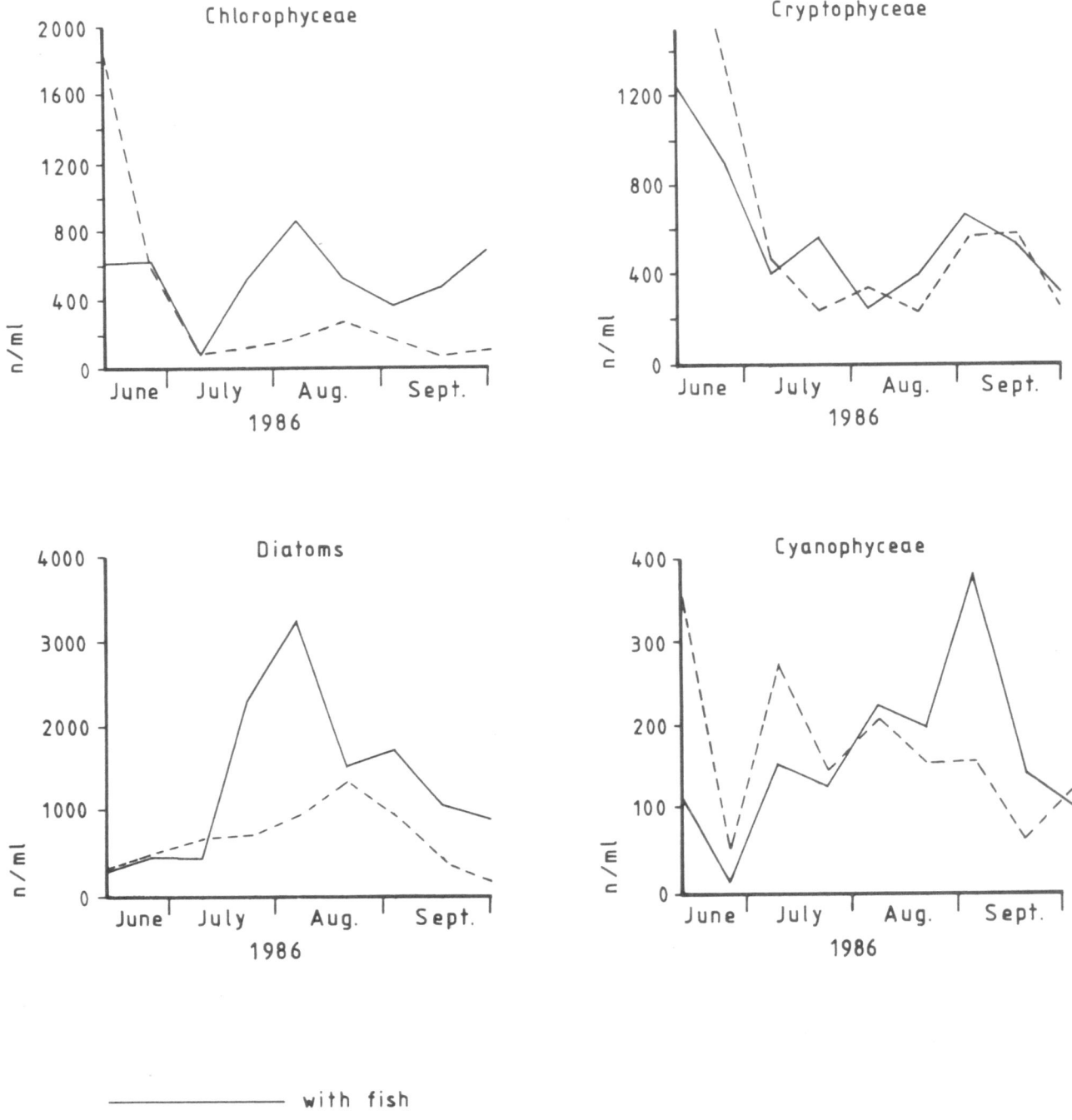

Fig. 4. Density (no. l^{-1}) of the main phytoplankton groups throughout the experiments. Average values of ten ponds.

60 to 120 cm. This turbidity pattern was found in all ponds (Table 2). The small differences in chlorophyll *a* concentrations would hardly contribute to this difference in turbidity. The brown-grey colour of the water in the fish compartments suggests that its turbidity was mainly caused by suspended sediment particles. Some observations indicate that in these experiments carp was mainly responsible for the turbidity caused by resuspension of the sediment. In August two non-fish compartments suddenly turned turbid. Carp were discovered in both compartments. After their removal the compartments cleared within 2 days.

Nutrients

Nutrient concentrations in the fish and non-fish compartments did not show significant differences (Table 2). In spite of the addition of fertilizer to the ponds, nutrient levels stayed fairly low. No apparent rise in nutrient concentrations after fertilisation could be observed. The total phosphate concentration was about 0.08 mg P l^{-1}, total nitrogen concentration was on average 1.2 mg N l^{-1}. The concentrations of soluble nutrients were low; the soluble nitrogen concen-

tration was about 0.02-0.08 mg N l^{-1}, orthophosphate was 0.005-0.015 mg P l^{-1}.

Macrophytes

In all ponds dense stands of submerged vegetation established themselves. *Chara* was the main genus in all ponds from July to September. In one pond *Potamogeton pectinatus* was dominant. The degree of cover seemed slightly higher in the non-fish compartments. The significance of these differences cannot be tested because of lack of data in the fish compartments due to high turbidity.

Gut contents

Figure 5 shows the diet of the fish in four ponds investigated. Bream mainly ate zooplankton during the whole period. Roach fed on zooplankton until the end of July and then switched to macrophytes. Carp ate zooplankton, chironomids, snails and macrophytes. Approximately 25% of the biomass of the diet of carp consisted of zooplankton.

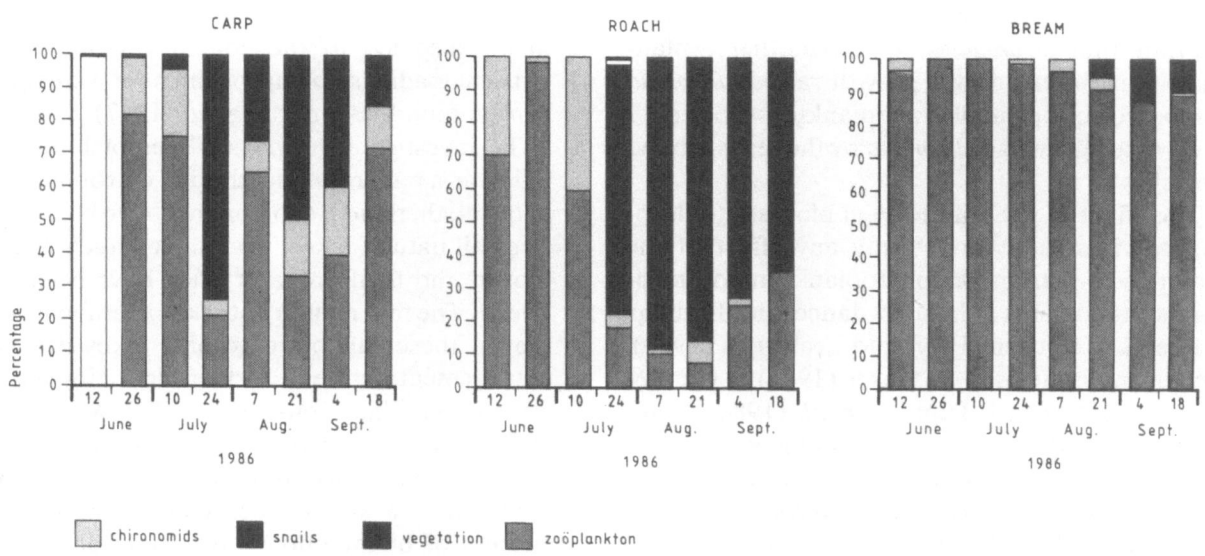

Fig. 5. Relative contribution of different food organisms (%) in the gut of 0 + bream, roach and carp.

An indication of the selectivity of fish when feeding on zooplankton is presented in Fig. 6. Nauplii are strongly negatively selected by all species. *Bosmina* on the other hand is consumed in disproportionally high quantities by bream and roach. It is striking that *Daphnia* seems to be negatively selected by all three fish species, although the time-series presented in Fig. 1 shows that population densities of *Daphnia* are suppressed by the presence of fish.

Discussion

The average figures for the ten ponds give a representative picture for each one of them. The observed pattern is generally alike for all ponds, but the periodicity is not the same. In accordance with the findings of several authors (de Bernardi & Guissani, 1982; Andersson *et al.*, 1978; Reinertsen & Olsen, 1984; Benndorf *et al.*, 1984) the largest zooplankton species are found in the absence of fish. The high densities of small zooplankton species found in the fish compartments are probably due to reduced competition from the larger species (Brooks & Dodson, 1965).

The negative preference for *Daphnia* found in the analysis of the gut content of the fish is not in accordance with the suppression of *Daphnia* by fish. This discrepancy may be caused by differences between the time of the fish sampling and the time that *Daphnia* is eaten. Another explanation may lie in a lower growth rate of *Daphnia* compared to the smaller zooplankton species, or a repressed growth of *Daphnia* by the resuspended particles.

The effect of fish on the algal biomass (chlorophyll *a*) is significant. Hardly any effect of the presence of fish on the phytoplankton composition is found. This is in accordance with findings of Post & McQueen (1987) and Leah *et al.* (1980). However, Reinertsen & Olsen (1984), Lynch & Shapiro (1981) and Benndorf *et al.* (1984) found no decrease of the phytoplankton biomass, but a shift to other algal species in the absence of fish.

The density of *Cryptomonas* and *Rhodomonas* is not influenced by the presence of fish. The equal density of these small flagellates in situations with and without fish was also found by Shapiro & Wright (1984), Benndorf *et al.* (1984) and Andersson *et al.* (1978). The lack of impact of zooplankton on the abundance of these species is probably not due to a negative preference of the zooplankton. Several authors showed that zooplankton grazes on these algae (Lynch & Shapiro, 1981; Porter, 1977). It is more likely that the high growth rate of these algae prevents substantial reduction of their density by zooplankton (Fott, 1975; Reynolds *et al.*, 1982).

Contrary to the results of other authors (Tatrai & Istvánovics, 1986; Anderson *et al.*, 1978; Henrikson *et al.*, 1980) the fish did not influence the total nutrient concentrations in the ponds. This is remarkable because of the large differences in particulate matter content of the water.

In spite of repeated fertilisation, the nutrient concentrations were too low to be sure that no nutrient depletion of the algal growth had occurred. Probably nutrient depletion occurred in August. No reason was found for the fairly low chlorophyll *a* concentrations in the fish compartments in the other months. Added nutrients were quickly buffered by the system. Probably the growth of macrophytes contributed to this.

According to several models (McQueen *et al.*, 1986; Scheffer, 1989) fish removal will be more successful at relatively low nutrient levels. A comparative study of several biomanipulation experiments also led to the conclusion that at high nutrient loadings no improvement of water quality can be found (Benndorf *et al.*, 1987).

To investigate the possibilities of biomanipulation as a restoration method in eutrophic waters in the Netherlands, experiments have been started in small natural lakes. Fish have largely been removed and then predatory fish have been introduced. The first results of these experiments show that in these shallow eutrophic lakes fish stock management can lead to low algal biomass and clear water (Van Donk *et al.*, 1990; Meijer *et al.*, 1989). The experiments will be continued for at least four years to study whether the obtained situation is a stable one. Until now long-term success of biomanipulation is only documented for an oligotrophic lake (Henrikson *et al.*, 1980).

283

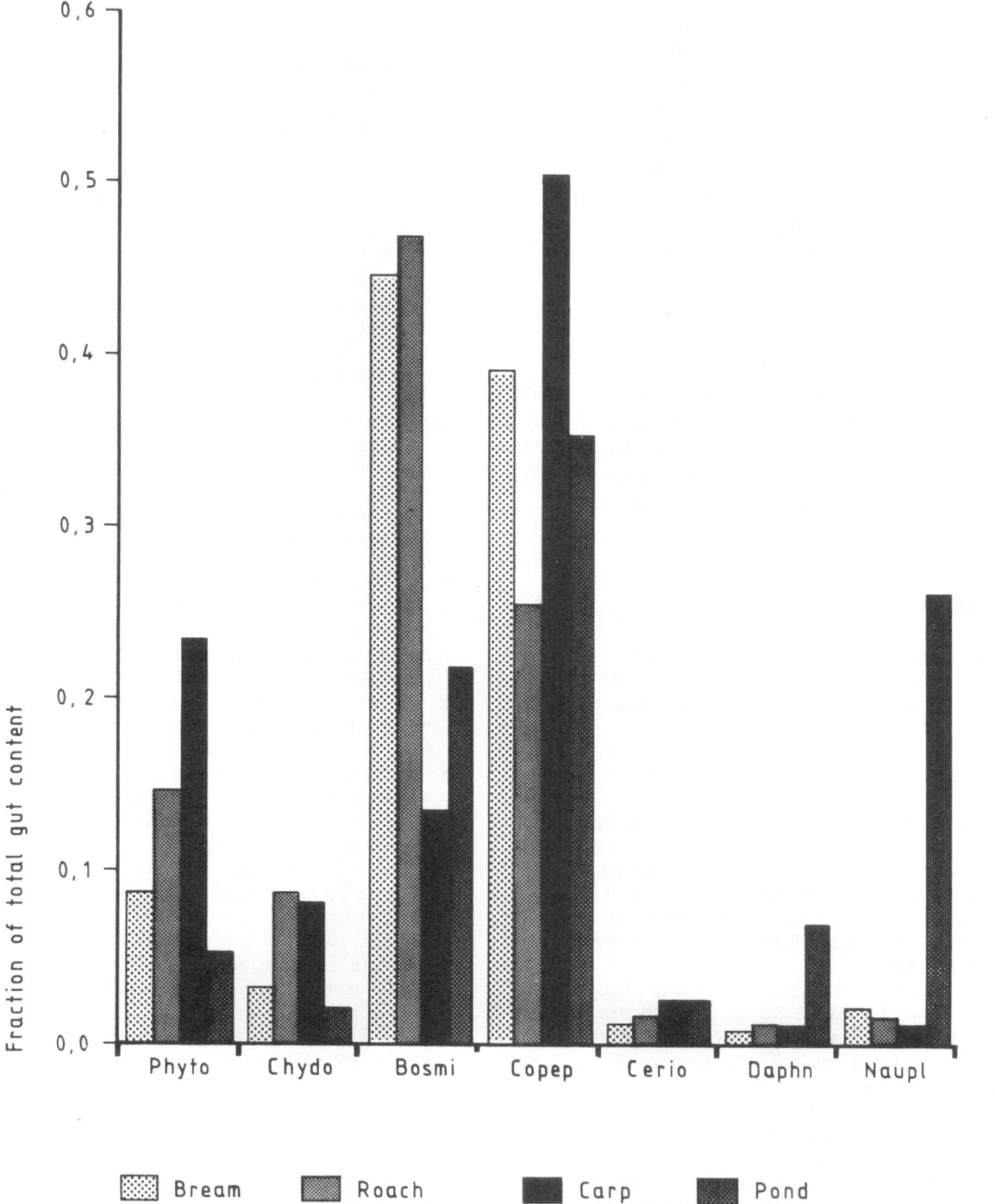

Fig. 6. Selection of zooplankton species by 0 + bream, roach and carp. Percentages of most important species or groups of the total number of zooplankton are given for the gut contents of these fishes and for the ponds.

To conclude, the results reported here indicate that in shallow mesotrophic waters removal of planktivorous and benthivorous 0 + fish may have a pronounced effect on turbidity. Resuspension of sediment by fish is an important factor for the increase of turbidity in the experiments. In addition, the density of large zooplankton species is reduced. A slightly higher chlorophyll *a* content of the water was observed in the presence of fish.

Acknowledgements

We thank F. Jacques and his co-workers for the management of the ponds; A. Klink for the analysis of the phytoplankton; J.A.W. de Wit and M. Scheffer for comment on the manuscript; and R. Helsdingen for preparing the figures.

References

Andersson, G., H. Berggren, G. Cronberg & C. Gelin, 1978. Effects of planktivorous and benthivorous fish on organisms and water chemistry in eutrophic lakes. Hydrobiologia 59: 9–15.

Benndorf, J., 1987. Food web manipulation without nutrient control: A useful strategy in lake restoration? Schweiz. Z. Hydrol. 49: 237–248.

Benndorf, J., H. Kneschke, K. Kossatz & E. Penz, 1984. Manipulation of the pelagic food web by stocking with predaceous fishes. Int. Revue ges. Hydrobiol. 69: 407–428.

Bernardi, R. de & G. Guissani, 1978. Effect of mass fish mortality on zooplankton structure and dynamics in a small Italian lake (Lago di Annone). Verh. int. Ver. Limnol. 20: 1045–1048.

Brooks, J. L. & S. I. Dodson, 1965. Predation, body size and composition of plankton. Science, 150: 28–35.

Fott, J., 1975. Seasonal succession of phytoplankton in the fish pond Smyslov near Blatna, Czechoslovakia. Arch. Hydrobiol./Suppl. 46, Algological Studies 12: 259–279.

Grimm, M. P., 1981. Intraspecific predation as a principal factor controlling the biomass of northern pike (*Esox lucius* L.). Fish. Management 12: 77–80.

Henrikson, L., H. G. Nyman, H. G. Oscarson & J. A. Stenson, 1980. Trophic changes without changes in the external nutrient loading. Hydrobiologia 68: 257–263.

Hosper, S. H., 1989. Biomanipulation, new perspective for restoring shallow, eutrophic lakes in the Netherlands. Hydrobiol. Bull. 23: 5–11.

Hrbáček, J. M. Dvořaková, V. Kořinek & L. Procházková, 1961. Demonstration of the effect of the fish stock on the species composition of zooplankton and the intensity of the metabolism of the whole plankton association. Verh. int. Ver. Limnol. 14: 192–195.

Lammens, E. H. R. R., 1986. Interactions between fished and the structure of fish communities in Dutch shallow eutrophic lakes. Thesis, Wageningen.

Lazarro, X., 1987. A review of planktivorous fishes: their evolution, feeding behaviours, selectivities, and impacts. Hydrobiologia 146: 97–167.

Leah, R. T., B. Moss & D. E. Forrest, 1980. The role of predation in causing major changes in the limnology of a hyper-eutrophic lake. Int. Rev. ges. Hydrobiol. 65: 223–247.

Lynch, M. & J. Shapiro, 1981. Predation, enrichment and phytoplankton community structure. Limn. Oceanogr. 26: 86–102.

McQueen, D. J., J. R. Post & E. L. Mills, 1986. Trophic relationships in freshwater pelagic ecosystems. Can. J. Fish. aquat. Sci. 43: 1571–1581.

Meijer, M.-L., A. J. P. Raat & R. W. Doef, 1989. Restoration by biomanipulation of the eutrophic shallow lake Bleiswijkse Zoom. First results. Hydrobiol. Bull. 23: 49–59.

Post, J. R. & D. J. McQueen, 1987. The impact of planktivorous fish on the structure of a plankton community. Freshwat. Biol. 17: 79–89.

Porter, K. G., 1977. The plant-animal interface in freshwater ecosystems. Amer. Scient. 65: 159–170.

Reinertsen, H. & Y. Olsen, 1984. Effects of fish elimination on the phytoplankton community of a eutrophic lake. Verh. int. Ver. Limnol. 22: 649–657.

Reynolds, C. S., J. M. Thompson, A. J. D. Ferguson & S. W. Wiseman, 1982. Loss processes in the population dynamics of phytoplankton maintained in closed systems. J. Plankton Res. 4: 61–600.

Scheffer, M., 1989. Alternative stable states in eutrophic freshwater ecosystems. A minimal model. Hydrobiol. Bull. 23: 73–85.

Shapiro, J., 1980. The need for more biology in Lake Restoration. In: Lake Restoration. Proc. of a Conference 22-24 Aug. 1978. U.S. Envir. Prot. Ag. 444/5-79-001. Minneapolis.

Shapiro, J. & D. I. Wright, 1984. Lake restoration by biomanipulation Round lake, Minnesota. Freshwat. Biol. 14: 371–383.

Sokal, R. R. & F. J. Rohlf, 1981. Biometry. Freeman and Company NY. 1–859.

Tatrai, I. & V. Istvánovics, 1986. The role of fish in the regulation of nutrient cycling in Lake Balaton, Hungary. Freshwat. Biol. 16: 520–525.

Ten Winkel, E. H. & J. T. Meulemans, 1984. Effect of fish upon submerged vegetation. Hydrobiol. Bull. 18: 157–158.

Van Donk, E., R. D. Gulati & M. P. Grimm, 1990. Restoration by biomanipulation in a small hypertrophic lake: first-year results. Hydrobiologia 191: 285–295.

Hydrobiologia **191**: 285–295, 1990.
P. Biró and J. F. Talling (eds), Trophic Relationships in Inland Waters.
© 1990 *Kluwer Academic Publishers.*

Restoration by biomanipulation in a small hypertrophic lake: first-year results

E. Van Donk[1], R.D. Gulati[2] & M.P. Grimm[3]
[1] *Provincial Waterboard of Utrecht, Postbox 80300, 3508 TH Utrecht, The Netherlands (correspondence address)*; [2] *Limnological Institute 'Vijverhof' Laboratory, Rijksstraatweg 6, 3631 AC Nieuwersluis, The Netherlands*; [3] *Organization for the Improvement of the Inland Fisheries, Postbox 433, 3430 AK Nieuwegein, The Netherlands*

Key words: biomanipulation, lake restoration, phytoplankton, zooplankton, grazing, fish

Abstract

Biomanipulation was carried out in order to improve the water quality of the small hypertrophic Lake Zwemlust (1.5 ha; mean depth 1.5 m). In March 1987 the lake was drained to facilitate the elimination of fish. Fish populations were dominated by planktivorous and benthivorous species (total stock c. 1500 kg) and were collected by seine- and electro-fishing. The lake was subsequently re-stocked with 1500 northern pike fingerlings (*Esox lucius* L.) and a low density of adult rudd (*Scardinius erythrophthalmus*). The offspring of the rudd served as food for the predator pike. Stacks of *Salix* twigs, roots of *Nuphar lutea* and plantlets of *Chara globularis* were brought in as refuge and spawning grounds for the pike, as well as shelter for the zooplankton.

The impact of this biomanipulation on the light penetration, phytoplankton density, macrophytes, zooplankton and fish communities and on nutrient concentrations was monitored from March 1987 onwards. This paper presents the results in the first year after biomanipulation.

The abundance of phytoplankton in the first summer (1987) after this biomanipulation was very low, and consequently accompanied by increase of Secchi-disc transparency and drastic decline of chlorophyll *a* concentration.

The submerged vegetation remained scarce, with only 5% of the bottom covered by macrophytes at the end of the season.

Zooplankters became more abundant and there was a shift from rotifers to cladocerans, comprised mainly of *Daphnia* and *Bosmina* species, the former including at least 3 species.

The offspring of the stocked rudd was present in the lake from the end of August 1987. Only 19% of the stocked pike survived the first year.

Bioassays and experiments with zooplankton community grazing showed that the grazing pressure imposed by the zooplankton community was able to keep chlorophyll *a* concentrations and algal abundance to low levels, even in the presence of very high concentrations of inorganic N and P. The total nutrient level increased after biomanipulation, probably due to increased release from the sediment by bioturbation, the biomass of chironomids being high.

At the end of 1987 Lake Zwemlust was still in an unstable stage. A new fish population dominated by piscivores, intended to control the planktivorous and benthivorous fish, and the submerged macrophytes did not yet stabilize.

286

Introduction

Most inland waters in The Netherlands suffer from a surfeit of nutrients, especially phosphorus. The increased nutrient loading has resulted in high algal biomass, and a decline or even disappearance of submerged macrophytes due to overshadowing by algae. For reducing this algal biomass various restoration techniques have been developed. In the past these included only chemical and physical manipulations like dilution, sediment dredging (Björk, 1985) and reduction of phosphorus inputs by precipitation techniques (Van Liere, 1986). However, despite some success of conventional methods, alternate restoration methods like biomanipulation have been developed (Shapiro *et al.*, 1975).

Biomanipulation includes a series of approaches aimed at improving the 'economy' of lakes by manipulation of their trophic structure and thus the functioning of the ecosystem (Shapiro *et al.*, 1975; Shapiro, 1980a, b; Shapiro *et al.*, 1982; Shapiro & Wright, 1984). These approaches are needed to overcome the high costs, delay in response and frequent inapplicability of conventional techniques. Only recently biomanipulation has been recognized as an additional restorative technique in The Netherlands (Hosper *et al.*, 1987).

Lake eutrophication in The Netherlands is accompanied by a decline of the predatory fish pike (*Esox lucius* L.) because of the decrease of submerged macrophytic vegetation (Grimm, 1981). The pike requires submerged vegetation as a refuge and spawning ground. Suppression of predatory fish resulted in an increase of planktivorous and benthivorous fish biomass formed mainly of bream, *Abramis brama* (Lammens, 1986). Several studies have shown that planktivorous fish can cause substantial changes in the zooplankton structure (De Bernardi & Giussani, 1975; Zaret, 1980), which in turn may affect the phytoplankton community (Shapiro *et al.*, 1975; Andersson *et al.*, 1978; Lynch & Shapiro, 1981; De Bernardi *et al.*, 1982; Lazzaro, 1987). Planktivorous fish feed on large zooplankton species and within one species the larger individuals are preferred (Hrbáček *et al.*, 1961; Brooks & Dodson, 1965).

Most biomanipulation experiments are based on increasing the biomass of large filter-feeding zooplankton and thus creating heavy grazing pressure on the phytoplankton. These experiments, although usually carried out in enclosures, confirm that biomanipulation is a useful technique for the rehabilitation of lakes. The technique has, however, not been widely applied (Stenson *et al.*, 1978; Benndorf *et al.*, 1984; Reinertsen & Olsen, 1984; Shapiro & Wright, 1984).

This paper describes the results of the first year after biomanipulation in the shallow, hypertrophic Lake Zwemlust at Nieuwersluis (The Netherlands).

Description of the lake

Lake Zwemlust is a small (1.5 ha) and shallow (mean depth, 1.5 m; maximum depth, 2.5 m) lake, located in the Province of Utrecht (Fig. 1). The lake receives nutrient-rich seepage water from the hypertrophic River Vecht, flowing some 50 m beside the lake; this river receives a high nutrient loading from the River Rhine, agricultural run-off and several sewage plants.

Until the mid-1960's the lake was characterized by high transparency, low algal densities and a large area covered by submerged vegetation (Le Cosquino de Bussy, 1968); for this period information on the fish population is not available. However, based on the area covered by surface vegetation, it is reasonable to assume that the population density of northern pike was high at that time (Grimm, 1983). Later the water quality deteriorated, high biomass of phytoplankton developed, submerged vegetation disappeared and the fish population structure changed when planktivores, mainly bream (*Abramis brama*), became dominant. At the onset of the biomanipulation vegetation-belts of *Phragmites* sp. were present.

The lake is important to the surrounding community because of its intensive recreational use, especially for swimming (100-3000 visitors a day).

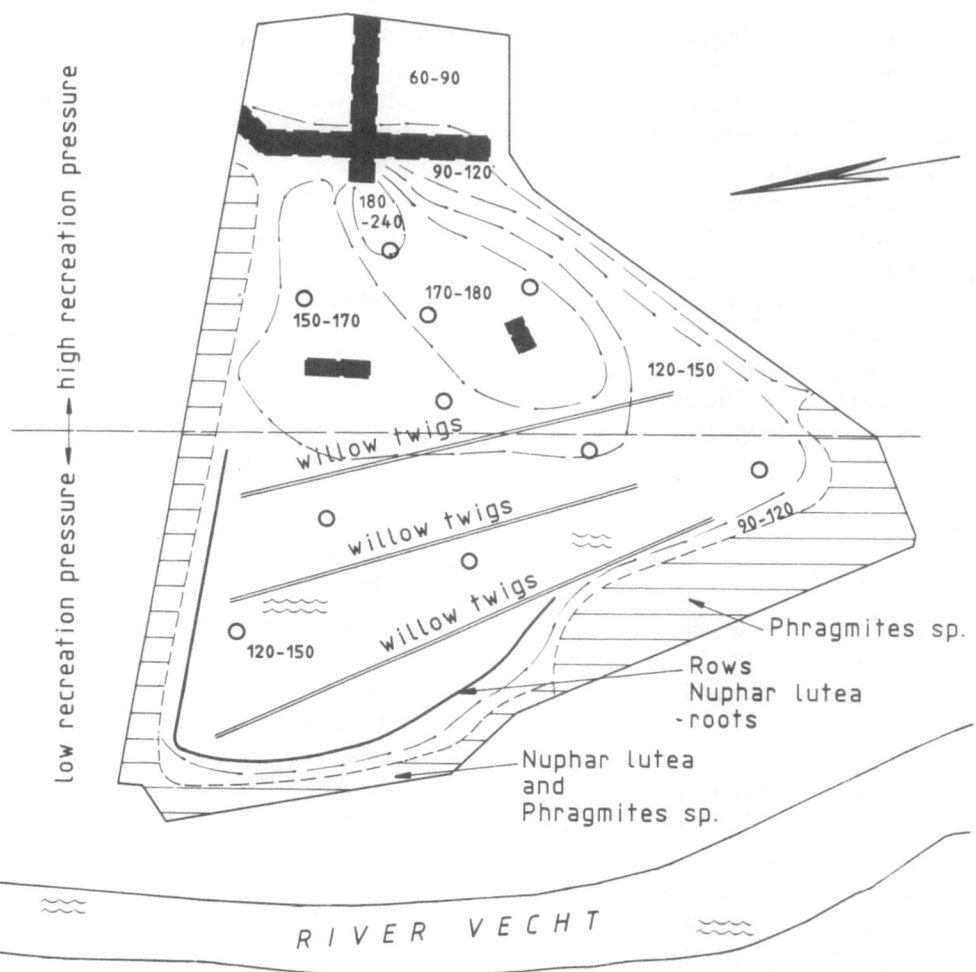

Fig. 1. Schematic map of Lake Zwemlust: sampling stations (○), arrangement of *Salix* (willow) twigs, marcophytes and depths (in cm) are given.

Problems of water quality occurred during the summer when blooms of cyanobacteria (especially *Microcystis aeruginosa*) developed. Secchi-disc transparency in summers before the present biomanipulation ranged from 0.1 to 0.3 m (Fig. 2). In The Netherlands, a Secchi-depth of ≥ 1 m is obligatory for swimming waters. In the past several restorative techniques have been attempted to reduce algal blooms in Lake Zwem-lust, including dredging of the sediments and the application in 1968 of the herbicide Karmex AA 80% (diuron) (Le Cosquino de Bussy, 1968). These techniques have had no success and the algal blooms persisted.

Methods

Biomanipulation operations

During 23-26 March 1987 the lake was drained by pumping. Before that, fishes were removed, as far as possible, by seine- and electro-fishing and the rest were pumped out. On 26 and 27 March, when the lake was empty, 280 *Salix*-twig stacks were placed (see Fig. 1), of which 170 were fixed to the bottom and 110 remained free, so they could float after refilling the lake. In three days after the lake was emptied, seepage-water from the River Vecht refilled the lake. The *Salix*-twig bundles are likely to provide shelter for pike fingerlings and spawning grounds for the adults. Fifteen hundred fingerlings of pike (4 cm) were stocked at the end of April. Further, 140 adult rudd (*Scardinius erythrophthalmus*) were introduced so the offspring could serve as food for the pike. The lake was divided in two parts by the floating structures: in one part a high recreation pressure was allowed but in the other there was no direct recreation activity (Fig. 1). Two hundred roots of yellow lily (*Nuphar lutea*) and plantlets of *Chara* were planted near the shore (Fig. 1). The zooplankters *Daphnia magna* and *Daphnia hyalina* were also introduced (c. 1 kg wet weight).

Sampling and sample analysis

In 1986, i.e. before biomanipulation, samples were taken fortnightly between May and September at one station in the deepest part of the lake. In 1987 the entire water column was sampled simultaneously with a transparent, perspex tube of 1.5 m length and 5 cm diameter. The samples were taken at ten stations (see Fig. 1) and mixed. The mixed sample from these stations was subsampled for the analysis of nutrients (nitrate, nitrite, ammonium and orthophosphate), total phosphorus, chlorophyll *a*, phytoplankton and zooplankton. Temperature, conductivity, pH, oxygen concentration and Secchi-disc transparency were measured *in situ*.

Methods employed for chemical analysis were according to the Dutch Standard Methods (NEN) which comply with The International Standards (ISO). The nutrients were determined by automated colorimetric methods. For chlorophyll *a* an ethanol extraction was used.

To determine the concentration of zooplankton, 25-l of lake water was filtered through 33 μm gauze. The samples were immediately preserved in 4% formalin and subsamples were counted. To determine phytoplankton 1 litre of water was fixed with Lugol solution and the cells allowed to settle for three days. The top 900 ml were then discarded by decanting carefully and the remaining 100 ml were mixed. From this ten-times con-

centrated sample, subsamples were taken for counting.

Fishes were caught fortnightly with a fyke-net. The fork- and total lengths were measured on live fish which were then set free. In June 1987, at 63 locations in the sediment, chironomid larvae were sampled with a perspex tube.

The growth of macrophytes was determined by monitoring monthly the composition and degree of coverage by the submerged vegetation in the lake.

Bioassay experiments

Bioassays with the phytoplankton community from the lake were carried out in June 1987 *in vitro* under the prevalent water temperature (16 °C) and light conditions (cool white fluorescent tubes at 25 W m^{-2}; 14 h light-10 h dark). Eight 1-litre Pyrex flasks were filled with lake water and filtered through 100 μm gauze. Two flasks were filled with unfiltered water. Five combinations (Reference (blank); + PO$_4$; + NO$_3$; + NH$_4$; and + zooplankton) were incubated in replicates (Table 1).

The phytoplankton growth was followed for seven days by counting (Coulter Counter, 70 μm orifice tube) daily the cells in subsamples. Because of the low phytoplankton concentrations, chlorophyll *a* measurements did not appear suitable to determine the growth rate. Three size fractions of phytoplankton were distinguished: 2.8-6 μm; 5.8-10.4 μm nd 10.0-28.0 μm.

Table 1. Combinations in the bioassay experiments.

Combination	Removal of zooplankton	Nutrient addition	
		P (mg l^{-1})	N (mg l^{-1})
reference	+	–	–
+ PO$_4$	+	0.32	–
+ NO$_3$	+	–	0.56
+ NH$_4$	+	–	0.30
+ zoopl.	–	–	–

Zooplankton grazing and phytoplankton primary production

The grazing and assimilation rates of the crustacean zooplankton were measured weekly in the laboratory by ^{14}C-tracer technique (Gulati *et al.*, 1982) using lake seston (<33 µm) as tracer food. The biomass of seston food (<33 µm) and zooplankton (>150 µm) was measured by the COD technique described by Golterman (1969) and modified by Gulati *et al.* (1982). The oxygen consumed was converted to carbon according to Winberg *et al.* (1971). The expression of data in carbon units also facilitates comparison with phytoplankton primary production rates (^{14}C-technique) (Van Liere *et al.*, 1986), which were measured only from 24 June 1987 onwards.

Results

Transparency

A marked increase of transparency occurred within 4 weeks after biomanipulation (Fig. 2). In 1986 Secchi-disc transparency in the summer was constantly low, averaging 0.3 m with little variation. However, a pronounced increase occurred in June-July 1987, when the Secchi-depth exceeded 2 m and even lake bottom was visible (Fig. 2).

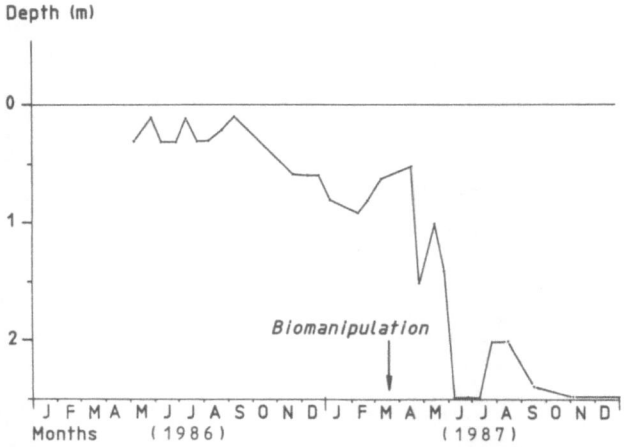

Fig. 2. Secchi-disc transparencies in Lake Zwemlust before (until mid-March 1987) and after biomanipulation.

Phytoplankton

A marked decrease was observed in chlorophyll *a* concentration in June-July 1987, when the low values coincided with the increase of Secchi-disc transparency. The summer concentrations in 1987 did not exceed 5 µg l^{-1} compared with 250 µg l^{-1} recorded in summer 1986 (Fig. 3).

Further, there was a change in phytoplankton species composition. In summer 1986 a bloom of cyanobacteria (*Microcystis aeruginosa*) was observed, but in 1987 green algae (*Chlamydomonas* sp., *Scenedesmus* sp.) and Cryptophyceae became dominant.

In 1987 no cyanobacteria were detected in the routine phytoplankton analyses.

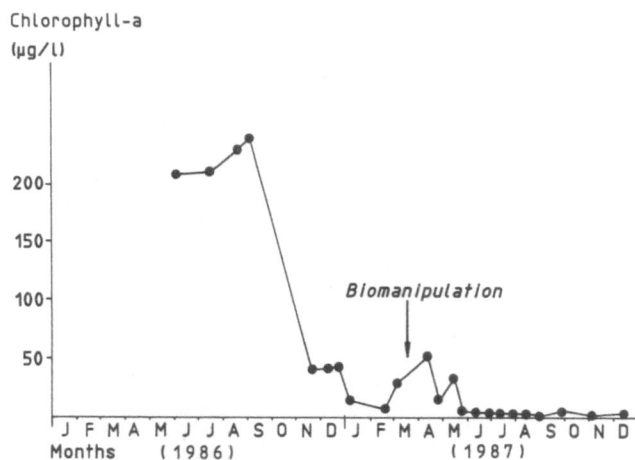

Fig. 3. Chlorophyll *a* concentration in Lake Zwemlust before and after biomanipulation.

Zooplankton

In 1986, as in the years before, rotifers (*Brachionus* sp., *Keratella* spp., *Polyarthra* sp.) comprised most of the zooplankton community. At the end of April 1987, i.e. 4 weeks after biomanipulation, larger rotifers (*Brachionus angularis*, *B. calyciflorus*, *Filinia longiseta*, *F. cornuta*) were first recorded. However, only *Polyarthra* sp. was important throughout the summer period, being the only persistent rotifer.

Among the crustacean zooplankters, the filter-feeding cladocerans (Fig. 4), namely *Daphnia hyalina* and *Bosmina longirostris*, became important already in mid-May 1987; i.e. about 6 weeks after the lake was filled. *Daphnia magna* appeared only in early July but rapidly achieved its maximum (190 ind. l^{-1}). Two other *Daphnia* species, *D. cucullata* and *D. longispina*, were also encountered but in relatively low densities.

The copepods were dominated by cyclopoids and their nauplii. The latter occurred in concentrations of 532-820 ind. l^{-1} in the period from early May-early July (Fig. 4).

Results from other sampling stations in the lakes, and bottom samples, show that the densities of crustacean zooplankton recorded by us in the open-water zone as described above are underestimates, particularly in case of *Daphnia magna*. This species tends to concentrate at the bottom and in the littoral zone during daytime and is, therefore, captured with a low efficiency using a tubesampler. This has also been confirmed in a recent preliminary diurnal study (E. Slim, pers. comm.). Areal concentrations of *D. magna* in nocturnal samples were clearly higher than those collected during daytime.

Chironomid larvae

No data on chironomids prior to biomanipulation are available. After the biomanipulation opera-

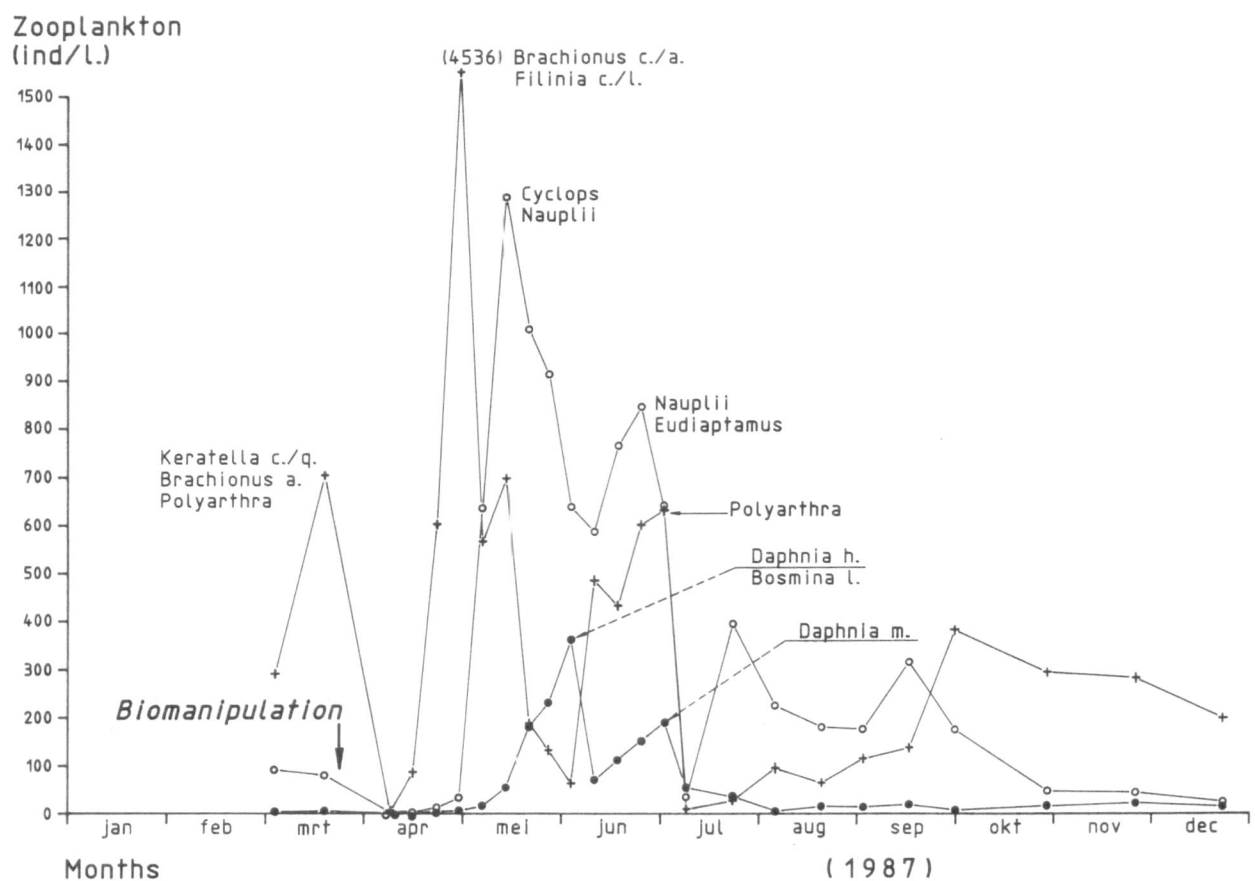

Fig. 4. The change before and after biomanipulation in the abundance of herbivorous zooplankton, Rotifera (+), Copepoda (o) and Cladocera (●). The dominant species for some of the sampling dates are indicated.

Table 2. Fish caught by seine- and electro-fishing in March 1987 (* measured by sampling).

Species	Weight (kg)	Number
Bream (*Abramis brama*) (> 20 cm)	99.4	116
(*Abramis brama*) (10–20 cm)	435.0	10000*
(*Abramis brama*) (< 10 cm)	50.6	6208*
Roach (*Rutilus rutilus*) (> 10 cm)	64.0	4025*
(*Rutilus rutilus*) (⩽ 10 cm)	24.7	2912*
(*Leucaspius delineatus*)	30.8	18547*
Pike (*Esox lucius*)	56.6	44
Rudd (*Scardinius erythophthalmus*) (> 20 cm)	1.0	4
(*Scardinius erythophthalmus*) (10–20 cm)	5.5	223
(*Scardinius erythophthalmus*) (< 10 cm)	0.6	115
White Bream (*Blicca bjoerkna*)	9.2	575
Eel (*Anguilla anguilla*)	15.7	24
Tench (*Tinca tinca*)	6.4	7
Carp (*Cyprinus carpio*)	6.4	1
Perch (*Perca fluciatilis*) (⩾ 20 cm)	3.2	9
(*Perca fluciatilis*) (< 20 cm)	1.7	357*
Pike-perch (*Stizostedion lucioperca*)	0.2	1

tions the biomass of chironomids in the sediment was high (in June 1987, 8000 chironomids m^{-2}), probably due to the decrease in predation by fish. More than 90% of the chironomid larvae were determined as *Chironomus plumosus*.

Fish

Results of the seine- and electro-fishing in March 1987 are given in Table 2. The total amount of fish removed, including those eliminated by the pumping, was estimated at 1200 – 1500 kg wet weight. The offspring of the stocked rudd were present in the lake from the end of August. Only 19% of the stocked pike survived the first year. Despite the absence of small prey-fish during most of the year, the length-weight relation of the pike from Zwemlust was similar to that of other waters. The diet of the pike consisted mainly of chironomid larvae.

Macrophytes

In 1986 there were virtually no submerged aquatic macrophytes. Of the 200 *Nuphar lutea* planted in 1987, 50 plants flowered. The submerged vege-

tation remained scarce, only 5% of the bottom being covered by macrophytes at the end of the season. *Chara globularis*, *Potamogeton crispus*, *P. berchtoldii* and *Elodea nuttallii* formed most of the vegetation.

In August 1987 the macroalgae *Hydrodictyon reticulatum* and *Enteromorpha* covered up to 25% of the lake area and were emergent in the littoral zones. In 1987 plant development was probably retarded due to low summer temperature.

Nutrients, oxygen and pH

In 1986, the phosphorus and nitrogen concentrations in Zwemlust were very high and very similar to the values in the River Vecht from which the lake receives water by seepage (Figs. 5, 6). After biomanipulation the phosphate and ammonium concentrations even increased (Figs. 5, 6) but oxygen concentrations decreased. The higher concentration of total phosphorus probably originated from the sediments due to bioturbation by the chironomids (Gallepp *et al.*, 1978). The pH declined from a maximum of 9.5 in summer 1986 to 7.7 in summer 1987.

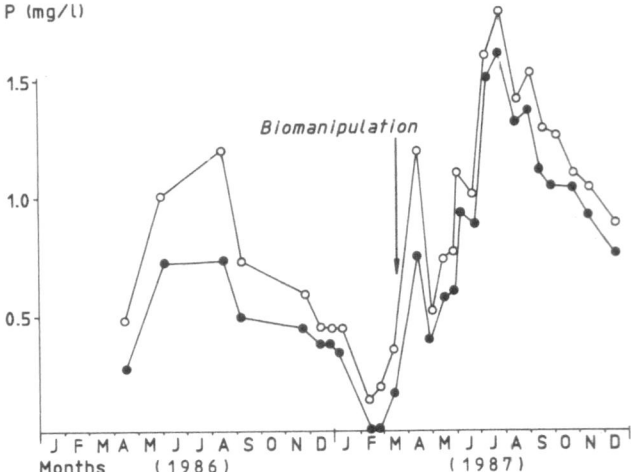

Fig. 5. The concentrations of ortho-P (●) and total-P (○) of Lake Zwemlust before and after biomanipulation.

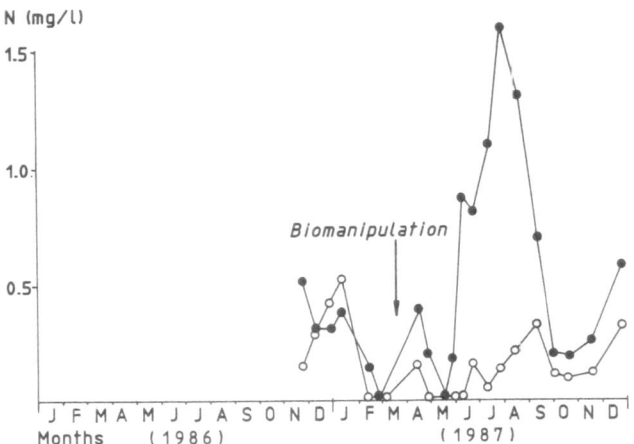

Fig. 6. The concentration of ammonium (●) and nitrate (○) of Lake Zwemlust before and after biomanipulation.

Bioassay experiments

The relative importance of nutrient limitation and increased grazing pressure was tested in the 7-day bioassay experiment. The reference (no nutrient addition, no zooplankton) and bioassays with NO_3, PO_4 or NH_4 enrichment, but no zooplankton, resulted in high growth rates which did not differ within one size fraction (Fig. 7). The highest growth rates were found in the two smallest

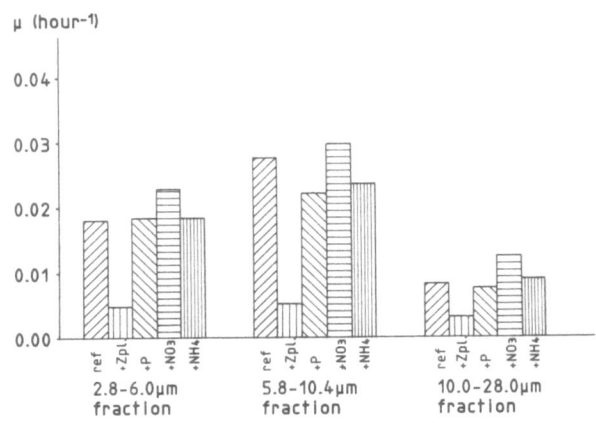

Fig. 7. Growth rates (μ) of the three phytoplankton fractions plotted for the five bioassay combinations (see text). The experiment was carried out in mid-June 1987.

fractions. In contrast, in the bioassays in which zooplankton was not removed and no nutrients were added, only a very small increase in phytoplankton numbers was observed (Fig. 7). For the size fraction of 2.8-10.4 μm the increase in numbers of phytoplankton in the different bioassays is plotted in Fig. 8. Microscopic investigation confirmed that the particles measured by the Coulter Counter were phytoplankton cells and not other organisms or detritus. The bioassays indicate that in June 1987 zooplankton grazing controlled the phytoplankton growth.

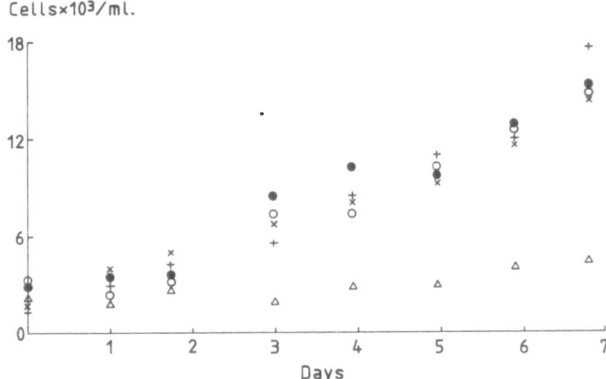

Fig. 8. Phytoplankton concentrations (size-fraction 2.8-10.4 μm) plotted against incubation time for the five bioassay combinations. Reference ●; + zooplankton, △; + P, ×; + NO_3, +; + NH_4, ○.

Table 3. Primary production, grazing and assimilation rates of the phytoplankton and crustacean zooplankton community and other relevant data of Lake Zwemlust in 1987. The zooplankton biomass (Z) and their seston food (S) are expressed as mg C l^{-1}; the primary production (PP), consumption (C) and assimilation rates (A) are expressed as mg C l^{-1} d^{-1}. For other codes, see footnote.

Date	T°C	S	Z	PP	SPP	G	SCR	C	A	A/C	SDA
15–05	13	1.91	0.62	–	–	73	1.17	1.39	0.19	14	31
25–05	18	1.28	1.16	–	–	170	1.47	2.18	0.95	44	82
03–06	18	0.66	0.87	–	–	121	1.39	0.79	0.22	28	26
10–06	17	0.69	0.56	–	–	195	3.48	1.35	0.63	47	112
17–06	17	0.66	0.48	–	–	55	1.14	0.36	0.11	29	22
24–06	17	0.62	0.32	0.35	62.8	80	2.50	0.50	0.21	43	67
01–07	21	0.70	0.66	–	–	112	1.69	0.78	0.32	40	48
08–07	24	0.75	0.34	–	–	65	1.91	0.49	0.15	31	44
22–07	21	0.68	0.32	0.07	142.7	–	–	–	–	–	–
05–08	18	0.71	0.20	0.09	142.2	–	–	–	–	–	–
19–08	21	0.52	0.22	0.07	31.8	14	0.65	0.08	0.04	51	18
02–09	20	0.63	0.30	0.20	49.0	46	1.53	0.29	0.15	53	50
16–09	18	0.45	0.28	0.03	20.0	33	1.18	0.15	0.08	54	29
30–09	14	0.25	0.28	0.04	56.7	226	8.08	0.58	0.19	33	68
14–10	11	0.27	0.15	0.01	28.7	125	8.31	0.34	0.09	25	60
28–10	10	0.10	0.16	0.02	16.0	–	–	–	–	–	–
11–11	8	0.24	0.18	0.01	4.1	22	1.24	0.05	0.02	44	11
25–11	6	0.25	0.23	–	–	24	1.04	0.06	0.02	29	9
22–12	7	–	–	0.01	12.5	22	–	–	–	–	–

SPP, specific primary production (mg C mg Chl^{-1} d^{-1}); G, grazing (% d^{-1});
SCR, specific clearance rate (l mg C^{-1} d^{-1}); A/C, assimilation efficiency (%) and
SDA, specific daily assimilation (A/Z in %).

Zooplankton grazing and phytoplankton primary production

Results of the grazing measurements carried out between May and December 1987 are presented in Table 3. The daily zooplankton grazing (G) was often very high, with values up to 226%. This is reflected in a marked reduction in seston concentration, i.e. from 1.9 to 0.1 mg C l^{-1}, the latter being one-twentieth of the spring level. The zooplankton biomass (Z) to their food (S) ratio (Z/S), chlorophyll *a* and Secchi transparency data all indicate the prevalence of 'oligotrophic' conditions under the high grazing pressure. The specific clearance rates (SCR) also reflect a high zooplankton filtering activity as is often found in some deep, oligo-mesotrophic Dutch lakes, with food concentrations reaching the incipient limiting level (Gulati, 1983). Both the consumption and assimilation rates of zooplankton appear to be much higher than can be supported by phyto-

plankton primary production. The high grazing and assimilation activities indicate the significant role of zooplankton in improving the light climate through high seston removal. Despite the relatively low assimilation efficiencies (14-54%), weight-specific zooplankton assimilation (SDA, Table 3) indicates generally favourable growth conditions for zooplankton dominated by larger cladocerans in the absence of planktivorous fish.

Discussion

The preliminary results of our biomanipulation experiment are comparable with those found in other similar studies in which planktivorous fish were eliminated. Larger species of herbivore zooplankton became dominant, accompanied by decline of chlorophyll *a* concentrations and algal densities and increasing transparency (Hrbáček *et al.*, 1961; Andersson *et al.*, 1978; Leah *et al.*,

1980; Shapiro *et al.*, 1982; De Bernardi *et al.*, 1982; Benndorf *et al.*, 1984; Shapiro & Wright, 1984).

We hypothesised that decrease in algal abundance and increase in water transparency would result mainly from the sharp increase in the herbivore zooplankton populations in the absence of planktivorous fish. The bioassay and zooplankton grazing experiments support this hypothesis. The grazing pressure exerted by the crustacean zooplankton community was able to keep chlorophyll *a* concentrations and algal abundance at very low levels, even in the presence of high concentrations of both inorganic N and P (Figs. 7, 8 and Table 3). The net daily grazing losses of seston appear to considerably exceed the daily primary production of phytoplankton, thus improving water clarity. Since practically all the phosphate in the lake is present as ortho-phosphate it may be assumed that algae are not P-limited, nor even light-limited because of the improved light climate.

In other lake restoration programmes a decline in nutrient concentrations was observed after removal of planktivorous fishes (Stenson *et al.*, 1978; Henrikson *et al.*, 1980; Shapiro & Wright, 1984). In these lakes fishes were eliminated by rotenone. Differences in water supply (seepage in Zwemlust versus supply from the watershed) and in bioturbation by chironomids (rotenone application is known to have a negative effect on benthos: Koksvik & Aagaard, 1984) might explain our results of nutrient enhancement.

Lake Zwemlust, however, is still in an unstable stage. A new fish population dominated by predatory fish, which will control the planktivorous fish, and a well-developed submerged vegetation, have not yet stabilized. Therefore, we placed artificial substrata and planted macrophytes as spawning grounds and refuges for the stocked pike.

In the other lakes restored by biomanipulation, all fish was removed but no measures were taken to improve the fish community by the introduction of other species and by construction of new habitats (Stenson *et al.*, 1978; Shapiro & Wright, 1984).

Leah *et al.* (1980) followed the effects of fish elimination by cormorants in Brundall Broad during one year only, and Shapiro & Wright (1984) studied the effect of fish removal by rotenone in Round Lake for only two years. Henrikson *et al.* (1980) reported that the beneficial effects of fish removal lasted for at least four years, but their lake was oligotrophic and kept without fish. According to an empirical model by McQueen *et al.* (1986) fish removal can only be successful in oligotrophic waters. An important question concerning the use of biomanipulation as a restoration technique in the hypertrophic Lake Zwemlust is the long-term effectiveness of such a measure. We intend to follow the effect of biomanipulation here for at least four years.

Acknowledgements

We thank F. Jaques and his co-workers for carrying out the biomanipulation operations. Further we thank K. Beneken Kolmer, E. Slim, P. Heuts and B. Vlink for their work in the field and laboratory, and L. M. Matulessya for typing the manuscript.

References

Andersson, G., H. Berggren, G. Cronberg & C. Gelin, 1978. Effect of planktivorous and benthivorous fish on organisms and water chemistry in eutrophic lakes. Hydrobiologia 59: 9–15.

Benndorf, J., H. Kneschke, K. Kossatz & E. Penz, 1984. Manipulation of the pelagic food web by stocking with predacious fishes. Int. Revue ges. Hydrobiol. 69: 407–428.

Björk, S., 1985. Lake restoration techniques. In: Proceedings of the International congress 'Lakes Pollution and Recovery' April 1985, Rome.

Brooks, J. L. & S. I. Dodson, 1965. Predation, body size and composition of plankton. Science 150: 28–35.

De Bernardi, R. & G. Giussani, 1975. Population dynamics of three cladocerans of Lago Maggiore related to the predation pressure by a planktiphagous fish. Verh. int. Ver. Limnol. 19: 2906–2912.

De Bernardi, R., G. Giussani & E. Lasso Pedretti, 1982. Select feeding of zooplankton with special reference to blue-green algae in enclosure experiments. Mem. Ist. Ital. Idrobiol. 40: 113–128.

Gallepp, G. W., J. F. Kitchell & S. M. Bartel, 1978. Phosphorus release from lake sediments as affected by chironomids. Verh. int. Ver. Limnol. 20: 458–465.

Golterman, H. L., 1969. Methods for chemical analysis of freshwater. I.B.P. Handbook, No. 8. Blackwell Sci. Publ., Oxford.

Grimm, M. P., 1981. The composition of northern pike (*Esox lucius* L.) populations in four shallow waters in The Netherlands, with special reference to factors influencing o$^+$ pike biomass. Fish. Management 12: 61–77.

Grimm, M. P., 1983. Regulation of biomasses of small (<41 cm) northern pike (*Esox lucius* L.) with special reference to the contribution of individuals stocked as fingerlings (4-6 cm). Fish. Management 14: 115–135.

Gulati, R. D., 1983. Zooplankton and its grazing as indicators of trophic status in Dutch lakes. Envir. Monit. Assess. 3: 343–354.

Gulati, R. D., K. Siewertsen & G. Postma, 1982. The zooplankton: its community structure, food and feeding, and role in the ecosystem of Lake Vechten. Hydrobiologia 95: 127–163.

Henrikson, L., H. G. Nyman, H. G. Oscarson & J. A. Stenson, 1980. Trophic changes without changes in the external nutrient loading. Hydrobiologia 68: 257–263.

Hosper, S. H., M.-L. Meijer & E. Jagtman, 1987. Biomanipulation a new perspective for restoring lakes in The Netherlands (in Dutch, English summary), H$_2$O (20) 12: 274–279.

Hrbáček, J., M. Dvořáková, V. Kořínek & L. Procházková, 1961. Demonstration of the effect of the fish stock on the species composition of zooplankton and the intensity of the metabolism of the whole plankton association. Verh. int. Ver. Limnol. 14: 192–195.

Koksvik, J. I. & K. Aagaard, 1984. Effects of rotenone treatment on the benthic fauna of a small eutrophic lake. Verh. int. Ver. Limnol. 22: 658–665.

Lammens, E. H. R. R., 1986. Interactions between fishes and structure of fish communities in Dutch shallow eutrophic lakes. Thesis, University of Agriculture, Wageningen 100 p.

Lazarro, X., 1987. A review of planktivorous fishes: their evolution, feeding behaviours, selectivities, and impacts. Hydrobiologia 146: 97–167.

Leah, R. T., B. Moss & D. E. Forrest, 1980. The role of predation in causing major changes in the limnology of a hyper-eutrophic lake. Int. Revue ges. Hydrobiol. 65: 223–247.

Le Cosquino de Bussy, L. J., 1968. Algiciden. Onderzoek over diuron als bestrijdingsmiddel van algen in de zwemvijver 'Zwemlust', Nieuwersluis, IG-TNO report, nr. A48.

Lynch, M. & J. Shapiro, 1981. Predation, enrichment and phytoplankton community structure. Limnol. Oceanogr. 26: 86–102.

McQueen, D. J., J. R. Post & E. L. Mills, 1986. Trophic relationships in freshwater pelagic ecosystems. Can. J. Fish. aquat. Sci. 43: 1571–1581.

Reinertsen, H. & Y. Olson, 1984. Effects of fish elimination on the phytoplankton community of a eutrophic lake. Verh. int. Ver. Limnol. 22: 649–657.

Shapiro, J., 1980a. The need for more biology in lake restoration. In: Lake Restoration. Proceedings of a National Conference, 22-24 August 1978. U.S. Environmental Protection Agency 444/5-79-001, Minneapolis.

Shapiro, J., 1980b. The importance of trophic-level interactions to the abundance and species composition of algae in lakes. In: Hypertrophic ecosystems (Eds. J. Barica & L. R. Mur). Junk, The Hague.

Shapiro, J., B. Forsberg, V. Lamarra, G. Lindmark, M. Lynch, E. Smeltzer & G. Zoto, 1982. Experiments and experiences in biomanipulation: studies of ways to reduce algal abundance and eliminate bluegreens. U.S. Environmental Protection Agency EPA-600/3-82-096. Also Limnological Research Center Interim Report no. 19.

Shapiro, J., V. Lamarra & M. Lynch, 1975. Biomanipulation: an ecosystem approach to lake restoration. In: Water quality management through biological control. (Eds. P. L. Brezonik & J. L. Fox). Rep. no. ENV-07-75-1, University of Florida, Gainesville.

Shapiro, J. & D. I. Wright, 1984. Lake restoration by biomanipulation: Round Lake, Minnesota, the first two years. Freshwat. Biol. 14: 371–383.

Stenson, J. A. E., T. Bohlin, L. Henrikson, B. I. Nilsson, H. G. Nyman, H. G. Oscarson & P. Larsson, 1978. Effects of fish removal from a small lake. Verh. int. Ver. Limnol. 20: 794–801.

Van Liere, E., 1986. Loosdrecht lakes, origin, eutrophication, restoration and research programme. Hydrobiol. Bull., Amsterdam 20: 9–15.

Van Liere, E., L. van Ballegooyen, W. A. de Kloet, K. Siewertsen, P. Kouwenhoven & T. Aldenberg, 1986. Primary production in the various parts of the Loosdrecht Lakes. Hydrobiol. Bull., Amsterdam 20: 77–85.

Winberg, G. G. et al., 1971. Symbols, units and conversion factors of freshwater productivity. IBP, London, 23 pp.

Zaret, T. M., 1980. Predation and freshwater communities. Yale University Press, New Haven and London. 187 pp.

Hydrobiologia **191**: 297–306, 1990.
P. Biró and J. F. Talling (eds), Trophic Relationships in Inland Waters.
© 1990 *Kluwer Academic Publishers.*

Operation of the Kis-Balaton reservoir: evaluation of nutrient removal rates

Ferenc Szilágyi, László Somlyódy & László Koncsos
Water Resources Research Center (VITUKI), H-1453 Budapest, Pf. 27

Key words: eutrophication, nutrient removal, wetlands, lake management

Abstract

As one of the major measures for controlling the man-made eutrophication of Lake Balaton, the Hidvég reservoir of 20 km² surface area was built near the mouth of River Zala, draining half the watershed of the lake, and representing the largest nutrient source for the lake. The reservoir, as the first element of the expected total system of 70 km² surface area (Kis-Balaton Control System), started to operate in June 1985, aiming at removing nutrients primarily through sedimentation, adsorption and uptake by macrophytes.

Detailed investigations began with the operation. These cover the observation of upstream and downstream nutrient loads and the water quality in the reservoir, the study of major phosphorus removal processes, and analysis of the nitrogen cycle and of the behaviour of phytoplankton, zooplankton, fish and macrophytes. The research programme is completed by the evaluation of observations (including the use of phosphorus budget models), with special emphasis on future operation modes of the reservoir.

The nutrient removal efficiencies in the reservoir came up to expectations. The removal rates for suspended solids, total-P, soluble reactive-P and nitrate-N exceeded 50% in the first full year of operation (1986). As a result of reservoir operation, nutrient loads in the western basin of Lake Balaton have been significantly reduced. However, the improvement in water quality can be expected only with a lag time due to the internal P load of the basin.

Introduction: eutrophication of Lake Balaton

Lake Balaton, the largest shallow lake in Central-Europe, is the most important recreational region of Hungary, providing about 50% of the total revenue from foreign tourism, and also playing a major role in the recreation of the local population.

Lake Balaton is 77.8 km long and 7.7 km wide on average, with surface area of 596 km². The watershed of the lake is 5775 km² (not including the lake surface). The lake is of elongate shape,

with an average depth of 3.14 m. Hydrologically it can be divided into four basins (Baranyi, 1974; Somlyódy & van Straten, 1986; see Fig. 1).

Significant socio-economic development has occurred during the last two decades in the watershed. This development resulted in increasing external loads and consequent algal blooms in the lake (Somlyódy & Tóth, 1987). Eutrophication was especially advanced in the most western basin (Keszthely Bay) representing only 4.3% of the total lake volume, as about one-third of the total external nutrient load is discharged

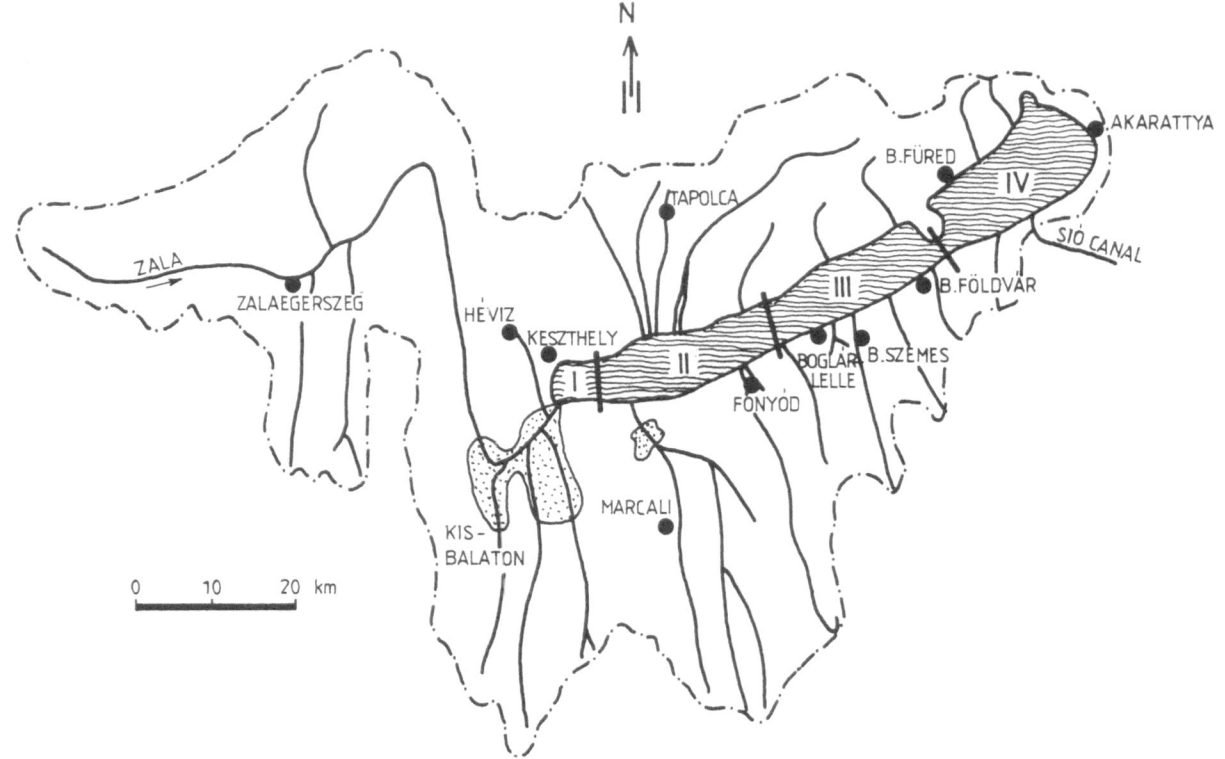

Fig. 1. Lake Balaton and its watershed.

into this basin by the River Zala. Consequently, the volume-related loads are an order of magnitude higher here than in the less polluted eastern basin (Somlyódy & van Straten, 1986). According to OECD classification (Vollenweider & Kerekes, 1981), the basins of the lake mostly fall into the mesotrophic to hypertrophic category (see Fig. 2). The highest observed chlorophyll *a* concentration has reached 250 mg m^{-3} in the Keszthely basin.

In order to control man-made eutrophication of Lake Balaton a project was initiated in 1982 to determine water quality target levels, required control alternatives and associated costs. The proposal born as a result of the project, and approved by the Council of Ministers in 1983, involved measures such as upgrading and extending existing wastewater treatment plants, introduction of chemical removal of phosphorus, construction of a regional sewage treatment system for exporting P from the watershed, establishment

of a reservoir system of Kis (Small-) Balaton in the vicinity of the mouth section of River Zala, and others (for details see Somlyódy & van Straten, 1986). The first element of the system of about 70 km^2 total surface area (Hidvég reservoir) was put into operation in 1985, while the construction of the second stage (Fenéki reservoir) is expected by the mid-'90s (Joó *et al.*, 1987).

The Kis-Balaton reservoir system

The objective of the Kis-Balaton reservoir system is to reduce nutrient loads of the River Zala by sedimentation, adsorption and uptake by algae and aquatic macrophytes. The major parameters of the two-stage reservoir system are shown in Table 1, and the completed layout of the Hidvég reservoir can be seen in Fig. 3.

The reservoirs of the Kis-Balaton system are extremely shallow (Table 1). The detention time

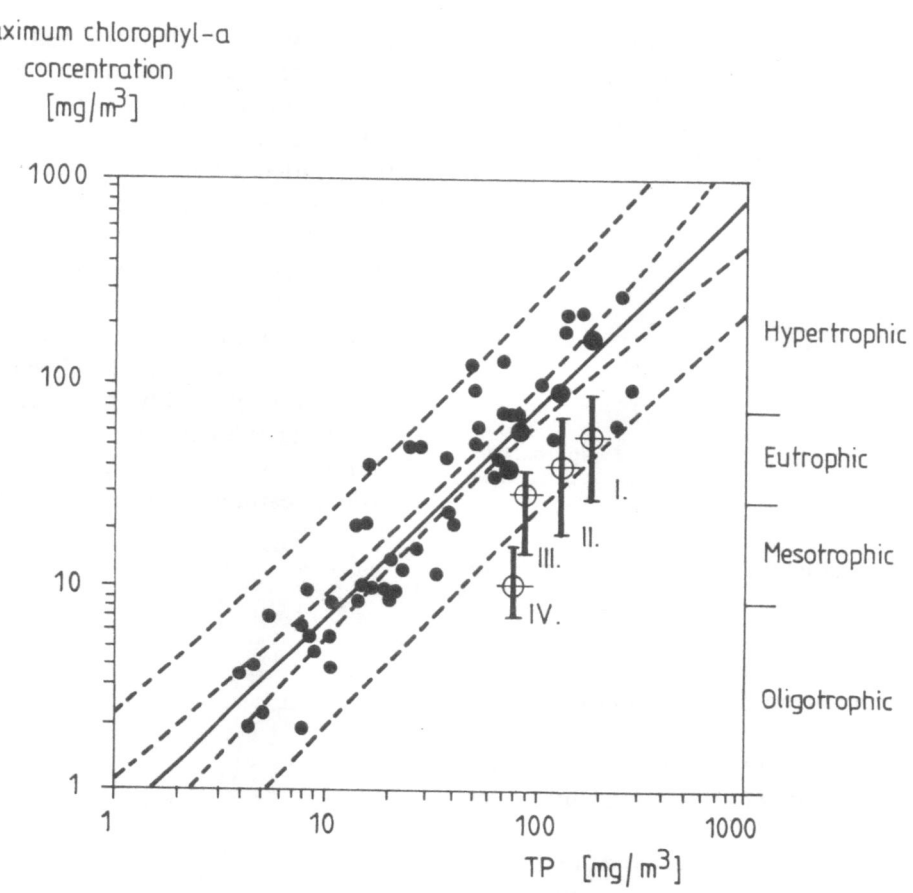

Fig. 2. Trophic conditions in Lake Balaton as compared to other OECD lakes. I, II, III, IV-Basins of Keszthely, Sziglitet, Szemes and Siófok, respectively ○ Average basin values for 1975-1979 I Range between minimum and maximum value ● Highest values in 1980 • Values of the OECD lakes (Vollenweider & Kerekes, 1981).

of c. 40 days of Zala water in the Hidvég reservoir is long enough for the processes of nutrient removal. The total phosphorus load of this reservoir is rather high (approximately 15 mg m^{-2} d^{-1}) and about 5-6 times higher than that forecast for the future downstream reservoir. The flow and water level of the Hidvég reservoir is regulated by dykes and three gates (Fig. 3: structures 2T, 3T and 4T). At the upper part of the reservoir, the Zala river was closed by a dam. The river flow enters the reservoir through several cuts made in the flood levee. The flow pattern in the reservoir shows two different routes (i) through the main reservoir, around the major diversion dyke (flow direction I, see Fig. 3), and/or (ii) through the internal (or sub-) reservoir (called casette) via two

regulating gates. The purpose of the casette is to retain and treat accidental pollution loads of the River Zala which may occur occasionally.

Research and monitoring of the Hidvég Reservoir

During the two years of operation of the Hidvég reservoir, a multidisciplinary research programme has been carried out with the participation of 11 institutes. The research was performed under the leadership of the Water Resources Research Centre, VITUKI, while the reservoir is operated by the Western Transdanubian District Water Authority at Szombathely.

The objective of the study was to determine

Table 1. Characteristics of the Kis-Balaton reservoir system.

	Reservoirs	
	Hidvégi	Fenéki
Water level (meter above Baltic sea level)	106.5	105.8
Surface area (km²)	18	51
Average depth (m)	1.14	1.2
Volume (10^6 m³)	20	64
Detention time (year)	0.12	0.24
Inflow (m³ s^{-1})	6	8
Total-P load (g m^{-2} yr^{-1})	5.28	approx. 0.98

logical variables investigated were different for each observation station. Daily nutrient load measurements are carried out at points 1 and 11 (in- and outflow). Major chemical components (suspended solids, chlorophyll *a*, chemical oxygen demand (COD_{Mn}) total-P, soluble reactive P, total-N, nitrate-N, nitrite-N, ammonia-N, pH, conductivity, dissolved oxygen) are measured twice a week in stations 1, 5, 8, 9, 10 and 11. Hydrobiological studies, referred to above, were carried out at each of the eleven stations monthly (winter) to bi-weekly (summer).

During 1986 and 1987 daily observations were carried out for the inflow section at Zalaapáti in order to estimate nutrient loads. In Balatonhidvég

nutrient removal efficiency, the exploration of processes and ecological changes affecting P and N removal rates, and the preparation of optimal operation guidelines for the reservoir.

The research involved hydrological (monitoring network, water budget and flow forecasting), hydraulic (calibration of the gates, determination of flow patterns and detention time), and water quality studies.

In addition to water budget and hydraulic studies, the main emphasis of the programme has been laid on observing and studying water quality and the processes influencing it. The investigations involved nutrient load measurements, water and sediment analyses and studies on the structural and functional changes in the ecosystem (bacterioplankton, phytoplankton, zooplankton, macrophytes, protozoa, macroscopic invertebrates and vertebrates). The major components of the phosphorus and nitrogen cycles (denitrification, nitrogen fixation, phosphorus uptake by phytoplankton, phosphorus adsorption of sediment particles, etc.) have been measured. A data-base has been established on an IBM PC for handling about 200 000 – 250 000 data collected and measured. Modelling of the nutrient budget has also been initiated, first with relatively simple models.

The different regions of the reservoir were monitored at 11 stations (Fig. 3, locations 1 to 11). Sampling frequencies and the number of bio-

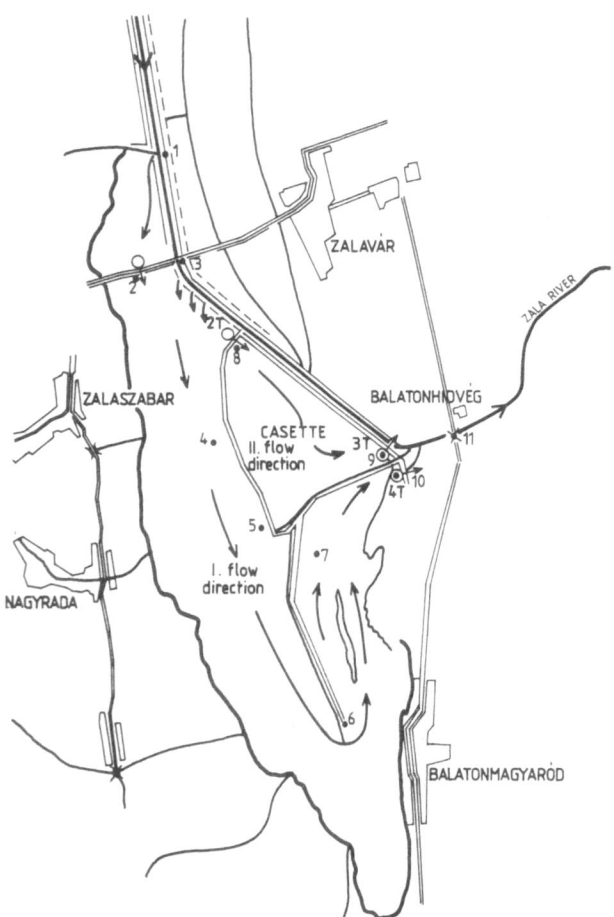

Fig. 3. Plan of the Hidvég reservoir of the Kis-Balaton Control System. ○→ Gate ● Regular sampling sites → Flow direction of the water in the reservoir = Dyke — Creek.

streamflow was measured daily during 1986, while water quality samples were taken only twice a week. From May 1987 daily sampling of the outflow at Balatonhidvég has also been performed. The removal rate of the reservoir can be calculated by analyzing the data of inflow and outflow sections at Zalaapáti and Balatonhidvég, respectively. The analysis involved such components as rate of flow, suspended solids, total-P, soluble reactive-P, total-N, and nitrate-N.

The removal efficiency (η) of the reservoir was calculated as $\eta = (L_{in} - L_{out})/L_{in}$ where L_{in} and L_{out} are the input and outlet loads, respectively; it was averaged for a period sufficiently long in comparison to the detention time.

The chemical analyses were carried out according to Felföldy (1980). Some of the research results are summarized below.

Results and discussion

Detention time
As follows from data of Table 1, about 40 days are needed to fill the reservoir at the average flow of the River Zala. The actual detention time was estimated by daily nitrate-N measurements at the in- and outflows, for a two-month winter period in 1986. The nitrate approximately behaved as a conservative component. Cross-correlation analyses of data from the inflow and outflow section were performed with different lag times. The correlation coefficients for flow and nitrate-N were plotted as a function of the lag time (Fig. 4). For water discharge the plot does not show any peaks, suggesting that the reservoir strongly attenuates the flow fluctuations of the River Zala. The two peaks of the nitrate-N curve that correspond to 3 and 7 days, respectively, indicate the expected detention time of the casette and the main reservoir under the rather high flow conditions. From the available observations, a pronounced variation in the retention time between 2 and 50 days can be estimated. The influence of this on water quality and removal efficiency is a subject of future studies.

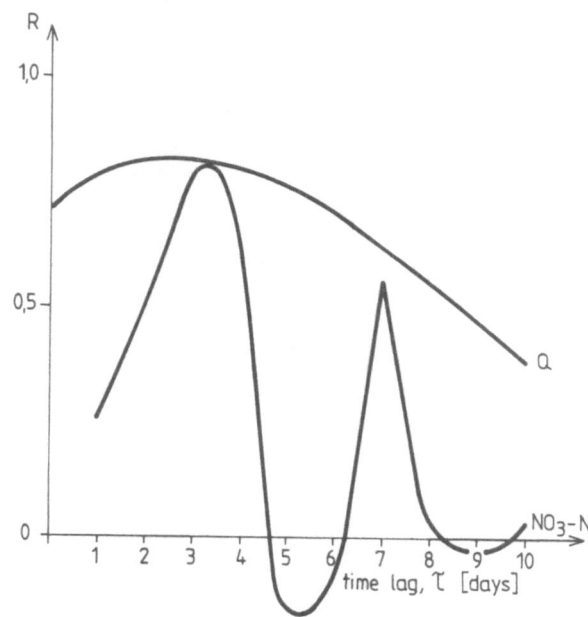

Fig. 4. Cross-correlations between parameters of the inflow and outflow sections of the Hidvég reservoir. R = coefficient of correlation Q = discharge rate of the water NO₃-N = Nitrate-nitrogen.

Water quality in the reservoir
The water quality of the River Zala is characterized by oxygen levels below saturation values, pH of 7.3-8.3, and conductivity of 500-800 μS cm^{-1}. Among the nitrogen compounds nitrate-N is the most important (2-4 g m^{-3}), while the concentration of ammonia-N varies between 0.3 and 2.0 g m^{-3}. Nitrite-N concentrations are below 0.3 g m^{-3}. The concentration of total nitrogen may reach 6.0 g m^{-3}. Total phosphorus varies from 0.1 to 2.0 g m^{-3}, and a significant portion of it is represented by soluble reactive-P (0.05-0.5 g m^{-3}). Chlorophyll a concentrations are below 20 mg m^{-3}, and suspended solids concentrations vary within wide ranges typical for Hungarian rivers (Major & Szilágyi, 1986; Szilágyi & Somlyódy, 1986). The reservoir has not yet formed its own bottom sediment. A high humic material content, up to 350 mg g^{-1}, characterizes the bed material. The CaCO₃ fraction is usually lower than 100 mg g^{-1}. The total phosphorus concentration in the bottom material amounts to 2 mg g^{-1}, while nitrogen is below 25 mg g^{-1}. Both nutrients are mostly in unavaila-

302

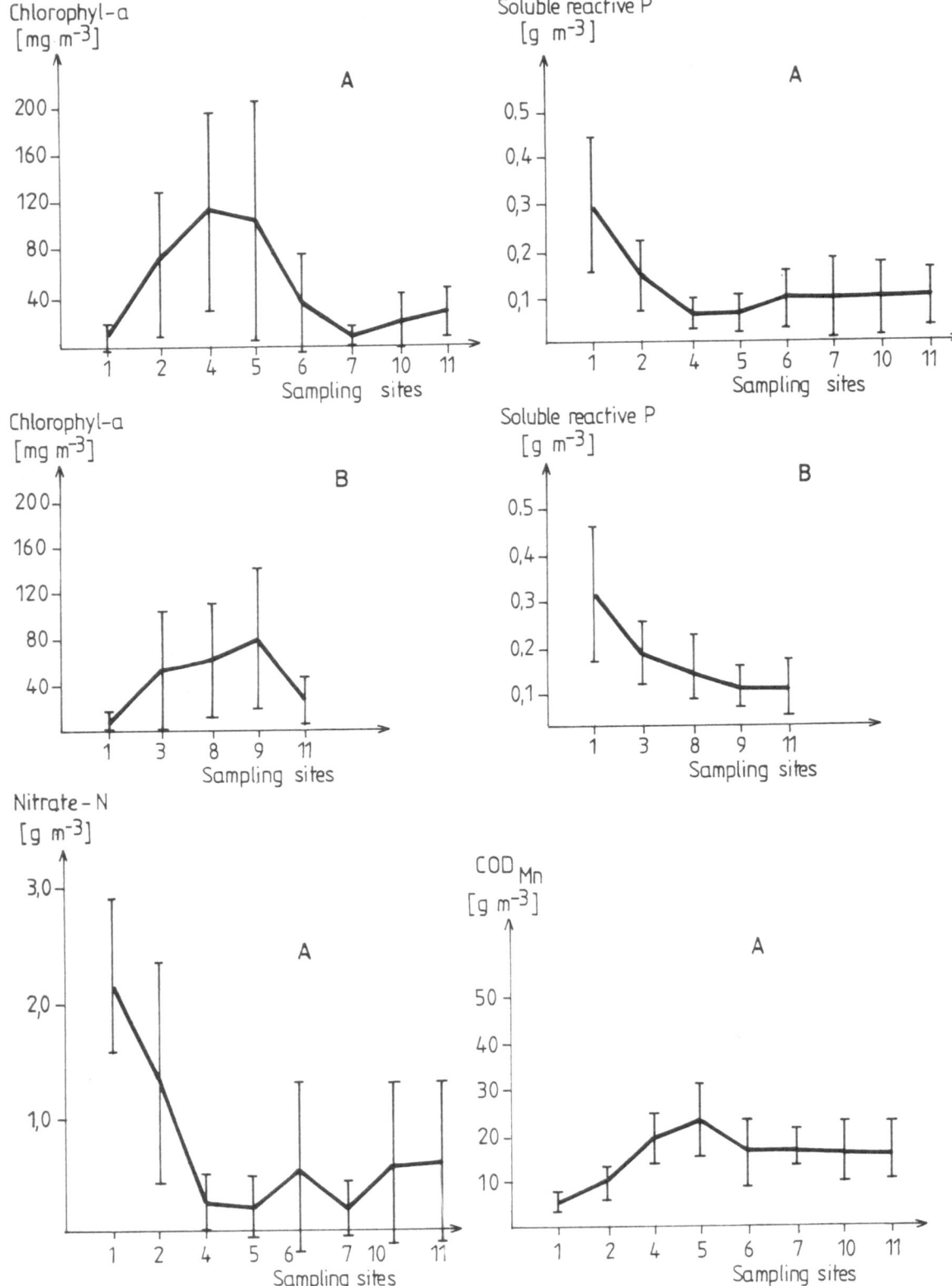

Fig. 5. See p. 303.

303

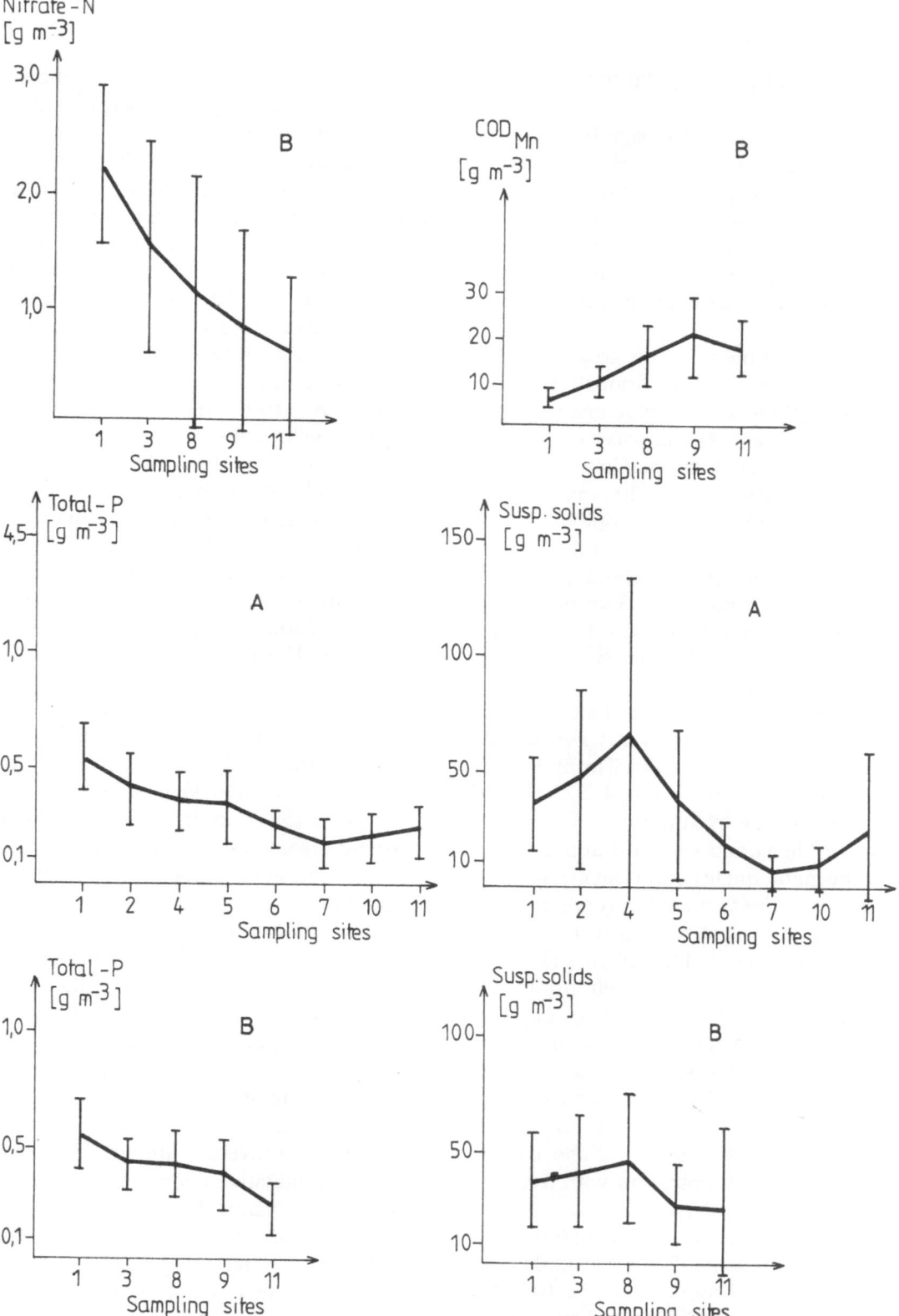

Fig. 5 (continued). Annual mean values (━━) and standard deviations of variables (───) along the two flow directions of the reservoir (data from regular sampling sites, 1986, see also Fig. 3). A = I flow direction B = II flow direction.

ble forms for plants (more than 80-90%: Szilágyi et al., 1987).

Nutrient load conditions vary significantly over the reservoir area, due to the variation of flow pattern and the chemical composition of the soil. As a result, water quality varies spatially as well, due to the different detention times. In the upper part of the reservoir, the water is relatively rich in nutrients due to the direct effect of the inflowing River Zala.

The middle part of the reservoir area is characterized by low nutrient concentrations and intensive summer blooms of blue-green algae (*Anabaena flos-aquae*, *Microcystis aeruginosa*, *Microcystis flos-aquae*), with maximum chlorophyll *a* concentrations reaching 400 mg m^{-3}. In the lower part of the reservoir macrophytes dominated in 1986. In relation to the total reservoir area the coverage amounted to about 50% with an annual specific biomass of 3.3 kg m^{-2}. The highest populations were represented by *Ceratophyllum submersum* and *Polygonum amphibium* (Pomogyi, 1987).

The chemical data measured at 11 stations can be used to characterize changes in water quality along the two main flow directions. The first direction is represented by stations 1, 2, 4, 5, 7, 10 and 11 and the second one by stations 1, 3, 8, 9 and 11 (see Fig. 3). The variations of the annual mean values of some major quality components are indicated in Fig. 5a-c. The vertical bar represents the standard deviation around a mean value.

The spatial variation of chlorophyll *a* (Fig. 5a) indicates that nutrients were taken up in the middle of the reservoir mostly by algae, and this resulted in frequent algal blooms and hypertrophic conditions. In the southern parts of the reservoir algal concentrations were reduced, presumable due to the competing effect of macrophytes. The chlorophyll *a* levels of the casette indicate planktonic eutrophication with frequent algal blooms (station 9).

Along both flow directions significant (60-70%) reduction of soluble reactive-P and nitrate-N concentrations could be observed (Fig. 5a, b). This process, however, was not linear and the rate of reduction was more marked in the upper part of the reservoir. The reduction of soluble reactive-P is attributed mostly to adsorption on solid particles but, in the summer warm-water periods, phosphorus uptake by phytoplankton was also significant. De-nitrification was the dominating process in reducing nitrate concentrations, with average de-nitrification rates of 41.1 mg m^{-2} d^{-1}, while the rate of nitrogen fixation was only 30.1 mg m^{-2} d^{-1} (Szilágyi et al., 1987).

The soil of the reservoir bottom is rich in humic materials, a feature that results in a high humic content of the water due to decomposition processes. This content is reflected by the COD_{Mn} values observed and the brownish colour of the water (Fig. 5b). The COD_{Mn} value of water increases by about four times during storage. It is also characterized by seasonal changes with high summer and low winter concentrations.

Contents of total phosphorus and suspended solids have also decreased – similarly to other nutrient forms – along both flow directions (Fig. 5c). The rate of reduction was higher for suspended solids than for total phosphorus. This indicates that P is bound to smaller fractions: the change in composition and particle size distribution of suspended solids during the storage can also play a role (Szilágyi et al., 1987).

Figures 5a-c indicate a marked scattering around the annual mean value. This is due to the high temporal variation in water quality characteristics.

In addition to the decreasing concentration of nutrients, some other favourable changes occurred in the quality of Zala water during storage in the reservoir. The bacterial counts decreased significantly (Szilágyi & Somlyódy, 1986; Szilágyi et al., 1987). The protozoan species of the fauna favouring less contaminated water became pre-dominant, the number of macroscopic invertebrate species has increased, and the qualitative composition of the fish stock improved (Csányi, 1986, 1987). These phenomena may indicate that the stabilization of the reservoir is progressing and a habitat is being formed with diversified populations.

Nutrient removal by the reservoir

During the planning and construction phase of the Hidvég Reservoir the nutrient removal efficiency of the system was estimated as approximately 60%. No seasonal changes of this removal efficiency were reported in the preliminary studies. in analysing the effectiveness of the reservoir, the years 1986 and 1987 should be dealt with separately, from the following reasons:

(a) 1986 was the first year of reservoir operation characterized by a relatively low sampling frequency at the outflow section (about 100 samples yr^{-1}). Thus the analysis of the removal rate is subject to an error of approximately 15% on an annual basis.

(b) Sampling frequency was increased in 1987, although data for the last quarter of the year are not yet available.

Table 2 illustrates the average removal rates for 1986 (full year) and the first three quarters of 1987. It is noteworthy that the reservoir operated with a removal efficiency of 50% or higher for most components, and that about 70% of the suspended solids and 61% of the soluble reactive-P load were removed in 1986. The low removal rate of total nitrogen is of little importance since the water of the River Zala is nitrogen-deficient.

In the first nine months of 1987 the loads of total-P, soluble reactive-P, total-N and nitrate-N to the reservoir were slightly higher than the total load in 1986, while the load of suspended solids was three times higher. Simultaneously with the increased loads, the removal efficiencies increased for suspended solids and total nitrogen, remained the same for soluble reactive-P and nitrate-N, and slightly decreased for total-P. Only rough estimations can be given for the loads from the watershed of the reservoir itself, due to lack of data. From the watershed of small creeks flowing into the reservoir a 6-8% contribution of the total nutrient load can be estimated. Consequently the removal efficiencies of Table 2 are somewhat underestimated.

Summarizing the removal efficiencies, the Hidvég reservoir works efficiently and meets the expectations made during the planning process, as far as nutrient removal efficiencies are concerned. As a result of reservoir construction the nutrient loads of the Keszthely Bay of Lake Balaton have been reduced significantly. However, the improvement in water quality of the bay can be expected only with a time lag, due to the observed internal P load.

Further decrease in nutrient loads of the River Zala would be expected after construction of the second reservoir (Fenéki reservoir) of the Kis-Balaton Control System.

As Table 2 shows, the nutrient load of Hidvég reservoir is very high. This is very unfavourable from a water quality point of view, especially for phosphorus because it is mostly accumulated in the sediment. Therefore it is necessary to reduce the phosphorus loading of the reservoir by intro-

Table 2. Input load (L_{in}) and mean removal efficiency (n %) of the Hidvég reservoir.

	Flow $10^6 m^3$	Suspended solids $Mg\ yr^{-1}$	Total-P $Mg\ yr^{-1}$	Soluble reactive-P $Mg\ yr^{-1}$	Total-N $Mg\ yr^{-1}$	Nitrate-N $Mg\ yr^{-1}$
1986						
L_{in}	214	12552	92	46	1097	531
n %		70	51	61	16	59
1987 months I–IX						
L_{in}	224	34599	104	52	1071	536
n %		84	37	53	26	55

ducing phosphorus removal from the wastewater of the largest town of the Zala watershed, Zalaegerszeg.

References

Baranyi, S., 1974. Composition of Lake Balaton Water in respect to its origin, and the analysis of water exchange processes (in Hungarian). Beszámoló a VITUKI 1971. évi munkájáról, VIZDOK, Budapest, 370 pp.

Csányi, B., 1986. Hydrobiological investigation of the Kis-Balaton Control System (in Hungarian). VITUKI Report No. 7622/3/10, Budapest, 48 pp.

Csányi, B., 1987. Hydrobiological investigation of the Kis-Balaton Control System (in Hungarian). VITUKI Report No. 7622/3/522, Budapest, 26 pp.

Felföldy, L., 1980. Biological water quality classification (in Hungarian). Vizügyi Hidrobiológia 9, VIZDOK, Budapest, 263 pp.

Joó, O., L. Déri & F. Laki, 1987. Construction of the Kis-Balaton Control System (in Hungarian). Vizügyi Közlemények 47: 331–354.

Major, J. & F. Szilágyi, 1986. Research and development work made for the determination of the efficiency of Kis-Balaton Control System: Optimization of Operation (In Hungarian). Proc. of the Conference of the Hungarian Hydrological Society, 17-19 June, 1986, Héviz, Hungary, p. 304–320.

Pomogyi, P., 1987. Results of botanical studies carried out in 1987 in the first stage reservoir of the Kis-Balaton Control System (in Hungarian). Western Transdanubian Water Authority Report, Keszthely, Hungary, 38 pp.

Somlyódy, L. & G. van Straten (eds.), 1986. Modeling and managing shallow lake eutrophication. Springer Verlag N.Y., 386 pp.

Somlyódy, L. & L. Tóth, 1987. Restoration of shallow lakes: Hungarian experiences. Proc. of the Symposium on Eco-system Redevelopment, Budapest, 1987. (in press)

Szilágyi, F. & L. Somlyódy, 1986. Chemical, biological and mass balance studies of the Kis-Balaton Control System (in Hungarian). VITUKI Report No. 7612/3, Budapest, 80 pp.

Szilágyi, F., L. Somlyódy & L. Koncsos, 1987. Chemical, biological and mass balance studies of the Kis-Balaton control System (in Hungarian). VITUKI Report No. 7612/3/8 Budapest, 137 pp.

Vollenweider, R. A. & J. J. Kerekes, 1981. Background and summary results of the OECD cooperative program on eutrophication. Int. Symposium on Inland Waters and Lake Restoration, Sept. 8-12, 1981 Portland, Main, USA, EPA, Washington D.C. Epa 440/5-81-110: 25–35.

Hydrobiologia **191**: 307–313, 1990.
P. Giró and J. F. Talling (eds), Trophic Relationships in Inland Waters.
© 1990 *Kluwer Academic Publishers.*

Enclosures study of trophic level interactions in the mesotrophic part of Lake Balaton

István Tátrai, László G.-Tóth, Vera Istvánovics & János Zlinszky
Balaton Limnological Research Institute, H-8237 Tihany, Hungary

Key words: bream, chain effect, bacteria, algae, nutrient release

Abstract

Enclosures, open to the bottom sediments and to the atmosphere, containing about 17 m³ of lake water in the mesotrophic area of Lake Balaton, were used to elucidate the role of the benthivorous fish bream (*Ambramis brama* L.) in the lake during 1984-1986.

Throughout the whole period water was less transparent in the enclosure containing fish, which strongly influenced the concentrations of suspended solids and chlorophyll *a*.

Both phytoplankton biomass and production readily responded to nutrient increase in the enclosure with fish. In 1985 diatoms were replaced by cyanobacteria whereas in 1986, at a lower fish stocking, a shift in algal structure towards chlorophytes was observed.

Egested organic substances and the resuspension of sediment particles by fish increased bacterial production.

Introduction

The importance of fish predation on zooplankton was first demonstrated by Hrbáček *et al.* (1961) and later by Brooks & Dodson (1965). In recent years the influence of fish predation on invertebrates, mainly crustaceans, in lake ecosystems was intensively studied. Most investigators have analyzed interactions between two trophic levels, especially between fish and zooplankton. Much less information is available on other processes, such as zooplankton predation and grazing, and strategies among crustaceans and phytoplankters to avoid being eaten. Relatively few studies deal with direct and indirect effects of fish on lower trophic levels.

Since 1982 a group of limnologists has been working on the functional role of fish in Lake Balaton (Tátrai *et al.*, 1985; Tátrai & Istvánovics, 1986). We conducted experiments, using enclosures open above and below, to study the effects of the dominant benthivorous fishes of the lake on the ecosystem. To demonstrate the ecosystem effect of these fishes we here present results from recent experiments carried out between 1984 and 1986.

Material and methods

The experiments were carried out between 1984 and 1986 in two enclosures (3 m diameter) open to the bottom sediments and to the atmosphere, each containing about 17 m³ of lake water (Tátrai & Istvánovics, 1986). In each year the enclosures were installed between June-July in the meso-

Table 1. The number and size of bream (*Abrham brama* L.) introduced into one of the enclosures.

Date of the experiments with fish	Average water temp. °C	Number and biomass of fish. g m^{-2}	Average weight of fish, g (range)	Total increment in weight, %
01.06–11.09. 1984	18.7	9 60.7	47.9 (10.0–106.5)	82.8
21.06–18.07. 1985	19.0	6 79.7	94.3 (8.0–158.0)	49.6
04.07–11.09. 1986	21.9	4 46.4	82.4 (69.0–107.0)	21.4

trophic area of the lake, 100 m offshore from the institute's jetty, at a water depth of 205-210 cm. One enclosure served as a control and was free of fish, while the other was stocked with bream (*Abramis brama* L.) of various size (Table 1).

Physical (Secchi-disc transparency, suspended matter), chemical (TN,TP) and biological (phyto- and bacterioplankton) analyses were done on samples collected by surface-mud length tube (12 cm in diameter). The suspended matter content of the water was determined by weight.

The concentration of chlorophyll *a* was determined photometrically following methanol extraction (Iwamura *et al.*, 1970). Algae were counted with an inverted microscope. Biomass as biovolume was estimated based on volumes of the individual species. Primary production was measured in triplicate from unfiltered samples using the ^{14}C method. The radioactivity was determined with a Rack-Beta-2 liquid scintillation counter. The production of bacterioplankton was measured after incorporation of thymidine, labelled with tritium (^{3}HT), into DNA of bacteria (Fuhrman & Azam, 1982).

Mixed, unfiltered water samples were used for measurements of total nitrogen (TN) and total phosphorus (TP) according standard analytical procedures (Mackereth *et al.*, 1978).

Results and discussion

Separation of enclosures from the open waters was predicted to reduce mixing inside, but sur- prisingly the water in our enclosures was well mixed. Transparency of the enclosed water, before fish introduction, was very similar to that of the external lake water. This indicates that the diameter of the enclosures was in good proportion to their depth. Due to intense mixing, temperature differences between surface and bottom water were always less than 1 °C in all three years.

Secchi-disc transparency was similar at all stations and enclosures before fish introduction. After introduction of fish (61 g m^{-2}) (Table 1) the transparency gradually decreased in the enclosure with fish, and remained lower than in the control or in the lake throughout the experiment in 1984 (Fig. 1). In 1985, at a higher stocking density (80 g m^{-2}), the Secchi-disc transparency averaged 10 cm less than in 1984 (Fig. 1). However, a lower stocked biomass of fish (46 g m^{-2}) resulted in a transparency like that of 1985 in the enclosure in 1986, being 10 cm less than in the control and lake.

In 1985 the concentration of suspended matter when fish were present was more than two times higher in the enclosure with fish than in the control (9.7 mg l^{-1}), while in the lake it lay between these two values (13.4 mg l^{-1}) (Fig. 1). In 1986 the content of suspended solids was more than three times higher in the enclosure containing fish (75.6 mg l^{-1}) than in the lake (29.8 mg l^{-1}). So far, concentrations of suspended matter were 2-5 times higher at all sampling stations in 1986 than in 1985.

The chlorophyll *a* concentration varied slightly both in the lake and in the control enclosure

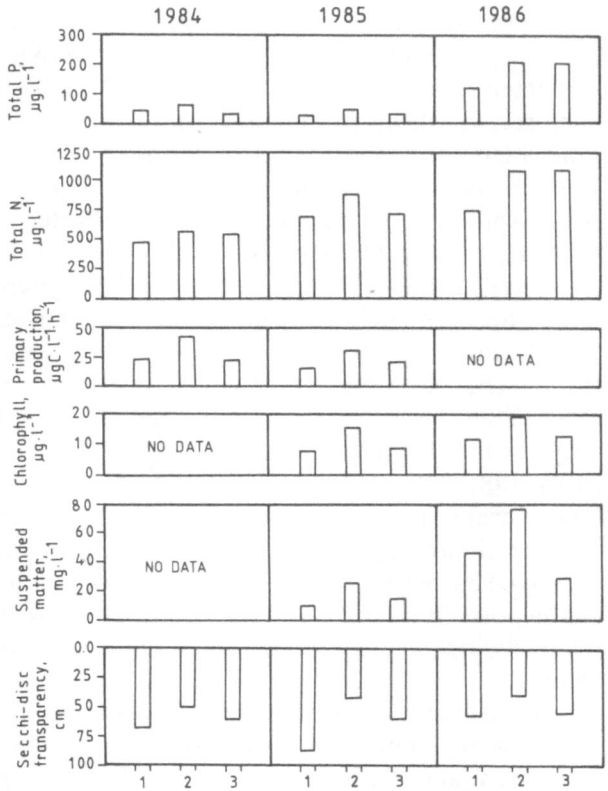

Fig. 1. The effects of adding various biomass of benthivorous fish to the enclosure of lake waters in different years (1- control enclosure, 2- enclosure with fish, 3- Lake Balaton).

(5-10 μg l^{-1}) in 1985). Otherwise, as expected, the chlorophyll content did not always follow changes in the algal biomass. The chlorophyll *a* content of algae oscillated, especially in the presence of fish in 1986, then exceeding two times the value measured in the fish-free enclosure and in the lake. After fish removal, however, the chlorophyll *a* concentration dropped below the level measured in the control enclosure.

Prior to fish introduction into the enclosures, no differences were observed in primary production between enclosures and lake during two years (Fig. 1). After fish introduction, phytoplankton production increased to the highest extent in the enclosures containing fish.

Concentrations of total nitrogen (TN) were very similar at all sampling points in 1984. The differences in primary production between dif-

ferent stations were not reflected in TN concentrations. A more intense contribution of small algae to the production could be the main reason for this. The total phosphorus (TP) content increased immediately after fish introduction and thereafter did not decrease in the enclosure with fish. At the same time, the peak in algal production coincided with more fluctuating TP values in the fish enclosure than in the control enclosure and lake (Fig. 1).

In 1985 the TN concentration ranged between 400 and 1150 μg l^{-1} at all stations, of which the larger part (average 74%) was in dissolved organic form. TN concentration was highest in the fish enclosure except in two cases. TP values ranged between 20 and 110 μg m^{-1} and in the fish enclosure significantly exceeded those in the control and lake. About 24-40% of the TP was in dissolved form. This proportion was highest in the control and lowest in the fish enclosure.

In 1986 the TN concentration ranged from 120 to 2300 μg l^{-1} at all stations, reaching maximum values in the enclosure with fish (Fig. 1). Similarly the TP values varied in a wide range (25-530 μg m^{-1}), and were higher in the presence of fish and in the lake.

In all years the presence of fish increased algal abundance and species diversity. In 1985 for example, 72 species were found in the lake, 71 in the control and 83 in the fish enclosure. After introduction of fish to the enclosure, diatoms were replaced by mixed populations of filamentous cyanobacteria with very high reproductive rate. The biomass (wet weight) of cyanobacteria exceeded 7 mg l^{-1} by the end of the experiments and comprised 93% of the total algal biomass in the fish enclosure (Fig. 2). Cyanobacteria also appeared in the lake and in the control enclosure, but their biomass reached only one-fifth of that found in the presence of fish. The biomass of algae was very similar at the beginning of the experiments at all sampling stations (2.66-2.96 mg l^{-1} in 1986). At this time mainly diatoms, *Cyclotella ocellata*, *Cymatopleura elliptica*, *C. elliptica* var. *constricta*, *C. elliptica* var. *nobilis*, *C. elliptica* var. *hibernica*, and *Campylodiscus noricus*, dominated the phytoplankton in both lake and enclosures. In

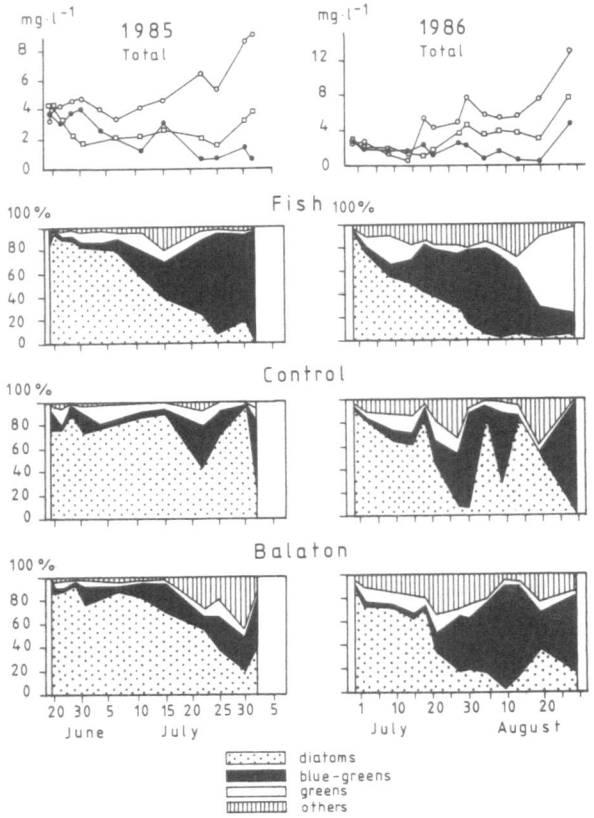

Fig. 2. Total phytoplankton biomass and the share of different types of algae in the total in the control enclosure (●————●), in the enclosure with fish (○————○), and in the lake (□————□) during two years.

the enclosure with fish the algal biomass increased throughout the experiments. In July mainly mixed populations of filamentous cyanobacteria, *Aphanizomenon flos-aquae* f. *klebahni* and *A. issatschenkoi*, dominated and their biomass reached 80% of the total by the end of this month in the fish enclosure (Fig. 2). In August cyanobacteria were replaced by *Closterium aciculare*. By the end of the experiments chlorophytes had increased to more than 10 mg l^{-1} and constituted 78% of the total biomass (13 mg l^{-1}). The biomass of cyanobacteria increased to 1.9 mg l^{-1} by the end of August (3.9 mg l^{-1}) in the lake. The dominance of chlorophytes did not appear in the control enclosure and the total biomass there increased to 4.5 mg l^{-1}, being three times lower than in the fish enclosure.

The production of bacteria in both enclosures and lake was measured only in 1985. The activity of bacterioplankton increased mostly in the fish enclosures, where production was almost two times higher than in the control and lake (Fig. 3). The bacterial production increased simultaneously with the enhancement of cyanobacteria, reaching a value of 2.8 nmol l^{-1} day^{-1}.

Averaged chlorophyll *a* concentration, determined for each enclosure and the lake, increased as a linear function of the average TP concentration (Fig. 4), except in the fish enclosure in 1986. The regression was significant ($p < 0.05$) only in the fish enclosure in 1985 and in the control in 1986. The enclosure with fish had relatively high average values of TP and chlorophyll *a*. The average chlorophyll *a* value in the fish enclosure during 1986 fell below the regression line, indicating possible inhibition. Control enclosures had relatively low average TP and high chlorophyll *a* values.

Several results during three years of experiments were inconsistent with both theoretical prediction and with results obtained in other years. Mixing processes of combined fish and wave action did increase the suspended solid content and decrease transparency as compared with the control. In all years transparency was less in the enclosure containing fish. However, its value seemed to be dependent on factors other than fish biomass since, for example, the water in 1985 was as transparent as in 1986, although fish stocking levels differed almost two-fold in the two years. It can be suggested that temperature indirectly affected the transparency. Average water temperature during the experiment in 1986 was almost 3 °C higher than in the previous year (Table 1). Higher temperature itself could be responsible for a higher 'base level' of chlorophyll *a* concentration in 1986, and the biomass of algae was similarly almost two times higher at that time in 1985.

Both phytoplankton biomass and production positively responded, without a lag-phase, to nutrients in the fish enclosure. In 1985, after the first algal peak, diatoms were replaced by mixed populations of filamentous cyanobacteria. At the

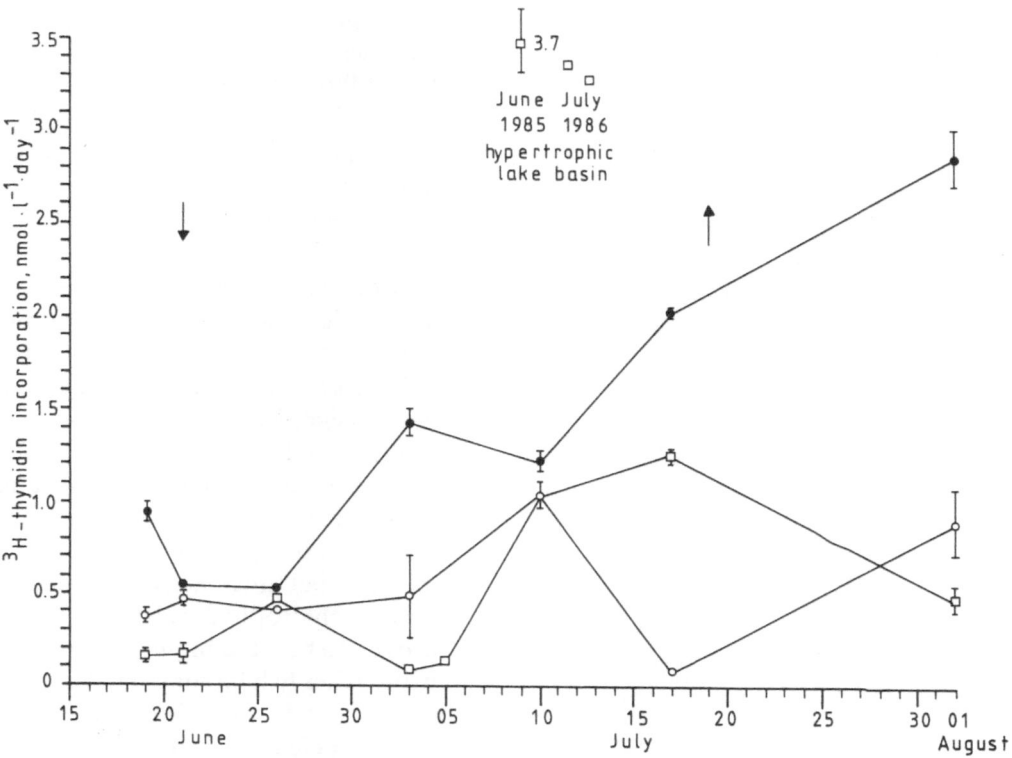

Fig. 3. Bacterial thymidine incorporation in the enclosures (o———o-control ●———●-with fish) and in the different basins of the lake □———□).

same time, the average TN : TP ratio suggested a double nutrient limitation in the fish enclosure (Vollenweider & Kerekes, 1980) through N_2-fixing species that became dominant during the second half of the experiment. In the hypertrophic area of Lake Balaton, blooms of N_2-fixing cyanobacteria have occurred irrespective of the TN : TP ratio since the middle of the 1970s, as a consequence of nitrogen loss through denitrification (Abdelmoneim, 1982).

A clear shift in algal structure also occurred in 1986 at a much lower biomass. The mixed populations of bluegreens, dominating during the first part of the experiment, were replaced by species of green algae. The domination of chlorophytes was observed neither in the control enclosure nor in the lake, but a spreading of cyanobacteria was rather typical. So far in 1986 the initial dominance of diatoms was replaced, as in preceding years, by cyanobacteria in the lake and control enclosure

and by chlorophytes, shortly after a dominance of cyanobacteria, in the fish enclosure. As an effect of fish, the biomass of algae in the fish enclosure was almost two times higher than in the lake and almost three times higher than in the control. Comparing the 1986 values of algal biomass with those obtained in 1985, the biomass in the enclosure with fish similarly exceeded that of the lake but was about 18 times higher than in the control.

The intensity of ^{3}H-thymidine incorporation by bacteria has indicated increased organic substrate in the fish enclosure. The higher bacterial production rate in the fish enclosure could result from egested organic substances, or alternatively from the resuspension of sediment particles, due to the mechanical mixing processes of fish searching for food. During such action organic substrates and bacteria, and particularly spores of heterocystic cyanobacteria (Gorzó, 1985), could be released to the water overlying the sediment. We concluded

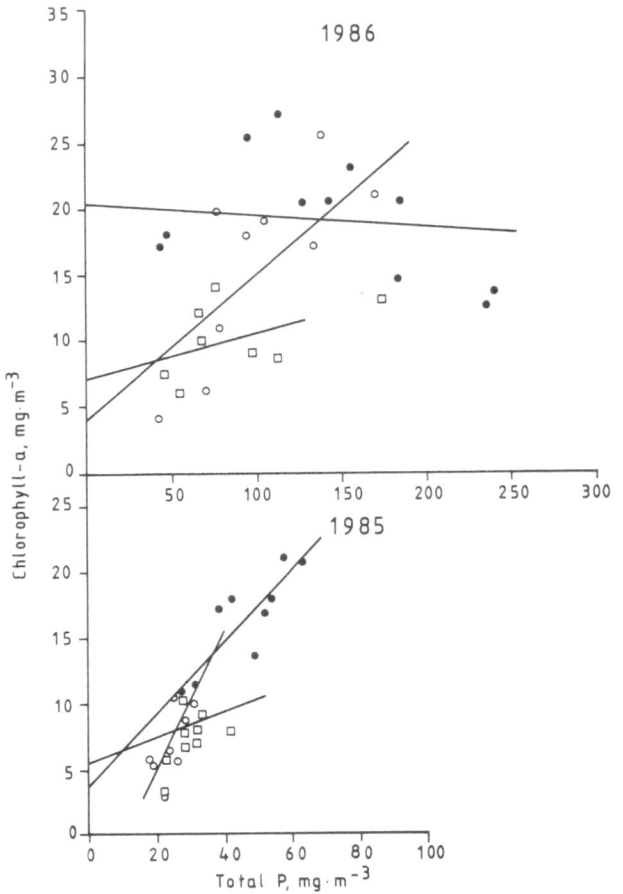

Fig. 4. Relationship between average concentrations of total phosphorus (TP) and chlorophyll *a* in the control enclosure (○- $y = 0.52 \times - 5.54$ $r = 0.70$, $n = 9$ in   1985; $y = 0.11 \times + 3.95$, $r = 0.71$, $n = 9$ in 1986), in the fish enclosure   (●- $y = 0.27 \times + 3.70$, $r = 0.86$, $n = 10$ in   1985; $u = - 0.01 \times + 20.44$, $r = - 0.21$, $n = 11$ in 1986), and in the lake   (□- $y = 0.10 \times + 5.46$, $r = 0.24$, $n = 9$ in   1985; $y = 0.034 \times + 7.04$, $r = 0.52$, $n = 7$ in 1986).

that fish in the mesotrophic area of Lake Balaton were able to generate a bacterial production typical of the hypertrophic area in summer.

The relationship between TP and chlorophyll *a* was insignificant. The reason is, undoubtedly, that P is only one of the several factors affecting algal abundance. One of the most important is the presence or absence of other organisms. Shapiro (1979) concluded that fish did not influence directly, but their activities did affect the phosphorus-chlorophyll relationship. His statement proved to be valid in the present experiments only with higher fish biomass in 1985. It is curious that it does not fit with data in the 1986 experiments with sufficiently less biomass in the enclosure.

Based on the results obtained during our enclosure experiments, the main pathways between different trophic levels affected by benthivorous fish are schematically presented in Fig. 5. We previ-

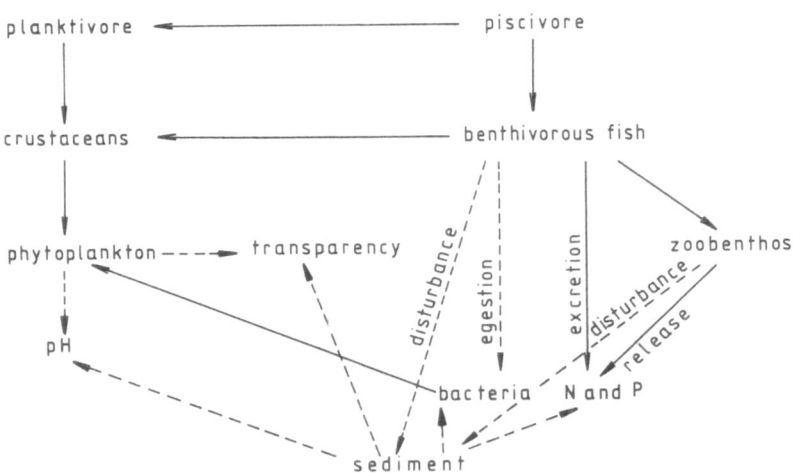

Fig. 5. The main pathways between different trophic levels affected by the activity of bream, a benthivorous fish. Direct influences are indicated by solid lines and indirect influences by broken lines.

ously found direct influence of bream on crustaceans – that is, they decreased the abundance of copepods and the size of cladocerans (Tátrai *et al.*, 1985) – and on zoobenthos. Here there is indirect influence on phytoplankton structure and production by direct nutrient excretion and by release of nutrients through sediment disturbance. By sediment-mixing processes as well as by egestion, fish might have indirect effects on bacterial production, pH and water transparency.

Acknowledgements

We are grateful to B. Arszovszky, I. Molnár, T. Németh and E. Kiss for building the enclosures, to the Diver's Club Amphora for the installation, and to T.F. Klein for technical assistance.

References

Abdelmoneim, M. A., 1982. Nitrogen fixation and denitrification in Hungarian shallow lake sediments. Ph.D. thesis, Fish. Res. Inst., Szarvas, Hungary, 109 pp.

Brooks, J. L. & S. I. Dodson, 1965. Predation, body size, and composition of plankton. Science 150: 28–35.

Fuhrman, J. A. & F. Azam, 1982. Thymidine incorporation as a measure of heterotrophic bacterioplankton production in marine surface waters: evaluation and field results. Mar. Biol. 66: 109–120.

Gorzó, Gy., 1985. Planktonic heterocystic cyanobacteria cultivated in the sediment of Lake Balaton. Hidrol. Közl. 6: 357–360.

Hrbáček, J., M. Dvořáková, V. Kořínek & L. Procházková, 1961. Demonstration of the effect of fish stock on the species composition of zooplankton and the intensity of metabolism of the whole plankton assemblage. Verh. int. Ver. Limnol. 14: 162–195.

Iwamura, T., H. Nagal & S. Ischimura, 1970. Improved methods for determining contents of chlorophyll, protein, ribonucleic acid and desoxyribonucleic acid in planktonic populations. Int. Revue ges. Hydrobiol. 55: 131–147.

Mackereth, F. J., J. Heron & J. F. Talling, 1978. Water analysis: some revised methods for limnologists. Freshwat. Biol. Assoc. Sci. Publ. No. 36.

Shapiro, J., 1979. The need for more biology in lake restoration. Contribution 183 from the Limnological Research Center, University of Minnesota, Minneapolis, EPA 440/5-79-001, 17 p.

Tátrai, I., L. G. Tóth & J. E. Ponyi, 1985. Effect of bream *Abramis brama* L. on the lower trophic level and on the water quality in Lake Balaton. Arch. Hydrobiol. 105: 205–217.

Tátrai, I. & V. Istvánovics, 1986. The role of fish in the regulation of nutrient cycling in Lake Balaton, Hungary. Freshwat. Biol. 16: 417–424.

Vollenweider, R. A. & J. J. Kerekes, 1980. The loading concept as a basis for controlling eutrophication: philosophy and preliminary results of the OECD program on eutrophication. Progr. Wat. Technol. 12: 5–38.

Hydrobiologia **191**: 315–318, 1990.
P. Giró and J. F. Talling (eds), Trophic Relationships in Inland Waters.
© 1990 *Kluwer Academic Publishers.*

Summary of the workshop on perspectives of biomanipulation in inland waters

M. Gophen
Kinneret Limnological Laboratory P.O. Box 345, Tiberias, Israel (14102)

Introduction

The term 'biomanipulation' was originally defined as management of aquatic communities by controlling natural populations of organisms aimed at water quality improvement (Shapiro *et al.*, 1975). In a broad sense, the term biomanipulation is similar to others presently used by several authors: top-down forces, trophic cascade or food-web manipulation (Carpenter *et al.*, 1985; McQueen *et al.*, 1986; Benndorf, 1987; Threlkeld, 1987; Carpenter, 1988). All these terms refer to manipulation of secondary or tertiary producers and its impact on the community structure of aquatic ecosystems. Recently, the complexity of the ecosystem response and the role of bottom-up forces (nutrients) and/or nutrient-mediated effects of planktivorous fish on plankton community structure was considered (Carpenter, 1988; McQueen & Lean, 1987; Threlkeld, 1987).

The classic or traditional limnological approach to lake ecosystem structure was oriented at the food-chain concept and went through nutrients – phytoplankton – zooplankton – fish. Nevertheless, fish as top consumers were not considered as a component which may significantly affect water-quality. Moreover, classic limnological science poorly included fishery biology (Rigler, 1982).

Pioneer experiments on biomanipulation were carried out by Hrbáček and his team during the late 50's and early 60's (Hrbáček *et al.*, 1961). They demonstrated a fish effect on water quality through zooplankton predation and consequent decline of its grazing capacity followed by algal enhancement. Shortly afterwards, Brooks &

Dodson (1965) and Hall *et al.* (1970) documented the effect of zooplanktivorous fish predation on lower throphic levels.

During the 70's studies on biomanipulation were focussed on the effect of high trophic level consumers on water quality (Hurlbert *et al.*, 1972; Zaret & Pain, 1973; Shapiro *et al.*, 1975; Andersson *et al.*, 1978; Hrbáček *et al.*, 1978; Stenson *et al.*, 1978).

Independently, an intensive research was carried out during the 70's on macrophyte removal by grass-carp (*Ctenopharyngodon idella*) (Shierman, 1979). The water quality accounted for by the grass-carp's researchers was mostly the absence of nuisance macrophytes. Transfer of nutrients (mostly P) from sediments through macrophytes and macrophytivorous fish (Hansson *et al.*, 1987) was not thoroughly studied.

Significant progress of biomanipulation studies was achieved during the 80's. Scientists increased the number of components that were studied within research on food-web manipulation and incorporated them into models: responses of phytoplankton and zooplankton to reduction or increase of predation pressure by planktivorous fish, and the effect of piscivory. Nutrients and fish mediated nutrient effects, bacteria, micro-zooplankton and picoplankton were also considered (Kitchell *et al.*, 1982; Shapiro *et al.*, 1982; Benndorf *et al.*, 1984; Drenner *et al.*, 1984; Goad, 1984; Reinertsen & Olsen, 1984; Shapiro & Wright, 1984; Carpenter *et al.*, 1985; Gophen, 1986; 1987; McQueen *et al.*, 1986; Threlkeld, 1987; Vinyard *et al.*, 1987; Carpenter, 1988; Stockner, 1988).

The Discussion

The following aspects were discussed:

Definition of biomanipulation

The definition of biomanipulation as a manipulation of food-web organisms and nutrient control to improve water quality was accepted. Water quality is partly defined, in a broad sense, by the kind of utilization – from drinking water standards to the removal of macrophytes to improve boat navigation. Participants agreed upon the impact of water utilization on quality standards: drinking water, fish farming, agricultural irrigation, recreation.

Water residence time

Biomanipulation operations should take into account the residence time of the treated waterbody. Biomanipulation strategy has to be different in water-bodies with long (years) retention time compared to those with shorter (days – months) period. In large water-bodies with retention periods of years the operations should be conducted gradually, including a low number of native species to ensure long-term stabilization and prevention of introduction of exotic species.

In small water-bodies with low residence time biomanipulation strategy may include more species, because of a lower level of risk due to the high rate of water renewal.

Biomanipulation and biological control

Elimination of water nuisances by fish to improve quality in water-supply systems was defined as 'biological control of water quality' (Leventer, 1978). The differences between biological control in relatively small reservoirs and biomanipulation strategy in larger lakes were discussed. The use of fish to eliminate nuisances from water-supply systems (reservoirs, canals) was considered as a different strategy from long-term food-web manipulation in larger lakes.

Artificial and natural water bodies

Artificial impoundments should be manipulated differently from natural large lakes. Crises due to a failure of biological treatment can be relatively easily repaired in small reservoirs but not in natural lakes. Ecosystem modification in natural large lakes might be slowly reversed. Therefore precautions are required when a strategy of food-web manipulation is implemented in large natural lakes. The management design of treatments involving natural lakes has to be based on thorough study of 'complex interactions' (Carpenter, 1988) between all components of the ecosystem, aimed at long-term stable relations. Natural lakes have not to be managed like a fish pond where energy is channelled mostly towards food production.

Two examples were discussed in which managers introduced predator fish to natural large lakes to improve fish production but did not adequately allow for ecosystem structure and trophic relations within the communities: Lake Gatum in Panama (Zaret & Pain, 1973) and Lake Victoria in Africa (Barel et al., 1985). The ecological crises as well as fishery destruction in these lakes were discussed.

Target species other than fish in biomanipulation strategy were also considered.

Participants concluded that there is much more to be studied in order to be able to predict: the response of community structure to food-web manipulation, as well as the complexity of the impacts on the whole ecosystem; long-term effects of biomanipulation treatments; the efficiency of food-web manipulation in large natural lakes; use of top trophic-level organisms other than fish for biomanipulation; reversibility and irreversibility of biological modifications in the ecosystems; the effects of secondary and tertiary consumers on plankton species diversity; the relative impact of top-down and bottom-up forces in ecosystems; long-term effects of excretions of secondary and tertiary consumers on the plankton assemblages; the role of 'microbial-loops' in biomanipulation treatments; and other topics.

References

Andersson, G., H. Berggren, G. Cronberg & C. Gelin, 1978. Effects of planktivorous and benthivorous fish on organisms and water chemistry in eutrophic lakes. Hydrobiologia 59: 9–15.

Barel, C. D. N., R. Dorit, P. H. Greenwood, G. Fryer, N. Hughes, P. B. N. Jackson, H. Kawanabe, R. H. Lowe-McConnell, M. Nagoshi, A. J. Ribbink, E. Trewavas, F. Witte & K. Yamaoka, 1985. Destruction of fisheries in Africa's lakes. Nature 315: 19–20.

Benndorf, J., 1987. Food web manipulation without nutrient control: a useful strategy in lake restoration? Schweiz. Z. Hydrol. 49: 237–248.

Benndorf, J., H. Kneschke, K. Kossatz & E. Penz, 1984. Manipulation of the pelagic food web by stocking with predaceous fishes. Int. Revue ges. Hydrobiol. 69: 407–428.

Brooks, J. L. & S. I. Dodson, 1965. Predation, body size and composition of plankton. Science 150: 28–35.

Carpenter, S. R. (ed), 1988. Complex interactions in lake communities. NSF Workshop, 21-26/3/1987, Notre-Dame, USA (in press).

Carpenter, S. R., J. F. Kitchell & J. R. Hodgson, 1985. Cascading trophic interactions and lake productivity. BioScience 35: 634–639.

Drenner, R. W., J. R. Mummert, F. DeNoyelles, Jr. & D. Kettle, 1984. Selective particle ingestion by a filter-feeding fish and its impact on phytoplankton community structure. Limnol. Oceanogr. 29: 941–948.

Goad, J. A., 1984. A biomanipulation experiment in Great Lake, Seattle, Washington, USA. Arch. Hydrobiol. 102: 137–154.

Gophen, M., 1986. Fisheries management in Lake Kinneret (Israel). In Proc. Ann. Mtg N. Amer. Lake Management Soc., Lake Geneva, Wisconsin, USA: 327–332.

Gophen, M., 1987. Fisheries management, water quality and economic impacts: a case study of Lake Kinneret. Proc. World Confer. on Large Lakes, Mackinac Island, Michigan, USA Vol. II: 5–24.

Hall, D. J., W. E. Cooper & E. E. Werner, 1970. An experimental approach to the production dynamics and structure of freshwater animal communities. Limnol. Oceanogr. 15: 839–928.

Hansson, L. A., L. Johansson & L. Persson, 1987. Effects of fish grazing on nutrient release and succession of primary producers. Limnol. Oceanogr. 32: 723–729.

Hrbáček, J., M. Dvořáková, V. Kořínek & L. Procházková, 1961. Demonstration of the effect of the fish stock on the species composition of zooplankton and the intensity of metabolism of the whole plankton assemblage. Verh. int. Ver. Limnol. 14: 192–195.

Hrbáček, J., B. Desortová & J. Popovský, 1978. The influence of the fish stock on the phosphorus-chlorophyll ratio. Verh. int. Verein. theoret. angew. Limnol. 20: 1624–1628.

Hurlbert, S. H., J. Zedler & D. Fairbanks, 1972. Ecosystem alteration by mosquitofish (Gambusia affinis) predation. Science 175: 639–641.

Kitchell, J. F., H. F. Henderson, E. Grygierek, J. Hrbáček, S. R. Kerr, M. Pedini, T. Petr, J. Shapiro, R. A. Stein, J. Stenson & T. Zaret, 1982. Management of lakes by foodchain manipulation. U.N.F.A.O., Rome.

Leventer, H., 1978. Biological control in reservoirs. Mekorot, Water supply Co. Jordan District, Nazareth, Israel, 80 pp.

McQueen, D. J., J. R. Post & E. L. Mills, 1986. Trophic relations in freshwater pelagic ecosystems. Can. J. Fish. aquat. Sci. 43: 1571–1581.

McQueen, D. J. & D. R. S. Lean, 1987. Influence of water temperature and nitrogen to phosphorous ratio on the dominance of blue-green alae in Lake St. George, Ontario. Can. J. aquat. Sci. 44: 598–604.

Reinertsen, H. & Y. Olsen, 1984. Effects of fish elimination on the phytoplankton community of a eutrophic lake. Verh. int. Ver. Limnol. 22: 649–657.

Rigler, F. H., 1982. The relation between fisheries management and limnology. Trans. Am. Fish. Soc. 111: 121–132.

Shapiro, J., V. Lammara & M. Lynch, 1975. Biomanipulation: an ecosystem approach to lake restoration. In P. L. Brezonik & J. L. Fox (eds), Proceedings of a symposium on water quality management through biological control. Univ. of Florida, Gainsville: 85–96.

Shapiro, J., B. Forsberg, V. Lammara, G. Lindmark, M. Lynch, E. Smeltzer & G. Zoto, 1982. Experiments and experiences in biomanipulation, pp. 1–125. Interim report No. 19 of the Limnological Research Center, Univ. of Minnesota, Minneapolis.

Shapiro, J. & D. I. Wright, 1984. Lake restoration by biomanipulation: Round lake, Minnesota, the first two years. Freshwat. Biol. 14: 371–383.

Shierman, J. V. (ed), 1979. Proceedings of the grass-carp conference. Gainsville, Fl: Aquatic Weeds Research Center, Univ. of Florida, Inst. of Food and Agricult. Sciences.

Stenson, J. A. E., T. Bohlin, L. Henrikson, B. I. Nilsson, H. G. Nyman, H. G. Oscarson & P. Larsson, 1978. Effects of fish removal from a small lake. Verh. int. Ver. Limnol. 20: 794–801.

Stockner, J. G., 1988. The role of autotrophic and heterotrophic picoplankton in structuring the food webs of large oligotrophic British Columbia lakes. Conference on Functional and Structural Properties of Large Lakes, Konstanz, 13-18/87 (abstract).

Threlkeld, S. T., 1987. Experimental evaluation of trophic-cascade and nutrient-mediated effects of planktivorous

318

fish on plankton community structure. In W. C. Kerfoot & A. Shi (eds.), Predation: direct and indirect impacts on aquatic communities. Univ. Press N. England, Hanover: 161–173.

Vinyard, G. L., R. W. Drenner, M. Gophen, U. Pollingher, D. L. Winkelman & K. D. Hambright, 1987. An experimental study of the plankton community impacts of two filter-feeding cichlids, the Galilee Saint Peter's fish (*Sarotherodon galilaeus*) and Blue Tilapia (*Tilapia aureus*). Can. J. Fish. aquat. Sci. 45(4): 685–690.

Zaret, T. M. & R. T. Pain, 1973. Species introduction in a tropical lake. Science 182: 449–455.

Hydrobiologia **191**: 319–320, 1990.
P. Biró and J. F. Talling (eds), Trophic Relationships in Inland Waters.
© 1990 *Kluwer Academic Publishers.*

Concluding remarks

H.L. Golterman
Station Biologique de la Tour du Valat, le Sambuc, 13200 Arles, France

This was a very satisfactory meeting, well organized, in a pleasant atmosphere, in short a very positive experience.

The title covered the contents well, which is not always the case. Apparently the organizers realized that the topic of 'trophic relations in inland waters' was ripe for an international discussion and the participants understood the message. So we broke through barriers of phytoplankton or zooplankton sessions, and looked at influences of one level of the food chain on the others. It was useful that even fish was discussed as part of the ecosystem. This has not always been accepted. Wesenberg Lund (1910) stated that 'many of the (freshwater biological) laboratories are bound by obligations to the fishery, which in my opinion neither promotes the fishery nor the limnology'. During the preparatory work for the International Biological Programme (IBP) there was still a lot of resistance. Fortunately IBP managed to bring the fish biologists into the ecosystem and after the IBP several institutes that owed their programmes to the impact of IBP continued to keep phytoplankton and fishery biology together. Here, in this meeting, we had some good results showing the influence of fish, direct or indirect, on phytoplankton populations.

Very many relationships were discussed indeed: all came together, hydrology, macrophytes, chemistry, bioperturbations, classic plankton work etc. Sometimes there were so many transparencies on top of each other, that no light could pass through the relationships!

This integrated work has several drawbacks: an algal physiologist cannot know sufficiently about fish as well; he simply does not have sufficient megabytes. So he will ask the fish biologist to do work in which the latter is perhaps not interested. And several of us feel that the best scientific work is always achieved to satisfy our own curiosity. For integrated work the workload becomes a very important factor: two different populations must be studied at the same moment; there are great logistic problems.

Nevertheless it was clear during this meeting that it can be done, sometimes. We were indeed impressed by the wide scope of variables analyzed.

The models in the P-flux begin to help us in our thinking – as we always said they would, but saw so seldom happen. The model discussed here will perhaps even be useful for the water manager; it seems that something like a compromise between the oversimplified models (such as the OECD one) and the too sophisticated ones begins to emerge – at least for shallow lakes. With a next step towards a better understanding of the phosphate equilibrium between sediments and water, this begins to be interesting. The sediments, however, remain very much a black box as long as we do not know more about their chemical composition. Operational extractions did not help very much in the past.

Another special topic was bio-manipulation. It is clear that bio-manipulation is no panacea against eutrophication. It may stimulate recovery when the essential measures are taken. Its effects are probably limited to the speeding up of the recovery process and do not change the pattern. I think it can be described by Fig. 1, which suggests that the way back may be quicker with than without bio-manipulation, but that this does not change the outcome of the restoration process qualitatively.

An important topic was work on measurement and understanding of algal growth; the problem

320

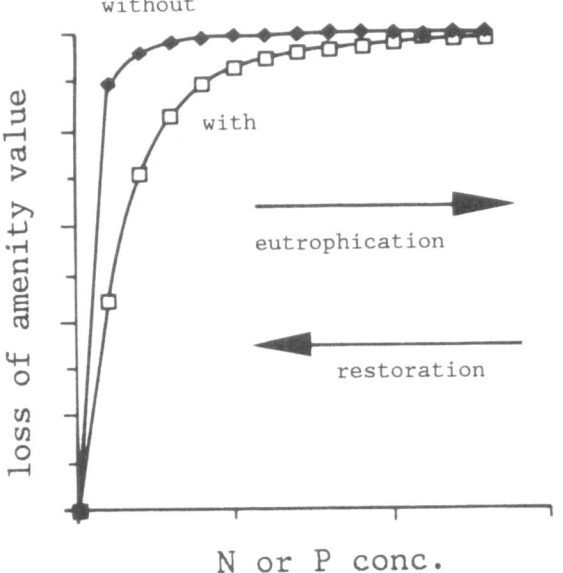

Fig. 1. Restoration process with or without biomanipulation (Loss of amenity value = 1/water quality).

of nutrient availability – especially of the nutrients attached to suspended matter or sediments – remains difficult but important. A critical point is, however, that most algal physiological constants so far used are derived from cultures. We know that they are probably not the same as in nature. Furthermore I should like to mention that iron is an important nutrient about the availability of which very little is known. Lund *et al.* (1975) have shown this in part of their work, by adding chelates to lake water and showing improved growth. It is a pity that the old work of Shapiro (1967) on iron availability, especially for blue-green algae, has never had a serious follow-up.

There are several points to be wished for: More experimental work in enclosures or small ecosystems should be carried out, but such experiments pose the question of the extrapolability of the results to the real lake system. The costs of these studies are impressive: the Loosdrecht Lakes study is paid for by a restoration programme, and falls far beyond the budget of a scientific institute – academic or otherwise. Often the two interests do not run parallel. Thinking about restoration programmes and assessing their effects poses a serious problem: If measures are taken, e.g. by precipitation of phosphate or ring canalization, how do we separate the effects of the removal of heavy metals from the effects of removing phosphate?

It is clear that there is an enormous data bank on phytoplankton – zooplankton interactions. There seems to be only a small data bank on fish – zooplankton interactions. I should like to express the opinion of several of the participants that we hope that the Tihany institute might undertake the responsibility of setting up a zooplankton – fish study group, comparable to the Plankton Ecology Group of SIL. We should like to stress the importance of close co-operation with local fishermen – although a caveat against their Latin must be observed.

Finally I should like to give a vote of thanks to the organizers: it was a good experiment especially as a pilot study for restoration projects.

References

Lund, J. W. G., G. H. M. Jaworski & C. Butterwick, 1975. Algal bioassays of water from Blelham Tarn, English Lake District and the growth of planktonic diatoms. Arch. Hydrobiol. (Suppl.) 49: 49–69.

Shapiro, J. 1967. Yellow organic acids of lake water: differences in their composition and behavior. In: Chemical environment in the Aquatic Habitat. (H. L. Golterman Ed). Proceedings of an I.B.P. symposium, held in Amsterdam, 10-15 October, 1966. North Holland Publishing Co.

Wesenberg-Lund, C. 1910. Summary of our knowledge regarding various limnological problems. In: Murray, J. & L. Pullar (Eds), Bathymetrical survey of the Scottish freshwater lochs. Report on the Scientific Results. Challenger Office, Edinburgh: 374–433.

General Index

324

257, 260, 262
structure 57, 160
trophogenic layer 47, 50
zone 120
trophy, lake 6
trout 6
Tsalmon reservoir 47, 48, 49, 53, 54
tufted duck 233, 235, 237, 239, 240
turbidity 249, 275, 278
turf 26
turnover 12
time 170
Typha 259, 260
angustifolia 258, 260

underwater chasers 230

Vaucheria dichotoma Ag. 23, 24, 25, 26
dry weight 25
growth rate 24
Vecht River 87, 90, 131, 175, 287
vegetable food 62
vendace 71
vertical attenuation 243
coefficient 245, 246
extinction 137
migration 11, 251
mixing 113
visiting birds 229
piscivores 225, 226, 228
Vistula river 169
Viviparus viviparus 235, 240
Volga river 169
Volvox spp. 253
vulnerability 40

Waddenzee 75
waste water 75, 87
discharge 87, 90, 92
Water Action Programme 78
water balance 77, 78
blooms 15, 17, 19
borax-buffered 152
budget 173
characteristics 139
chemistry 68
circulation 93
colour 78
discharge 90, 166, 169
exchange rate 234
– fowl 233
input 33, 34
intake 29, 31, 34, 35, 36
– level fluctuations 169
lilies 227, 230
lily zone 227, 229

management 90
oxygen-saturated 152
– plants 92, 93, 94
purification 195
quality 53, 75, 78, 103, 105, 107, 189, 254, 275, 315, 316, 320
and quantity model 77
deterioration 189, 197
gradient 189
management 77, 84
standard 78
renewal time 1, 131
reservoirs of the USSR 49
residence time 10, 140, 266, 316
– supply systems 316
temperature 20, 29, 31, 32, 33, 34, 113, 114, 133, 137, 142, 145, 154, 156, 157, 158, 166, 186, 199, 201, 210, 228, 265
transparency 113, 114
velocity 77
white bream 291
whitefish 59, 61, 65
wind effect 129, 249
roles of 129
velocity 250, 251
Winkler-technique 152, 250
winter overturn 2, 3
stratification 14
wintering period 237
Woloszynskia ordinata 11

yield 53, 61, 213, 214, 215, 216, 217, 218, 220
from gillnet catches 62
pelagic fish species 6

Zala river 297, 298, 299, 301, 304, 305
Zala watershed 297, 306
zoobenthos 40, 54
zooflagellates 177
zooplankters 120
zooplankton 6, 29, 31, 36, 40, 47, 50, 52, 55, 57, 61, 62, 63, 65, 66, 67, 71, 89, 90, 106, 123, 125, 126, 149, 151, 159, 165, 166, 177, 200, 204, 209, 218, 234, 253, 262, 275, 278, 289
abundance 93, 105, 108, 109, 110, 174, 184
analysis 49
assimilation 88
biomass 9, 40, 41, 44, 49, 65, 71, 139, 152, 153, 174, 181, 182, 184, 185, 206, 220, 293
clearance rate 154, 155
community 4, 6, 36, 39, 43, 57, 58, 71, 161, 200, 206, 220, 293
composition 60, 62, 71, 126, 149, 170, 174, 184, 199
consumption 120
control 53
crustacean number 50